Inorganic Chemistry

PRENTICE-HALL INTERNATIONAL SERIES IN CHEMISTRY

BOUDART Kinetics of Chemical Processes
DEMITRAS et al. Inorganic Chemistry
FERGUSON The Modern Structural Theory of Organic Chemistry
HARVEY Introduction to Nuclear Physics, 2nd ed.
JOLLY Synthesis and Characterization of Inorganic Compounds
KISER Introduction to Mass Spectrometry and Its Applications
MCGLYNN et al. Molecular Spectroscopy of the Triplet State
MORRISON AND MOSHER Asymmetric Organic Reactions
PASTO AND JOHNSON Organic Structure Determination
SANDORFY Electronic Spectra and Quantum Chemistry

PRENTICE-HALL INTERNATIONAL, INC., *London*
PRENTICE-HALL OF AUSTRALIA, PTY. LTD., *Sydney*
PRENTICE-HALL OF CANADA, LTD., *Toronto*
PRENTICE-HALL OF INDIA PRIVATE LIMITED, *New Delhi*
PRENTICE-HALL OF JAPAN, INC., *Tokyo*

Inorganic Chemistry

GREGORY C. DEMITRAS, F.S.C.
LA SALLE COLLEGE, PHILADELPHIA

CHARLES R. RUSS
UNIVERSITY OF MAINE, ORONO

JAMES F. SALMON, S.J.
LOYOLA COLLEGE, BALTIMORE

JAMES H. WEBER
UNIVERSITY OF NEW HAMPSHIRE, DURHAM

GERALD S. WEISS
MILLERSVILLE STATE COLLEGE, PENNSYLVANIA

Coordinator: Gregory C. Demitras, F.S.C.

PRENTICE-HALL, INC.
ENGLEWOOD CLIFFS, NEW JERSEY

10 9 8 7 6 5 4 3 2 1

ISBN: 0–13–466359–4

Library of Congress Catalog Card Number: 74–39589

Printed in the United States of America

Contents

Part III. The Behavior of Atoms and Molecules

B. THE STUDIED BEHAVIOR OF SELECTED ELEMENTS

Preface

Inorganic chemistry is perhaps the most difficult area of chemistry to define. Until its renaissance in the 1950's, inorganic chemistry consisted of the descriptive chemistry of all the elements other than carbon. During the 1950's it became a subject in which many loose ends of physical and analytical chemistry were tied together. Inorganic chemistry finally evolved into a course containing qualitative aspects of bonding theory, structural concepts, and the new explanations in rapidly developing areas such as coordination chemistry. This evolution relegated descriptive chemistry to a second class status if it were taught at all.

The authors of this text feel there is a present-day need for a book which is directed pointedly at the undergraduate chemistry major that is flexible, modern, and a challenge to the student. This basic premise led us to adopt the following guidelines to direct our efforts with regard to this text: (1) it should be aimed at the undergraduate level, (2) it should contain timely topics which are important to modern inorganic chemistry, and (3) the importance of descriptive chemistry should be recognized by its systematic treatment based on previously presented theoretical concepts. This treatment, then, should provide the advanced undergraduate with a fundamental treatment of inorganic chemistry that will stimulate his interest and initiative and provide him with a substantial foundation for further progress in chemistry.

It is felt that certain topics have unique coverage in this text. These include: (1) the inorganic solid state, (2) inorganic reaction mechanisms, (3) electrochemistry, (4) acid-base theory, and (5) organometallic chemistry. These topics lend balance to a modern treatment of modern inorganic chemistry and will surely be welcomed by both students and teachers. An additional feature of this text is the challenging comprehensiveness of many of the questions at the end of each chapter. The questions are representative of several concepts discussed in the text, and their challenge lies in the students tying these concepts together.

The influence of our past teachers and particular chemists will be evident in this text. Four of us, G.C.D., C.R.R., J.F.S., and G.S.W. were profoundly influenced by Alan MacDiarmid at the University of Pennsylvania. J.H.W. is particularly indebted to Daryle S. Busch at Ohio State University, as is G.C.D. to Earl L. Muetterties for his advice and inspiration. The influence of these men is rich and deep, and we are indebted to them.

Appreciation must also be expressed to our many colleagues who have encouraged us and generously have provided us with constructive criticism. A word of thanks is due to the authors and editors who have kindly granted us permission to reproduce diagrams from their works. A final word of thanks is due to Miss Mary Park for her persevering professional preparation of the manuscript.

Inorganic Chemistry

PART I

The Atom

CHAPTER 1

The Electronic Structure of the Atom

Part I. The Atom

INTRODUCTION

The results of the research of physicists working with light have shown us that there is a relationship to spectra and the structure of matter. This relationship has been the source of our knowledge of the electronic structure of the atom. Still more fascinating is the fact that the study of the nature of light has shown that light has the properties of both a propagated wave and a stream of particles. A hypothesized relationship by de Broglie,

$$p = \frac{h}{\lambda} \tag{1.1}$$

where p is the momentum, h is Planck's constant (6.6256×10^{-27} erg-sec), and λ is the wavelength of the light under discussion, has been shown to be correct. If this concept were extended to all moving species, one could see that it would be quite possible to calculate the wavelength of a student jaunting down a school corridor, and there would be a wavelength associated with a racing car speeding around the course.

If one looks on an object irradiated with white light, he sees that the object absorbs some of the light and reflects the remainder. The reflected light gives the image and color of the object to the eye. Therefore, white light is considered to be a probe. Electrons are much smaller than this probe, so when one irradiates an electron, the light hitting an electron may very well alter the path, position, or energy of the electron. One may very well be able to say something about the energy of the electron from a consideration of the light absorbed, but at the same time the information on the position of the electron will not be reliable. This example of the determination of the energy (or momentum) of the electron giving an unreliable determination of its position is an illustration of the Heisenberg uncertainty principle. The Heisenberg uncertainty relationship is

$$(\Delta x)(\Delta p) \geqslant \frac{h}{2\pi} \qquad [1.2]$$

where Δx and Δp represent uncertainty in position and momentum, respectively, and h is Planck's constant. This relationship warns us that in the measurement of position and momentum there is an uncertainty that is equal to or greater than $h/2\pi$. Of course, the smaller the system under measurement, the more significant is the uncertainty.

The purpose of the presentation of these two items is to give the reader a feeling for the *uncertainty physics* of the day. We shall also call upon these concepts to explain certain phenomena.

THE WAVE EQUATION

The famous wave equation presented by Schrödinger is an application of the classical wave concepts to matter. A simplistic form of the equation is

$$\mathcal{H}\psi = E\psi \qquad [1.3]$$

where $\mathcal{H}$ is the Hamiltonian operator, ψ is an eigenfunction or wave function, and E is the eigenvalue or energy associated with a given wave function. The Hamiltonian operator contains all the energy terms, both kinetic and potential, of the system under study. It has been said that if you know the Hamiltonian of a system, you know all that there is to know about that system. The solution of the wave equation is very difficult, if not impossible, to solve for most systems.

The symbol ψ represents a wave function or eigenfunction. The wave function is a function that is finite, single valued, and continuous at all points. The wave function is open to a few seemingly different interpretations. First, the wave function may be considered the amplitude of a de

Broglie wave. The square of the amplitudes is the intensity of a wave; hence, the square of the wave function, ψ^2, is the intensity of the wave. In terms of the hydrogen atom, we may say that the wave function defines the motion of the electron in the hydrogen atom; hence, ψ^2 will give the intensity of the electron at a given point. By intensity is meant the relative density of the electron (charge density) if the electron is considered to be in motion; however, if the electron is assumed to be stationary, the intensity may be defined as the probability of finding an electron at a given point. It may be well to recall at this point the Heisenberg principle—one can never know the exact position of an electron.

We should note that equation [1.3] is time independent. This fact makes the preceding two interpretations of intensity identical—one interpretation is the density interpretation and the other is a position interpretation. If we made short, instantaneous studies of each interpretation at different times, we would obtain fuzzy accounts of both interpretations. However, if we collect an innumerable number of these fuzzy accounts and superimpose them, a full picture of electron density and electron position probability would develop. Both pictures would be identical. This, of course, means that we may use the terms electron density and electron probability interchangeably.

Probability, P, may be defined as

$$P = \int \psi^2 \, d\tau \qquad [1.4]$$

where ψ is the wave function and $d\tau$ is an element of volume of space ($d\tau = dx\, dy\, dz$). This definition implies that

$$\int_{-\infty}^{+\infty} \psi^2 \, d\tau = 1 \qquad [1.5]$$

which means that the probability of finding an electron associated with a given hydrogen nucleus (if one is considering the hydrogen atom) is unity, or, simply, that there is an electron somewhere to be found. If the condition that is expressed by equation [1.5] is true for a given integration of a squared wave function, then that expression is said to be normalized. If, on the other hand, the expression is not normalized, then we may include a normalization constant such that

$$N^2 \int_{-\infty}^{+\infty} \psi_a^2 \, d\tau = 1 \qquad [1.6]$$

where N is the normalization constant.

It may be as well at this point to introduce the reader to an alternate notation—the *bra* and *ket* symbolism of Dirac. Some consider this system easier to write and find it more compact. In this notation, the wave function ψ_x is represented by the *ket* $|\rangle$; ψ_x is now written as $|x\rangle$ and ψ_{1s} is now written as $|1s$. The *bra* symbol $\langle|$ is used in conjunction with the ket to indicate the integration of the product of the wave functions over all space in the bra and the ket. Thus, if ψ_a is $|a\rangle$ and ψ_b is $|b\rangle$, then $\int \psi_a \psi_b d\tau$ is written as $\langle a|b\rangle$. If the Hamiltonian is operating on ψ_b, then $\int \psi_a H\psi_b d\tau$ may be written as $\langle a|H|b\rangle$. Now equation [1.5] may be written as

$$\langle a|b\rangle = 1 \qquad [1.5]$$

and equation [1.6] may be written as

$$N^2\langle a|a\rangle = 1 \qquad [1.6]$$

Another property of the wave function is expressed by the following equation:

$$\langle a|b\rangle = 0 \qquad [1.7]$$

When the integration of the product of two different wave functions over all space is zero, then the wave functions are said to be orthogonal. One interpretation of orthogonality as applied to an atom might be that the probability of finding an electron in two orbitals of the same atom at the same time is zero.

The Schrödinger equation may be expressed in alternate forms. In Cartesian coordinates, the form is

$$\frac{\partial^2\psi}{\partial x^2} + \frac{\partial^2\psi}{\partial y^2} + \frac{\partial^2\psi}{\partial z^2} + \frac{8\pi^2 m}{h^2}(E - V)\psi = 0 \qquad [1.8]$$

where ψ is the wave function, h is Planck's constant, m is the mass of the electron, E is the energy of the electron, and V is the potential energy of the electron. In spherical coordinates, equation [1.8] has the following form:

$$\frac{1}{r^2}\frac{\partial}{\partial r}\left(r^2\frac{\partial\psi}{\partial r}\right) + \frac{1}{\sin\theta}\frac{\partial}{\partial\theta}\left(\sin\theta\frac{\partial\psi}{\partial\theta}\right) + \frac{1}{\sin^2}\frac{\partial^2\psi}{\partial\phi^2} + \frac{8\pi^2 mr^2}{h^2}(E - V)\psi = 0 \qquad [1.9]$$

where r is the distance of the electron from the nucleus, ϕ is the azimuthal angle (angular deviation from the XZ plane), and θ is the polar angle (angular deviation from the x axis). To give the student some idea of the complexity of solving the Schrödinger equation, we list a number of wave functions in Table 1.1.

TABLE 1.1 *Normalized Wave Functions of Hydrogenlike Atoms*

($\rho = Zr/a_0$; Z = atomic number; $a_0 = 0{\cdot}5$ A, the Bohr radius)

Orbital	Polar Coordinates	Cartesian Coordinates
$1s$	$\frac{1}{\sqrt{\pi}}\left(\frac{Z}{a_0}\right)^{3/2} e^{-\rho}$	$\frac{1}{\sqrt{\pi}}\left(\frac{Z}{a_0}\right)^{3/2} e^{-\rho}$
$2s$	$\frac{1}{4\sqrt{2\pi}}\left(\frac{Z}{a_0}\right)^{3/2}(2-\rho)e^{-\rho/2}$	$\frac{1}{4\sqrt{2\pi}}\left(\frac{Z}{a_0}\right)^{3/2}(2-\rho)e^{-\rho/2}$
$2p_x$	$\frac{1}{4\sqrt{2\pi}}\left(\frac{Z}{a_0}\right)^{3/2}\rho e^{-\rho/2}\sin\theta\cos\phi$	$\frac{1}{4\sqrt{2\pi}}\left(\frac{Z}{a_0}\right)^{3/2}\rho e^{-\rho/2}\frac{x}{r}$
$3d_{xy}$	$\frac{1}{81\sqrt{\pi}}\left(\frac{Z}{a_0}\right)^{3/2}\rho^2 e^{-\rho/3}\sin^2\theta\sin 2\phi$	$\frac{\sqrt{2}}{81\sqrt{\pi}}\left(\frac{Z}{a_0}\right)\rho^2 e^{-\rho/3}\frac{xy}{r^2}$

When the Schrödinger equation is solved, ψ is resolved into three terms called quantum numbers: n, l, and m_l. If our model is the hydrogen atom, we may say that these quantum numbers describe the behavior of an electron. There is a fourth quantum number, a spin quantum number (m_s), appearing in a later refinement of the theory.

The Principal Quantum Number, n

This quantum number defines the energy of an electron. The energy of an electron is directly related to the principal quantum number,

$$E = f\left(-\frac{1}{n^2}\right) \qquad [1.10]$$

where E is the energy of the electron and n is the principal quantum number. The quantum number, n, is quantized; the values that n may assume are 1, 2, 3, 4 . . . . When $n = 1$, the electron in the hydrogen atom is said to be in its lowest energy state, which is called the *ground state*. In general, the term ground state is used to describe any system in its lowest energy state.

Any system not in the ground state is said to be in an *excited state*, as when $n = 2, 3, 4 \ldots$, etc., for the hydrogen atom. Note that as the energy of the electron is increased, so is its distance from the nucleus. If an atom is excited, say, by some electrical means, the excited atom normally returns to the ground state. The energy that was absorbed by the atom in becoming excited is given off as quanta of light ($h\nu$). This is the origin of an emission spectrum. In summation, the principal quantum number indicates the relative energy of the electron and its relative distance from the nucleus.

The Azimuthal Quantum Number, l

This quantum number gives a measure of the angular momentum of an electron in its orbital motion about the nucleus; it is generally referred to as orbital angular momentum. Spectroscopy has used letters to identify this quantum number. The letters used are *s, p, d, f, g, h, i, . . .* when the values of l are 0, 1, 2, 3, 4, 5, 6, 7 . . . . As with n, the higher the value of l, the higher is the energy of the electron. The quantum number l is quantized by taking the integral values 0, 1, 2, 3, . . . to a maximum value of $n - 1$. Therefore, when $n = 1$, the only value that l may assume is $l = 0$. When $n = 3$, $l = 2$, 1, and 0.

The Magnetic Quantum Number, m_l

If an excited atom is placed in a strong magnetic field, some of the spectral lines that appear as single lines in the absence of the magnetic field will appear as several lines. This is known as the Zeeman effect. Since an electron is a charged particle, its motion in the orbit causes it to set up its own magnetic field, which may detract or enhance an applied, external, magnetic field. Therefore, the plane of the electron motion may take several orientations with respect to the applied magnetic field. These orientations are of different energies. In the absence of a magnetic field, the planes of electron motion are degenerate—i.e., they are of equal energy. Quantum mechanics has shown there are $2l + 1$ orientations of the planes of electronic motions; therefore, the values of m_l are $-l \ldots 0 \ldots +l$. An electron having $n = 3$, $l = 2$, may have m_l values $-2, -1, 0, 1, 2$. When $l = 2$, m_l assumes five values ($2l + 1$).

The Spin Quantum Number, m_s

Experimental observations of spectra of atoms with only one electron in the valence shell revealed that the number of energy states was twice as great as was predicted. It was shown that the number of energy states expected would be doubled if the electron were spinning about its own axis. The electron would then possess both orbital momentum and spin momentum. Dirac introduced an extra energy term into the wave equation and expressed the equation in a relativistic invariant form. As a result, the spin quantum number appeared in the solution.

It was found that m_s may assume two values, $+\frac{1}{2}$ or $-\frac{1}{2}$. These two possibilities indicate that a spinning electron may interact in two ways with an applied magnetic field—it may enhance it or it may oppose it.

At this point, we should write the quantum numbers for the electrons in a number of atoms. For the hydrogen atom in the ground state, $n = 1$. Therefore, the only value that l may have is 0; m_l must be zero; and, m_s may be either $+\frac{1}{2}$ or $-\frac{1}{2}$. Table 1.2 contains the four quantum numbers of electrons having $n = 1$. Or, Table 1.2 may be used to determine the number of elements in the first row of the periodic table.

TABLE 1.2 ***Four Quantum of Electrons Having n = 1***

n	l	m	m_s
1	0	0	+*
1	0	0	−

* $+ = +\frac{1}{2}$; $- = -\frac{1}{2}$

This short table offers the following information. When $n = 1$, only two variations in quantum numbers are permitted, and they lie in the spin number of the electron. Now, if one were considering the helium atom ($Z = 2$) in the ground state, it may be said that the two electrons of helium may occupy the 1s orbital, but they must have opposite spins. This is an application of the Pauli principle, which states that in a given atom no two electrons may have the same set of quantum numbers. One final interpretation of Table 1.2 may be made in terms of the periodic table: Only two elements may exist in the first row of the periodic table. This point will become clearer.

Table 1.3 contains the quantum number possibilities for $n = 2$. 'This tabulation is also open to various interpretations. In terms of the hydrogen atom, the tabulation tells us that if the lone electron associated with the hydrogen nucleus is on the second energy level, there are eight orientations or eight excited states of the atom. In terms of the total number of electrons, the Pauli principle dictates that the maximum number of electrons on the

TABLE 1.3 ***The Four Quantum Numbers of Electrons Having n = 2***

n	l	m_l	m_s
2	1	−1	−
2	1	−1	+
2	1	0	−
2	1	0	+
2	1	1	−
2	1	1	+
2	0	0	+
2	0	0	−

second energy level is eight. In terms of the periodic table, the tabulation argues the fact that there are eight elements in the second row.

The tabulation of all quantum numbers where $n = 3$ is given in Table 1.4. The reader should rationalize this tabulation.

TABLE 1.4 ***The Four Quantum Numbers of Electrons Having n = 3***

n	l	m_l	m_s
3	2	−2	−
3	2	−2	+
3	2	−1	−
3	2	−1	+
3	2	0	−
3	2	0	+
3	2	1	−
3	2	1	+
3	2	2	−
3	2	2	+
3	1	−1	−
3	1	−1	+
3	1	0	−
3	1	0	+
3	1	1	−
3	1	1	+
3	0	0	−
3	0	0	+

At this point, it might be well to return to some concepts related to the wave function. The wave function describes the electron in the hydrogen atom or any atom. There is a given energy (eigenvalue) associated with a given wave function (eigenfunction), and this is obtained from the wave equation. Now the wave function is also related to the electron density or the probability of locating the electron. Since the quantum numbers are directly involved in the wave equation, they are also directly involved with energy and probability. In other words, the quantum numbers are directly related to volumes of space at given distances from the nucleus (energy) where we may reliably find the electron (probability). These volumes of space are called orbitals. An orbital may be defined as a geometric volume where one has a high degree of probability of finding an electron.

The orbitals in multielectron atoms are described by referring to their n and l quantum numbers, using the letter representation for the l quantum number. On the first energy level, $n = 1$ and $l = 0$; this is the only possible combination. Therefore, one says that the first energy level has or contains a 1s orbital. The second energy level ($n = 2$; $l = 0, 1$) contains 2s and 2p orbitals. On the third energy level ($n = 3$; $l = 0, 1, 2$), 3s, 3p, and 3d orbitals exist.

Closer scrutiny of the tabular representations of the quantum numbers will give more information. In all cases where $l = 0$, $m_l = 0$, m_s is + or −. This indicates that there can be only one s orbital on each energy level, and the maximum number of electrons that an s orbital may contain is two. When $l = 1$, we have six possibilities of sets of quantum numbers. The number of different combinations of l and m_l is the key to the number of orbitals. On both the second and third energy levels, when $l = 1$, there are three different combinations of l and m_l: 1, −1; 1, 0; and 1, 1. Extrapolating, one can then say that in each of the energy levels above the first there are three p orbitals. In each of these orbitals, m_s has a + and a − value, indicating that each of the p orbitals can contain two electrons and all of the p orbitals can contain a maximum of six electrons. The third energy level has a 3s orbital, three 3p orbitals, and five 3d orbitals. For a complete understanding of these ideas, the student should rationalize the number of 3d orbitals, the number of electrons in each of the 3d orbitals and the total number of electrons in all of the 3d orbitals, and the maximum number of electrons that the third energy level ($n = 3$) may contain.

We now focus upon the relative energies of the orbitals. This discussion is summarized in Fig. 1.1. If we were to excite the lone electron in the hydrogen atom by stepwise, quantum energy increments, stopping at each possible quantum jump, the electron would rise from the first energy level, to the second, to the third, and so on. In the second energy level, the electron would go into the 2s orbital and then the 2p orbital before going on to the

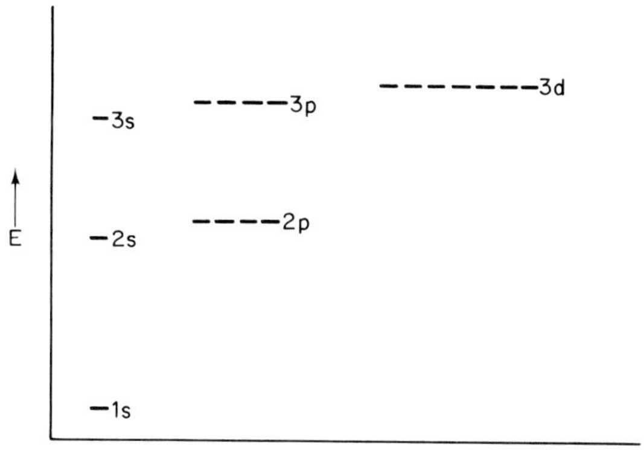

Figure 1.1. Relative energy levels of the orbitals in the hydrogen atom.

third energy level; in the third energy level, the electron would pass from the 3*s* to the 3*p* to the 3*d* orbital. In general, the energy levels increase

$$n = 1 < 2 < 3 < 4 < 5 \ldots \qquad [1.11]$$

and for the same value of *n*, the energy of the orbitals increases as

$$l = s < p < d < f < g \ldots \qquad [1.12]$$

It is necessary to state that the energy of a given orbital is very strongly dependent on *Z*, the atomic number, and that the preceding schemes are an oversimplification of the problem. The energy of a given orbital decreases with an increase in atomic number. The relationship is not a simple one because as the number of protons increases, so does the number of electrons, and the electron–electron interaction changes and becomes more complicated. So if we leave our simple model using the hydrogen atom and go on to the multielectron atom, we should expect discrepancies.

At this point, a mnemonic will aid in determining the ground state electronic configuration of the multielectron atom. This plan of *building*

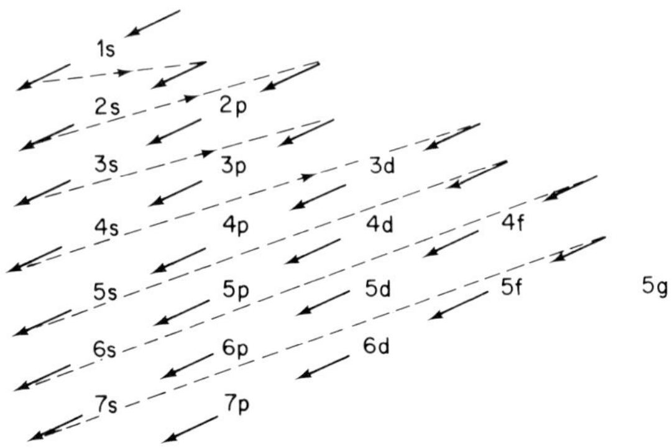

electrons around the nucleus is called the *Aufbau principle*. This principle is based on experimentation.

In placing the electrons in the orbitals around a given atom, just follow

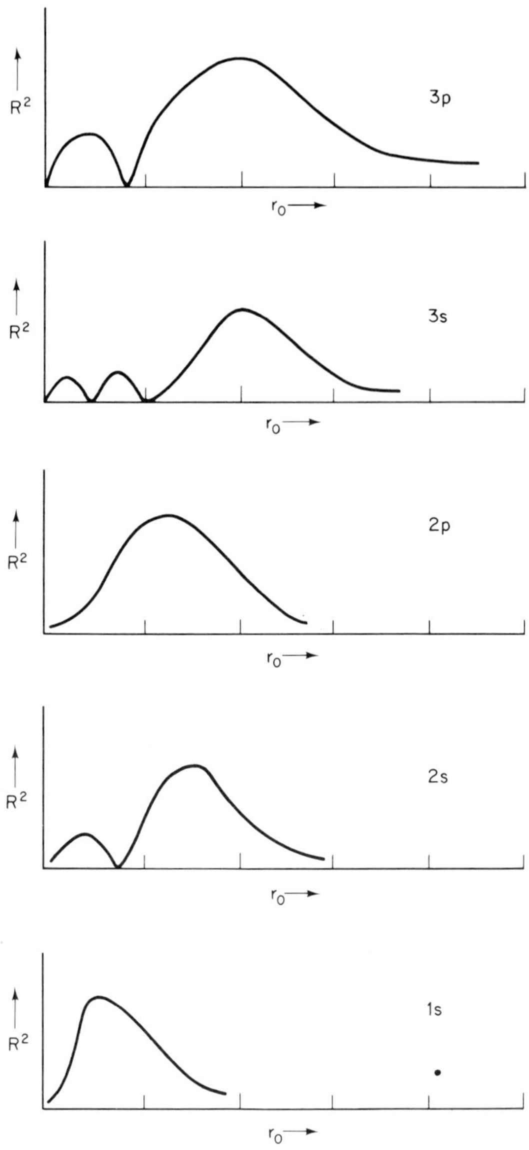

Figure 1.2. Plot of the radial distribution functions for some of the orbitals. $R^2 = 4\pi r^2 \psi_r^2$ and r_0 = units of Bohr radii.

the arrow. Once again, this diagram is sufficient to determine the ground state electronic configuration of most of the known atoms.

There is an explanation offered as to why the 4*s* orbital is filled before the 3*d* orbital; it lies in the concept of the radial distribution function. The wave function for the $2p_x$ orbital in hydrogenlike atoms is $\frac{1}{4}\sqrt{2\pi}(Z/a_0)^{3/2}\ \rho e^{-\rho/2}\ (x/r)$. In the preceding wave function, $\frac{1}{2}\sqrt{6}(Z/a_0)^{3/2}\ \rho e^{-\rho/2}$ is the radial part of the function and $\sqrt{\frac{3}{2}}/\pi(x/r)$ is the angular part of the wave function. The radial distribution function is defined as $4\pi r^2\ \psi_r^2\ dr$, where ψ_r^2 is the radial part of the wave function for a given orbital. The radial part of the wave function is multiplied by $4\pi r^2\ dr$ to obtain the probability of electron density per spherical shell. In general, the radial distribution function is used to give the variation of electron density with distance from the nucleus for a given orbital. Figure 1.2 contains the plot of $4\pi r^2\psi_r^2\ dr$ for various orbitals.

In Fig. 1.3, we have superimposed the plot of radial distribution functions vs. r for the 3*d* and 4*s* orbitals. You must realize that the inner electrons in an atom screen the nuclear charge and prevent it from having a full attractive force on the outer electrons. The comparison of the radial distribution plots in Fig. 1.3 shows that the 4*s* orbital has a greater probability nearer the nucleus than does the 3*d* orbital. In other words, the 4*s* orbital can more effectively penetrate the screen of the inner electrons than can the 3*d*. Therefore, it may be said that 4*s* orbital may be lower in energy than the 3*d* orbital, and the electron would prefer this lower energy state.

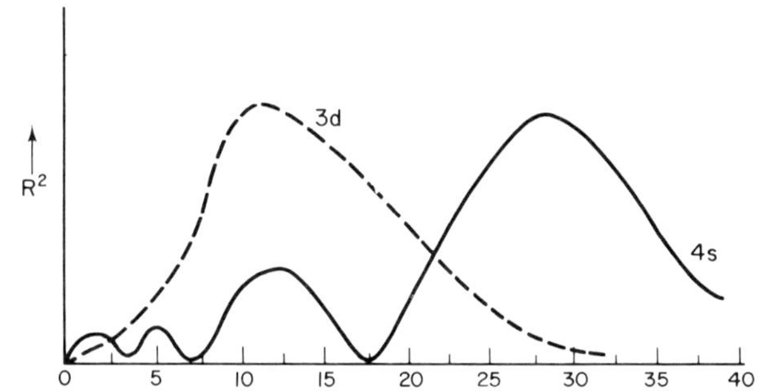

Figure 1.3. The radial distribution functions for the 4*s* and 3*d* orbitals superimposed.

We shall now consider the ground state electronic configurations of the atoms, remembering that the atomic number determines the number of electrons in a neutral atom. Table 1.5 contains the electronic configurations of the atoms in the first row of the periodic table.

TABLE 1.5 ***Ground State Electronic Configuration of Atoms in First Row of the Periodic Table***

Z	Symbol	Electronic Configuration	Valence Electrons $1s$
1	H	$1s^1$	↑
2	He	$1s^2$	↑↓

The symbol $1s^1$ indicates that the electronic ground state consists of one electron in the $1s$ orbital. The symbol $1s^2$ indicates that the $1s$ orbital contains two electrons. Pauli's principle is operative here and the electrons must have opposite spins. This is indicated by the arrow notation, ↑↓.

The reader should note the filling of the $2p$ orbitals with reference to boron, carbon, and nitrogen in Table 1.6. The p orbitals are filled with

TABLE 1.6 ***Ground State Electronic Configuration of Atoms in the Second Row of the Periodic Table***

Z	Symbol	Electronic Configuration	Valence Electrons $2s$	$2p$		
3	Li	$1s^22s^1$	↑	_	_	_
4	Be	$1s^22s^2$	↑↓	_	_	_
5	B	$1s^22s^22p^1$	↑↓	↑	_	_
6	C	$1s^22s^22p^2$	↓↑	↑	↑	_
7	N	$1s^22s^22p^2$	↑↓	↑	↑	↑
8	O	$1s^22s^22p^4$	↑↓	↑↓	↑	↑
9	F	$1s^22s^22p^5$	↑↓	↑↓	↑↓	↑
10	Ne	$1s^22s^22p^6$	↑↓	↑↓	↑↓	↑↓

electrons that are parallel in spin until each orbital has one electron. There is no distinction among the p orbitals because they are degenerate—i.e., they are of equal energy. This preferential placing of electrons with parallel spins in orbitals that are degenerate has been shown, experimentally, to be a state of lower energy than that where the spins are paired. Formally, this rule is known as Hund's rule of *maximum multiplicity*.

The long form of the periodic table is constructed so that each row of the table represents the energy level of the valence electrons. There are only two elements in the first row of the periodic table because the $1s$ orbital has a capacity of two electrons. In the second row of the periodic table there are eight elements, because the second energy level has a capacity of eight electrons.

TABLE 1.7 *Ground State Electronic Configurations of Elements in the Third Row of the Periodic Table*

Z	Symbol	Electronic Configuration	Valence Electrons $3s$	$3p$		
11	Na	(Ne)*$3s^1$	↑			
12	Mg	(Ne) $3s^2$	↑↓			
13	Al	(Ne) $3s^2 3p^1$	↑↓	↑		
14	Bi	(Ne) $3s^2 3p^2$	↑↓	↑	↑	
15	P	(Ne) $3s^2 3p^3$	↑↓	↑	↑	↑
16	S	(Ne) $3s^2 3p^4$	↑↓	↑↓	↑	↑
17	Cl	(Ne) $3s^2 3p^5$	↑↓	↑↓	↑↓	↑
18	Ar	(Ne) $3s^2 3p^6$	↑↓	↑↓	↑↓	↑↓

* (Ne) refers to a *Ne core*, that is, the inner electrons have the identical configuration as Ne – $1s^2 2s^2 2p^6$.

Tables 1.7 and 1.8 offer abundant material for discussion. Notably, there are only 8 elements in the third row of the periodic table (Na to Ar) and there are 18 elements in the fourth row. It might be expected, from first predictions, that these rows might contain 18 and 32 elements, respectively. Remember that the atomic number is determined experimentally and that the chemical properties of the elements are experimental facts. We showed previously that the $4s$ orbital may be filled with electrons before the $3d$ orbital. After the argon atom is *constructed*, the next element is potassium. It has been determined that the ground state of potassium's valence electron is $4s^1$. Its chemical properties also indicate that it belongs to the same family as does sodium. The same argument may be extended for positioning calcium under magnesium and elements 31 to 36 in their respective positions. Since the elements 21 to 30 have electrons in the $4s$ orbital, they rightly belong in the fourth row of the periodic table because the row number of the periodic table indicates the maximum energy level of the valence electrons. This should rationalize the positioning of elements 11 to 18 and elements 19 to 36 in rows 3 and 4 of the periodic table.

A comparison of Tables 1.5 through 1.8 should also reveal a vertical relationship among the ground state electronic configurations of the

TABLE 1.8 *Ground State Electronic Configurations of Elements in the Fourth Row of the Periodic Table*

Z	Symbol	Electronic Configuration	Outer Electrons 4s	3d					4p		
19	K	(Ar) $4s^1$	↑								
20	Ca	(Ar) $4s^2$	↑↓								
21	Sc	(Ar) $4s^23d^1$	↑↓	↑	—	—	—	—	—	—	—
22	Ti	(Ar) $4s^23d^2$	↑↓	↑	↑	—	—	—	—	—	—
23	V	(Ar) $4s^23d^3$	↑↓	↑	↑	↑	—	—	—	—	—
24	Cr	(Ar) $4s^13d^5$	↑	↑	↑	↑	↑	↑	—	—	—
25	Mn	(Ar) $4s^23d^5$	↑↓	↑	↑	↑	↑	↑	—	—	—
26	Fe	(Ar) $4s^23d^6$	↑↓	↑↓	↑	↑	↑	↑	—	—	—
27	Co	(Ar) $4s^23d^7$	↑↓	↑↓	↑↓	↑	↑	↑	—	—	—
28	Ni	(Ar) $4s^23d^8$	↑↓	↑↓	↑↓	↑↓	↑	↑	—	—	—
29	Cu	(Ar) $4s^13d^{10}$	↑	↑↓	↑↓	↑↓	↑↓	↑↓	—	—	—
30	Zn	(Ar) $4s^23d^{10}$	↑↓	↑↓	↑↓	↑↓	↑↓	↑↓	—	—	—
31	Ga	(Ar) $4s^23d^{10}4p^1$	↑↓	↑↓	↑↓	↑↓	↑↓	↑↓	↑	—	—
32	Ge	(Ar) $4s^23d^{10}4p^2$	↑↓	↑↓	↑↓	↑↓	↑↓	↑↓	↑	↑	—
33	As	(Ar) $4s^23d^{10}4p^3$	↑↓	↑↓	↑↓	↑↓	↑↓	↑↓	↑	↑	↑
34	Se	(Ar) $4s^23d^{10}4p^4$	↑↓	↑↓	↑↓	↑↓	↑↓	↑↓	↑↓	↑	↑
35	Br	(Ar) $4s^23d^{10}4p^5$	↑↓	↑↓	↑↓	↑↓	↑↓	↑↓	↑↓	↑↓	↑
36	Kr	(Ar) $4s^23d^{10}4p^6$	↑↓	↑↓	↑↓	↑↓	↑↓	↑↓	↑↓	↑↓	↑↓

elements in each group of the periodic table. In the first column of the periodic table (hydrogen and the alkali metals), one should notice that these elements all have a ground state configuration ns^1. The second group has the ground state configuration ns^2. The boron family of elements has the ground state valence configuration ns^2np^1; the oxygen family has the ns^2np^4 ground state valence configuration; and, with the exception of He, the noble gases have the ns^2np^6 configuration. Of course, the value of n is determined by the row of the periodic table in which the element resides.

In the preceding paragraph, we discussed the elements in which the electrons are filling the ns and np orbitals. These elements are called *representative elements*. On long forms of the periodic chart, they are usually referred to as the A elements, and their groups are called I-A, II-A, III-A, etc. There is another group of elements exemplified by elements 21 to 30 in Table 1.8. They fill the d orbitals and possess s electrons. They have the outer orbital configuration ns^2 $(n-1)d^x$, where x is the number of electrons in the d orbitals. These elements are called transition elements.

They are found in the middle portion of the long form of the periodic table and are usually labeled the B elements; I-B, II-B, etc.

Also very noticeable in Table 1.8 are the electronic configurations of Cr[(Ar)$4s^1$ $3d^5$] and Cu[(Ar)$4s^1$ $3d^{10}$]. At first glance, it might be expected that Cr should be (Ar)$4s^2$ $3d^4$ and Cu should be (Ar)$4s^2$ $3d^9$. The given ground state electronic configurations are rationalized by the fact that a half-filled set of degenerate orbitals is a very stable species, as is a completely filled set of orbitals. Experimentally, this fact is supported by the ionization potential (work necessary to remove the electron most distant from the nucleus) of some of the elements. If one would remove a single p electron from boron, carbon, and nitrogen, the energy required to remove an electron from a singly occupied p orbital increases due to the addition of a proton to the nucleus of the atoms involved. In Table 1.9, we see that the ionization potential of a ground state oxygen atom is less than that of a ground state nitrogen atom. The difference in electronic configuration lies in the fact that the electron removed from the oxygen atom is paired with another electron. No doubt repulsion exists between electrons, and this repulsion energy is related to oxygen having a lower ionization potential than nitrogen. The correlation of these ideas with the fact that the noble gases are relatively inert leads to the concept that half-filled orbitals (p^3, d^5, f^7, etc.) and filled orbitals have the property of some extra stability. Analogously, we may rationalize the ground state electronic configurations

TABLE 1.9 *Ionization Potentials of Some Elements (eV/atom)*

Li	5.390	N	14.54	Na	5.138	P	10.484
Be	9.320	O	13.614	Mg	7.644	S	10.357
B	8.296	F	17.418	Al	5.984	Cl	13.01
C	11.264	Ne	21.559	Si	8.149	Ar	14.339

of chromium and copper by assuming that the energy required to uncouple and transfer an electron from the $4s$ to the $3d$ orbital is more than compensated for by the added stability gained by the $3d^5$ and $3d^{10}$ configurations.

Table 1.10 contains the ground state electronic configurations for the remainder of the elements of the periodic table. There are two groups of elements—elements 58 through 71 and elements 90 through 103—that may be characterized by the fact that in building up their electronic structure they are filling the $(n-2)f$ orbitals. Here, n represents the maximum energy level of the valence electrons. These elements are called inner transition elements. On the long form of the periodic table, these elements are usually

TABLE 1.10 *Ground State Electronic Configurations of Elements 37 through 103*

Z	Symbol	Electronic Configuration	Z	Symbol	Electronic Configuration
37	Rb	(Kr) $5s^1$	70	Yb	$—4f^{14}6s^2$
38	Sr	$—5s^2$	71	Lu	$—4f^{14}5d^16s^2$
39	Y	$—4d^15s^2$	72	Hf	$—4f^{14}5d^26s^2$
40	Zr	$—4d^25s^2$	73	Ta	$—4f^{14}5d^36s^2$
41	Nb	$—4d^45s^1$	74	W	$—4f^{14}5d^46s^2$
42	Mo	$—4d^55s^1$	75	Re	$—4f^{14}5d^56s^2$
43	Tc	$—4d^55s^2$	76	Os	$—4f^{14}5d^66s^2$
44	Ru	$—4d^75s^1$	77	Ir	$4f^{14}5d^76s^2$
45	Rh	$—4d^85s^2$	78	Pt	$4f^{14}5d^96s^1$
46	Pd	$—4d^{10}$	79	Au	$4f^{14}5d^{10}6s^1$
47	Ag	$—4d^{10}5s^1$	80	Hg	(Au) $6s^2$
48	Cd	$—4d^{10}5s^2$	81	Te	$—6s^26p^1$
49	In	(Kr) $4d^{10}5s^25p^1$	82	Pb	$—6s^26p^2$
50	Sn	$—4d^{10}5s^25p^2$	83	Bi	$—6s^26p^3$
51	Sb	$—4d^{10}5s^25p^3$	84	Po	$—6s^26p^4$
52	Te	$—4d^{10}5s^25p^4$	85	At	$—6s^26p^5$
53	I	$—4d^{10}5s^25p^5$	86	Rn	$—6s^26p^6$
54	Xe	$—4d^{10}5s^25p^6$	87	Fr	(Rn) $7s^1$
55	Cs	(Xe) $6s^1$	88	Ra	$—7s^2$
56	Ba	$—6s^2$	89	Ac	$—6d^17s^2$
57	La	$—5d^16s^2$	90	Th	$—6d^27s^2$
58	Ce	$—4f^26s^2$	91	Pa	$—5f^26d^17s^2$
59	Pr	$—4f^36s^2$	92	U	$—5f^36d^17s^2$
60	Nd	$—4f^46s^2$	93	Np	$—5f^57s^2$
61	Pm	$—4f^56s^2$	94	Pu	$—5f^67s^2$
62	Sm	$—4f^66s^2$	95	Am	$—5f^77s^2$
63	Eu	$—4f^76s^2$	96	Cm	$—5f^76d^17s^2$
64	Gd	$—4f^75d^16s^2$	97	Bk	$—5f^86d^17s^2$
65	Tb	$—4f^96s^2$	98	Cf	$—5f^{10}7s^2$
66	Dy	$—4f^{10}6s^2$	99	Es	$—5f^{11}7s^2$
67	Ho	$—4f^{11}6s^2$	100	Fm	$—Ff^{12}7s^2$
68	Er	$—4f^{12}6s^2$	101	Md	$—5f^{13}7s^2$
69	Tm	$—4f^{13}6s^2$	102	No	$—5f^{14}7s^2$
			103	Lw	$—5f^{14}6d7s^2$

placed in two rows below the body of the chart. Elements 58–71 should immediately follow the lanthanum square, hence the name lanthanides (rare earth elements); elements 90–103 should immediately follow the actinium square, hence the name actinides. Their positioning is a matter of convenience.

Finally, we might point out that the electronic configuration may be utilized to distinguish between a metal and a nonmetal. The distinction lies in the value of n, the principal quantum number of the valence shell.

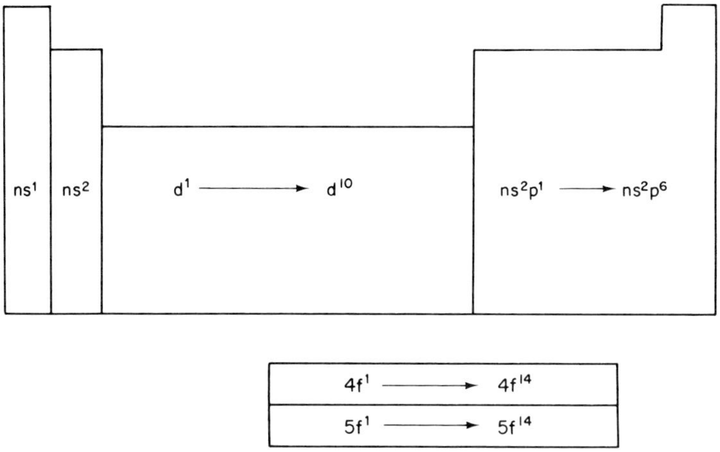

Figure 1.4. Relationship of the positions of the elements in the periodic table to their ground state electronic structure.

If the number of electrons in the valence shell is less than *n*, then the element is a metal; if the number of electrons in the valence shell is greater than *n*, then the element is a nonmetal. This distinction between metal and nonmetal is not sharp, especially when *n* and the number of valence electrons are equal or nearly equal. The elements along the silicon, arsenic, and tellurium line on the periodic chart have some metallic properties and are usually called metalloids. But as we depart from this line, the distinction between metal and nonmetal based on electronic configuration is sharp.

THE GEOMETRY OF THE ORBITALS

Because the wave functions are directly related to the probability of finding an electron in space, it follows that if one sets boundaries on the space in which there is a 99% chance of finding an electron, a three-dimensional geometric shape would result. This *diaphanous space* is the orbital. The orbital is not a solid thing but merely a volume of space where there is a very high probability of finding an electron. The memorization of the geometrical shapes of these orbitals is essential for understanding the theories of bonding.

The boundary surface of an *s* orbital describes a sphere. The wave functions of the 2*s* orbitals given in Table 1.1 show that these wave functions have no angular dependence. This fact implies that the orbital is spherically symmetrical. Comparison of the 1*s* and 2*s* orbitals shows that they are both spherically symmetrical, but the 2*s* orbital is more extended in space or its boundary surface is farther from the nucleus. Quantum

In the Russell-Saunders scheme, L is the resultant azimuthal quantum number, S is the resultant spin number, and J is the resultant of the combination of L and S. For example, if we are considering an atom with an electron configuration p^1d^1, L would have a maximum value of 3(p, $l = 1$; d, $l = 2$) but could also have the values 2 and 1. This vector addition is illustrated in Fig. 1.6.

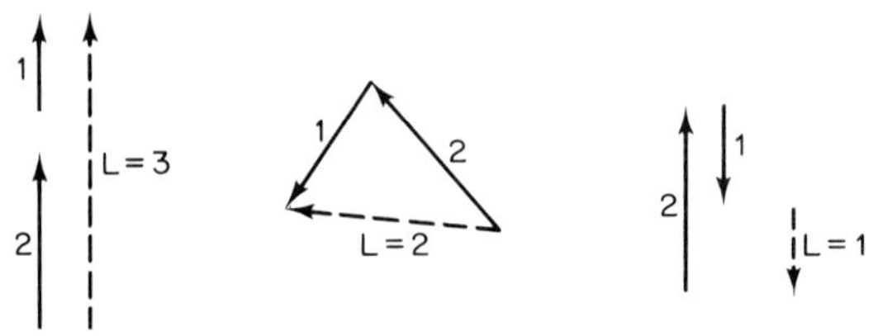

Figure 1.6. Addition of vectors having values of, 2 and 1.

Now the value of S is determined in a similar fashion. Since there are two electrons and the m_s value of these may be $+\frac{1}{2}$ or $-\frac{1}{2}$, S may have the values 1, 0, or -1 when two electrons are considered.

The symbol of an energy term representing a given electron arrangement is ^{2S+1}L where S and L have the preceding definitions. The $2S+1$ part of the symbol is called the *multiplicity*, which may be considered as being indicative of the number of unpaired electrons in a given atom. The L part of the symbol is represented by a capital letter, similar to orbital notation. When $L = 0, 1, 2, 3, 4, 5 \ldots$, it is represented by the corresponding letter S, P, D, F, G, H . . . . Thus, the symbol ^4F implies $S = \frac{3}{2}$, or 3 unpaired electrons having parallel spins exist in an atom described by this symbol and that $L = 3$. Now if the J value is considered, J may assume the values $\frac{9}{2}$, $\frac{7}{2}$, $\frac{5}{2}$, and $\frac{3}{2}$. The complete set of symbols representing the ^4F term are $^4F_{9/2}$, $^4F_{7/2}$, $^4F_{5/2}$, and $^4F_{3/2}$. These are called states of an atom which exhibit the effect of spin-orbit coupling that removes the degeneracy of a given term to produce multiplet (J) levels. In reality, the multiplicity ($2S + 1$) reveals the number of orientations the spin vector may assume in a magnetic field. In other words, if $L > S$, then the multiplicity is indicative of the number of J levels in a given term.

In determining the relative energy of given J values, remember the following points. (1) The lowest value of J has the lowest energy when the term is derived from a less than half-filled subshell. (2) The highest value of J has the lowest energy when the subshell is more than half full. The reader should also, at this time, convince himself that for completely filled shells, $L = 0$, $S = 0$, and $J = 0$.

Let us now consider atoms with one electron in the valence shell, say, a single s electron. For this electron, $L = 0$ and $S = \frac{1}{2}$, which generates a *doublet-S term*, ^2S. The multiplicity is 2, indicating that we have one unpaired electron and that it may assume two orientations with respect to an external magnetic field.

An atom with a single p electron would have $L = 1$ and $S = \frac{1}{2}$, giving a *doublet-P term*, ^2P. There are three p orbitals in which the electron may reside; the electron may be said to have threefold orbital degeneracy. The electron may assume two spin states; therefore it has a twofold spin degeneracy. Its total degeneracy is sixfold.

An atom containing a single d electron may be described by the term *doublet D*, ^2D. This electron has fivefold orbital degeneracy, a twofold spin degeneracy, and tenfold total degeneracy.

Let us now consider an atom with two electrons, say, one in the p orbital and one in the d orbital. In this system, $L = l_1 + l_2 = 3$. Therefore, $L = 3$, 2, or 1 (see Fig. 1.6) and $S = 1$ or 0. There are six possible terms for this atomic system: ^1P, ^1D, ^1F, ^3P, ^3D, and ^3F, all of which differ in energy. This simple approach to terms does not always work, because if we had two equivalent electrons, some terms or electronic arrangements would be forbidden because of the Pauli principle. Therefore, a more complicated procedure is required to distinguish the terms that are Pauli permitted.

Both L and S are space quantized. The components of L will be termed M_L and $M_L = L, L-1, L-2, \ldots, -L$. Analogously, M_S will represent the components of S where $M_S = S, S-1, S-2, \ldots, -S$, with respect to a given field direction. When capital letters are used, such as M_L and M_S, it indicates the total orbital and total spin components. For individual electrons, small letters are used; m_l and m_s represent orbital and spin components for individual electrons. Hence,

$$M_L = \Sigma m_l \qquad [1.16]$$

$$M_S = \Sigma m_s \qquad [1.17]$$

The preceding concepts will be utilized in the derivation of terms for the s^2, p^2, and d^2 cases.

The s^2 case is really trivial; $L = 0$, $S = 0$, and the term is ^1S or $^1S_0$. There is only one orientation of the electrons: they must be paired. This is regulated by the Pauli principle, since there is only one s orbital on each energy level. The term ^3S is valid for two nonequivalent s electrons—i.e., $1s^12s^1$.

For the p^2 case, in practice, it is easier to determine our values for L and S from M_L and M_S. For a p electron, $m_l = +1, 0,$ or -1, and $m_s = \pm\frac{1}{2}$. A microstate, $(\overset{+}{1}, \overset{-}{0})$, will be defined where the numeral—1 or 0—represents the m_l for a given electron and the $+$ and $-$ represents the electron's spin orientation. The reader is asked to rationalize the following statements relative to microstates:

(1) $(\overset{+}{1}, \overset{-}{0})$ has $M_L = 1 + 0 = 1$ and $M_S = \frac{1}{2} + (-\frac{1}{2}) = 0$.

(2) $(\overset{+}{1}, \overset{+}{1})$ is forbidden by the Pauli principle.

(3) $(\overset{+}{1}, \overset{-}{0})$ and $(\overset{-}{1}, \overset{+}{0})$ represent distinct microstates.

(4) $(\overset{+}{0}, \overset{-}{0})$ and $(\overset{-}{0}, \overset{+}{0})$ do not represent distinct microstates because of our inability to distinguish between electrons.

M_L ↓ \ M_S →	1	0	−1
2		$(\overset{+}{1}, \overset{-}{1})$	
1	$(\overset{+}{1}, \overset{+}{0})$	$(\overset{+}{1}, \overset{-}{0})$ $(\overset{-}{1}, \overset{+}{0})$	$(\overset{-}{1}, \overset{-}{0})$
0	$(\overset{+}{1}, -\overset{+}{1})$	$(\overset{+}{1}, -\overset{-}{1})$ $(\overset{-}{1}, -\overset{+}{1})$ $(\overset{+}{0}, \overset{-}{0})$	$(-\overset{-}{1}, -\overset{-}{1})$
−1		$(-\overset{+}{1}, \overset{-}{0})$ $(-\overset{-}{1}, \overset{+}{0})$	$(-\overset{-}{1}, \overset{-}{0})$
−2		$(-\overset{+}{1}, -\overset{-}{1})$	

Figure 1.7. Listing or microstates for a p^2 configuration.

The next step is to set up a listing of microstates according to M_L and M_S values, as illustrated in Fig. 1.7. The correct number of microstates may be determined by the following formula:

$$\text{Number of microstates} = \prod_{N=l}^{N=m} \frac{2(2l+1) - N + 1}{N} \qquad [1.18]$$

where l is the azimuthal quantum number of the set of orbitals under consideration and Π stands for the product of all the evaluated terms from $N = 1$ to $N = m$; m is the number of electrons in the system being considered. For a p^2 case, the number of microstates is equal to $6 \times \frac{5}{2} = 15$.

This is the number of microstates in Fig. 1.7. For a d^2 case, the number of microstates is $10 \times \frac{9}{2} = 45$.

The microstates are now arranged in a table, as in Fig. 1.7. The possible M_S values are listed as column heads across the top of the table and the possible M_L values are listed as row headings down the left-hand side of the table. The number of microstates may be calculated beforehand, so that one may know how many microstates should be represented in the table. The student should realize that no microstate can be placed in the square $M_L = 2$, $M_S = 1$, $(\overset{+}{1}, \overset{+}{1})$, because it would violate Pauli's principle.

There now remains the determination of L and S values for the p^2 case. From the table, one determines the highest value of M_L, which in this case is 2. This must be from a term where $L = 2$ or a D term. The only M_S value associated with this term is zero $(\overset{+}{1}, \overset{-}{1})$. Therefore, the microstate $(\overset{+}{1}, \overset{-}{1})$ belongs to a ^1D term. This term is orbitally fivefold degenerate, since $M_L = 2, 1, 0, -1, -2$. Since $S = 0$ and contains no degeneracy, the total total spin orbital degeneracy is five; that is, a ^1D term comprises five microstates. Five microstates are removed from the list, one from each M_L value of 2, 1, 0, -1, -2, and each with the M_S value of 0, or from the center column of the list in Fig. 1.7. The next highest M_L value is chosen, which is $M_L = 1$. When $M_L = 1$, $M_S = 1$, 0, and -1. Now $M_L = 1$, 0, and -1; therefore, these terms are derived from a ^3P term. This term is threefold degenerate orbitally and threefold degenerate spinwise. The total spin orbital degeneracy is nine; therefore, of the ten microstates remaining, nine of them are derived from ^3P state. There is only one microstate remaining with $M_L = 0$ and $M_S = 0$. This clearly is a ^1S term, which is nondegenerate and comprises only one microstate. The p^2 electronic configuration gives rise to the ^1S, ^3P, and ^1D terms.

Hund's rules, which follow, are employed to determine the relative energies of these terms: (1) Of the terms given by equivalent electrons, those with the greatest multiplicity are of lowest energy. (2) Of the terms with greatest multiplicity, the term with the greatest L value is lowest in energy. (3) If a subshell is more than half filled, the maximum J value represents the state of lowest energy. For the p^2 configuration the ^3P term has the greatest multiplicity and represents the p^2 ground state. The terms in increasing order of energy are ^3P, ^1D, and ^1S.

What has been done up to this point in reference to term symbols is summarized in Fig. 1.8. As previously described, the J values are obtained from L and S. The number of microstates is obtained from the degeneracy of J. When $J = 2$, then J may assume the following values: $J = 2, 1, 0, -1, -2$. As evident from Fig. 1.8, the total number of microstates is 15. The

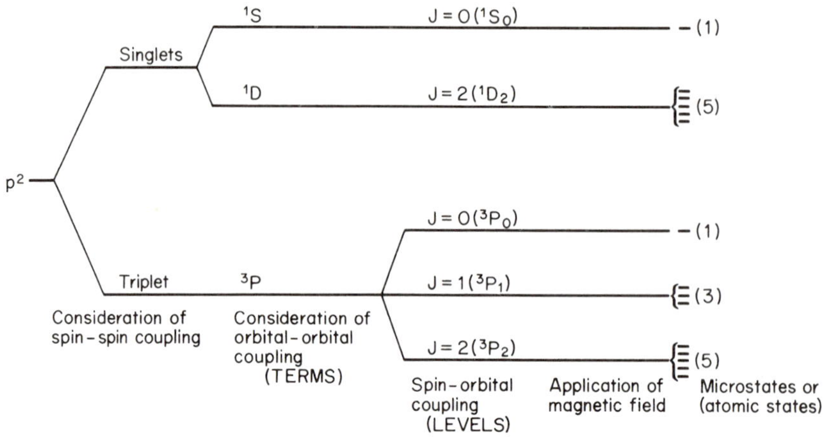

Figure 1.8. Summary of Russell-Saunders scheme for determining the term symbol of atoms with a p^2 configuration. The words in parentheses name the symbols used.

d^2 case is given as an exercise at the end of the chapter.

Table 1.11 is a summary of the terms arising from the different s, p, and d configurations of equivalent electrons. Note that the terms for p^1 and p^5 or p^2 and p^4 are identical. This identity offers the opportunity to introduce the concept of *hole formalism*. In general, l^m will produce the same terms

TABLE 1.11 *Terms for Various Electron Configurations*

Electronic Configuration	Term
s^1	^2S
s^2	^1S
p^1	^2P
p^2	$^1S, {}^1D, {}^3P$
p^3	$^2P, {}^2D, {}^4S$
p^4	$^1S, {}^1D, {}^3P$
p^5	^2P
p^6	^1S
d^1, d^9	^2D
d^2, d^8	$^1S, {}^1D, {}^1G, {}^3P, {}^3F$
d^3, d^7	$^2P, {}^2D, {}^2F, {}^2G, {}^2H, {}^4P, {}^4F$
d^4, d^6	$^1S, {}^1D, {}^1F, {}^1G, {}^1I, {}^3P, {}^3D$ $^3F, {}^3G, {}^3H, {}^5D$
d^5	$^2S, {}^2P, {}^2D, {}^2F, {}^2G, {}^2H, {}^2I, {}^4P, {}^4D,$ $^4F, {}^4G, {}^6S$

as l^{x-m}, where x is the total number of electrons a given orbital (l) may hold and m is the actual number of electrons present. The *holes* in a given electronic configuration may be regarded as positive electrons. Note the arrangement of d^m terms in Table 1.11.

THE ELECTRONIC CONFIGURATION OF POSITIVE IONS

For the first three periods of the periodic table, the prediction of the electronic structure is rather straightforward and is obtained by removing the required number of electrons from the outermost orbital to produce the appropriate positive charge. For example, Mg^{2+} has the configuration $1s^2\ 2s^2\ 2p^6$, which is also the configuration for Na^+, Al^{3+}, and Si^{4+}.

The electronic configuration for titanium is $1s^2\ 2s^2\ 2p^6\ 3s^2\ 3p^6\ 4s^2\ 3d^2$. The correct electronic configuration for Ti^{2+} is $1s\ 2s^2\ 2p^6\ 3s^2\ 3p^6\ 3d^2$. Although the $4s$ is filled before the $3d$, the $4s$ electrons are removed first. The process of removing an electron is identical to increasing effective nuclear charge. The binding energy of the electron is then determined by the positions of the maxima in their radial distribution curves. The extra penetration of the $4s$ orbital is insufficient to outweigh the stabilization caused by the $3d$ electrons being closer to the nucleus in their radial distribution maxima. Therefore, the $4s$ electrons are removed before the $3d$ electrons.

In general, in predicting ground state electronic configurations for positive ions, remove the electrons with highest principal quantum number first.

Suggested Readings

1. Coulson, C. A., *Valence*, 2nd ed., Oxford University Press, London, 1965.
2. Durant, P. J., and B. Durant, *Introduction to Advanced Inorganic Chemistry*, Longmans, Green & Company, Ltd., London, 1962.
3. Gray, H. B., *Electrons and Chemical Bonding*, W. A. Benjamin, Inc., New York, 1964.
4. Hertzberg, G., *Atomic Spectra and Atomic Structure*, Dover Publications, New York, 1944.
5. Hochstrasser, R. M., *Behavior of Electrons in Atoms*, W. A. Benjamin, Inc., New York, 1954.
6. Lever, A. B. P., *Inorganic Electronic Spectroscopy*, Elsevier Publishing Co., Amsterdam, 1968.

7. Phillips, L. F., *Basic Quantum Chemistry*, John Wiley & Sons, Inc., New York, 1965.

Study Questions

1.1. Determine the degeneracy of the nth level of the hydrogen atom, neglecting spin.

1.2. Draw the probability density (radial density) vs. radius diagram for a $3s$ orbital. What significance is the shape of the diagram?

1.3. From the atomic spectrum of a first transition series element, the ground state was calculated to have the term symbol $^2S_{1/2}$. Which element is this and what do you know about its electronic configuration?

1.4. Niobium has the electronic structure d^4s^1. What is its ground state term symbol?

1.5. A wave function, ψ, is said to be normalized if $\int \psi\psi^* \, d\tau = 1$. Show that ψ_{1s} for the hydrogen atom is normalized. Assume

$$\psi_{1s}\psi^*_{1s} = \psi^2_{1s}; \qquad \psi_{1s} = \frac{1}{\sqrt{\pi}}\left(\frac{1}{a_0}\right)^{3/2} e^{-r/a_0}$$

1.6. Two wave functions are said to be orthogonal if $\int \psi_1 \psi_2 \, d\tau = 0$. Show that $\int \psi_{1s}\psi_{2s} \, d\tau = 0$ for the hydrogen atom.

$$\psi_{2s} = \frac{1}{4}\sqrt{2\pi}\left(\frac{Z}{a_0}\right)^{3/2}\left(2 - \frac{Zr}{a_0}\right)e^{-r/2a_0} \quad \text{where} \quad Z = 1$$

1.7. Show that the most probable distance of the electron from the nucleus in the $1s$ state (orbital) of the hydrogen atom is equal to the Bohr radius, a_0.

1.8. Prove that the terms given in the right-hand column are the only ones that arise from the configurations on the left.

	Configurations	*Terms*
(a)	$ns^1(n+1)s^1$	^1S, ^3S
(b)	$np(n+1)p$	^1S, ^1P, ^1D, ^3S, ^3P, ^3D
(c)	$nd^1(n+1)d^1$	^1S, ^1P, ^1D, ^1F, ^1G, ^3S, ^3P, ^3D, ^3F, ^3G
(d)	np^4	^1S, ^1D, ^3P

1.9. Construct a table of microstates for an atom having an inner core of electrons and a d^2 configuration outside.

1.10. What are the chief differences between a $1s$ and a $2s$ orbital?

1.11. What is the electronic configuration of palladium ($Z = 46$) in the absence of a magnetic field? Pictorially indicate the electron orbital occupancy in the n and $n - 1$ energy levels of palladium under field free conditions, where $n =$ the valence energy level. What significant differences exist among the quantum electronic energy levels in palladium when under field free and when under magnetic field conditions? What experimental evidence would you predict in support of your answer?

1.12. What is the significance of the square of the eigenfunction which is a solution of the Schrödinger equation?

1.13. By the absorption of an appropriate amount of energy, the hydrogen atom can be raised to an excited state having the electronic configuration $3d^1$. Derive the complete term symbol(s) for such a configuration.

1.14. Show that $N^{-2} = \int \psi^2 \, d\tau$ for a given wave function; N is the normalization constant.

1.15. In terms of the hydrogen atom, rationalize why ψ should be finite, single valued, and continuous.

1.16. An excited state of hydrogen is given by the term symbol ^2P. What is the electronic configuration of this atom?

CHAPTER 2

Some Periodic Properties of Atoms

There is no reason to believe, nor is there any reason to guarantee, that an atom can be assigned a radius which will be characteristic of it in all compounds. Different environments may well have a measurable effect on the radius of the atom. The best effort consists in measuring the radius of an atom in a large number of compounds and then taking an average value when the atom is involved in the formation of a certain number and type of bonds. Tables of *atomic radii* in reference books sometimes differ considerably from year to year as more measurements are made on more compounds and more data are available. Since the wave function for an atom does not indicate a definite distance from the nucleus at which the probability of finding electrons drops suddenly to zero—the probability decreases gradually—a measured atomic radius must depend on the limits of approach of the atom with respect to other identical atoms. These limits are determined by experimental measurement of the distances between nuclei of adjacent atoms, to an accuracy usually within a few hundredths of an angstrom or better. Electron diffraction, X-ray techniques, and various spectroscopic methods are employed to determine the distances between atoms. The radius of the atom is then taken as half the internuclear distance when both bonded atoms are identical. The radii of atoms so measured can then be assigned to one of four categories: covalent, van der Waals, metallic, or ionic.

COVALENT RADII

The covalent radius of an atom may be defined as one-half the interatomic distance between two atoms of the same element when they are joined by a covalent bond. The bond may be a single, double, or triple bond, necessitating a further distinction in the radius description since multiple bonds are invariably shorter than are single bonds.

Single Bond Covalent Radii (SBCR)

The SBCR for nonmetals can be obtained from bond length measurements on their nonpolar compounds. For example, the SBCR for chlorine (0.99 A) is obtained from the Cl—Cl bond length (1.98 A) in the chlorine molecule, Cl_2. The SBCR for carbon (0.77 A) can be obtained from diamond, in which carbon is covalently bonded in a tetrahedral network with a uniform C—C distance of 1.54 A. This is very nearly the same as the value for saturated hydrocarbons (1.52 to 1.55 A).

Since the nitrogen (N_2) and oxygen (O_2) molecules are not singly bonded, the SBCR values for nitrogen and oxygen cannot be taken from measurements on these molecules. The SBCR for nitrogen is not taken as one-half the N—N distance in hydrazine, N_2H_4, since this value is slightly larger than that obtained by indirect methods on other compounds, such as subtracting the SBCR for carbon (0.77 A) from the C—N distance in methylamine. The longer distance in hydrazine can be attributed to the repulsion effects of the nonbonding electron pair on each nitrogen atom (see Chapter 5; see also effect on bond energies in Chapter 3). A similar situation exists for the SBCR for oxygen. The SBCR value for oxygen as obtained from H_2O_2 is significantly larger (0.73 A) than the values deduced from many compounds such as CH_3OH, which average out to about 0.66 A. Repulsion effects are undoubtedly important for oxygen also, so that the best values in these instances appear to be as indicated previously.

The SBCR for two atoms of different elements can be added to give the approximate length of the single covalent bond. Thus, the SBCR of carbon added to the SBCR of silicon yields the interatomic distance in the silicon–carbon bond from the following relationship:

$$D_{AB} = \tfrac{1}{2}D_{AA} + \tfrac{1}{2}D_{BB} \qquad [2.1]$$

$$D_{\mathrm{Si-C}} = \tfrac{1}{2}S_{\mathrm{SiSi}} + \tfrac{1}{2}D_{\mathrm{CC}} \qquad [2.2]$$

$$= \tfrac{1}{2}(2.3) + \tfrac{1}{2}(1.54)$$

$$= 1.94 \text{ A}$$

$$D_{\text{SiC}} \text{ (expt)} = 1.93 \text{ A}$$

Thus, the interatomic distance between the carbon and silicon atoms in a molecule such as H_3CSiH_3 could be calculated using the SBCR of carbon and silicon as determined in other species, such as diamond and elemental silicon.

This relationship works well, however, only when the bonds are predominantly covalent. As the amount of ionic character in the bonds increases, the difference between the calculated and experimental values also frequently increases. Schömacher and Stevenson have formulated an empirical correction for this increased ionic character in many bonds by making the bond distance depend on the electronegativity difference between the atoms in a bond, where the electronegativity difference is used as an index of degree of ionic character in the bond. This correction often gives improved agreement between calculated and experimental values:

$$D_{AB} = \tfrac{1}{2}D_{AA} + \tfrac{1}{2}D_{BB} - 0.09|X_A - X_B| \qquad [2.3]$$

where X_A and X_B are the electronegativity values of elements A and B, respectively, and $|X_A - X_B|$ is the absolute difference in their electronegativity values. For example,

$$\tfrac{1}{2}D_{\text{NN}} + \tfrac{1}{2}D_{\text{CC}} = 1.51 \text{ A}$$

$$\tfrac{1}{2}D_{\text{NN}} + \tfrac{1}{2}D_{\text{CC}} - 0.09|X_{\text{N}} - X_{\text{C}}| = 1.47 \text{ A}$$

$$D_{\text{NC}} \text{ in } (CH_3)_3N \text{ (expt)} = 1.47 \text{ A}$$

Unfortunately, in certain cases, the Schömacher-Stevenson correction leads to an incorrect value:

$$\tfrac{1}{2}D_{\text{CC}} + \tfrac{1}{2}D_{\text{ClCl}} = 1.76 \text{ A}$$

$$\tfrac{1}{2}D_{\text{CC}} + \tfrac{1}{2}D_{\text{ClCl}} - 0.09|X_{\text{Cl}} - X_{\text{C}}| = 1.71 \text{ A}$$

$$D_{\text{CCl}} \text{ in } CCl_4 \text{ (expt)} = 1.76 \text{ A}$$

Undoubtedly, a major reason why the Schömacher-Stevenson correction does not work in all cases is that the amount of ionic character in a bond depends on factors other than the electronegativity difference between the atoms in the bond. The predicted values using the Schömacher-Stevenson equation are generally high. In the case of BF_3, for example, the bond shortening found is greater than that calculated by the Schömacher-Stevenson relationship (uncorrected value −1.54 A, corrected value −1.36 A, measured value 1.30 A). The difference between the corrected and measured values can be attributed to B—F double bonding, in which a fluorine atom with a filled p orbital donates a pair of electrons into the vacant p orbital of boron:

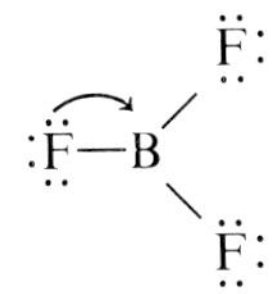

Similarly, in silicon tetrafluoride, this type of π bonding, using the $2p$ orbitals on fluorine and $3d$ orbitals on silicon, contributes to the Si—F bond shortening (uncorrected value −1.83 A, corrected value −1.62 A, measured value −1.54 A). It is thus evident that multiple bonding in certain types of compounds makes an important contribution to the interatomic distances in addition to the influence of electronegativity differences on the percent of ionic character in a bond.

In addition, hybridization effects are important. Thus, the C—C distance of 1.54 A refers to a bond between two sp^3 hybridized carbon atoms, and the C=C distance refers to a bond between two sp^2 hybridized carbon atoms, whereas the C≡C bond generally occurs between two sp hybridized carbon atoms. Since the $2s$ orbital of carbon has a smaller mean radius than the $2p$ orbitals, it would be expected that the greater the degree of s character in the hybrid orbitals used, the shorter would be the internuclear distance at which the best balance of overlap and repulsion would occur in the σ bonds. Therefore, the hybrid orbitals used in bonding to hydrogen in methane, ethylene, and acetylene will have different radii, and at least part of the decreases in bond lengths just discussed is attributable to this effect rather than to the π bonding. The single bond radii for the carbon atom in the several states of hybridization have been estimated as sp^3—0.77 A, sp^2—0.74 A, sp—0.70 A.

The covalent radii for metals can be obtained from measurements on their covalent compounds. For example, dimethyl zinc, $Zn(CH_3)_2$, is a covalent liquid as distinct from zinc perchlorate, $Zn(ClO_4)_2$, which is an ionic solid. The covalent radius of zinc can be obtained from the zinc-carbon distance in dimethyl zinc, by use of the relationship

$$R_{Zn} = D_{ZnC} - R_c \quad [2.4]$$

TABLE 2.1 ***Single Bond Covalent Radii and Metallic Radii (Coordination Number 12) of the Elements (in A)***

KEY: 14 Si; SBCR — 1.17; METALLIC RADIUS — 1.32

Periods	Group I-A	II-A	III-B	IV-B	V-B	VI-B	VII-B	VIII	VIII	VIII	I-B	II-B	III-A	IV-A	V-A	VI-A	VII-A	
1	1 **H** 0.32																	2 **He**
2	3 **Li** 1.23 1.55	4 **Be** 0.89 1.12											5 **B** 0.82 0.98	6 **C** 0.77 0.91	7 **N** 0.70 0.92	8 **O** 0.66	9 **F** 0.64	10 **Ne**
3	11 **Na** 1.54 1.90	12 **Mg** 1.36 1.60											13 **Al** 1.18 1.43	14 **Si** 1.17 1.32	15 **P** 1.10 1.28	16 **S** 1.04 1.27	17 **Cl** 0.99	18 **Ar**
4	19 **K** 2.03 2.35	20 **Ca** 1.74 1.97	21 **Sc** 1.44 1.65	22 **Ti** 1.32 1.47	23 **V** 1.22 1.36	24 **Cr** 1.18 1.30	25 **Mn** 1.17 1.27	26 **Fe** 1.17 1.26	27 **Co** 1.16 1.25	28 **Ni** 1.15 1.25	29 **Cu** 1.17 1.28	30 **Zn** 1.25 1.37	31 **Ga** 1.26 1.41	32 **Ge** 1.22 1.37	33 **As** 1.21 1.39	34 **Se** 1.17 1.40	35 **Br** 1.14	36 **Kr**
5	37 **Rb** 2.16 2.48	38 **Sr** 1.91 2.15	39 **Y** 1.62 1.78	40 **Zr** 1.45 1.60	41 **Nb** 1.34 1.47	42 **Mo** 1.30 1.39	43 **Tc** 1.27 1.35	44 **Ru** 1.25 1.34	45 **Rh** 1.25 1.34	46 **Pd** 1.28 1.37	47 **Ag** 1.34 1.44	48 **Cd** 1.48 1.54	49 **In** 1.44 1.66	50 **Sn** 1.40 1.62	51 **Sb** 1.41 1.59	52 **Te** 1.37 1.60	53 **I** 1.33	54 **Xe**
6	55 **Cs** 2.35 2.67	56 **Ba** 1.98 2.22	57–71 **La-Lu**	72 **Hf** 1.44 1.62	73 **Ta** 1.34 1.49	74 **W** 1.30 1.41	75 **Re** 1.28 1.37	76 **Os** 1.26 1.35	77 **Ir** 1.27 1.36	78 **Pt** 1.30 1.39	79 **Au** 1.34 1.46	80 **Hg** 1.44 1.57	81 **Tl** 1.48 1.71	82 **Pb** 1.47 1.75	83 **Bi** 1.46 1.70	84 **Po** 1.46 1.76	85 **At** 1.45	86 **Rn**
7	87 **Fr**	88 **Ra**	89–103 **Ac-Lr**															
Lanthanide series				57 **La** 1.69 1.88	58 **Ce** 1.65 1.82	59 **Pr** 1.64 1.83	60 **Nd** 1.64 1.82	61 **Pm** 1.63	62 **Sm** 1.62 1.81	63 **Eu** 1.85 1.99	64 **Gd** 1.62 1.81	65 **Tb** 1.61 1.80	66 **Dy** 1.60 1.80	67 **Ho** 1.58 1.79	68 **Er** 1.58 1.78	69 **Tm** 1.58 1.77	70 **Yb** 1.94	71 **Lu** 1.58 1.75
Actinide series				89 **Ac**	90 **Th** 1.65 1.80	91 **Pa**	92 **U** 1.42	93 **Np**	94 **Pu**	95 **Am** 1.73	96 **Cm**	97 **Bk**	98 **Cf**	99 **Es**	100 **Fm**	101 **Md**	102 **No**	103 **Lr**

The SBCR for various elements are presented in Table 2.1.

Multiple Bond Radii

Multiple bond radii are less than single bond radii, and, in general, the interatomic distances are less as the multiplicity of the bond increases. Here, again, the multiple bond radius is equal to half the distance between two identical atoms that are joined by a double or triple bond. The double bonded radius for oxygen, as obtained from the oxygen molecule (0.60 A), is larger due to repulsion effects than that obtained from measurements of many carbonyl compounds in which the average value for the oxygen double bond radius is about 0.55 A. However, the triply bonded radius of nitrogen is about the same if taken from the N_2 molecule as from HCN, or other molecules in which nitrogen is triply bonded. In nitrogen gas, repulsion effects are negligible due to the presence of only a single lone pair, which is disposed at an angle of 180° to the bond pairs.

TABLE 2.2 *Various Covalent Radii of Carbon, Nitrogen, and Oxygen (in A)*

	Carbon	Nitrogen	Oxygen
Single bond radius	0.77 ($H_3C{-}CH_3$)	0.70 ($CH_3{-}NH_2$)	0.66 ($CH_3{-}OH$)
Double bond radius	0.665 ($H_2C{=}CH_2$)	0.60 ($RN{=}NR$)	0.55 ($R_2C{=}O$)
Triple bond radius	0.60 ($HC{\equiv}CH$)	0.55 ($N{\equiv}N$)	

TABLE 2.3 *Multiple Bond Radii of Some Nonmetallic Elements (in A)***

	B	C	N	O	F
Double bond radius	0.71	0.665	0.60	0.55	0.60
Triple bond radius	0.64	0.60	0.55		
		Si	P	S	Cl
Double bond radius		1.07	1.00	0.94	0.89
Triple bond radius		1.00	0.93	0.87	
		Ge	As	Se	Br
Double bond radius		1.12	1.11	1.07	1.04
		Sn	Sb	Te	I
Double bond radius		1.30	1.31	1.27	1.23

* Pauling, L., *Nature of the Chemical Bond*, 3rd ed., Cornell University Press, 1960.

As with single bond covalent radii, the multiple bond radii can be used to calculate interatomic distances in other molecules. For example, consider the C≡N distance in HCN; the triple bond radii of carbon and nitrogen are 0.60 and 0.55 A, respectively, and the calculated C≡N distance is 1.15 A. The experimentally determined C≡N distance in HCN is 1.154 A, and that of C≡N in CH_3CN is 1.157 A. Multiple bond radii of some elements with known values are presented in Table 2.3.

If the experimental bond lengths in a certain compound are found to be intermediate between single and double, or between double and triple, it is usually assumed that delocalization of the bonding π electrons exists. This is usually rationalized in terms of resonance structures, two or more of which contribute to the actual structure of the compound, which is referred to as a resonance hybrid. For example, all the carbon–carbon bond lengths in the benzene molecule are identical, but intermediate between the values predicted for single and double bonds:

$$D_{C-C} \text{ (calc'd)} = 1.54 \text{ A}$$

$$D_{C=C} \text{ (calc'd)} = 1.34 \text{ A}$$

$$\text{Measured } D_{C-C} \text{ in benzene} = 1.39 \text{ A}$$

Thus, the prediction is made that the following resonance forms (among others) contribute to the true structure of benzene:

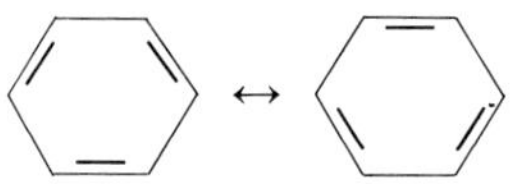

To appreciate more fully the *delocalization of electrons* concept for benzene, we usually represent the structure as

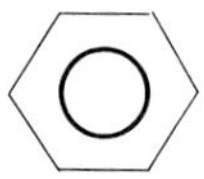

to indicate that the π bond formed by the overlap of the six P_Z carbon orbitals on benzene is not confined between any two carbon atoms but is instead delocalized over the entire ring.

Similarly, in the nitrate ion, all the N—O bond lengths are identical, but intermediate between the predicted values for single and double bonds:

N—O (calc'd) = 1.36 A; N=O (calc'd) = 1.15 A;

measured N—O distance in NO_3^- = 1.21 A.

The postulated resonance forms for the NO_3^- ion are

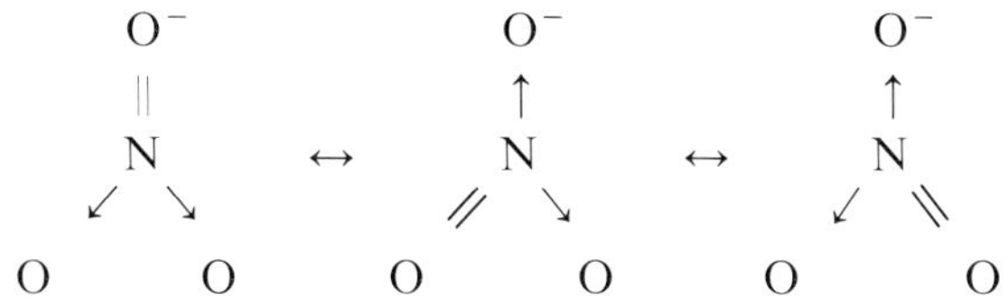

Since the bond lengths for such compounds in which electron delocalization occurs vary considerably from compound to compound, it is unrealistic to speak of a *covalent resonance radius* of atoms in such compounds which can be tabulated and used to predict the interatomic distances in other compounds involving resonance structures.

Van der Waals Radii

When the internuclear distances between nonmetal atoms are measured by diffraction techniques in the solid, crystalline state, it is found that two main types of distances are observed. The closer of these is the intramolecular distance between two atoms joined by a covalent bond; it is called the covalent bond length. The SBCR is derived from such bond length measurements. The longer distance, between adjacent atoms on neighboring molecules, is known as the van der Waals distance. For example, in solid chlorine, each atom is found to have one close neighbor at a distance of 1.98 A (bond length) and another at a distance of 3.60 A (van der Waals distance). Thus, the single bond covalent radius of chlorine is 0.99 A, whereas the van der Waals or nonbonding radius is 1.80 A. The van der Waals radii are generally much larger than the covalent radii, because the orbital overlap within a bond brings the atoms much closer together. The term van der Waals radii is used because the principal bonding forces between molecules in solid crystals of the nonmetals are the weak van der Waals forces of attraction. Similarly, for the inert gases, the crystal radii are actually van der Waals radii, since there exist forces of attraction other than the van der Waals forces of attraction. A list of van der Waals radii is given in Table 2.4.

Metallic Radii

The metallic radius may be defined as one-half the distance between the atoms of a metal in the metallic close-packed crystal lattice (in which the metal exhibits a coordination number of 12). Metallic radii are generally

TABLE 2.4 *Van der Waals Radii of Some Common Elements (in A)*

		H 1.2	He 0.93
N 1.5	O 1.40	F 1.35	Ne 1.31
P 1.9	S 1.85	Cl 1.80	Ar 1.74
As 2.0	Se 2.00	Br 1.95	Kr 1.89
Sb 2.2	Te 2.20	I 2.15	Xe 2.09
			Rn 2.14

some 10 to 15% larger than the single bond covalent radii (see Table 2.1) but at the same time are considerably smaller than the nonbonding van der Waals radii. The metallic bond (see Chapter 7) that governs the proximity of metal atoms to one another in the metallic crystal is *not* a localized bond between any two pairs of atoms as is the covalent bond. This results in the two metal atoms not being drawn quite as close to each other in the metallic crystal as they are in forming a covalent bond. At the same time, however, the bonding forces are appreciable in governing interatomic distances, as evidenced by the comparatively much larger van der Waals radii for the same atom. Metallic radii are quite often referred to as *atomic radii* in reference tables.

Ionic Radii

Cations are smaller, considerably, than the parent atom, whereas anions are considerably larger. This can be attributed to interelectronic repulsion, which is less for the cation and larger for the anion relative to the parent atom. If the interelectronic repulsion is reduced, the remaining electrons may be drawn closer to the nucleus (cation), and the more electrons lost, the smaller is the radius (see Table 2.5). Compare Fe^0, 1.26 A; Fe^{2+}, 0.76 A; and Fe^{3+}, 0.64 A. For anions, the increased interelectronic repulsion caused by the increased electron density causes the electrons to move apart and the atom to expand. For a series of isoelectronic ions, those with the same electronic configuration, the ionic radii decrease as atomic number increases due to the greater attractive force exerted by the increased number of protons in the nucleus. Compare the following: N^{3-} (1.71 A), O^{2-} (1.40 A), F^- (1.36 A), Na^+ (0.95 A), Mg^{2+} (0.65 A), Al^{3+} (0.50 A), and Si^{4+} (0.41 A).

PERIODIC TRENDS IN ATOMIC RADII

Whether one considers the covalent radii, metallic radii, or van der Waals radii, significant trends appear as one traverses a period of the periodic

TABLE 2.5 ***Ionic Radii of the Elements in A (Oxidation Number for Ion is in Parentheses)***

Periods	Group I-A	II-A	III-B	IV-B	V-B	VI-B	VII-B	VIII	VIII	VIII	I-B	II-B	III-A	IV-A	V-A	VI-A	VII-A	
1	1 **H**																	2 **He**
2	3 **Li** 0.68	4 **Be** 0.35											5 **B** 0.23	6 **C** 0.16	7 **N** 1.71(−3)	8 **O** 1.40(−2)	9 **F** 1.36(−1)	10 **Ne**
3	11 **Na** 0.95	12 **Mg** 0.65											13 **Al** 0.50	14 **Si** 0.41	15 **P** 0.44(3) 0.35(5)	16 **S** 1.83(−2) 0.37(4) 0.30(6)	17 **Cl** 1.81(−1) 0.34(5) 0.37(7)	18 **Ar**
4	19 **K** 1.33	20 **Ca** 0.99	21 **Sc** 0.81	22 **Ti** 0.90(2) 0.76(3) 0.68(4)	23 **V** 0.88(2) 0.74(3) 0.59(5)	24 **Cr** 0.84(2) 0.63(3) 0.52(6)	25 **Mn** 0.84(2) 0.66(3) 0.60(4)	26 **Fe** 0.76(2) 0.64(3)	27 **Co** 0.72(2) 0.63(3)	28 **Ni** 0.69(2) 0.62(3)	29 **Cu** 0.96(1) 0.72(2)	30 **Zn** 0.74	31 **Ga** 0.62	32 **Ge** 0.73(2) 0.53(4)	33 **As** 0.58(3) 0.46(5)	34 **Se** 1.98(−2) 0.50(4) 0.42(6)	35 **Br** 1.96(−1) 0.47(5)	36 **Kr**
5	37 **Rb** 1.47	38 **Sr** 1.12	39 **Y** 0.92	40 **Zr** 0.79(4)	41 **Nb** 0.74(4) 0.69(5)	42 **Mo** 0.70(4) 0.62(6)	43 **Tc** 0.56(7)	44 **Ru** 0.67(4)	45 **Rh** 0.68(3)	46 **Pd** 0.80(2) 0.65(4)	47 **Ag** 1.26(1) 0.89(2)	48 **Cd** 0.97	49 **In** 0.81	50 **Sn** 0.93(2) 0.71(4)	51 **Sb** 0.76(3) 0.62(5)	52 **Te** 2.21(−2) 0.56(6)	53 **I** 2.20(−1) 0.50(7)	54 **Xe**
6	55 **Cs** 1.67	56 **Ba** 1.98	57–71 **La-Lu**	72 **Hf** 0.78(4)	73 **Ta** 0.68(5)	74 **W** 0.70(4) 0.62(6)	75 **Re** 0.70(4) 0.56(7)	76 **Os** 0.69(4)	77 **Ir** 0.68(4)	78 **Pt** 0.80(2) 0.65(4)	79 **Au** 1.37(1) 0.9(3)	80 **Hg** 1.10	81 **Tl** 1.47(1) 0.95(3)	82 **Pb** 1.20(2) 0.84(4)	83 **Bi** 0.96(3) 0.74(5)	84 **Po** 0.95(3)	85 **At** 2.25(−1)	86 **Rn**
7	87 **Fr** 1.76	88 **Ra** 1.43	89–103 **Ac-Lr**															
Lanthanide series				57 **La** 1.14	58 **Ce** 1.07(3) 0.94(4)	59 **Pr** 1.06(3) 0.92(4)	60 **Nd** 1.04(3)	61 **Pm** 1.06(3)	62 **Sm** 1.00(3)	63 **Eu** 1.12(2) 0.98(3)	64 **Gd** 0.97(3)	65 **Tb** 0.93(3) 0.81(4)	66 **Dy** 0.92(3)	67 **Ho** 0.91(3)	68 **Er** 0.89(3)	69 **Tm** 0.87(3)	70 **Yb** 1.13(2) 0.86(3)	71 **Lu** 0.85
Actinide series				89 **Ac** 1.18	90 **Th** 1.02(4)	91 **Pa** 0.98(4) 0.9(5)	92 **U** 0.97(4) 0.80(6)	93 **Np** 1.10(3) 0.95(4)	94 **Pu** 1.08(3) 0.93(4)	95 **Am** 1.07(3) 0.92(4)	96 **Cm**	97 **Bk**	98 **Cf**	99 **Es**	100 **Fm**	101 **Md**	102 **No**	103 **Lr**

table or compares these properties in a periodic group. The importance of the radius lies in the fact that the size of the atom, as much as any other single property of the atom, determines its chemical and physical properties. The chemical reactivity of an atom is a function of the ability of the atom to lose, gain, or share electrons, and this ability is very largely controlled by the electrostatic attractive force exerted by the nucleus on the valence electrons. In turn, this attractive force is governed largely by the distance of the valence electrons from the nucleus, i.e., the radius. At the same time, many physical properties, such as density, are dependent on the radius of the atom.

Change in Radius Moving Down a Periodic Group

A general increase in radius is observed moving down each periodic group (Table 2.1). With each successive element, the added electrons occupy new principal energy levels, which are a function of the distance from the nucleus. In comparison, the increased attraction of more electrons by a more highly charged nucleus is a relatively minor effect.

Change in Radius Across a Period from Left to Right

There is a general, overall decrease in radius observed across any period of the periodic table from left to right (Table 2.1). Each successive element has one more electron added to the same principal energy level and one proton added to the nucleus. The result is an increase in electron density. The corresponding nuclear charge exerts a greater attraction for the electrons, with the result that the atom contracts. Slater's rules (p. 44) indicate that the efficiency of screening the nuclear charge will decrease at the same time, allowing the effective atomic number to increase with the result that the valence electrons are attracted more effectively to the nucleus. This increased attractive force more than compensates for the interelectronic repulsive force caused by higher electron density.

If one compares the changes in radii across the nontransition metal periods with those of the transition metals and inner-transition metals, one notes that the average percentage decrease in radius from element to element varies in the following order: nontransition > transition > inner-transition. The reason for this sequence lies in the fact that the percentage increase in nuclear charge follows the same sequence. However, another factor controlling size changes is the filling of the *d* orbitals of transition metals and the filling of the *f* orbitals of the inner-transition metals. The filling of the *d* orbitals, which occupy the $(n - 1)$ energy level, cannot effectively influence the size of the atom, which is determined by the distance from the nucleus to the outermost *ns* electrons. This statement is even truer for the inner-transition elements, which fill the $(n - 2)$ *f* level. Remember also that the attractive force exerted by the nucleus for the outermost electrons is always offset to some degree by the repulsion effects between those same outermost electrons and the other electrons occupying orbitals lying closer to the nucleus. It is likely that transition metals would exhibit repulsion effects of this type more than would the nontransition elements. As more and more *d* electrons enter the $(n - 1)$ level, this repulsive effect becomes more important relative to the attractive power exerted by the nucleus, resulting in expansion of the atoms toward the end of the series. This latter argument is borne out by the fact that each of the transition metal series exhibits a slight increase in radius toward the end of the series. The same effect is not noticed with the inner-transition elements. Presumably, the $(n - 2)$ *f* electrons are too far removed from the outermost *ns* electrons to exert a significant repulsion effect. Thus, increased nuclear charge exhibits its greatest effect for the inner-transition metals, resulting in an overall decrease in radius to the end of the series.

Lanthanide Contraction

The *lanthanide contraction* is the decrease in radius that occurs across the lanthanide series of elements; it is the only variation in radii that is given a specific title. The significance of the lanthanide contraction is contained in the following two points.

(1) The change in radius (Table 2.1) across this series of 15 elements is very small. There is an average decrease of less than 0.01 A between successive elements. This very small size change between successive elements explains the historically great difficulty that exists in the chemical separation of these elements. The chemistry of the lanthanides as a group of elements is strikingly uniform, more so than any other group of elements in the periodic table. For many years, the tedious separation techniques of repeated fractional crystallization or precipitation were the only methods available. Since the late 1940's, ion exchange techniques have rendered the previous separation techniques obsolete.

(2) The second effect of the lanthanide contraction is on the properties of the elements in the third transition series. An examination of the radii of the lanthanides and the elements immediately following them leads to the observation that if the lanthanide elements did not intervene, the radii of

hafnium and the following elements in the third transition series would be on the order of 0.15 A larger than they actually are. Instead, the lanthanide contraction nearly erases completely this expected increase, with the result that the radii for hafnium and zirconium, niobium and tantalum, and molybdenum and tungsten are nearly identical. In fact, the radii of the fifth period elements from the third transition series on are more similar to those of the sixth period elements below them in the same group than they are to the radii of the fourth period elements. This results in the properties of the fifth and sixth period elements (beyond lutetium) being more similar than any other two periods in the periodic table. The chemical behavior of zirconium and hafnium is extremely similar, more so than for any other pair of congeners known.

IONIZATION POTENTIAL

The energy exerted to convert a neutral atom to a positively charged ion is one of the most fundamental and important properties of an atom. When this ionization energy is measured in solution under certain restrictive conditions of temperature and concentration, it is known as the oxidation potential. When the energy is measured in the gaseous state, however, the atom is considered essentially independent of its environment and the energy necessary to effect the ionization is exactly the energy that binds the electron to the atom. When the ionization energy of an atom is measured in this way, it is known commonly as the ionization potential and its magnitude is a quantitative measure of the stability of the electronic structure of the isolated atom. Since more than one electron can be removed from the atom, a precise definition of ionization potential must necessarily be specific for each electron. Thus, the first ionization potential (I_1) is the energy required to remove the most loosely bound electron from the neutral gaseous atom in its lowest energy state.

A number of factors can be expected to contribute significantly to the magnitude of the ionization potential for a given atom. Since the attractive force holding the electron in the atom is a coulombic one, both the distance of the electron from the nucleus and the effective nuclear charge that the electron *sees* are important. This can be understood readily from Coulomb's law: *The attractive force between two oppositely charged particles is proportional to the product of the charges divided by the square of the distance between them.* The effective nuclear charge is always much less than the sum total of protons in the nucleus due to the screening effect exerted on the nuclear charge by the electrons in the intervening energy levels between the

TABLE 2.6 ***Ionization Potentials of the Elements at 25°C in kcal/mole***

(Note: Values in parentheses are doubtful.)

	1st	2nd	3rd	4th	5th	6th	7th	8th
H	315	—	—	—	—	—	—	—
He	568	1256	—	—	—	—	—	—
Li	126	1745	2822	—	—	—	—	—
Be	216	421	3547	5019	—	—	—	—
B	193	581	874	5978	7843	—	—	—
C	261	564	1103	1486	9037	(11240)	—	—
N	337	684	1097	1784	2256	(11660)	(15300)	—
O	316	812	1271	1781	2624	(3170)	(16960)	(20000)
F	403	808	1444	2009	2633	(3606)	(4244)	(21880)
Ne	499	949	—	—	—	—	—	—
Na	120	1092	—	—	—	—	—	—
Mg	178	348	1847	—	—	—	—	—
Al	140	436	657	2766	—	—	—	—
Si	189	378	773	1040	3838	—	—	—
P	255	455	697	1184	1499	—	—	—
S	240	541	808	1090	(1670)	(2020)	—	—
Cl	301	550	920	1255	1563	(2232)	(2622)	—
Ar	365	639	945	1380	—	—	—	—
K	102	735	1100	1405	—	—	—	—
Ca	142	275	1180	1550	—	—	—	—
Sc	(153)	(299)	572	1704	—	—	—	—
Ti	160	(316)	(650)	997	2301	—	—	—
V	156	329	(687)	(1100)	(1475)	(3061)	—	—
Cr	158	382	(715)	(1140)	(1682)	(2075)	(3867)	—
Mn	173	362	(779)	(1232)	(1752)	(2323)	(2738)	(4759)
Fe	183	375	708	—	—	—	—	(3470)
Co	182	495	774	—	—	—	—	—
Ni	178	420	812	—	—	—	—	—
Cu	180	469	851	—	—	—	—	—
Zn	218	416	917	—	—	—	—	—
Ga	139	475	710	1482	—	—	—	—
Ge	184	369	791	1053	2154	—	—	—
As	228	468	655	1162	1443	(2926)	—	—
Se	226	498	740	989	(1565)	(1878)	(3818)	—
Br	275	(500)	829	(1154)	(1387)	(2010)	(2345)	(4814)
Kr	324	568	853	—	—	—	—	—
Rb	98	636	921	—	—	—	—	—
Sr	133	256	—	1300	—	—	—	—
Y	148	284	475	—	—	—	—	—
Zr	160	304	532	793	—	—	—	—
Nb	160	332	581	(885)	1153	—	—	—
Mo	165	374	627	1071	—	—	—	—
Tc	169	353	—	—	—	—	—	—
Ru	171	(388)	658	—	—	—	—	—
Rh	174	(418)	(716)	—	—	—	—	—
Pd	193	(449)	(761)	(1125)	—	—	—	—
Ag	176	497	805	—	—	—	—	—
Cd	209	391	866	—	—	—	—	—
In	135	436	648	1252	—	—	—	—
Sn	171	339	705	941	1870	—	—	—
Sb	201	381	584	1021	1284	(2471)	—	—
Te	209	431	721	882	1389	(1661)	(3149)	—
I	243	442	(723)	(960)	(1200)	(1772)	(2081)	(3901)
Xe	281	490	741	—	—	—	—	—
Cs	91	580	—	—	—	—	—	—
Ba	122	232	—	—	—	—	—	—
La	130	265	444	—	—	—	—	—
Hf	161	345	—	—	—	—	—	—
Ta	183	375	—	—	—	—	—	—
W	185	409	—	—	—	—	—	—
Re	183	384	—	—	—	—	—	—
Os	201	391	—	—	—	—	—	—
Ir	201	—	—	—	—	—	—	—
Pt	211	445	—	—	—	—	—	—
Au	214	474	—	—	—	—	—	—
Hg	242	434	790	—	—	—	—	—
Tl	142	472	688	1171	—	—	—	—
Pb	172	348	738	(977)	1607	—	—	—
Bi	(170)	386	591	1044	1291	(2168)	—	—
Rn	249	—	—	—	—	—	—	—
Ra	123	236	—	—	—	—	—	—

nucleus and the electron being removed. The effective nuclear charge (Z^*) on a given electron can be determined by subtracting from the atomic number (Z) the screening constant (S) for the other electrons. According to Slater, this screening constant on a given electron is estimated by summing the following contributions:

(a) Zero for each electron farther away from the nucleus
(b) 0.35 for each electron in the same $n(sp)$ group (except $1s$ where 0.30 is used)
(c) 0.85 for each electron in the $n-1$ group
(d) 1.00 for each electron closer than the $n-1$ group
(e) For an electron in a d or f orbital, every electron in a lower group contributes 1.00 to the screening constant.

The preceding rules are based on considering ns and np electrons together as a single group, whereas nd and nf electrons each form separate groups. The order of grouping the electrons thus becomes: ($1s$); ($2s$, $2p$); ($3s$, $3p$); ($3d$); ($4s$, $4p$); ($4d$); ($4f$); etc. A few examples will serve to illustrate the use of Slater's rules.

(1) Potassium: $1s^22s^22p^63s^23p^64s$. The effective nuclear charge *seen* by the $4s$ electron of potassium is

$$\begin{aligned} Z^* &= Z - S \\ &= 19 - [0.85\,(\text{number of } n = 3 \text{ electrons}) + \\ &\quad 1.00\,(\text{number of electrons with } n < 3)] \qquad [2.5] \\ &= 19 - [(0.85 \times 8) + 1.00(10)] \\ &= 19 - 16.8 \\ &= 2.20 \end{aligned}$$

(2) Tungsten: Kr $4d^{10}4f^{14}5s^25p^65d^46s^2$. The effective nuclear charge on one of the $6s$ electrons in tungsten is

$$\begin{aligned} Z^* &= Z - S \\ &= 74 - [0.35\,(\text{number of } n = 6 \text{ electrons}) + \\ &\quad 0.85\,(\text{number of } n = 5 \text{ electrons}) + 1.00 \\ &\quad (\text{number of electrons with } n < 5)] \qquad [2.5] \\ &= 74 - [0.35(1) + 0.85(12) + 1.00(60)] \\ &= 74 - 70.55 \\ &= 3.45 \end{aligned}$$

The fact of the existence of the *last* electron in the $4s$ orbital of potassium, rather than the $3d$, can be understood readily by comparing the effective nuclear charge on the *last* electron if it were actually in the $3d$ orbital rather than the $4s$.

(3) Potassium: $1s^2 2s^2 2p^6 3s^2 3p^6 3d^1$.

$$\begin{aligned} Z^* &= Z - S \\ &\doteq 19 - 1.00 \text{ (number of electrons in a group lower than } 3d) \qquad [2.5] \\ &= 19 - 18 \\ &= 1.00 \end{aligned}$$

Thus, the electron *sees* a significantly larger Z^* if it is in the $4s$ orbital, 2.20, compared with that seen while in the $3d$ orbital, 1.00. It is reasonable, therefore, that the *last* electron in potassium would reside in the lower energy $4s$ orbital in the ground state of the atom.

If, on the other hand, the Z values are calculated for an outer electron of an element in the first transition series, such as vanadium, a different result is found.

(4) Vanadium: $1s^2 2s^2 2p^6 3s^2 3p^6 3d^3 4s^2$.

(a) The effective nuclear charge on one of $4s$ electrons is

$$\begin{aligned} Z^* &= Z - S \\ &= 23 - [0.35 \text{ (number of } n = 4 \text{ electrons)} \\ &\quad + 0.85 \text{ (number of } n = 3 \text{ electrons)} \\ &\quad + 1.00 \text{ (number of } n < 3 \text{ electrons)}] \qquad [2.5] \\ &= 23 - [0.35(1) + 0.85(11) + 1.00(10)] \\ &= 23 - 19.70 \\ &= 3.30 \end{aligned}$$

(b) The effective nuclear charge on one of the $3d$ electrons is

$$\begin{aligned} Z^* &= Z - S \\ &= 23 - [0.35(2) + 18(1.00)] \qquad [2.5] \\ &= 4.30 \end{aligned}$$

Therefore, note that for the vanadium atom, the $3d$ electrons are more tightly bound than the $4s$ because of the higher effective nuclear charge. Hence, in ionization of a neutral vanadium atom the first electron removed is from a $4s$ orbital. On the other hand, the preceding calculations have shown that for potassium, the $4s$ electron is bound more tightly than the $3d$.

Related to the screening effect and differences just observed for $4s$ and $3d$ electrons is the degree to which the electron being removed can penetrate underlying levels and, in effect, reduce the screening effect of the inner electrons. This degree of penetration varies in the order $ns > np > nd > nf$, i.e., an ns orbital has more of its charge density disposed nearer the nucleus than an np orbital, an np more than an nd, etc. These differences in penetration can be attributed to the inherently different shapes of the various atomic orbitals and are supported by the order of energies of the electrons in the same orbitals.

Although there is some question about the success of using Slater's rules in certain theoretical interpretations, they have been used quite successfully in the concept of electronegativity, as defined by Allred and Rochow (see p. 71). Recently, Clementi and Raimondi have formulated a set of rules for the evaluation of the screening constant S, which purportedly gives better agreement in theoretical interpretations for a greater number of elements. In the Clementi and Raimondi rules, the factors of numbers of electrons, their distance from the nucleus, angular momentum, nuclear charge, and spin multiplicity are all considered in a set of rather involved equations for the calculations of S.

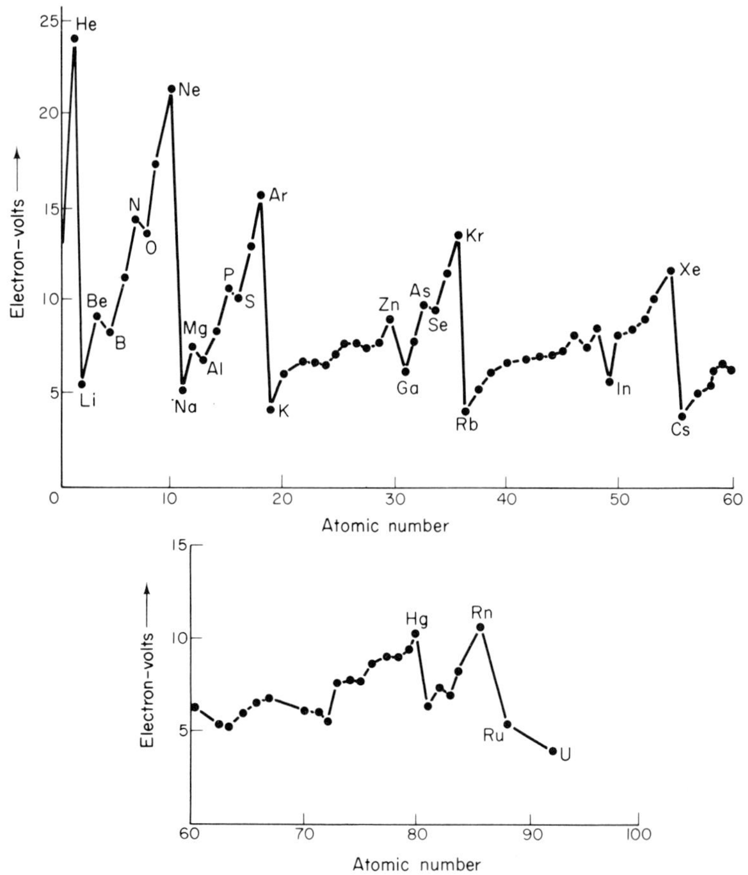

Figure 2.1. Plot of ionization potential I_1 (eV) against atomic number.

There are periodic trends in ionization potentials (see Fig. 2.1). To obtain meaningful correlations in discussing periodic trends in ionization potential, it is necessary to be consistent and use the same definition throughout the discussion. As is customary, first ionization potentials will be used exclusively in the following discussion.

The changes in ionization potential that occur going across a period from left to right or down a group in the periodic table are, with some exceptions, the exact opposite of the corresponding changes in the radii of the elements. Thus, the radii of the elements increase down a nontransition element group, and the ionization potentials decrease in accordance with Coulomb's law. There are very few exceptions to this trend considering the large percentage increase in radii among successive elements down a group. This radius effect apparently more than compensates for the increase in nuclear charge as a group is descended.

In the case of the transition metal groups, the trend is not regular, i.e., the ionization potentials generally do decrease from the first to the second member of the group, but then increase again from the second to the third member. The initial decrease from the first to the second member can be attributed to the corresponding radius change, but the change in radius from the second to the third member is at best minimal and often nonexistent. Thus, the increase in ionization potential from the second to the third member of the group can be attributed to the fact that there is an increase in nuclear charge *without* a corresponding increase in radius (see Tables 2.1 and 2.6). This effect is exhibited past the third transition series to lead, which has a higher value than that of tin immediately above it. After each inert gas there is a sharp decrease in ionization potential as the first electron being removed originates in a new principal energy level, is much farther away from the nucleus, and is thus less tightly held.

The inverse correlation of ionization potential values with atomic radius is less fully realized across any period than down a group, and there are some significant exceptions to the trend. In each period, there is a sharp drop in ionization potential at Group III-A as the filling of *p* orbitals begins. Due to the inherently higher potential energy of a *p* orbital (as opposed to an *s* orbital), the *p* electrons are held less tightly and less energy is required for their removal than would be *expected* purely on the basis of radius changes.

Note also that the ionization potentials for Group V-A elements are generally larger than those of the corresponding elements in Group VI-A. This exception to the trend can be attributed to the repulsion effect between two electrons in the same *p* orbital of the VI-A element. This is not present for the preceding V-A element. The electron is thus removed more readily for the VI-A element even though its nuclear charge is greater. The result is an apparent increase in stability of the half-filled p^3 energy level. A similar phenomenon is observed with d^5 and f^7 levels. This has been compared to the high stability of filled levels as exemplified by the unusually high ionization potentials of the inert gases—p^6 configuration, alkaline earth metals—s^2 configuration, and Group II-B elements—$d^{10}s^2$ configuration.

On traversing a transition or inner-transition metal series from left to right, the ionization potential increases very slightly overall but there is no regular change from element to element, particularly among the Group VIII elements. This is consistent with the preceding observations on radius change.

ELECTRON AFFINITY

The property of ionization potential is a measure of the tendency of elements to lose electrons and requires an input of energy in all cases. However, the process in which an element acquires a single electron is often exothermic, particularly for the nonmetals. This is the electron affinity, A, and is defined as the energy released when a neutral gaseous atom in its ground state acquires an electron to form the gaseous negative ion in its ground state:

$$X\,(g) + e^-\,(g) \;\rightarrow\; X^-\,(g), \qquad \Delta H = \text{electron affinity} \qquad [2.6a]$$

The electron affinity for the neutral atom, X, possesses the same magnitude but opposite sign to that of the ionization potential for the negatively charged ion, X^-.

$$X^-\,(g) \;\rightarrow\; X\,(g) + e^-\,(g), \qquad \Delta H = \text{ionization potential} \qquad [2.6b]$$

From the two equations (2.6a and 2.6b) we see that the sign convention used is similar to that of thermodynamic processes: a positive sign represents an endothermic process and a negative sign is indicative of an exothermic process. Be warned that not all chemists agree on this sign convention, and you should establish a consistent practice of your own.

Table 2.7 also reveals that electron affinities follow expected trends; there is an overall increase across a period from left to right and an overall decrease as a group is descended. In general, as radii decrease, the atom

TABLE 2.7 ***Electron Affinities of Gaseous Atoms of the Elements, in kcal/mole at 25°C***

Group / Periods	I-A	II-A	III-B	IV-B	V-B	VI-B	VII-B	VIII-B	VIII-B	VIII-B	I-B	II-B	III-A	IV-A	V-A	VI-A	VII-A	VIII-A
1	1 **H** (−17.3)*																	2 **He**
2	3 **Li** (−13.4)	4 **Be** (4.4)											5 **B** (−7.6)	6 **C** (−25.8)	7 **N** (6.2)	8 **O** −33.8 168†	9 **F** −81.0	10 **Ne**
3	11 **Na** (−17.9)	12 **Mg** (7.4)											13 **Al** (−11.9)	14 **Si** (−32)	15 **P** (−17.9)	16 **S** −47.6 78.2	17 **Cl** −84.8	18 **Ar**
4	19 **K** (−11.5)	20 **Ca** (36.8)	21 **Sc** (9.2)	22 **Ti** (−3.46)	23 **V** (−15.0)	24 **Cr** (−19.6)	25 **Mn** (27.7)	26 **Fe** (−2.3)	27 **Co** (−16.1)	28 **Ni** (−25.4)	29 **Cu** −34.6	30 **Zn**	31 **Ga**	32 **Ge**	33 **As**	34 **Se** 96.8	35 **Br** −79.1	36 **Kr**
5	37 **Rb** (−13.8)	38 **Sr** (11.5)	39 **Y** (−6.9)	40 **Zr** (−23.1)	41 **Nb** (−29.8)	42 **Mo** (−29.8)	43 **Tc** (−23.1)	44 **Ru** (−33.4)	45 **Rh** (−31.1)	46 **Pd** (−32.3)	47 **Ag** −46.1	48 **Cd**	49 **In**	50 **Sn**	51 **Sb**	52 **Te**	53 **I** −72.1	54 **Xe**
6	55 **Cs**	56 **Ba**	57 **La-Lu**	72 **Hf**	73 **Ta**	74 **W**	75 **Re**	76 **Os**	77 **Ir**	78 **Pt**	79 **Au** −64.4	80 **Hg** −35.4	81 **Tl**	82 **Pb**	83 **Bi**	84 **Po**	85 **At**	86 **Rn**
7	87 **Fr**	88 **Ra**	89 **Ac-Lr**															
	Lanthanide series			57 **La**	58 **Ce**	59 **Pr**	60 **Nd**	61 **Pm**	62 **Sm**	63 **Eu**	64 **Gd**	65 **Tb**	66 **Dy**	67 **Ho**	68 **Er**	69 **Tm**	70 **Yb**	71 **Lu**
	Actinide series			89 **Ac**	90 **Th**	91 **Pa**	92 **U**	93 **Np**	94 **Pu**	95 **Am**	96 **Cm**	97 **Bk**	98 **Cf**	99 **Es**	100 **Fm**	101 **Md**	102 **No**	103 **Lr**

* Values in parentheses are calculated by various methods.

† Total value for both electrons.

becomes more compact and the effect of the nuclear charge becomes more pronounced. This causes a corresponding increase in the ability of the atom to attract an electron of itself. Thus, more energy is released in ion formation. It should be noted that the same factors govern both electron affinity and radii changes.

A notable exception to the trend of electron affinities decreasing on descending a group is found with the first members of nontransition element groups. In all cases, the electron affinity for the first element in each of those groups is less than that for the second. The values for the second member are, as a result, the highest in the group as the expected trend continues down the group. This irregularity is attributed to the extreme interelectronic repulsive forces in operation in the first member elements, which, because they are so compact, exhibit a high electron density. This extreme interelectronic repulsion decreases the energy released in the formation of X^-. The second and succeeding members of these groups are larger and have vacant d orbitals of the same principal quantum number, which allow the interelectronic repulsive forces to be reduced considerably.

For all elements, the processes of gaining the second and succeeding electrons are endothermic. This is not unexpected, since the second electron, for example, must be brought toward an ion already negatively charged and the resulting repulsive forces require an input of energy. In the case of the elements in the oxygen groups, for example, this second electron affinity, A_2, is numerically much larger than A_1, with the result that the overall process is endothermic for the formation of X^{-2}. The student should reconcile this fact with the known experimental reactivity of oxygen and its congeners.

The principal drawback in the use of electron affinities is the great difficulty in obtaining values experimentally for more than the more electronegative elements. Since reliable experimental values of electron affinities as determined by absorption spectroscopy are available for a few elements, the use of electron affinities is limited. One of the alternate methods of determining electron affinities is by use of the Born-Haber cycle, which will be discussed in different sections of this text.

USE OF IONIZATION POTENTIALS AND ELECTRON AFFINITIES IN BORN-HABER CYCLES

The important relationships between electron affinity, ionization potential, and other energy factors underlying the driving force in chemical reactions can be understood quite clearly with the use of a Born-Haber cycle. If one uses the approximation that $\Delta H = \Delta G$, more data are available for a greater variety of reactions without any serious loss in significance to the correlations that become evident with consistent use of ΔH values, accurate calculations are still possible. As an example, we shall use the following reaction,

$$Na\ (s) + \tfrac{1}{2}Cl_2\ (g) \xrightarrow{25^\circ C} Na^+Cl^-\ (c) \qquad [2.7]$$

for which the cycle in Fig. 2.2 is obtained.

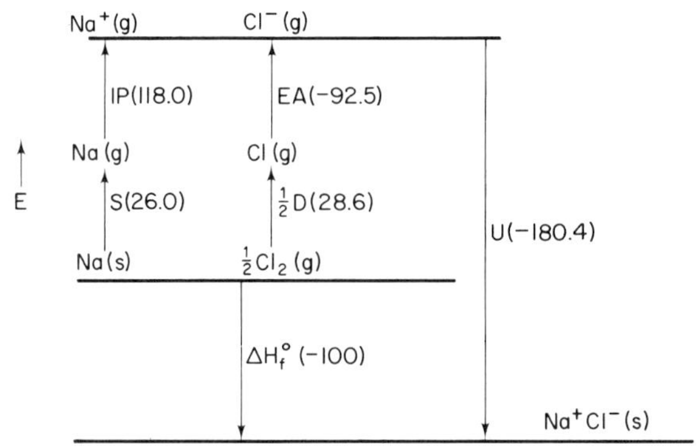

Figure 2.2. The Born-Haber energy cycle respresenting the formation of Na^+Cl^-(c) from Na(s) and $\frac{1}{2}Cl_2$ at 25°C. The numbers in parentheses are the energy values in kcal/mole for each process.

If one follows the path *up the ladder* in Fig. 2.2, the initial step represents conversion of the elements in their standard states to gaseous atoms. This requires, on a per mole basis, the addition of the sublimation energy of sodium, 2.60 kcals, and one-half the bond dissociation energy of Cl_2, 28.9 kcals. The second step involves conversion of gaseous atoms to gaseous ions and requires the addition of the ionization potential of sodium, 118 kcals, and the release of 92.5 kcals (electron affinity) during the formation of Cl^-. The energy level of the gaseous ions is now 80.4 kcals above that of the elements in their standard states. If the gaseous ions are now allowed to react to form solid Na^+Cl^-, 180.4 kcals or the crystal lattice energy, U, will be released. The cycle is completed by including the standard enthalpy of formation of $Na^+Cl^-_{(s)}$ from $Na^0_{(s)}$ and $\frac{1}{2}Cl_{2(g)}$, which is -100 kcals. Obviously, the crystal lattice energy plays a very important role in determining the driving force for reactions of this type. In a more practical sense, such Born-Haber cycles can be used to evaluate one of the various energy terms if each of the others is known. This can be especially useful for quantities that are difficult to evaluate experimentally, such as sublimation energies or electron affinities.

MAGNETIC SUSCEPTIBILITIES OF ATOMS AND IONS

Of all the properties that may be used to characterize the structure of atoms and ions, perhaps none is so readily obtained, nor so revealing a structural feature, as is the magnetic susceptibility of a substance. The cause of this feature of the atom is the spinning electron. The results of measuring magnetic properties are useful in determining the number of unpaired electrons present in various transition metal complexes. The method is quite applicable for determining the number of unpaired electrons in the oxygen molecule, nitric oxide, or other stable free radicals.

Magnetic susceptibility refers to the change in density of a magnetic field as it passes by some substance. A vacuum will have no effect on magnetic lines of force. A paramagnetic substance will be attracted to a magnetic field or draw the magnetic lines of force to itself. A diamagnetic material will be repelled by a magnetic field or repel the magnetic lines of force.

All materials are diamagnetic. Diamagnetism is caused by paired electrons, which all atoms except the hydrogen atom possess. This property does not exhibit itself very strongly; even the weakest paramagnetism is a thousand times stronger than any diamagnetism. This characteristic is not practically important to the chemist.

Paramagnetic properties and their measurements are important to the chemist because these measurements can elucidate some structures. The origin of the paramagnetic moment or magnetic susceptibility lies in the spinning electron, which establishes a magnetic spin moment and the orbital moment established by the electron. Substances that are paramagnetic have unpaired electrons, and these unpaired electrons cause the paramagnetism.

Magnetic susceptibilities are determined by measuring the force on a substance while it is in an inhomogeneous magnetic field. The paramagnetic substance, which is attracted to the magnetic field, will weigh more than it does normally, and the increased weight can be related to the magnetic susceptibility. However, since all materials are diamagnetic to some degree, the paramagnetism must be corrected for this. This correction is contained in the relationship

$$X_m = N\alpha + \frac{N}{3kT}B^2\mu_m^2 \qquad [2.8]$$

where X_m is the molar magnetic susceptibility, N is Avogadro's number, α is the negatively signed diamagnetic constant, k is the Boltzmann constant, T is the absolute temperature, B is the Bohr magneton,[1] and μ_m is the total magnetic moment expressed in Bohr magnetons. The diamagnetic constant, α, is temperature independent, whereas the paramagnetic term, the second term of equation [2.8], varies inversely with the temperature. These temperature relationships can be explained in conjunction with the causes of magnetic susceptibility. We have already mentioned that the moving spinning electron possesses a spin moment and an orbital moment and that the two moments combine to give a paramagnetic moment. The tendency for unpaired electrons to have their moments aligned with the applied field, as is the case in paramagnetism, is opposed by thermal agitation of the magnetic dipoles. Thus, the inverse temperature relationship with paramagnetic substances, as indicated in equation [2.8], is seen. In a diamagnetic substance, all electrons are paired, and the magnetic field that is induced in the sample opposes the applied field. This results in a tendency for the sample to be repelled by the applied field. The diamagnetic susceptibility, which carries a negative sign, will become larger as the size of the atom or ion increases and as the number of electrons increases, but there is no temperature dependency.

The total magnetic moment for a substance used in equation [2.8] which is dependent on both spin and orbital moments is given by

$$\mu_m = \mu_{S+L} = \sqrt{4S(S+1) + L(L+1)} \qquad [2.9]$$

where S is the resultant spin angular momentum quantum number and L is the resultant orbital angular momentum quantum number, as defined in Chapter 1.

The application of magnetic susceptibility data is generally limited to those substances containing one or more unpaired electrons, although the simple knowledge that a substance is not paramagnetic is often quite revealing as to its structure. Thus, any free radical, if stable enough for susceptibility data to be obtained, could be detected by this technique. A number of such stable radicals exist (O_2, NO, ClO_2) and have been studied. Generally, it is not applicable to those free metallic elements with unpaired

[1] The Bohr magneton $= eh/4\pi mc = 9.27 \times 10^{-21}$ erg per gauss (erg/G), where e is the electronic charge, h is Planck's constant, m is the rest mass of the electron, and c is the speed of light.

electrons, since the outer electrons in metals are not closely associated with any one atom. Thus, the orbital angular momentum component can vary considerably from metal to metal, and no regular change in magnetic susceptibility is observed as the number of unpaired electrons changes.

The most important application of magnetic susceptibility data is in determining the presence and number of unpaired electrons in the *d* orbitals of the metal in transition metal complexes. In this context, paramagnetic transition metal ions may be divided into three types. The first type is that in which the *L* component of μ_m may be neglected, and a regular increase of μ_m with the increase of the number of unpaired electrons is observed. This group includes most of the familiar ions of the first transition series. Experimental results (Table 2.8) indicate that the orbital angular momentum contribution to the magnetic moment is small enough to be neglected for most of these ions, producing a direct correlation between the spin angular momentum and the number of unpaired electrons in the metal ion. Since the unpaired electrons in these ions occupy the outermost orbitals, orbital momentum effects in one ion are quenched wholly or partially by the electric fields of neighboring ions. Thus, equation [2.9] reduces to

$$\mu_m = \mu_S = \sqrt{4S(S+1)} \qquad [2.10]$$

Since the number of unpaired electrons, n, is equal to twice the sum of the spin quantum numbers, i.e., $n = 2S$, an equivalent relationship is

$$\mu_S = \sqrt{n(n+2)} \qquad [2.11]$$

TABLE 2.8 ***Magnetic Moments of Some Transition Elements in Bohr Magnetons***

Ion	Electronic Configuration	Term Symbol	U_s	U_{S+L}	Observed U
V^{+4}	$(Ar)3s^2 3p^6 3d^1$	^2D	1.73	3.0	1.7–1.8
Cu^{+2}	$(Ar)3s^2 3p^6 3d^9$	^2D	1.73	3.0	1.7–2.2
V^{+3}	$(Ar)3s^2 3p^6 3d^2$	^3F	2.83	4.47	2.6–2.8
Ni^{+2}	$(Ar)3s^2 3p^6 3d^8$	^3F	2.83	4.47	2.8–4.0
Cr^{+3}	$(Ar)3s^2 3p^6 3d^3$	^4F	3.87	5.20	~ 3.8
Co^{+2}	$(Ar)3s^2 3p^6 3d^3$	^4F	3.87	5.20	4.1–5.2
Fe^{+2}	$(Ar)3s^2 3p^6 3d^6$	^5D	4.90	5.48	5.1–5.5
Co^{+3}	$(Ar)3s^2 3p^6 3d^6$	^5D	4.90	5.48	~ 5.4
Mm^{+2}	$(Ar)3s^2 3p^6 3d^5$	^6S	5.92	5.92	~ 5.9
Fe^{+3}	$(Ar)3s^2 3p^6 3d^5$	^6S	5.92	5.92	~ 5.9

For example, for the V^{+2} ion, which has a $3d^3$ configuration, $\mu_S = 3(3 + 2) = 3.88$, since $n = 3$; μ_m(expt'l) $= 3.7 - 3.9$ for various V^{+2} salts. For the first transition series, the difference between the calculated values of μ_S and the experimental values can be attributed to some contribution of the orbital moment. This is usually additive to μ_S but, in some cases, may act in opposition to the spin contribution and moments slightly below the μ_S value are observed, as for Cr(III).

The second type of transition metal ion is found among the lanthanide and actinide ions. For these ions, the energy separation between the first excited state and the ground state, with a single, well-defined J value, is appreciably large, so that very few ions are found in the first excited state. At the same time, the unpaired electrons are well shielded from external field effects by the electrons in intervening energy levels. Under these circumstances, L-S coupling between the orbital and spin moments is significant, so that the resultant J moment must be used in the equation for magnetic moment,

$$\mu_m = \mu_J = g\sqrt{J(J + 1)} \quad \text{(in Bohr magnetons)} \qquad [2.12]$$

where g is the gyromagnetic ratio or Landé splitting factor and is the ratio of magnetic moment to angular momentum; its value is approximately 2.00.

This equation gives good agreement with experimental values for all lanthanide ions except Sm(III) and Eu(III), which do not have the large separation between the first excited and ground J states. Thus, the excited states are somewhat populated and the ions yield magnetic moments larger than those calculated assuming only ground state population (Fig. 2.3).

The third type of transition metal ion includes those in which the separation of J energy levels is small. The L-S coupling, as a result, is negligible, and L and S each contribute independently to μ_m, so that the entire equation [2.9] must be used. This type of magnetic behavior is important when the orbital contributions to μ_m are not quenched, and it includes those ions having D or F ground states. Quenching is complete for an ion with an S ground state, and excellent agreement is obtained between μ_S and observed values for such ions. For ions such as Co^{+2} and Fe^{+2}, however, orbital contributions are significant (see Table 2.8).

For metal ions of the second and third transition series, observed magnetic moments corresponding to one electron are found when the number of electrons is odd, whereas the ions are diamagnetic when there is an even number of electrons. Thus, there is a tendency for these ions *not* to adhere to Hund's rule but instead to have maximum pairing of electrons within the $4d$ and $5d$ orbitals. Because of the diffuse nature of the $4d$ and $5d$ orbitals, interelectronic repulsion effects are reduced with the result that the tendency of the d electrons to occupy the d orbitals singly is lessened. Also important, in all likelihood, is the crystal field splitting of these levels, which is larger in the order $5d > 4d > 3d$ (see Chapter 9) and which would

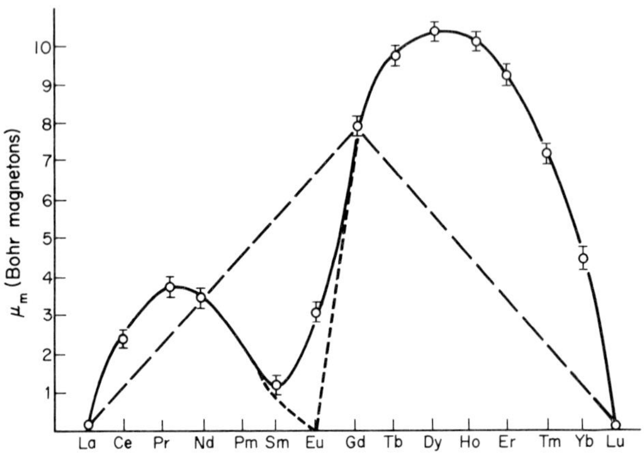

Figure 2.3. Experimental and calculated magnetic moments of the trivalent lanthanide ions at 300°K. ○, experimental values and those calculated from appropriate J ground states, including complications for S_m and Eu (discussed in text). |, vertical bars represent experimental ranges of μ_m. – – –, calculated values, ignoring complications for S_m and Eu. ——, calculated values using μ_S only.

produce the same effect: a lesser tendency to adhere to Hund's rules. Even considering these factors, however, we find that values obtained at room temperature often fall below μ_S (spin only) moments, and with only a few exceptions susceptibility data on these elements cannot be used to determine relative d orbital energy levels and oxidation states, as well as the number of unpaired electrons.

Earlier we mentioned that the magnetic susceptibility of a material is temperature dependent. The studies of Pierre Curie resulted in the Curie law, which is

$$X_m = \frac{C}{T} \qquad [2.13]$$

where X_m is the molar magnetic susceptibility, C is the Curie constant, and T is the absolute temperature. If X_m is plotted against absolute temperature, the normal parabolic curve will result as is illustrated in Fig. 2.4(a). However, if the reciprocal of X_m is plotted against absolute temperature, a

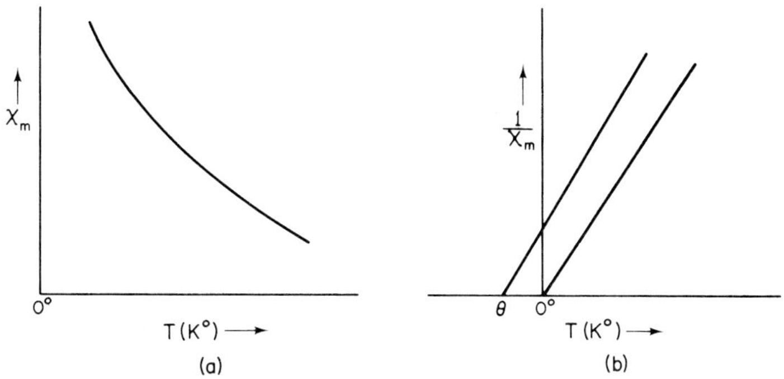

Figure 2.4. Graphs of magnetic susceptibility. (a) Expected relationship; (b) illustration of Curie-Weiss law.

straight line curve passing through absolute zero should result [see Fig. 2.4(b)]. In many cases, the line does pass through the ordinate above absolute zero and crosses the absolute temperature line below absolute zero! This observation has led to the Curie-Weiss law,

$$X_m = \frac{C}{T - \theta} \qquad [2.14]$$

where all the terms are defined as previously and θ is the Weiss constant, which provides the necessary correction to cause the line to pass through absolute zero [see line in Fig. 2.4(b)].

Two additional types of magnetic properties remain to be mentioned: ferromagnetism and antiferromagnetism. A substance is said to be ferromagnetic when at lower temperatures it behaves according to the Curie or Curie-Weiss law (section *AB* of the curve in Fig. 2.5) and then at a given

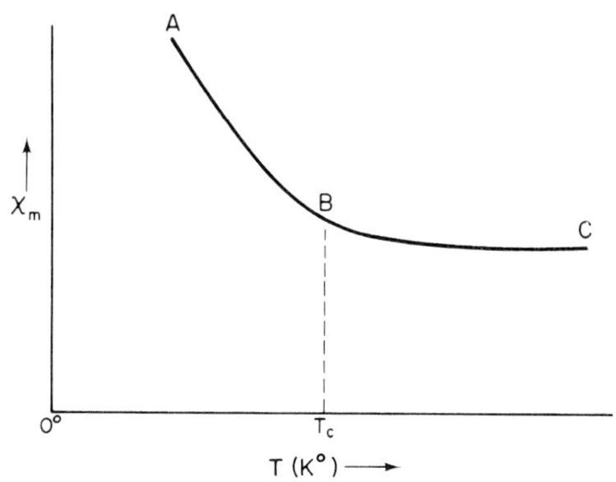

Figure 2.5. Illustration of Curie temperature.

temperature, a discontinuity appears in the curve and then the magnetic susceptibility follows a new relationship. The temperature at which this discontinuity occurs is called the Curie temperature, T_C. Antiferro-

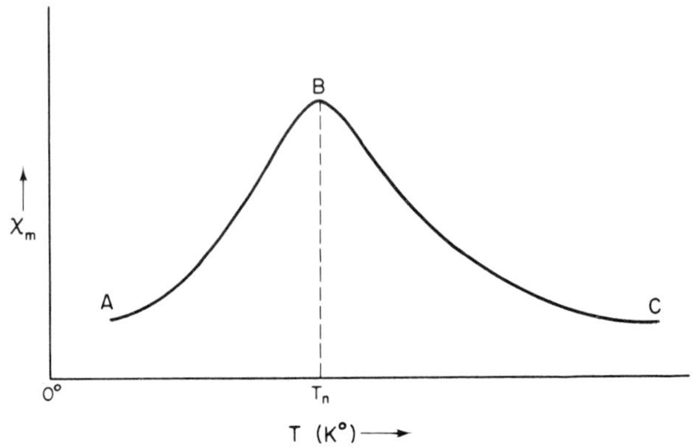

Figure 2.6. Illustration of Neel temperature.

magnetism is exhibited by a substance that follows the Curie or Curie-Weiss law from a given temperature upward. Below this temperature, called the Néel temperature, T_N, the magnetic susceptibility decreases with a decrease in temperature. Ferromagnetism may be explained by the fact that thermal agitation of the ions prevents the proper decrease in magnetic susceptibility. Ion alignment also explains antiferromagnetism. Here, the ions in the lattice arrange themselves so their magnetic moments tend to cancel one another.

Suggested Readings

1. Clementi, E., and D. L. Raimondi, *J. Chem. Phys.*, **38**, 2686 (1963).
2. Coulson, C. A., *Valence*, Clarendon Press, Oxford, 1952, p. 41.
3. Cubicotti, D., *J. Chem. Phys.*, **31**, 1646 (1959); **34**, 2189 (1961).
4. Figgis, B. N., and J. Lewis, *Progr. Inorg. Chem.*, **6**, 37 (1964).
5. Nyholm, R. S., *Record Chem. Progr.*, **19**, 45 (1958).
6. Pauling, L., *The Nature of the Chemical Bond*, 3rd ed., Cornell University Press, Ithaca, N.Y., 1960.
7. Sanderson, R. T., *Inorganic Chemistry*, Reinhold Publishing Corp., New York, 1967.

8. Schomaker, V., and D. P. Stevenson, *J. Am. Chem. Soc.*, **63**, 37 (1941).

9. Sutton, L. E., *Tables of Interatomic Distances and Configurations in Molecules and Ions*, Special Publication 11, The Chemical Society, London, 1958.

10. Wells, A. F., *Structural Inorganic Chemistry*, 3rd ed., Oxford University Press, Oxford, 1962, Chap. 2.

General References

1. Cotton, F. A., and G. Wilkinson, *Advanced Inorganic Chemistry*, 2nd ed., Interscience Publishers, New York, 1966.

2. Day, M. C., and J. Selbin, *Theoretical Inorganic Chemistry*, 2nd ed., Reinhold Publishing Corp., New York, 1968.

3. Douglas, B. E., and D. H. McDaniel, *Concepts and Models of Inorganic Chemistry*, Blaisdell Publishing Co., New York, 1965.

4. Harvey, K. B., and G. B. Porter, *Introduction to Physical Inorganic Chemistry*, Addison-Wesley Publishing Co., Inc., Reading, Mass., 1963.

5. Heslop, R. B., and P. L. Robinson, *Inorganic Chemistry*, 3rd ed., Elsevier, Amsterdam, 1967.

6. Phillips, C. S. G., and R. J. P. Williams, *Inorganic Chemistry*, vols. 1 and 2, Oxford University Press, Oxford, 1965.

Study Questions

2.1. Discuss the relationships among covalent radius, electron affinity, and ionization potential. Take particular note of the factors affecting them and integrate the discussion as much as possible.

2.2. A number of different radii values can be assigned to the tellurium atom which characterize the atom in various environmental conditions. The values are (in A): 1.27, 1.35, 1.60, and 2.20. Without referring to a reference, identify each of these values and give the basis for your answer.

2.3. From what measurements on what compounds, could you obtain:
(a) The single bond covalent radius for nitrogen?
(b) The double bond radius for oxygen?

2.4. Which species in each of the following possesses the indicated property (give the structural basis for each answer in addition to the generalization or periodic trend which applies)?
(a) Smallest radius of (i) O, F, Ne; (ii) Cr, Mo, W; (iii) Li^+, Be^{2+}, B^{3+}; (iv) N^0, N^{3+}, N^{3-}; (v) S^{2-}, Ar, or Ca^{2+}.

(b) Smallest ionization potential of (i) N, O, F; (ii) Mg, Al, Si; (iii) Cu, Ag, Au.
(c) Smallest magnetic susceptibility of Cr^{3+}, Fe^{2+}, or Ni^{2+}.
(d) Largest electron affinity of (i) N, P, As; (ii) B, C, N.

2.5. List and explain the factors that affect ionization potential.

2.6. Given the following data, calculate the crystal lattice energy of LiF: $I_{Li} = 124$ kcal/mole; $S_{Li} = 38$ kcal/mole; $A_F = 81$ kcal/mole; $D_{F_2} = 135$ kcal/mole; $\Delta H_f^\circ = -146$ kcal/mole.

2.7. What particular combination of hydration energy of the gaseous ion and crystal lattice energy of the corresponding solid favors a high heat of solution for an ionic compound in water?

2.8. The atomic radii of titanium, zirconium, and hafnium are 1.47, 1.60, and 1.62 Å, respectively. (a) Account for the closeness of the values for zirconium and hafnium. (b) The ionic radius of Ti(II) is 0.90 Å. What other metals form dipositive ions of similar size? Can any valid comparisons be made?

2.9. Estimate the magnetic moment (spin contribution only) of each of the following ions:

(a) Cr^{2+} (c) Cu^{2+} (e) FeF_6^{3-}
(b) Fe (d) $Ni(CN)_4^{2-}$ (f) $Fe(CN)_6^{3-}$

2.10. The magnetic moments of the tripositive ions of samarium, dysprosium, and thulium are 1.5, 10.5, and 7.3 Bohr magnetons, respectively. Calculate the magnetic moments of these ions based on the spin-only contribution of the $4f$ electrons. Account for any differences.

2.11. Calculate the electron affinity of chlorine from the following data:

Heat of formation of KCl (s) from K (s) and Cl_2 (g)	104.9 kcal/mole
Heat of sublimation of K	21.7 kcal/mole
Heat of dissociation of Cl_2	57.8 kcal/mole
Ionization potential of K	99.6 kcal/mole
Crystal lattice energy of KCl	164.4 kcal/mole

2.12. In the following graph, the first ionization potential is plotted against the atomic number for some elements in the second row of the periodic table. The term symbols for the ground state are next to the elements and are in parentheses. Devise some explanation as to why the first ionization potential for nitrogen is higher than the first ionization potential for oxygen.

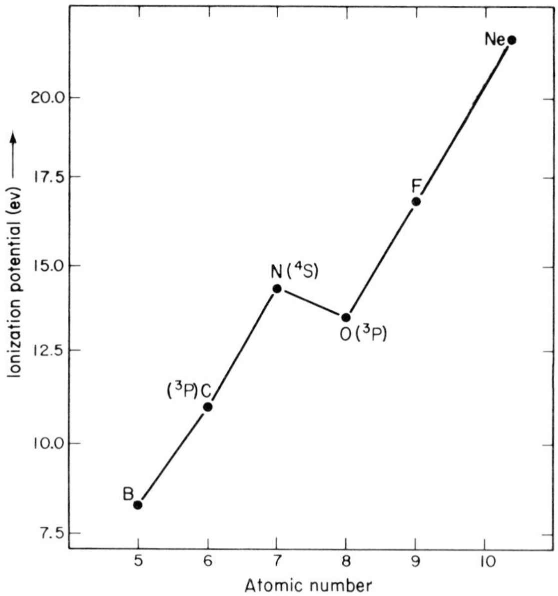

2.13. Calculate the sulfur–fluorine bond distance in SF_2 (expt'l = 1.58 A). Comment.

PART II

Bonded Atoms

CHAPTER 3

Some Properties of Covalent Bonds

Part II. Bonded Atoms

In the same fashion that one described the properties of atoms in their free or uncombined state, speaking of atomic radius, ionization potential, magnetic susceptibility, etc., one may also speak of the properties of the combined atom. In so speaking of the properties of a combined atom, one necessarily describes the bond holding that atom to another or several atoms. When one speaks of bond strength, bond energy and force constant, and the stability of the atom in its combined state, one describes these properties in a manner similar to the way he describes the properties of the uncombined atom. When one speaks of electronegativity, percent ionic character, and dipole moment, one is actually describing the dispersal of electronic charge among chemically combined atoms. Each of the properties of the bonded atoms and its bonds is related to each of the other properties in ways that may be described mathematically, which is the purpose of the presentations in this chapter.

BOND STRENGTH

Bond strength refers to the energy required to break only one bond in a molecule. This is also called the bond dissociation energy. It is illustrated by the following equation:

$$CH_4 + \text{energy} \rightarrow CH_3\cdot + H\cdot \qquad [3.1]$$

It is difficult to remove the first hydrogen atom—103 kcal/mole. The remaining atoms are removed more easily, and only 81 kcal/mole are needed to remove the last hydrogen atom. It might be noted that in the case of carbon dioxide, the energy required to remove one oxygen atom, viz.,

$$CO_2\,(g) \rightarrow CO\,(g) + O\,(g) \qquad [3.2]$$

is 127 kcal/mole, whereas 257 kcal/mole are required for reaction [3.3]:

$$CO\,(g) \rightarrow C\,(g) + O\,(g) \qquad [3.3]$$

The second bond dissociation energy in this case is larger than the first, although admittedly the types of bonds in the two species are not held constant.

The relative strengths of various bonds have been calculated by Pauling in his wave mechanical studies, and his general conclusion was as follows: The greater the degree of overlap of the atomic orbitals of the atoms forming the bond, the stronger is the bond. Assuming that the overlap of two *s* orbitals would have a relative bond strength of 1.000, Pauling calculated other bond strengths, listed in Table 3.1.

TABLE 3.1 ***Relative Bond Strengths of Typical Bonds***

s—s	1.00	s—sp^3	2.000
s—p	1.73	s—dsp^2	2.694
s—sp	1.93	s—d^2sp^3	2.923
s—sp^2	1.99	p—p	3.00

BOND ENERGY

A compound's heat of formation, or, more correctly, its standard enthalpy of formation, ΔH_f°, is usually defined as the heat energy involved at constant pressure in the formation of 1 mole of a compound in its standard state from the elements in their standard states. The sign of the ΔH_f° term in this context is negative for heat lost during the formation of the compound (exothermic) and positive for heat absorbed. In the present discussion of bond energy, it is more convenient to define the enthalpy of formation of a compound as the heat energy at constant pressure involved in the formation of 1 mole of gaseous molecules from gaseous atoms. In this definition, the sign of ΔH_f will also be considered negative for energy released during formation of the compound. Figure 3.1 diagrams the energy relationships for this reaction.

$$C\ (g) + 4H\ (g) \rightarrow CH_4\ (g) \tag{3.4}$$

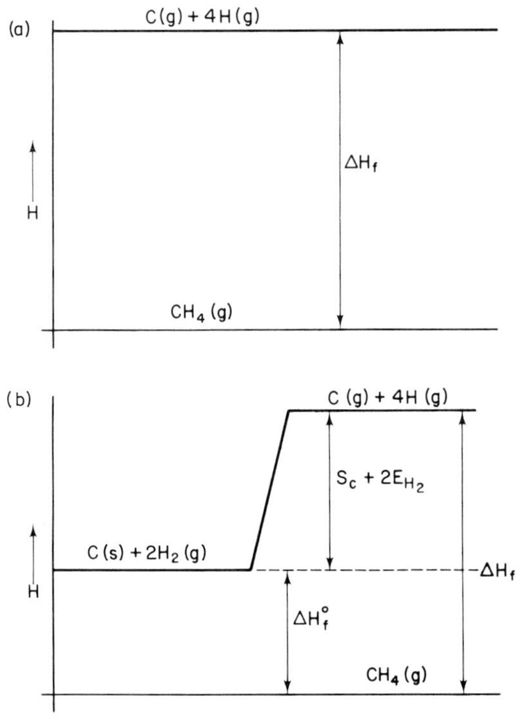

Figure 3.1. Energy relationships for the reaction C (g) + 4 H (g) → CH_4 (g).

In Fig. 3.1, S_c refers to the heat of sublimation of carbon, and D_{H_2} is the bond dissociation energy of the hydrogen molecule.

The energy of a bond can now be defined as the average amount of heat energy liberated per bond when gaseous atoms combine to form gaseous molecules. Thus, if one takes the absolute value of ΔH and ΔH_f°, the carbon–hydrogen bond energy is given by

$$E_{C-H} = \tfrac{1}{4}(\Delta H_f) = \tfrac{1}{4}(\Delta H_f^\circ + S_c + 2D_{H_2}) \qquad [3.5]$$

Bond energy is distinct from bond strength. Bond energy refers to the average energy per bond, whereas bond strength refers to the energy required to dissociate the first bond or a specific bond in the molecule. Since our definition of bond energy involves the enthalpy of a specific reaction, it is a thermochemical quantity that can be measured by any of the techniques available for measuring enthalpies of reaction. These can be divided generally into two groups—equilibrium measurements and nonequilibrium measurements—of which the former are ideally the only precise method. Equilibrium measurements include evaluation of ΔH° for a reaction from ΔG°; ΔS° and equilibrium constant data; electronic absorption or emission spectroscopic analysis, which is limited to diatomic molecules; mass spectroscopic analysis, in which threshold electron energies required to produce a known set of fragments from a given molecule are measured; and other techniques involving magnetic susceptibility, magnetic resonance spectroscopy, dilatometry, molar refraction, etc. The large majority of our currently available bond energy data has been provided by kinetic studies, which depend on determination of accurate rate constants and correlation of activation energies to enthalpies of reaction.

Several pitfalls must be avoided in calculating bond energies. The examples that follow shed light on some. The N—H bond energy is equal to $\frac{1}{3}\,\Delta H_f(NH_3)$; the N—N bond energy would be equal to $\Delta H_f(N_2H_4)$ – 4N—H bond energies, since the reaction

$$H_2NNH_2\,(g) \quad \rightarrow \quad 2N\,(g) + 4H\,(g) \qquad [3.6]$$

involves the breaking of one N—N bond and four N—H bonds.

It has been found experimentally that bond energies can often be used in an additive sense to estimate the energy released when a gaseous molecule is produced from gaseous atoms. For example, ΔH_f of CH_3Cl (g)

$$3H\,(g) + C\,(g) + Cl\,(g) \quad \rightarrow \quad CH_3Cl\,(g) \qquad [3.7]$$

may be determined by summing bond energy values, i.e.,

$$\begin{aligned} \Delta H_f &= -\,(3E_{C-H} + E_{C-Cl}) \\ &= -\,(3 \times 99 + 78) \\ &= -\,375\ \text{kcal/mole} \end{aligned} \qquad [3.8]$$

Determined experimentally, ΔH_f is -376 kcal/mole. The small discrepancy between the calculated and experimental values demonstrates that although an average bond energy is only an approximation to the true dissociation energy of each bond in a molecule, some excellent correlations can be obtained.

Some representative bond energies are presented in Table 3.2. Multiple bond energies are usually *not* the product of single bond energies, since the bonds in a multiple bond are of a different type (σ and π). The average energy per bond may increase or decrease with multiplicity (Table 3.3).

TABLE 3.2 ***Some Typical Covalent Single Bond Energies, in kcal/mole at 25°C****

	H	C	N	O	F	Si	P	S	Cl	Br	I	Ge	As
H	104												
C	99	83											
N	93	73	38										
O	111	84	48	33									
F	135	117	66	44	38								
Si	76	76	*78*	108	135	53							
P	75	63	*50*	*80*	117	*51*	51						
S	88	65	*49*	*64*	68	54	*55*	51					
Cl	103	78	46	49	61	91	78	61	58				
Br	88	58	*58*	*50*	61	74	63	51	52	46			
I	71	57	*42*	48	*89*	51	44	*43*	50	43	36		
Ge	55	*73*	*74*	*104*	*142*	*49*	*50*	*59*	81	66	51	45	
As	71	*59*	*39*	*86*	111	*44*	*43*	48	70	58	43	41	35

TABLE 3.3 ***Some Multiple Bond Energies, in kcal/mole at 25°C***

	C—C	C=C	C≡C	N—N	N=N	N≡N
Bond energy	83	100	123	38	100	226
Average energy/bond	83	50	41	30	50	75

PARTIAL IONIC—COVALENT CHARACTER OF BONDS AND ELECTRONEGATIVITY

Covalent and ionic bonds are extreme theoretical types; all bonds in compounds are mixtures of these extreme types and are intermediate between them. For the sake of convenience, bonds that are more than 50% covalent

* Estimated values as determined from electronegativity differences are italicized.

are referred to as covalent bonds, whereas those that are less than 50% covalent are referred to as ionic bonds. Both types will actually vary considerably from molecule to molecule in the amount of ionic or covalent character. These differences can be attributed to the unequal attraction exerted by the two atoms in a bond for the electron pair shared between them. Since the different atoms attract electrons to different degrees, the resulting bond formed will depend, in its degree of ionic character, upon the unique set of atoms involved in bond formation. A measure of the degree of attraction for a pair of electrons in a covalent bond is electronegativity.

Pauling Concept of Electronegativity

The concept of electronegativity was originally devised by Pauling in 1932 to explain differences in bond energies in certain molecules. If we consider a bond between two different atoms A and B, from the viewpoint of VB theory the bond may be regarded as a resonance hybrid:

$$\underset{\text{(I)}}{A - B} \quad \leftrightarrow \quad \underset{\text{(II)}}{A^{+} \overset{\times}{\bullet} B^{-}} \quad \leftrightarrow \quad \underset{\text{(III)}}{A^{-} \overset{\times}{\bullet} B^{+}}$$

If B is considerably more electronegative than A, the latter structure (III) can be ignored, as it will not make an important contribution. The contribution of II can be related to the difference in electronegativities (Table 3.4).

TABLE 3.4 ***Bond Energy Data Used to Estimate Electronegativities by Pauling's Method (kcal/mole)***

A	B	E_{AA}	E_{BB}	$\sqrt{E_{AA}E_{BB}}$	E_{AB}	$\Delta(E_{A-B} - \sqrt{E_{AA}E_{BB}})$	$\Delta^{1/2}$
F	Br	38	46	42	61	19	4.4
Si	F	53	38	45	135	90	9.5
Si	Br	53	46	49	73	24	4.9

In Table 3.4, E_{AA} refers to the dissociation energy of the $A—A$ bond. The various bond dissociation energies in Table 3.4 were obtained from thermodynamic data for F_2, Br_2, FBr, SiF_4, and $SiBr_4$. Note in each case that E_{AB} (experimental value) is greater than the geometric mean, $\sqrt{E_{AA}E_{AB}}$. The geometric mean energy is taken as the hypothetical $A—B$ bond energy if $A—B$ were the pure covalent bond, as represented in structure (I). The structure (I). The differences, Δ's, are identified as *ionic resonance energies* because of the contribution of the ionic resonance structures (II) and (III). In the case of SiBr or any primarily covalent bond, the ionic resonance energy contributes additional stability to the actual structure over and above that of the 100% covalent species, structure (I). It is assumed that the greater is the electronegativity difference, the greater will be Δ, where $|X_A - X_B|$ represents the absolute difference between any two Pauling electronegativities; Δ itself is not proportional to $|X_A - X_B|$.

$$\begin{array}{ccccc} \Delta(\text{FSi}) & - & \Delta(\text{SiBr}) & \neq & \Delta(\text{FBr}) \\ (90) & - & (24) & \neq & (19) \end{array} \qquad [3.9]$$

But the quantities $\Delta^{1/2}$ are approximately additive

$$\begin{array}{ccccc} \Delta^{1/2}(\text{FSi}) & - & \Delta^{1/2}(\text{SiBr}) & = & \Delta^{1/2}(\text{FBr}) \\ (9.5) & - & (4.9) & = & (4.6) \end{array} \qquad [3.10]$$

Pauling used such data to devise a table of relative electronegativities, using the relationship

$$|X_A - X_B| = K\Delta^{1/2} \qquad [3.11]$$

where $K = (23)^{-1/2}$. With K equal to $(23)^{1/2}$, the values of Δ are converted in effect from kcal/mole to eV/mole, utilizing the factor $\frac{1}{23}$ eV mole/kcal. By setting the electronegativity values of the second period elements in the range of 1.0 to 4.0, we get a convenient scale for the electronegativities of the various elements (see Table 3.4).

The Mulliken Concept of Electronegativity

Mulliken used theoretical arguments to show that the tendency of an atom in a molecule to compete for electrons with the atom to which it is bound should be proportional to $(I + EA)/2$, or the average of the ionization potential and the electron affinity of the atom. The physical picture here is that the tendency to attract shared electrons should be the average of the tendency of the atom to hold its electrons and its tendency to acquire more. When Mulliken values are expressed in electron volts, they can be adjusted to the Pauling scale by dividing them by 2.8:

$$E_M = \frac{I + A}{2} \qquad [3.12]$$

and

$$X = \frac{E_M}{2.8} \qquad [3.13]$$

where E_M is a Mulliken electronegativity, I and A are ionization potential and electron affinity, and X is the Pauling electronegativity. When the values of E and I are expressed in kilocalories per mole, the relationship is

$$X = \frac{I + A}{130} \qquad [3.14]$$

Allred and Rochow Concept of Electronegativity

The Allred and Rochow concept is based on the electrostatic force of attraction existing between an electron in an atom and the nucleus of that atom. This electrostatic force is taken as the electronegativity. The force of attraction of the electron and the nucleus is governed by Coulomb's law:

$$\text{Force} = \frac{Z^*e^2}{r^2} \qquad [3.15]$$

where e is the charge on the electron, r the covalent radius, and Z the effective nuclear charge. This Z^* is the same as that calculated using Slater's rules. The forces obtained in this way can be adjusted to the Pauling scale by plotting them against Pauling's values. From the slope and intercept of the straight-line plot, the Allred-Rochow electronegativities (Pauling scale) are obtained by the relationship

$$X = 0.359\frac{Z^*}{r^2} + 0.744 \qquad [3.16]$$

Sanderson's Stability Ratios

Sanderson has formulated still another unique concept of electronegativity. In his concept, the electronegativity is considered to be a function of the compactness of the atom. The stability ratio or electronegativity, S, is defined as

$$S = \frac{D}{D_i} \qquad [3.17]$$

where D is the electronic density of the atom, $3Z/4\pi r^3$, and D_i is the *ideal* electronic density of the hypothetical isoelectronic inert gas atom. The term D_i is needed to correct the average electronic densities for those variations with Z (atomic number) that are unrelated to chemical reactivity and thus to electronegativity. The term D_i is determined by linear extrapolation between D values for the next higher and next lower inert gases. These S values are adjusted to the Pauling scale by

$$X^{1/2} = 0.21S + 0.77 \qquad [3.18]$$

None of the preceding methods of defining electronegativity has a distinct advantage over the others, with the possible exception that some are more widely applicable than others. In the Mulliken concept, for example, many values cannot be determined for many metallic elements simply because electron affinity data are unavailable. Similarly, the Pauling values are somewhat limited because the necessary accurate bond energy data are not available for all the elements. The Sanderson and Allred-Rochow concepts are more nearly inclusive, in that they depend in an experimental way only on covalent radii which are quite generally known. The agreement among the four concepts is not immediately outstanding, but when one considers the uniquely different approaches used, the agreement that does exist is quite remarkable. Electronegativity values for the elements calculated by the various methods are given in Table 3.5.

PERCENTAGE IONIC CHARACTER, ELECTRONEGATIVITIES, AND DIPOLE MOMENTS

As a necessary consequence of the shared electron pair in a covalent bond being attracted to the more electronegative atom, an unequal distribution of electronic charge will result in the bond. This unequal distribution of charge can be measured experimentally by measuring the electric dipole moment of the molecule. Such measurements then lead to an equation relating the electronegativity difference between the two atoms in a bond to the percent ionic character of the bond.

The dipole moment of a molecule is a vector quantity, which gives a direct indication of the degree of charge separation in a molecule. It is determined experimentally by measuring the dielectric constant of a substance. The definition of dipole moment is

$$\mu = q \times l \qquad [3.19]$$

TABLE 3.5 ***Electronegativities of the Elements as Determined by Different Methods (All on Pauling Scale)***

KEY: 14 Si — 1.8 (PAULING), 1.74 (ALLRED-ROCHOW), 2.44 (MULLIKEN), 1.74 (SANDERSON)

Periods \ Group	I-A	II-A	III-B	IV-B	V-B	VI-B	VII-B	VIII	VIII	VIII	I-B	II-B	III-A	IV-A	V-A	VI-A	VII-A	
1	1 **H** 2.1 2.2 2.3																	2 **He**
2	3 **Li** 1.0 0.97 0.94 0.86	4 **Be** 1.5 1.47 1.46 1.61											5 **B** 2.0 2.01 2.01 1.88	6 **C** 2.5 2.5 2.63 2.47	7 **N** 3.0 3.07 2.33 2.93	8 **O** 3.5 3.5 3.17 3.46	9 **F** 4.0 4.1 3.91 3.92	10 **Ne**
3	11 **Na** 0.9 1.01 0.93 0.85	12 **Mg** 1.2 1.23 1.32 1.42											13 **Al** 1.5 1.47 1.81 1.54	14 **Si** 1.8 1.74 2.44 1.74	15 **P** 2.1 2.06 1.81 2.16	16 **S** 2.5 2.44 2.41 2.66	17 **Cl** 3.0 2.83 3.0 3.28	18 **Ar**
4	19 **K** 0.8 0.91 0.8 0.74	20 **Ca** 1.0 1.04 1.06	21 **Sc** 1.3 1.2 1.09	22 **Ti** 1.5 1.32 1.13	23 **V** 1.6 1.45 1.24	24 **Cr** 1.6 1.56 1.35	25 **Mn** 1.5 1.6 1.44	26 **Fe** 1.8 1.64 1.47	27 **Co** 1.8 1.7 1.47	28 **Ni** 1.8 1.75 1.47	29 **Cu** 1.9 1.75 1.36 1.74	30 **Zn** 1.6 1.66 1.49 1.86	31 **Ga** 1.8 1.82 1.95 2.1	32 **Ge** 1.8 2.02 2.31	33 **As** 2.0 2.2 1.75 2.53	34 **Se** 2.4 2.48 2.23 2.76	35 **Br** 2.8 2.74 2.76 2.96	36 **Kr**
5	37 **Rb** 0.8 0.89 0.7	38 **Sr** 1.0 1.0 0.96	39 **Y** 1.2 1.11 0.98	40 **Zr** 1.4 1.22 1.0	41 **Nb** 1.6 1.23 1.12	42 **Mo** 1.8 1.30 1.24	43 **Tc** 1.9 1.36 1.33	44 **Ru** 2.2 1.42 1.4	45 **Rh** 2.2 1.45 1.47	46 **Pd** 2.2 1.35 1.57	47 **Ag** 1.9 1.42 1.36 1.72	48 **Cd** 1.7 1.46 1.4 1.73	49 **In** 1.7 1.49 1.8 1.88	50 **Sn** 1.8 1.72 2.02	51 **Sb** 1.9 1.82 1.65 2.19	52 **Te** 2.1 2.01 2.10 2.34	53 **I** 2.5 2.21 2.56 2.5	54 **Xe**
6	55 **Cs** 0.7 0.86 0.69	56 **Ba** 0.9 0.97 0.93	57–71 **La-Lu**	72 **Hf** 1.3 1.23 0.98	73 **Ta** 1.5 1.33 1.04	74 **W** 1.7 1.40 1.13	75 **Re** 1.9 1.45 1.19	76 **Os** 2.2 1.52 1.26	77 **Ir** 2.2 1.55 1.33	78 **Pt** 2.2 1.44 1.36	79 **Au** 2.4 1.42 1.72	80 **Hg** 1.9 1.44 1.92	81 **Tl** 1.8 1.44 1.96	82 **Pb** 1.8 1.55 2.01	83 **Bi** 1.9 1.67 2.06	84 **Po** 2.0 1.76	85 **At** 2.2 1.96	86 **Rn**
7	87 **Fr** 0.7 0.86	88 **Ra** 0.9 0.97	89–103 **Ac-Lr**															
Lanthanide series				57 **La** 1.1 1.08 0.92	58 **Ce** 1.1 1.06 0.92	59 **Pr** 1.1 1.07 0.92	60 **Nd** 1.1 1.07 0.93	61 **Pm** 1.1 1.07 0.94	62 **Sm** 1.1 1.07 0.94	63 **Eu** 1.1 1.01 0.94	64 **Gd** 1.1 1.11 0.94	65 **Tb** 1.2 1.11 0.94	66 **Dy** 1.2 1.10 0.94	67 **Ho** 1.2 1.10 0.96	68 **Er** 1.2 1.11 0.96	69 **Tm** 1.2 1.11 0.96	70 **Yb** 1.2 1.06 0.96	71 **Lu** 1.2 1.14 0.96
Actinide series				89 **Ac** 1.1 1.0 11	90 **Th** 1.3 1.11 13	91 **Pa** 1.5 1.14 15	92 **U** 1.7 1.22 14	93 **Np**	94 **Pu**	95 **Am**	96 **Cm**	97 **Bk** ← 1.2 (est) →	98 **Cf**	99 **Es**	100 **Fm**	101 **Md**	102 **No**	103 **Lr**

where μ is the dipole moment in Debye units, 10^{-18} esu-cm units; q is the magnitude of the charge on one end of the dipole measured in electrostatic units, esu; and l is the distance of charge separation, i.e., bond length, in centimeters. Bond dipole moments are caused by the partial ionic character of bonds due to the different electronegativity values of the atoms in the bond.

TABLE 3.6 ***Electronegativity Differences and Dipole Moments of Some Common Covalent Bonds***

Bond	ΔX	$\mu(D)$	Bond	ΔX	$\mu(D)$
As—H	0.1	0.1	HCl	0.9	1.0
C—H	0.4	0.38	HBr	0.7	0.79
C—O	1.0	0.8	ICl	0.5	0.65
H—F	1.9	1.98	H—I	0.4	0.65

A rough empirical relationship has been found to exist between the dipole moment of a primarily covalent bond and the electronegativity difference of the bonded atoms. For a few molecules, especially the hydrogen halides, the electronegativity difference is approximately equal to the dipole moment of the bond in Debye units (D). There are many exceptions to this relationship, but the correlations of many bonds are surprisingly close (Table 3.6). Any chemical bond involving atoms with different electronegativities will have a bond dipole moment. In a diatomic molecule, this means that the molecule must also have a dipole moment. For a polyatomic molecule, however, whether the molecule has a net dipole moment will depend on its geometrical configuration, even if the electronegativities of the involved atoms are quite different. Thus, carbon dioxide has no net dipole moment because of its linear structure, in which the two bond dipoles, in effect, cancel each other. On the other hand, the water molecule, has a net dipole moment because of its angular structure. Although water is actually a tripole, because of its small bond angle it is approximately a dipole and has become known as such. Thus, the observation that a triatomic molecule of type AB_2 possesses a permanent dipole moment is definite evidence as to its nonlinearity. Similarly, evidence as to structure can be obtained from dipole moment data for more complicated molecules.

ELECTRONEGATIVITY AND PERCENTAGE IONIC CHARACTER

The relationship between the electronegativity difference between two atoms in a bond and the percentage ionic character in the bond was originally derived by Pauling from dipole moment data on the hydrogen halides. Thus, if HCl were 100% ionic and the hydrogen and chlorine atoms at the same distance from each other as they are in the molecule, as observed experimentally, hydrogen chloride would have a dipole moment of 6.08 D:

$$\begin{array}{ccc} H^{+1} & & Cl^{-1} \\ |\longleftarrow & 1.27\ \text{A} & \longrightarrow| \end{array}$$

(1.27 A is the experimental bond length in HCl). That is, if HCl were 100% ionic, its dipole moment would be

$$\begin{aligned} \mu &= q \times l \\ &= 4.77 \times 10^{-10}\ \text{esu} \times 1.27 \times 10^{-8}\ \text{cm} \\ &= 6.08 \times 10^{-18}\ \text{esu-cm} \\ &= 6.08\ \text{D} \end{aligned} \qquad [3.20]$$

Since the observed dipole moment of HCl is only 1.03 D, the ratio of the observed moment to the 100% ionic value can be used as a measure of the degree of ionic character in the bond. That is, the percentage ionic character of the HCl molecule is $(1.03/6.08) \times 100 = 17\%$. Now, since the electronegativity difference between hydrogen and chlorine is 0.9 (Pauling values), it can be stated that when the electronegativity difference is 0.9 the bond is approximately 17% ionic. Carrying out a similar calculation for the other hydrogen halides, Pauling was able to obtain, from a plot of ionic character vs. electronegativity difference, the following equation which fits the curve:

$$\text{Percentage ionic character} = 18|X_A - X_B|^{1.4} \qquad [3.21]$$

More recently, with more data available, Hannay and Smyth were able to formulate a relationship using the same method as Pauling, but which afforded a better correlation with a greater variety of molecules. The Hannay and Smyth relationship is

$$\text{Percentage ionic character} = 16|X_A - X_B| + 3.5|X_A - X_B|^2 \qquad [3.22]$$

Thus the carbon–chlorine bond, in which $X_{Cl} - X_C = 0.5$, is 8.9% ionic, whereas a bond in which the electronegativity difference is 2.1 corresponds to 50% ionic character. The latter case fits the elements sodium and chlorine and is rather surprising considering the observed highly ionic character of sodium chloride. This situation can be rationalized in the following way. In a crystal of sodium chloride each sodium atom is surrounded by six chlorine atoms and each chlorine atom is surrounded by six sodium atoms. In this environment, the sodium atom can be regarded, hypothetically, as attached to one chlorine atom by a bond that is 50% covalent in character, and to the remaining five chlorine atoms by bonds that are 100% ionic. However, since all the bonds in crystalline sodium chloride are identical, the amount of covalent character in each sodium–chloride bond averages out to $\frac{1}{6}$ of 50% or 8.35%. Thus, a sodium–chlorine bond in a crystal of sodium chloride may be regarded as approximately 8.4% covalent or 91.6% ionic. Similar arguments may be applied to other ionic crystals, although it is obvious that the percentage ionic character concept as related to electronegativity differences becomes cumbersome and less meaningful for ionic compounds. Even for primarily covalent compounds, the preceding relationships between electronegativity differences, dipole moments, and percentage ionic character are at best inexact and are unreliable for other than qualitative work.

INTERRELATIONSHIPS OF BOND ENERGY, RESONANCE ENERGY, ELECTRONEGATIVITY, IONIC CHARACTER, AND DIPOLE MOMENTS FOR CHEMICAL BONDS

As mentioned earlier, the bond energy for the 100% covalent structure of a given bond, such as hydrogen–chlorine bond, was considered to be the geometric mean of the bond energies of the hydrogen–hydrogen bond and the chlorine–chlorine bond:

$$\begin{aligned} E(\mathrm{HCl};\ 100\%\ \text{covalent}) &= \sqrt{E_{\mathrm{H_2}} \cdot E_{\mathrm{Cl_2}}} \\ &= \sqrt{(104)(58)} = 76\ \text{kcal/mole} \end{aligned} \qquad [3.23]$$

Since the experimental bond energy of the hydrogen chloride molecule is 103 kcal/mole, the difference (Δ) of 27 kcal/mole between the experimental bond energy and geometric mean value (100% covalent bond) is considered to be additional stability imparted to the hydrogen chloride through ionic resonance forms.

$$\Delta = E(\mathrm{HCl};\ \text{experimental}) - \sqrt{E_{\mathrm{H_2}} \cdot E_{\mathrm{Cl_2}}} = 27\ \text{kcal/mole} \qquad [3.24]$$

More generally,

$$\Delta_{AB} = E_{AB} - \sqrt{E_{AA} \cdot E_{BB}} \qquad [3.25]$$

Therefore, from a table of bond energies (Table 3.2), the corresponding ionic resonance energy may be calculated for each bond in the molecule. For a diatomic molecule, such as hydrogen chloride, this resonance energy is also the enthalpy change for the reaction.

$$\tfrac{1}{2}H_2\,(g) + \tfrac{1}{2}Cl_2\,(g) \rightarrow HCl\,(g), \qquad \Delta H_f^\circ = -27\ \text{kcal/mole} \qquad [3.26]$$

The energy liberated in reaction [3.25] should be distinguished from that for the reaction

$$H\,(g) + Cl\,(g) \rightarrow HCl\,(g), \qquad \Delta H_f = -103\ \text{kcal/mole} \qquad [3.27]$$

For a polyatomic molecule, the resonance energies are additive to produce the enthalpy change of formation of the gaseous molecule from gaseous molecules of the elements. For example, the resonance energy of the phosphorus–hydrogen bond is (in kcal/mole)

$$\begin{aligned} \Delta_{P-H} &= E_{P-H} - \sqrt{E_{P-P} \cdot E_{H-H}} \\ &= 75 - 51 - 104 \\ &= 75 - 73 = 2\ \text{kcal/mole} \end{aligned} \qquad [3.28]$$

The enthalpy change for the formation of 1 mole of phosphine from phosphorus (P_4) and hydrogen is, then,

$$2\ \text{kcal/mole bond} \cdot 3\ \text{bonds} = 6\ \text{kcal/mole} \qquad [3.29]$$

and the balanced equation is

$$\tfrac{1}{4}P_4\,(g) + \tfrac{3}{2}H_2\,(g) \rightarrow PH_3\,(g) \qquad [3.30]$$

These relationships are valid only when single bonds are involved in all species, since the energy of a double or triple bond is not an integral multiple of the single bond energy. For the ammonia molecule, the nitrogen–hydrogen bond energy is not $\frac{1}{3}$ of the enthalpy change for the reaction

$$\tfrac{1}{2}N_2\,(g) + \tfrac{3}{2}H_2\,(g) \rightarrow NH_3\,(g), \qquad \Delta H_f = -11\ \text{kcal/mole} \qquad [3.31]$$

in which three nitrogen–hydrogen bonds are formed. The reason lies in the fact that the energy of two triply bonded nitrogen atoms, 226 kcal/mole, is considerably larger than three times the nitrogen–nitrogen single bond energy, 38 kcal/mole. The extra stability of each bond in molecular nitrogen over simple addition of three single bonds is

$$\frac{226 - 3(38)}{3} = 37 \text{ kcal/mole} \qquad [3.32]$$

Consequently, some of the energy which would otherwise be given off as heat-of-reaction energy is expended in breaking up the very stable, triply bonded nitrogen atoms. Since this extra stability of the nitrogen molecule encompasses two nitrogen atoms, the resonance of the nitrogen–hydrogen bond could be calculated by dividing the change in enthalpy for reaction [3.30] by three and adding the correction factor $\frac{37}{2}$ kcal/mole.

$$\begin{aligned} \Delta(\text{N—H}) &= \tfrac{\Delta H}{3} + \tfrac{37}{2} \text{ kcal/mole} \\ &= (\tfrac{11}{3} + 18.5) = 22 \text{ kcal/mole} \end{aligned} \qquad [3.33]$$

Table 3.7 lists ionic resonance energies for some common bonds.

BOND DATA USED IN THE BORN-HABER CYCLE

All the previously discussed properties of bonds—bond energy, bond resonance energy, electronegativity, percentage ionic character, and dipole moment—can be correlated and their interrelationships understood more readily from a problem involving a Born-Haber cycle such as the following. Using bond energy data (Table 3.1) and the standard enthalpy change of formation for silane from the elements

$$\text{Si (s)} + 2\text{H}_2\text{ (g)} \rightarrow \text{SiH}_4\text{ (g)} \qquad \Delta H_f^\circ = -11 \text{ kcal/mole} \qquad [3.34]$$

(1) Calculate a value for the heat of sublimation of elemental silicon.
(2) Calculate the enthalpy change for the reaction

$$\text{Si (g)} + 4\text{H (g)} \rightarrow \text{SiH}_4\text{ (g)}, \qquad \Delta H_f' = ? \qquad [3.35]$$

(3) Calculate the enthalpy change for the reaction

$$\text{Si (g)} + 2\text{H}_2\text{ (g)} \rightarrow \text{SiH}_4\text{ (g)}, \qquad \Delta H_f'' = ? \qquad [3.36]$$

(4) Given the electronegativity of hydrogen [2.1], calculate the electronegativity of silicon.

(5) Calculate the percentage ionic character of the silicon–hydrogen bond.

(6) What is the approximate dipole moment of the silicon–hydrogen bond?

(7) Given the ionization potential of silicon (188 kcal/mole), calculate its electron affinity.

TABLE 3.7 ***Ionic Resonance Energies for Some Typical Single Covalent Bonds, in kcal/mole at 25°C****

	H	C	N	O	F	Si	P	S	Cl	Br	I	Ge	As
H	0												
C	6	0											
N	22	17	0										
O	53	32	13	0									
F	71	61	28	8	0								
Si	2	10	*33*	66	90	0							
P	2	−2	6	39	72	−1	0						
S	15	0	*5*	*23*	24	2	4	0					
Cl	25	8	−1	5	14	35	24	7	0				
Br	19	−4	16	*11*	19	24	15	3	0.3	0			
I	10	3	*5*	14	*52*	7	1	*0*	4	2	0		
Ge	*−13*	*11*	*33*	*65*	*41*	*0*	*2*	*11*	30	20	11	0	
As	*11*	*5*	*23*	*52*	74	*0.9*	*0.2*	*5*	25	18	8	0.9	0

The solution to this problem lies in the relationships among several energy terms, and these relationships may be related diagrammatically in a Born-Haber cycle (Fig. 3.2).

Reaction [3.32] may be considered to take place via either of two pathways. If one proceeds up the ladder, starting at the lower left with Si (s) and $2H_2$ (g) in their standard states, the first step requires the addition of the sublimation energy of silicon, S. The second addition of energy causes the dissociation of two hydrogen molecules, $2E_{H_2}$. The reactants are now converted to gaseous atoms—Si (g) and 4H (g)—and they may now combine to form gaseous silane, SiH_4. This union of atoms releases four silicon–hydrogen bond energies. The alternate pathway involves the conversion of solid silicon and gaseous hydrogen to silane directly and will have associated with it the standard enthalpy change of formation, ΔH_f°. The enthalpy of silane must lie below that of solid silicon and gaseous

*Estimated values as determined from electronegativity differences are italicized.

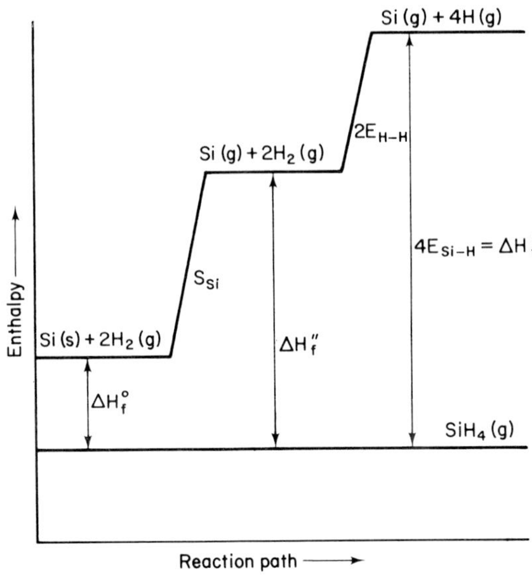

Figure 3.2. Born-Haber cycle for the preparation of SiH_4 from its elements.

hydrogen since the change in enthalpy, ΔH_f°, is negative in sign. The relationship of the energy terms in Fig. 3.2 is expressed by the following equation:

$$|\Delta H_f^\circ| + S_{Si} + 2E_{H^2} = 4E_{Si-H} \qquad [3.37]$$

The corresponding answers to the previously enumerated questions follow:

(1) Substituting values of E_{H-H} and E_{S-H} from Table 3.1 and the value of ΔH_f° from equation [3.34] leads to a value of 85 kcal/mole for S_{Si}, the sublimation energy of silicon, which is in fair agreement with literature values.

(2) The enthalpy change for reaction [3.35], $\Delta H_f'$, is equivalent to four silicon–hydrogen bond energies, i.e., $\Delta H_f' = 4E_{Si-H} = 4 \times 76 = 304$ kcal/mole.

(3) The enthalpy change for reaction [3.36] $\Delta H_f''$, can be calculated as follows:

$$\begin{aligned} \Delta H_f'' &= \Delta H_f' - 2E_{HH} \\ &= (304 - 208) \text{ kcal/mole} \qquad [3.38] \\ &= 96 \text{ kcal/mole} \end{aligned}$$

(4) Often, the condition of gaseous molecules is found at the first step of the ladder, and the energy term, equivalent to $\Delta H_f''$, can then be related to the resonance energy of the heteronuclear bond. In this case, it cannot, since silicon is already in the atomic state. Thus, the silicon–hydrogen bond resonance energy, Δ_{SiH}, which can be related to the silicon–hydrogen electronegativity difference, must be calculated from the relationship

$$\begin{aligned} \Delta_{\text{SiH}} &= E_{\text{SiH}} - \sqrt{E_{\text{SiSi}} \times E_{\text{HH}}} \\ &= 76 - \sqrt{51 \times 104} \\ &\simeq 3 \text{ kcal/mole} \end{aligned} \tag{3.39}$$

Now, this value of Δ_{SiH} can be related through equation [3.11] to the electronegativity difference $X_{\text{H}} - X_{\text{Si}}$:

$$\begin{aligned} X_{\text{H}} - X_{\text{Si}} &= \sqrt{\tfrac{3}{23}} \\ &\simeq 0.4 \end{aligned} \tag{3.40}$$

Given the electronegativity of H = 2.1, the electronegativity of Si is thus calculated to be approximately 1.7, which is in the range of values reported in Table 3.5 for X_{Si}.

(5) The percentage ionic character can now be calculated from equation [3.21]:

$$\begin{aligned} \text{Percentage ionic (Si–H bond)} &= 16(X_{\text{H}} - X_{\text{Si}}) + 3.5(X_{\text{H}} - X_{\text{Si}})^2 \\ &= 16(0.4) + 3.5(0.16) \\ &= 6.4 + 0.56 \\ &= 7.1\% \end{aligned}$$

(6) The approximate dipole moment of the silicon–hydrogen bond is given by the electronegativity difference, 0.4, and is thus on the order of 0.4 D. Keep in mind that, because of symmetry, the SiH_4 molecule has no net dipole moment.

(7) The electron affinity of silicon may be calculated by equation [3.14]:

$$\begin{aligned} X_{\text{Si}} &= \frac{I + A}{130} \\ A &= 130X_{\text{Si}} - I \\ &= [130(1.7) - 188] \text{ kcal/mole} \\ &= 33 \text{ kcal/mole} \end{aligned} \tag{3.41}$$

MOLECULAR VIBRATIONS AND FORCE CONSTANTS OF CHEMICAL BONDS

Perhaps the most important property of a chemical bond to those inorganic chemists who have an interest in molecular structure, particularly as determined by the techniques of absorption spectroscopy, is the force constant of the bond. Force constants are obtained by various methods from an analysis of the vibrational absorption spectrum of the molecule. A knowledge of the force constant of a chemical bond is important since it affords an additional means of measuring the strength of the bond.

The total energy of a molecule can be regarded as a sum of the kinetic energy and the internal potential energy:

$$E_T = E_{\text{kinetic}} + E_{\text{internal}} \qquad [3.42]$$

In turn, E_{internal} is a sum of energies:

$$E_{\text{internal}} = E_e + E_v + E_r \qquad [3.43]$$

where E_e is the electronic energy of the molecule, E_v is the vibrational energy of the molecule, and E_r is the rotational energy of the molecule.

The relationship between these contributions to the internal potential energy of a molecule may best be seen from a plot of potential energy vs. internuclear distance for a simple diatomic molecule (Fig. 3.3). In Fig. 3.3, the minimum in the potential energy diagram corresponds to the equilibrium internuclear distance, r_0. In the context of this discussion, it is important to recognize the energy differences associated with electronic, vibrational, and rotational transitions. Thus, $n = 1$ and $n = 2$ represent the ground and excited electronic states, respectively. Associated with each electronic energy state is a series of vibrational energy states, and we can see from Fig. 3.3 that a transition of the molecule from $n = 1$ to $n = 2$ may have associated with it a vibrational transition from $n = 1$, $v = 0$ to $n = 2$, $v = 1$. In addition, associated with each vibrational energy level is a series of rotational energy levels (shown expanded at the right of Fig. 3.3), in which transitions between rotational energy levels may take place in conjunction with the vibrational transition shown: $n = 1$, $v = 0$, $J = 0$, to $n = 1$, $v = 1$, $J = 1$. Both processes, vibrational and rotational transitions take place with an absorption of energy. The rotational energy of a molecule changes as the molecule rotates about its center of gravity. Therefore, changes in the electronic energy of a molecule (e.g., $n = 1$ to $n = 2$) involve

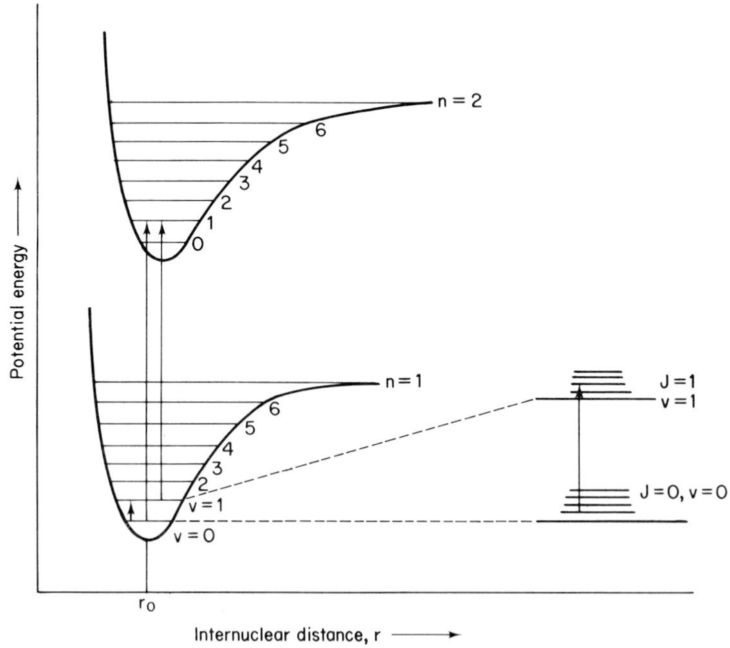

Figure 3.3. Internal electronic vibrational and rotational levels in a diatomic molecule.

vastly greater amounts of energy ($10^5 - 10^4$ cm^{-1}) than changes between vibrational levels (e.g., $n = 1$, $v = 0$, to $n = 1$, $v = 1$), and they usually fall in the ultraviolet–visible region of the spectrum. Vibrational transitions, involving energies on the order of $10^4 - 10^2$ cm^{-1}, are observed in the infrared region. The much smaller separations between rotational levels ($10^2 - 10^{-1}$ cm^{-1}) lead to transitions found in the microwave region of the spectrum.

Changes in electronic energy are particularly important in the application to coordination chemistry, as in the determination of crystal field stabilization energies (Chapter 9). Rotational transitions can be related directly to the moments of inertia of a molecule and thus to equilibrium bond distances and angles, which can be determined very precisely from microwave spectra. Of primary importance to this discussion, however, are vibrational transitions in molecules, which can be related to the force constants of the bonds in the molecule. To exemplify this relationship, we shall limit the discussion to diatomic molecules. Force constants can be determined from vibrational spectra of polyatomic molecules, although not directly, as for diatomic molecules, because there are often strong interactions between the vibrations of different bonds. Well-established methods are used for dealing with polyatomic molecules, but these are beyond the scope of this discussion.

For a diatomic molecule, AB, the single vibration must consist in a stretching and contracting of the bond relative to the equilibrium separation, r_e:

$$\begin{array}{ccc} \leftrightarrow & & \leftrightarrow \\ A & \text{———————} & B \\ |\leftarrow & r_e & \rightarrow| \end{array}$$

To a good approximation, the molecule behaves like a pair of weights attached to a spring that obeys Hooke's law [which says that to stretch or contract the spring by Δr requires work equal to $\frac{1}{2}k(\Delta r)^2$]. The potential energy of the spring is thus given by

$$V = \tfrac{1}{2}k(\Delta r)^2 \tag{3.44}$$

where k is the force constant of the spring (or of a bond for a molecule). Here, k is the restoring force in millidynes per angstrom (md/A) resisting displacement from the equilibrium position, r_e. It is a measure of the *stiffness* of the spring, or, for a molecule, the strength of the chemical bond.

Such a vibrating system is a harmonic oscillator, following the laws of simple harmonic motion. The expression that relates the frequency of vibration, ν, k, and the masses of a harmonic oscillator, is

$$\nu_{\text{cm}^{-1}} = \frac{1}{2\pi c}\sqrt{\frac{k}{\mu}} \tag{3.45}$$

where $\nu_{\text{cm}^{-1}}$ is the vibrational frequency in wave numbers, cm^{-1}, c is the speed of light, k is the force constant of the harmonic oscillator, and μ is the reduced mass of the system (in grams) and is defined by

$$\frac{1}{\mu} = \frac{1}{m_A} + \frac{1}{m_B} \tag{3.46}$$

or

$$\mu = \frac{m_A m_B}{m_A + m_B} \tag{3.47}$$

Thus, if we can measure the vibrational frequency of the molecule, AB, we can calculate the force constant of the AB bond.

As an example, consider the HCl molecule, which absorbs infrared radiation with a frequency of 2890 cm^{-1}. Substitution of this value into equation [3.45], along with the appropriate values for π and c and a value of $\mu_{HCl} = m_H m_{Cl}/(m_H + m_{Cl}) = 1.63 \times 10^{-24}$ g, leads to the value of 4.84 md/A for k. Similar calculations may be made for other diatomic molecules and values obtained for their force constants (Table 3.8). As may be seen from Table 3.8, definite qualitative relationships exist between the force constant of a bond and other bond properties.

The most obvious relationship that we can observe for homonuclear diatomic molecules in Table 3.8 is that the force constant decreases as the

TABLE 3.8 ***Fundamental Vibrational Frequencies (v = 0 to v = 1) of Some Diatomic Molecules and Corresponding Bond Properties***

Molecule	(cm^{-1})	k(md/A)	Bond Order*	Bond Energy (kcal/mole)	Bond Length (A)
H_2	4159†	5.2	1	104	0.74
D_2	2990	5.3	1	105	0.74
HF	3958‡	8.8	1	135	0.92
HCl	2886	4.8	1	103	1.27
HBr	2559	3.8	1	88	1.41
HI	2230	2.9	1	71	1.61
CO	2143	18.7	3	256	1.13
NO	1876	15.5	2.5	162	1.15
O_2	1556	11.4	2	118	1.21
N_2	2331	22.6	3	225	1.10
F_2	892	4.5	1	38	1.42
Cl_2	557	3.2	1	58	1.99
Br_2	321	2.4	1	46	2.28
I_2	213	1.7	1	36	2.67
Li_2	246	1.3	1	25	2.67
Na_2	158	1.7	1	17	3.08

masses of the atoms increase. This is particularly true and valid for the halogen molecules, and is associated with a decrease in frequency of vibration. Similarly, the force constants of the heteronuclear hydrogen halide molecules decrease as the vibrational frequency decreases and the

* As predicted from molecular orbital theory.

† All values for homonuclear diatomic molecules are taken from Raman data.

‡ All values for heteronuclear diatomic molecules are taken from infrared data.

mass of the halogen atom increases. In each of these examples the bond order is one, so that the observed changes can be related to the bond energy and to the bond length of the molecule.

Thus, by analysis of the data in Table 3.8, we may say that (1) as the bond length increases the force constant decreases, and (2) as the bond energy decreases the force constant decreases. Both of these observations are consistent with the concept that the force constant is also a measure of the strength of the chemical bond.

Also, the force constants for double and triple bonds are significantly larger than the range of values found for single bonds. This fact is also consistent with the idea that as the bond order increases the bond length decreases and the bond energy increases.

Although none of these relationships is linear over any appreciable range of values, such qualitative relationships are often useful in predicting or interpreting physical and chemical properties of bonds in molecules.

The force constants of bonds have been related quantitatively to the bond lengths by Badger, using the equation

$$k^{-1/3} = a_{ij}(D - b_{ij}) \qquad [3.48]$$

where k is the force constant of the bond, D is the equilibrium bond distance, and a_{ij} and b_{ij} are constants whose values depend on the nature of the bonded atoms (see Table 3.9).

TABLE 3.9 ***Constants for Gas Molecules for Use in Equation [3.48]***

Type of Bond				
Atom i in period	Atom j in period	Example	a_{ij}	b_{ij}
1	1	H_2	2.32Å	0.025Å
1	2	HF	2.32	0.335
1	3	HCl	2.32	0.585
1	4	HBr	2.32	0.65
2	2	CO	1.75	0.68
3	2	PN	1.87	0.94
4	2	TiO	2.00	1.06
5	2	SnO	2.04	1.18
6	2	PbO	2.04	1.26
3	3	Cl_2	2.04	1.25
5	3	ICl	1.98	1.48
4	4	Br_2	1.98	1.48
5	5	I_2	2.04	1.76

Suggested Readings

1. AC_3 Resource Paper III, *J. Chem. Educ.*, **42**, 502 (1965).
2. Allred, A. L., and E. G. Rochow, *J. Inorg. Nucl. Chem.*, **5**, 264, 269 (1958).
3. Badger, R. M., *J. Chem. Phys.*, **2**, 128 (1934); **3**, 710 (1935).
4. Barrow, G., *Molecular Spectroscopy*, McGraw-Hill Book Company, New York, 1962, Chap. 2.
5. Cottrell, T. L., *The Strengths of Chemical Bonds*, 2nd ed., Butterworth & Co., Ltd., London, 1958.
6. Gaydon, A. G., *Bond Dissociation Energies*, Chapman & Hall, Ltd., London, 1953.
7. Hannay, N. B., and C. P. Smyth, *J. Am. Chem. Soc.*, **68**, 171 (1946).
8. Mulliken, R. S., *J. Chem. Phys.*, **2**, 782 (1934); **46**, 497 (1949).
9. Sanderson, R. T., *Chemical Periodicity*, Reinhold Publishing Corp., New York, 1960; *Science*, **114**, 670 (1951).
10. Swarc, M., *Chem. Rev.*, **47**, 75 (1950).

General References

1. Cotton, F. A., and G. Wilkinson, *Advanced Inorganic Chemistry*, 2nd ed., Interscience Publishers, New York, 1966.
2. Day, M. C., and J. Selbin, *Theoretical Inorganic Chemistry*, 2nd ed., Reinhold Publishing Corp., 1968.
3. Douglas, B. E., and D. H. McDaniel, *Concepts and Models of Inorganic Chemistry*, Blaisdell Publishing Co., New York, 1965.
4. Harvey, K. B., and G. B. Porter, *Introduction to Physical Inorganic Chemistry*, Addison-Wesley Publishing Co., Inc., Reading, Mass., 1963.
5. Heslop, R. B., and P. L. Robinson, *Inorganic Chemistry*, 3rd ed., Elsevier, Amsterdam, 1967.
6. Moeller, T., *Inorganic Chemistry*, John Wiley & Sons, Inc., New York, 1952.
7. Pauling, L., *The Nature of the Chemical Bond*, 3rd ed., Cornell University Press, Ithaca, N.Y., 1960.
8. Phillips, C. S. G., and R. J. P. Williams, *Inorganic Chemistry*, vols. 1 and 2, Oxford Publishing Co., Oxford, 1965.
9. Sanderson, R. T., *Inorganic Chemistry*, Reinhold Publishing Corp., New York, 1967.

Study Questions

3.1. Criticize in detail the statement: *Elements combine because in so doing, the energy of the system is decreased due to the attainment of stable inert gas configurations by the atoms.*

3.2. Offer an explanation of the lower bond energy of F_2 as compared with Cl_2.

3.3. Explain why the B—F bond length in BF_3 is less than that in BF_4^-.

3.4. Calculate the electron affinity of iodine, given the following data:

Reaction	ΔH (kcal/mole)
I_2 (s) → I_2 (g)	15.0
I_2^0 (g) → $2I^0$ (g)	70.8
I^- (g) → I^- (aq)	−85.8
$\frac{1}{2}I_2$ (s) + e → I^- (aq)	−114.3

Sketch a Born-Haber diagram, showing the preceding transformations.

3.5. (a) Given the following data, calculate the C—Hg bond energy.
(b) Sketch a Born-Haber diagram to illustrate all energy terms involved.

Thermochemical Process	ΔH (kcal/mole)
(i) Hg (l) + 2C (graphite) + $3H_2$ (g) → $(CH_3)_2Hg$ (l)	37.8
(ii) $(CH_3)_2Hg$ (l) → $(CH_3)_2Hg$ (g)	7.7
(iii) C (graphite) → C (g)	171.0
(iv) Hg (l) → Hg (g)	14.6

3.6. (a) Calculate the heat absorbed or evolved per mole of PH_3 formed in the reactions

(i) P_4 (g) + $6H_2$ (g) → $4PH_3$ (g)
(ii) 4P (g) + 12H (g) → $4PH_3$ (g)

Given:

H—H bond energy = 104.2 kcal/mole
P—P bond energy = 51.3 kcal/mole
P—H bond energy = 78.4 kcal/mole

and ΔH_f of PH_3 (g) from P_4 (s) and H_2 (g) is 6.00 kcal/mole (heat absorbed during formation of PH_3). State clearly whether heat is absorbed or evolved in each reaction.

(b) Calculate the value for the heat of sublimation of P_4.
(c) Sketch a Born-Haber diagram for the preceding processes including all energy terms.
(d) Given the electronegativity of hydrogen [2.1], calculate the electronegativity of phosphorus.
(e) Calculate the percentage ionic character of a P—H bond.
(f) Would you expect PH_3 or NH_3 to have the higher dipole moment? Explain.
(g) Given the ionization potential of phosphorus (254 kcal/mole), calculate the electron affinity of phosphorus.

FROM **PRENTICE-HALL, INC.**
ENGLEWOOD CLIFFS, N.J. 07632

829122F

We are interested in the opinion of —

DR. J. A. BERAN
DEPT. OF CHEMISTRY
TEXAS A&I UNIVERSITY
KINSVILLE, TEXAS 78363

concerning these texts —

1 466359-4 INORGANIC CHEMISTRY - Demitras

FUTURE

1 TOTAL BOOKS	C	D
	XX	

2

bn

Sent with the compliments of your P-H representative

L. Baudean

THANK YOU FOR CONSIDERING OUR TEXTS

3.7. Account for the fact that the dipole moment of NH_3 is considerably larger than the dipole moment of NF_3, even though the electronegativity differences between the atoms in the respective bonds are almost identical.

3.8. Account for the value of the ionic resonance energy of the N—H bond, as obtained from the following data:

	N_2	NH	HH
E (kcal/mole)	225	93	104

3.9. The fundamental vibrational frequencies, ν, and the internuclear distances, r, of the hydrogen halides are

HX	HF	HCl	HBr	HI
cm^{-1}	3960	2885	2560	2230
r (A)	0.92	1.27	1.41	1.60

(a) Calculate the force constants of the HX bonds.
(b) Estimate the infrared stretching frequencies of the deuterium halides, assuming the force constants to be the same as those of the corresponding hydrogen halides. Compare your calculated values with the observed values and account for any differences.

3.10. The enthalpy of formation of gaseous XeF_4 from xenon and fluorine is reported to be -54.8 kcal/mole. Calculate the Xe—F bond energy in XeF_4.

3.11. Given the following electronegativity values: B = 2.1, S = 2.5, P = 2.1, Cl = 3.0; state which compound—H_2S or PCl_3—might be expected to decompose into molecules of its constituent elements more readily on heating. Show all the necessary equations and assumptions used to arrive at your answer. If the bond energies of H—S = 88.0 kcal/mole and P—Cl = 78.0 kcal/mole, which compounds, H_2S or PCl_3, would be more readily dissociated into atoms of its constituent elements? How much energy would be needed to break PCl_3 and H_2S into atoms?

3.12. Calculate the P—H bond energy in PHF_2 (g), given that

$$P_4\,(g) + 2H_2\,(g) + 4F_2\,(g) \rightarrow 4PHF_2\,(g) + A \text{ kcal}$$
$$P_4\,(g) + B \text{ kcal} \rightarrow 4P\,(g)$$
$$D_{H_2} = C \text{ kcal/mole}, \quad D_{F_2} = D \text{ kcal/mole}, \quad E_{PF} = E \text{ kcal/mole}$$

3.13. Distinguish clearly among (1) bond energy, (2) bond dissociation energy, and (3) bond resonance energy.

3.14. Using the following information, derive an expression that would give the P—B bond energy in the compound $(CH_3)_3P—BH_3$. Show your reasoning.

Heat of dissociation of P_4 (g)	= A kcal/mole
Heat of sublimation B (s)	= B kcal/mole
Heat of sublimation of graphite	= C kcal/mole
Heat of sublimation of $(CH_3)_3PBH_3$	= D kcal/mole
Heat of sublimation of P_4 (s) to P_4 (g)	= E kcal/mole
Heat of formation of $(CH_3)_3PBH_3$ (s) from elements in standard states	= F kcal/mole
C—H bond energy	= G kcal/mole
B—H bond energy	= H kcal/mole

C—P bond energy = I kcal/mole
H—H bond energy = J kcal/mole

3.15. Calculate the heat of formation of BF_3; include the use of the following information: calculated B—F bond length – 1.50 A, expt'l = 1.30 A.

3.16. Given the (N≡N) bond energy is 225 kcal/mole and N—N is 38 kcal/mole, calculate the heat of formation (ΔH_f) of NF_3 from N_2 and F_2 from these data and electronegativity data.

3.17. Explain the sharp drop in single bond energy between C—C (83 kcal/mole) and N—N (38 kcal/mole).

CHAPTER 4

Theories of the Covalent Bond

THE VALENCE BOND APPROACH RESONANCE

The valence bond (VB) method is a quantum mechanical approach to the understanding of chemical bonding that employs the linear combination of atomic orbitals. This presentation will be restricted to the formation of covalent bonds in molecules using the characteristic concept of the VB treatment, which is that of resonance.

One must recall that a wave function, ψ, describes an electron or a number of electrons in a given system at a given energy. The system under question may be an atom, molecule, or ion. Therefore, for the nth quantum state, there exists an energy, E_n, that is associated with this state. If E_n is minimal, then the nth quantum state is the lowest energy state and is called the ground state of the system. The quantum mechanical efforts to calculate this state of lowest energy involve various approximation methods that are beyond the scope of this text. However, some qualitative concepts will be presented.

If ψ_0 is the wave function for the ground state, it is fundamental that the energy, E_0, is less than that associated with any other wave function, ψ_n, that might be proposed. The structure of the ground state is the one of all conceivable structures that results in maximum stability. So if we consider two structures of a given system, I and II, that contribute to its stability, the VB method would represent the system by a general wave function, ψ, in the following manner:

$$\psi = a\psi_{\text{I}} + b\psi_{\text{II}} \qquad [4.1]$$

In equation [4.1], a and b are arbitrary numerical coefficients; the ratio a/b is important in energy calculations. If a/b is large, then ψ_{I} is the important species contributing to the stability of the system; if a/b approaches unity, then ψ_{I} and ψ_{II} are about equally influential in contributing to the stability of the system; and if a/b is small, the species most responsible for the system's stability is ψ_{II}. The VB method has conventualized this concept by saying that the real structure of the system is a resonance hybrid of structures I and II. The structure that best represents the system is neither I nor II, but is derived from both with the best ratio of a/b. The a/b ratio that yields the minimum energy that is below both E_{I} and E_{II} is chosen. This added stability is referred to as *resonance energy*.

The application of the foregoing ideas to a real system should help clarify them. Consider the hydrogen molecule, $H_a - H_b$, where the subscripts a and b serve only as a method of artificial distinction between the two hydrogen atoms. If the two hydrogen atoms were infinitely distant from one another, the wave function for the system would be

$$\psi = \psi_a \psi_b \qquad [4.2]$$

Equation [4.2] indicates that neither atom affects the other and that the energy of the system is the sum of the energy of each component. If the two atoms approach each other until a covalent bond between the two atoms is formed, the energy of the system changes continuously until it reaches a minimum. Since the hydrogen molecule, H_2, is a stable molecule, more stable than the separate atoms, the energy of the combined atoms must be less than the separate atoms.

The wave function for the system will now be written as

$$\psi_{\mathrm{I}} = \psi_a(1)\psi_b(2) \qquad [4.3]$$

where (1) represents the electron associated with H_a and (2) the electron associated with H_b. Figure 4.1 shows the plot of energy vs. R_{ab}, the distance between the two hydrogen nuclei. Curve (A) represents the energy plot of the system described by equation [4.3]. Curve (E) represents the experimental energy curve for the system. Comparison shows poor agreement between curves (A) and (E). The comparison may be improved if the electrons on the atoms are exchanged, because in reality they are shared. Actually, this would be an accurate description of a covalent bond. Therefore, the total wave function would now be

$$\psi_{\mathrm{cov}} = \psi_a(1)\psi_b(2) + \psi_a(2)\psi_b(1) \qquad [4.4]$$

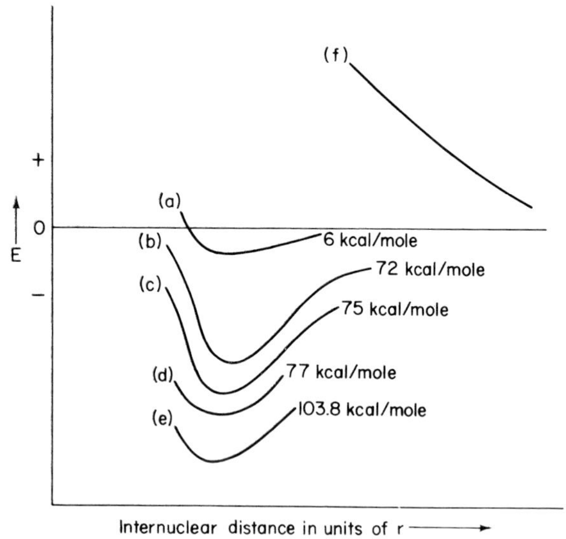

Figure 4.1. Potential energy curves for two hydrogen nuclei approaching one another to form the hydrogen molecule.

or

$$\psi_{\text{cov}} = \psi_{\text{I}} + \psi_{\text{II}} \quad [4.5]$$

where ψ_{cov} is the covalent contribution to chemical bond and $\psi_{\text{I}} = \psi_a(1)\psi_b(2)$ and $\psi_{\text{II}} = \psi_a(2)\psi_b(1)$. The energies calculated using ψ_{cov} plotted against R_{ab} give curve (*B*), which is a distinct improvement over curve (*A*). Curve (*C*) results from the same wave function, but the effective nuclear charge, Z^*, is used instead of the atomic number for hydrogen in the calculations. In this situation, one would be taking into account repulsive effects in the system, which improve agreement between curves (*C*) and (*E*). It is also quite possible that the two electrons of this system have a finite moment on each atom, resulting in ionic forms of bonding. These species would be $[H_a^- {:}H_b^+]$ and $[H_a^+ {:}H_b^-]$. The wave function for the ionic system would have the following form:

$$\psi_{\text{ionic}} = \psi_a(1)\psi_a(2) + \psi_b(1)\psi_b(2) \quad [4.6]$$

The chemistry of the hydrogen molecule announces the fact that the bond in the molecule is more covalent than it is ionic. Therefore, moderation of the amount of ionic character that is put into the total wave function is necessary. This moderation is accomplished by placing the correct numerical coefficients into the wave function. We now have a wave equation representing the hydrogen molecule in the form

$$\psi = a\psi_{\text{cov}} + b\psi_{\text{ionic}} \qquad [4.7]$$

The best value a/b is $\frac{6}{1}$. The plot of the resultant energy values against R_{ab} gives curve (D).

In all the above considerations, we assumed that the electron spins were paired. However, if the equation [4.7] is considered with parallel electron spins, energy vs. R_{ab} plots result in curve (F). This is a repulsive state, since there is no minimum energy and consequently no bonding.

From the foregoing one may draw the resonance forms for the hydrogen molecule in the following manner:

$$\text{H:H} \quad \leftrightarrow \quad \text{H}^+\text{:H}^- \quad \leftrightarrow \quad \text{H}^-\text{:H}^+ \qquad [4.8]$$

The double headed arrow (↔) indicates species that are *resonance forms* or *canonical forms*. Remember that within the confines of the resonance concept, each of these forms does not have independent existence or equivalent influence; they all contribute to the stability of the hydrogen molecule. The resonance forms actually originate in the mathematical approach to the lowest energy of the system.

This penetration also highlights the fact that the pure covalent and pure ionic forms of the hydrogen molecule are of a higher energy than a linear combination of the two forms, i.e., the *mixed* state is more stable than either *pure* state. The amount of energy or stability that a system has gained over the more stable of the two individual states (ionic or covalent) is called the resonance energy. This concept is illustrated diagrammatically as follows:

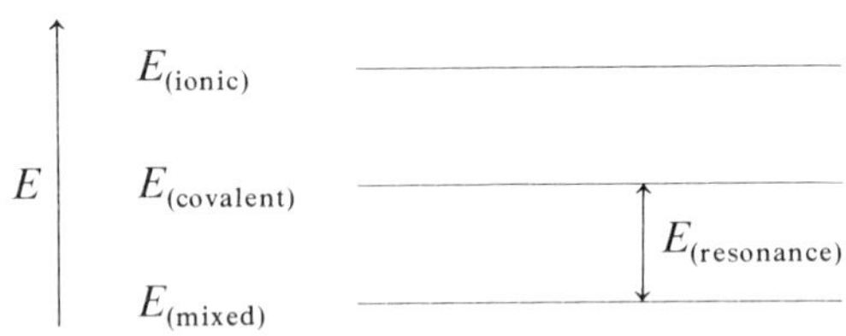

More will be said about arriving at resonance forms of various molecules, including a formulation of rules for doing so.

HYBRIDIZATION

The ground state electronic configuration for the carbon atom is $1s^22s^22p^2$. Although it is theoretically possible for carbon to form CX_2 type compounds through the loss of the $2p^2$ electrons, the most stable and unreactive compounds formed have the empirical formula CX_4. Therefore, it would seem that at the time of bond formation, carbon must have an electronic configuration with four unpaired electrons to bond four halogen atoms or four hydrogen atoms. The lowest energy configuration would be $1s^22s^12p_x^12p_y^12p_z^1$.

Figure 4.2 represents the energetics of methane formation. The carbon atom is excited from the solid state to the vapor state and on to its proper valence state. The gaseous hydrogen molecules are dissociated into gaseous hydrogen atoms. Then each gaseous carbon atom unites with four gaseous

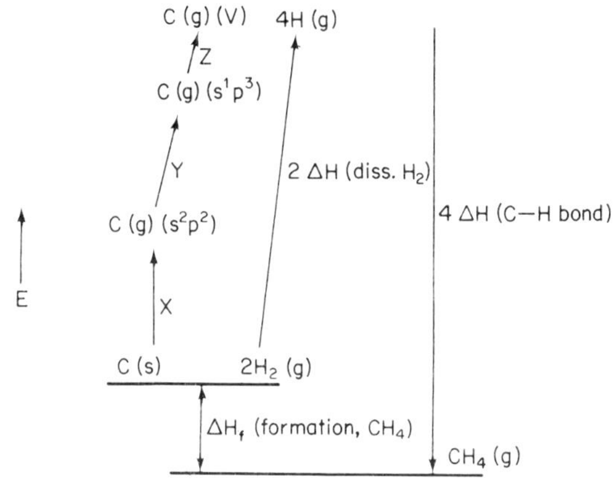

Figure 4.2. Various promotional energies in the formation of CH_4 (g).

hydrogen atoms and methane is formed. From Fig. 4.2, one can determine that X represents the energy to vaporize graphite, Y represents the energy needed to convert gaseous carbon (s^2p^2) to carbon (s^1p^3), and Z represents the energy necessary to promote carbon to its valence state, V.

It now remains to justify the process of absorbing energy Z to promote carbon to valence state V. The basis for the justification lies in the geometry of the orbitals and experimental data concerning the structure and reactivity of the C—H bond. The three p orbitals are directed in space along the coordinate axes, subtending angles of 90°. Each of these orbitals has a maximum capacity of two electrons. The p orbitals would then account for three C—H bonds subtending angles of 90°. The fourth C—H bond formed by hydrogen with the $2s$ carbon orbital would arrange itself in space in the lowest energy configuration. This bond would not form an angle of 90° with the bonds constructed from the p orbitals, the implication being that one C—H bond was different from the other three. It has been shown experimentally that all C—H bonds in methane are equivalent and all H—C—H bond angles are 109°28′. Therefore, the orbitals of carbon in methane are not pure p or pure s orbitals; they are equivalent. Therefore, they must be orbitals composed of s and p atomic orbitals. This type of orbital is called a *hybrid orbital*. This is the basis for distinguishing between the sp^3 state and the valence state V. The energy needed to pass from the sp^3 state to the valence state V is used in hybridizing the carbon orbitals. In methane, the carbon orbitals are called sp^3 hybrid orbitals, or, simply, sp^3 hybrids.

A finite number of valence electrons exist in the atomic orbitals of different atoms, and since we are considering only the covalent bond, it is reasonable, then, to assume that there are a limited number of hybridized orbitals that can be formed. With regard to the s and p orbitals, the possibilities are sp, sp^2, and sp^3. The d orbitals may also be included in hybridization schemes; they are dsp^2, dsp^3, and d^2sp^3.

The bright aspect of the concept of hybridized orbitals is that due to their directional character an a priori judgment on the structure of the molecule may be made. Of course, structural considerations based on experiment are difficult, and structural judgments without experimental support are valueless. However, it is an academic consolation that a theory has some predictive power.

The systematic presentation of the geometries of directed valence orbitals (hybridized orbitals) now remains. The most simple case is the formation of sp hybrid orbitals. An atom forming sp hybrid orbitals has the potentiality of forming two equivalent bonds. In Fig. 4.3, we see that

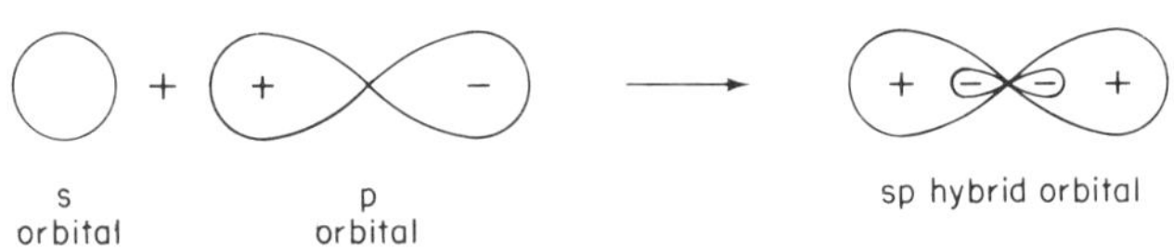

Figure 4.3. The formation of an sp hybrid orbital.

when an s and a p orbital combine, they form a set of linear orbitals, each of which has a large positive (symmetry) lobe and a smaller negative (symmetry) lobe. Therefore, any atom that hybridizes one of its s and one of its p orbitals forms two sp hybrid orbitals that subtend an angle of 180°.

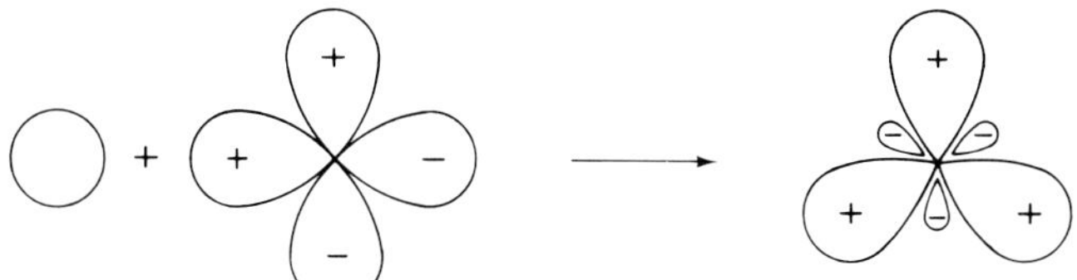

Figure 4.4. The formation of sp^2 hybrid orbitals.

The formation of sp^2 hybrid orbitals is illustrated in Fig. 4.4. The sp^2 hybrid orbitals result from the mixing of one s and two p orbitals, which form a trigonal planar structure, each lobe being directed to the vertex of an equilateral triangle. The angle subtended by the directed valence orbitals is 120°.

Hybridization of the sp^3 type, as already discussed, results in a tetrahedral structure. Here, the bonding lobes are directed to the vertices of a regular tetrahedron subtending angles of $109^\circ 28'$. See Fig. 4.5.

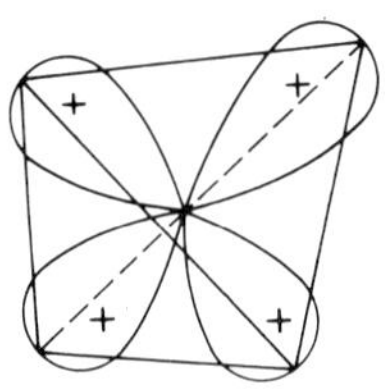

Figure 4.5. The sp^3 hybrid orbitals in a regular tetrahedron. The negative lobes of the hybrid orbitals are not shown.

These simple structures should give the reader an insight into the hybridization process. The previous section on resonance should be correlated with this section to conceptualize the energetics associated with the hybridization process. Finally, remember that in the hybridization process, the number of orbitals is conserved. If three orbitals are mixed to form hybrid orbitals, three hybrid orbitals are formed. Table 4.1 contains the more common types of hybrids and their geometries.

TABLE 4.1 ***Valence Bond Notations and Hybrid Structures***

Hybridization (Structure)	Examples	Electron Configurations (Valence Electrons Only)*
		$2s$ $2p$
sp	BeH_2	Be .. _ _ _ g.s.
Linear	All gaseous halides of Group II-A, mercuric halides, and	·× ·× _ _ h.s. sp
$X—M—X$	pseudohalides ($6s$, $6p$)	× = electron from the hydrogen atom

TABLE 4.1 *(Continued)*

Hybridization (Structure)	Examples	Electron Configurations (Valence Electrons Only)*
sp^2 Trigonal planar 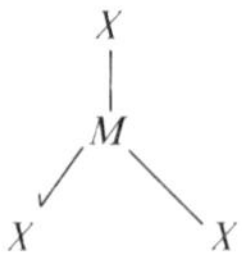	BF_3 All elements in Group III-A having a g.s. configuration ns^2np. X = halogen or organic radical. Some of the molecules dimerize [B_2H_6, Al_2Cl_6] with change in hybridization.	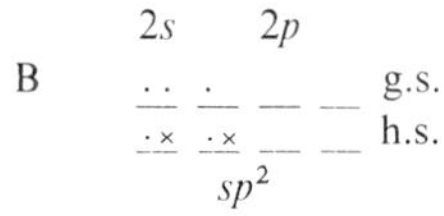 × = electrons from F
sp^3 Tetrahedral 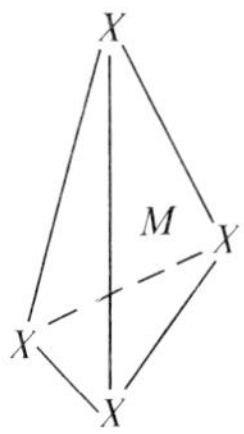	CF_4 All elements of Group IV-A, where individual molecules of the type MX_4 exist. X = hydrogen, halogen, organic radical.	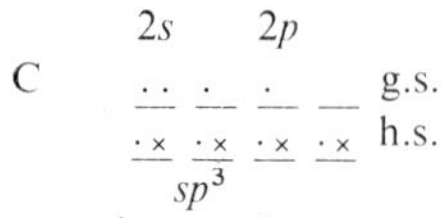 × = electron from F
dsp^2 Square planar $d_{x^2-y^2}, s, p_x, p_y$ 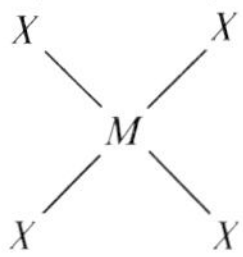	$Ni(CN)_4^{2-}$ Ni^{2+}, Pd^{2+}, Pt^{2+}, Cu^{2+}, Au^{3+} are known to form square planar complexes when bonded with four ligands.	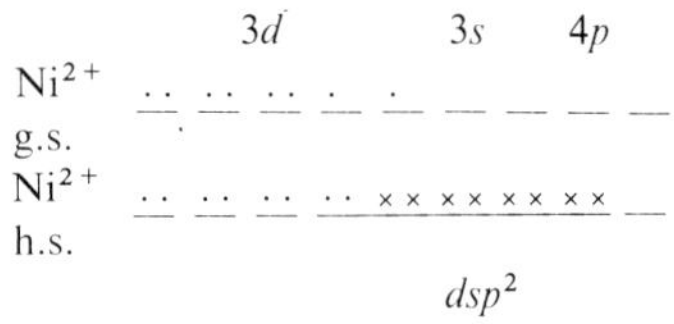 ×× = electron pairs from CN^-
dsp^3 Trigonal pyramid $d_{z^2}, s, p_x, p_y, p_z$ 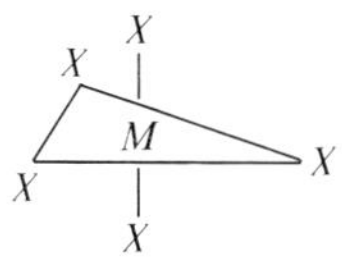	PF_5 SF_4, $Fe(CO)_5$, $Fe(PF_3)_5$, $SbCl_5$	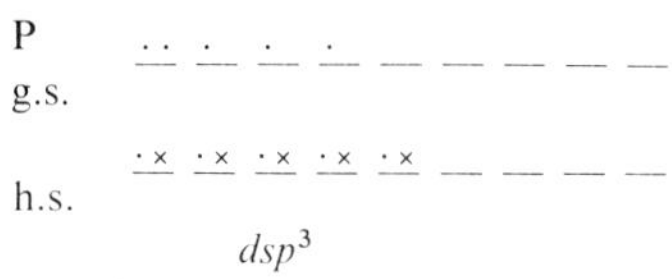 × = electrons from F
dsp^3 Square pyramid structure $d_{x^2-y^2}, s, p_x, p_y, p_z$ 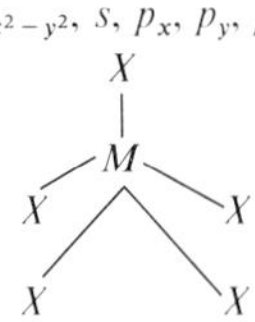	$Sb(C_6H_5)_5$ $Ni(CN)_5^{3-}$ (This structure is not as common as the trigonal bipyramid.)	Ni^{2+} (h.s.) 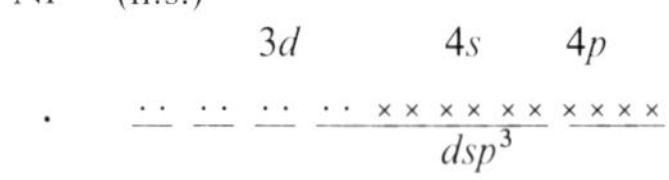 ×× = electron pairs from CN^-

TABLE 4.1 ***(Continued)***

Hybridization (Structure)	Examples	Electron Configurations (Valence Electrons Only)*
d^2sp^3	SF_6	S $3s$ $3p$ $3d$
Octahedral 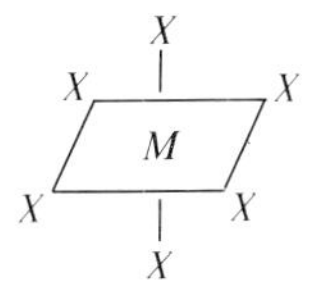	SeF_6, $Fe(CN)_6^{3-}$, $Fe(CN)_6^{4-}$; almost all hexacoordinated species	g.s. ·· ·· · · _ _ _ _ _ h.s. ·× ·× ·× ·× ·× ·× _ _ _ (d^2sp^3) ×× = electron from F

*Note that g.s. is ground state and h.s. represents hybridized state.

MOLECULES WITH LONE PAIRS OF ELECTRONS

The electron configuration of the ground states of nitrogen and oxygen are

	$1s$	$2s$	$2p$
N	··	··	· · ·
O	··	··	·· · ·

In view of these electronic configurations, let us now consider the structures of the hydrides of these elements, ammonia and water. Since the p orbitals subtend an angle of 90°, we would expect that the H—N—H and H—O—H bond angles would be near 90° if no hybridization occurs. If sp^3 hybridization is in effect, one would expect H—N—H and H—O—H bond angles near 109°28′. The bond angles in NH_3 and H_2O, respectively, are 106 and 105°. The first observation is that these angles are nearer to the tetrahedral angle than they are to 90°, the angle subtended by p orbitals. In accounting for this angle, one could argue that the ligands (hydrogen atoms), being positively charged, repel one another, and therefore open up the 90° angle of the p orbital. Calculations show that the angle may be opened up to 92° and possibly but unlikely to a maximum opening of 97°. This value is too far from the angle of unhybridized orbitals to offer a satisfactory explanation. These facts lead to the consideration of electron repulsive forces and hybridized orbitals. Ammonia has one lone pair of electrons on the nitrogen, whereas oxygen in water has two lone pairs of electrons. Lone pairs of electrons as well as bonding pairs of electrons may occupy hybrid orbitals. Now the lone pair of electrons is a very large entity because it is a pair of electrons associated with only one nucleus and has a relative freedom not enjoyed by a bonding pair of electrons, which is under the influence of two nuclei and consequently is more restricted. Therefore, it would seem plausible that the large lone pairs of electrons would tend to repel the bonding pairs of electrons. This electronic repulsion would naturally make the tetrahedral angle smaller. Consideration of the H—E—H bond angle in methane, ammonia, and water (where E would be C, N, and O) leads to the conclusion that as the number of lone pairs increases, the H—E—H bond angle decreases or the greater is the deviation from 109° in this series of compounds.

Extension of this argument one step leads to the establishment of a trend: repulsion between two lone pairs of electrons is greatest; repulsion between a lone pair and a bonding pair is next in intensity; and the repulsion between bonding pairs is relatively the weakest repulsion in this group.

Another extension of the correlation between bond angle and hybridization may be made. If we compare a series of compounds in a given series of elements, say, the elements of the first row of the periodic table (BeF_2, BF_3, CF_4, . . .), we find that the hybridization of the central element changes from *sp* to sp^2 to sp^3. The X—E—X bond angle decreases in size from 180 to 120 to 109°. In each case, the percentage of *s* character decreases or the percentage of *p* character increases. Therefore, the trend is as follows: As the percentage *s* character in a hybrid orbital decreases, the X—E—X bond angle decreases.

The hydrides of Group VI-A of the periodic table also present a trend. As one descends the Group VI-A hydrides, the H—E—H bond decreases: water (105°), hydrogen sulfide (92°), hydrogen selenide (91°), hydrogen telluride (90°). The Group VI-A atoms increase as we descend the group. This fact is important. Four ligands around an oxygen atom would experience more mutual repulsion than four ligands around a selenium atom. The tetrahedral arrangement of four point charges in space is said to be the arrangement of lowest energy. Considering the lone pairs of electrons as ligands, then, hybridization is very important in water. As we pass down to sulfur and selenium, we find that their *p* orbitals extend farther into space; they are at higher energy levels, the repulsion between ligands is not as influential as perceived, and hybridization is not necessary for the central atom. Similarly, since the electron pairs are farther from one another in space, they repel each other less. This is one explanation for the smaller angles as we descend the group.

CRITERION OF BOND STRENGTH: ORBITAL OVERLAP

When two atoms form a covalent bond, their atomic orbitals overlap. The greater is the overlap region between the two atoms, the stronger is the bond. This is an alternate verbalization of the definition of a covalent bond. Since the orbitals may represent electron density, a great deal of overlap indicates a high electron density between the two atoms. The amount of overlap is given by the expression

$$S = \int \psi_a \psi_b \, d\tau \qquad [4.9]$$

where S is called the overlap integral (see the Chapter 4 section on molecular orbital theory), ψ_a and ψ_b are the wave functions for the overlapping orbitals, and $d\tau$ is an element of space. Quantum mechanics shows that this integral is included in the consideration of resonance, and it can also be shown that the resonance energy increases in magnitude with an increase in overlapping of the two atomic orbitals. Finally, the bond formed by a given orbital will tend to lie in the direction in which the orbital is concentrated. Consequently, directional properties are important in the consideration of overlap.

Through an application of the VB method, Pauling calculated bond-forming powers, i.e., a relative set of values that can be given to the *s*, *p*, and *d* orbitals. Pauling concluded that the relative bond strengths of bonds formed by different orbitals would simply be related to the produce of these bond-forming powers (see Table 3.1). Remember that Pauling's calculations are merely a qualitative judgment on the effectiveness of the overlap of a given bond.

Since VB structures are relatively rigid, the direction of overlap is very important. The comparison of Fig. 4.6(a) and 4.6(b) shows that there is more overlap (shaded areas) in a *s*–*p* bond than there is in an *s*–*s* bond. Comparison of Fig. 4.6(b) and 4.6(c) shows that the direction of approach of *s* and *p* orbitals is important in achieving effective overlap. If the *s* orbital meets the *p* orbital head-on, there is effective overlap; however, if the *s* orbital approaches the *p* orbital at its side, there is no effective overlap and consequently no bonding. Orbitals must be of the same symmetry to overlap effectively, and in the case just cited [Fig. 4.6(c)], there is as much antibonding force (+ and − symmetries) as there is bonding force (+ and +, or − and − symmetries). The bonding power in this case is essentially zero. The reader must be aware, therefore, that the structure, the direction of the orbitals, and the signs of the lobes of the orbitals must be known to judge overlap and bond formation.

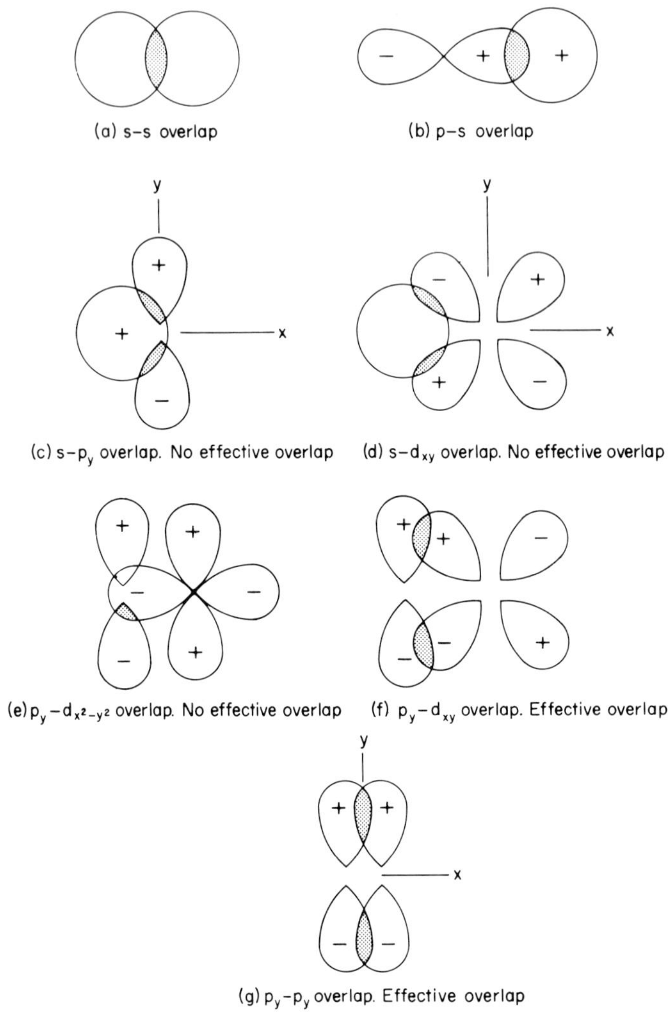

Figure 4.6. Diagrammatic representation of selected orbitals overlapping, and the net result.

Recall at this point our hybrid orbital structures—*sp*, sp^2, and sp^3 configurations. Since the *s* orbitals have a bonding power of 1 and the *p* orbitals a bonding power of 1.73, it would seem that the hybrid orbital with the greatest amount of *p* character, the sp^3 hybrid orbital with 75% *p* character, would have the greatest bonding power and form the strongest bonds. The diamond, the hardest substance known to exist, composed of carbon with sp^3 hybrid bonds, is supporting evidence.

BONDS FORMED BETWEEN HYBRID ORBITALS, SIGMA BONDS, PI BONDS

Some of the slightly more complex hydrides of carbon will be considered in terms of the VB theory. The most simple catenated hydride of carbon is ethane (C_2H_6 or $H_3C—CH_3$). Each carbon atom has its orbitals hybridized tetrahedrally, each orbital being an sp^3 hybrid orbital. The bond joining the two carbon atoms together is called an sp^3-sp^3 sigma bond. A sigma bond is defined as a bond having its electron density cylindrically symmetrical around the line or axis joining the two nuclei. The bond joining the carbon and hydrogen atoms is called an s-sp^3 sigma bond because the hydrogen 1s orbital overlaps with the sp^3 hybrid orbital of the carbon.

In ethylene, $H_2C{=}CH_2$, the experimentally determined H—C—H bond angle is 120°, which would lead us to believe that carbon possesses sp^2 hybrid orbitals. The carbon–hydrogen bond would then be described as an s–sp^2 sigma bond. Only two of the p orbitals in carbon participate in hybridization, so the description of the carbon–carbon multiple bond is slightly more complicated. First of all, the two carbon atoms are joined together by an sp^2–sp^2 bond. For purposes of discussion, we shall assume that the p_z orbital is the one that is not hybridized and that it is perpendicular to the plane of the molecule. Now the second bond formed in ethylene's carbon–carbon bond may be described as the two p_z orbitals approaching each other to give a cloud of electron density above and below an axis joining the two carbon atoms with zero electron density on this internuclear axis. This is a pi bond. Therefore a double bond consists of a sigma bond and a pi bond. This is illustrated in Fig. 4.7.

The linear acetylene molecule, H—C≡C—H, may be described in the same fashion. Here, the carbon–hydrogen bond is composed of an s–sp sigma bond. The carbon–carbon triple bond consists of a sigma bond. Two mutually perpendicular p orbitals on each carbon atom form two pi bonds. Therefore, a triple bond consists in one sigma bond and two pi bonds.

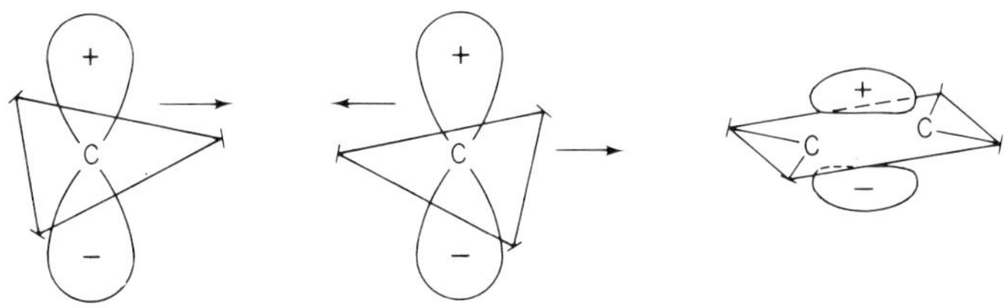

Figure 4.7. Formation of a double bond.

DETERMINATION OF RESONANCE STRUCTURES

Carbon monoxide is said to have four resonance structures:

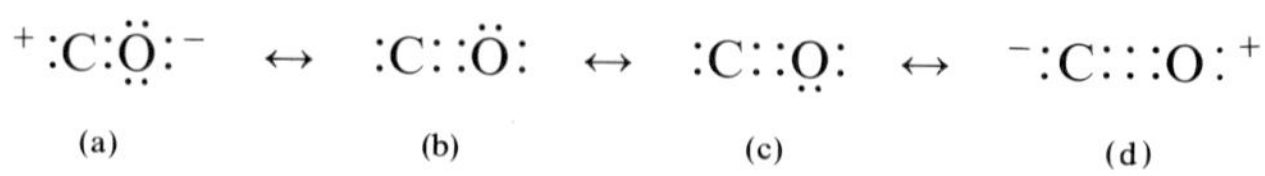

[4.10]

Each of these structures makes a contribution to the general stability of the carbon monoxide molecule. The determination of resonance structures of any molecule is based on one's intuition and Lewis structures in combination with the following basic rules:

(1) Similar atoms should have the same position in contributing resonance structures.
(2) All contributing resonance structures of a particular species must have the same number of unpaired electrons.
(3) Formal charges on neighboring atoms should not be like charges, and unlike charges should not be greatly separated.[1]

Implied in these rules is the fact that the structures that comply with them are the ones that are the most important contributors to the stability afforded by resonance energy. Any structure that violates these rules contributes very little to stabilize the system.

Let nitrous oxide serve as an example. The resonance structures with the formal charges of the respective atoms shown are

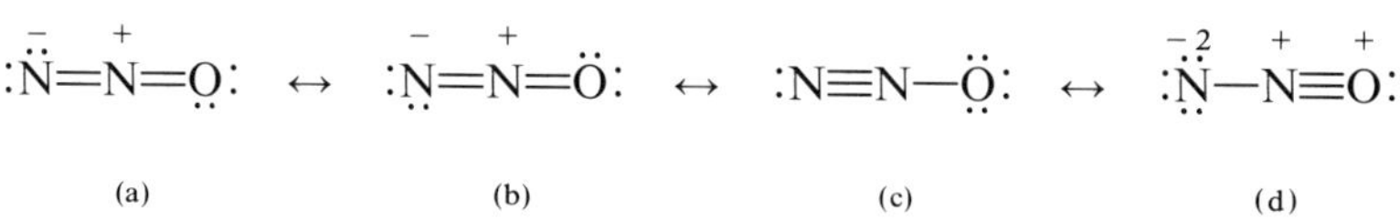

[4.11]

Example (d) is of little importance with regard to the total resonance energy because of its charge distribution. Pauling states that structures that place electric charges of the same sign on adjacent atoms make very little contribution to the ground state of the molecule. The dipole moment of nitrous oxide has been measured to be 0.116 D. If structure (d) made a significant contribution to the real structure of nitrous oxide, it would seem that the actual dipole moment should be appreciably larger. In general, the electronic structure of substances is such as to cause each

[1]Formal charge = Periodic group number of the atom − Number of unshared electrons around the atom − $\frac{1}{2}$ Number of shared electrons

atom to have a minimal electric charge. Pauling refers to this as the *electroneutrality rule*.

The heavier elements also exhibit multiple bond character. Consider sulfur dioxide,

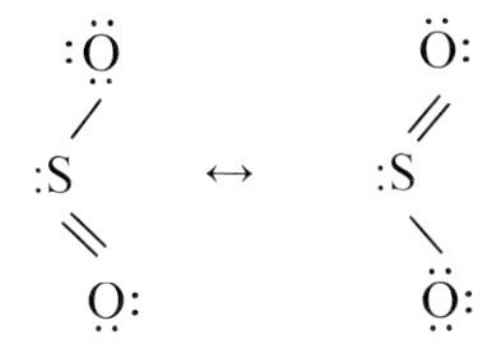

[4.12]

The sulfur-oxygen bond distance is 1.432 A and the O—S—O bond angle is 119.5°. The observed dipole moment is 1.59 D, whereas the expected dipole moment is 3.1 D, assuming 22% ionic character in the sulfur-oxygen bond. The acceptable sulfur–oxygen single covalent bond length is 1.7 A, and the acceptable double bond distance is 1.49 A. The shortened bond distances are indicative of a great deal of multiple bonding.

In general, note that the unexpectedly short bond length and decrease in dipole moment lead to the conclusion that unexpected bond multiplicity does exist. Trimethyl amine, $(CH_3)_3N$, has been found to have a pyramidal structure, whereas, trisilylamine, $(SiH_3)_3N$, has been found to be planar with shorter than expected silicon–nitrogen bonds. The mechanism for this bond shortening has been rationalized by saying that the overlap of the properly oriented empty silicon *d* orbitals with nitrogen *p* orbitals provides π bonding, specifically $(p \rightarrow d)\pi$ bonding. See Fig. 4.8. This multiple bond character shortens the bond. There is no direct experimental evidence for this type of bonding; thus, that concept of back bonding or $(p \rightarrow d)\pi$ bonding provides an area of contemporary research.

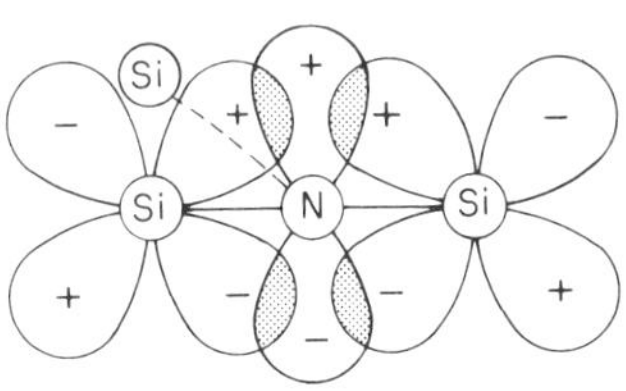

Figure 4.8. Representation of $(p \rightarrow d)\pi$ bonding between silicon and nitrogen in trisilyl amine. Bonding is shown between the nitrogen and only two silicon atoms for the sake of clarity.

THE MOLECULAR ORBITAL (MO) APPROACH TO THE COVALENT BOND

The qualitative approach of the MO theory contained herein will be better understood if the reader recalls the theoreticians' approach to the electronic configurations for atoms. The hydrogen atom was studied and all the energy levels and sublevels were determined, along with the geometry of the orbitals. The relative energies of these orbitals were also ascertained. These concepts were then extended to the multielectron atom, and the electronic configuration for the ground state was obtained by placing the electrons of a given atom in these energy levels, beginning with the lowest level. This is the application of the *Aufbau principle*.

The MO approach to bonding in molecules is analogous to the preceding concept. We place the nuclei of a given molecule at their equilibrium distances and calculate the orbitals that are spread over the entire molecule (molecular orbitals!); we then place our electrons in these orbitals, beginning with the MO of lowest energy. We can thus arrive at the electronic ground state configuration of a molecule.

The MO's are arrived at by the linear combination of atomic orbitals. This mathematical approach also aids us in determining the relative energies of the MO's. Use of these concepts will make the preceding analogy more meaningful and sensible.

The most simple species that we could choose to study is the H_2^+ ion. This species consists in two protons with *one electron bonding the protons together*. The two protons, positively charged, would ordinarily repel one another. With the bonding electron, the energy of the system reaches a minimum when the protons are about 1.06 A away from each other. If the protons are moved nearer or are removed from one another, the energy of the system increases. This is illustrated in Fig. 4.9, where the energy of the system is plotted against internuclear distance.

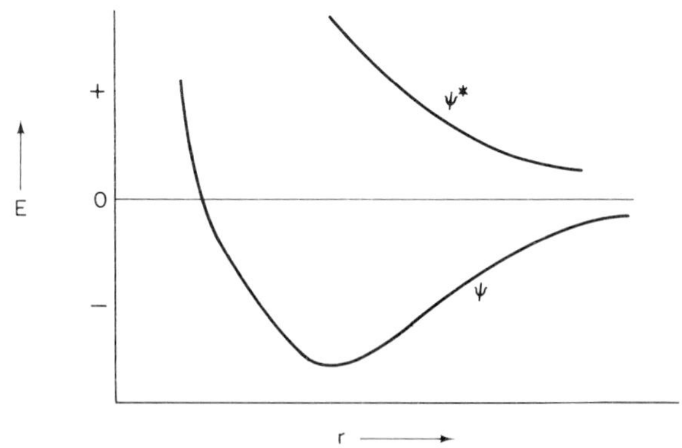

Figure 4.9. The plot of energy vs. internuclear distance for the H_2^+ species using the bonding and antibonding wave functions.

Let us assume that we can distinguish between the two hydrogen nuclei and label them *A* and *B*. The orbital of lowest energy associated with each of these nuclei is the 1*s* orbital. Each of these atomic orbitals may be represented by a wave function, ψ_A and ψ_B. Now each molecular orbital may also be represented by a wave function, ψ. This wave function is determined by a linear combination of the atomic orbitals. Hence, ψ may be expressed in two ways:

$$\psi = \phi_A + \phi_B \quad [4.13]$$

$$\psi^* = \phi_A - \phi_B \quad [4.14]$$

The wave function for the MO's must have all the properties of an atomic wave function, especially the property of being normalized and orthogonal. The expressions for the normalized wave functions are:

$$\psi = N_1(\phi_A + \phi_B) \quad [4.15]$$

$$\psi^* = N_2(\phi_A - \phi_B) \quad [4.16]$$

Normalization simply means

$$\int \psi_{\text{MO}}^2 \, d\tau = 1 \quad \text{or} \quad \langle \text{MO}|\text{MO} \rangle = 1 \quad [4.17]$$

Normalization informs us that the probability of finding an electron in a given molecular orbital is unity, or the electron is there somewhere if we want to look for it.

The normalization constant may be evaluated in the following fashion:

$$\langle \text{MO}|\text{MO} \rangle = N_1^2 \langle A + B|A + B \rangle \quad [4.18]$$

$$= N_1^2(\langle \text{A}|\text{A} \rangle + \langle \text{B}|\text{B} \rangle + 2\langle \text{A}|\text{B} \rangle) \quad [4.19]$$

Since the atomic wave functions are already normalized,

$$\langle A|A \rangle = \langle B|B \rangle = 1 \quad [4.20]$$

There remains $\langle A|B \rangle$, which is called the overlap integral and is represented by *S*. Equation [4.19] now becomes

$$\langle \text{MO}|\text{MO} \rangle = N_1^2(2 + 2S) \quad [4.21]$$

and

$$N_1 = \pm \frac{1}{\sqrt{2 + 2S}} \quad [4.22]$$

We shall assume that $S \ll 1$ and that the wave function demands only the positive value of N. The expression for the normalization constant is

$$N_1 = \frac{1}{\sqrt{2}} \qquad [4.23]$$

and the final expressions for the MO's are

$$\psi = \frac{1}{\sqrt{2}}(\phi_A + \phi_B) \qquad [4.24]$$

$$\psi^- = \frac{1}{\sqrt{2}}(\phi_A - \phi_B) \qquad [4.25]$$

The confirmation of the value of N_2 is left to the reader.

It has already been stated that

$$\mathscr{H}\Psi = E\Psi \qquad [4.26]$$

Now if each side of equation [4.26] is multiplied by ψ and then integrated, the result is

$$\langle\Psi|\mathscr{H}|\Psi\rangle = E\langle\Psi|\Psi\rangle \qquad [4.27]$$

but

$$\langle\Psi|\Psi\rangle = 1 \qquad [4.28]$$

Then

$$E = \langle\Psi|\mathscr{H}|\Psi\rangle \qquad [4.29]$$

Substituting [4.24] in this expression, we obtain

$$E_1 = \tfrac{1}{2}(\langle A + B|\mathscr{H}|A + B\rangle) \qquad [4.30]$$

or

$$E_1 = \tfrac{1}{2}\langle A|\mathscr{H}|A\rangle + \tfrac{1}{2}\langle B|\mathscr{H}|B\rangle + \tfrac{1}{2}\langle A|\mathscr{H}|B\rangle + \tfrac{1}{2}\langle B|\mathscr{H}|A\rangle \qquad [4.31]$$

No evaluation of these integrals will be attempted at this point, but the following representations will be made:

$$q_A = \langle A|\mathscr{H}|A\rangle = q_B = \langle B|\mathscr{H}|B\rangle \qquad [4.32]$$

and

$$\beta = \langle A|\mathscr{H}|B\rangle = \langle B|\mathscr{H}|A\rangle \qquad [4.33]$$

The final expression that results is

$$E_1 = q + \beta \tag{4.34}$$

In a similar manner, we can show for ψ^* that $E = q - \beta$.

The *Coulomb integral*, q, may be interpreted in several ways. (1) It may represent the energy necessary to remove an electron from the orbital. (2) It may represent the energy of a hydrogen atom disturbed by a proton at a distance, r. The energy of the system should be very near that of the hydrogen atom. The *exchange integral*, β, is sometimes referred to as the *resonance integral*, because it refers to the amount of extra stability added to a bond. In Fig. 4.9, the plot of these two wave functions is shown; energy is plotted against internuclear distance. It should be apparent to the reader from this plot that β must be a negative quantity, leading to the obvious conclusion than that ψ is of lower energy than ψ^*. The orbital associated with ψ is referred to as the *bonding orbital*, whereas the orbital associated with ψ^* is the *antibonding orbital*. The energy difference between these two orbitals is 2β. Figure 4.10 illustrates the preceding concepts.

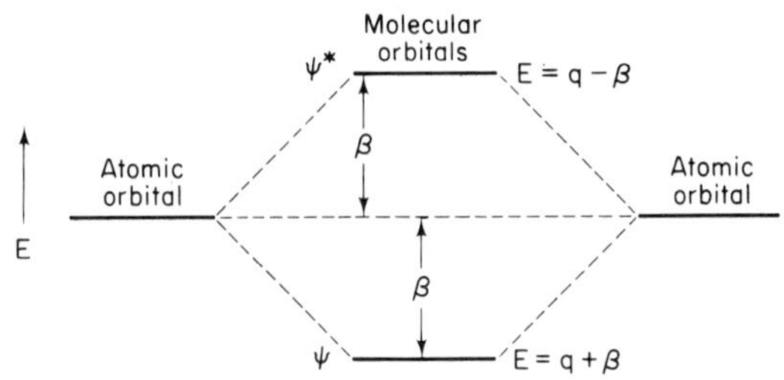

Figure 4.10. Energy level diagram relating the bonding and antibonding orbitals in the H_2^+ species.

Remember that for every bonding orbital that is found, a corresponding antibonding orbital is also formed.

At this point it is imperative to assume a new tack and look at the configurations of the molecular orbitals formed by the linear combination of the atomic orbitals. First, the combination of two *s* orbitals will be considered. These two orbitals combine to form a bond that is cylindrically symmetrical around the internuclear axis. This is the normal sigma bond. The electron density is concentrated between the two nuclei and it contributes to their bonding. When the wave functions for the two orbitals

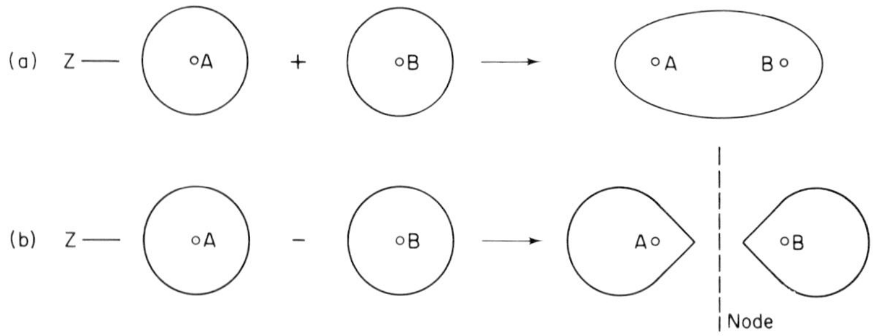

Figure 4.11. (a) Formation of a σ_s bonding orbital; (b) formation of σ_s^* antibonding orbital.

are subtracted (or if the wave functions of two different orbitals of different symmetries are added), an antibonding orbital results [Fig. 4.11(b)]. There is a node, a plane of zero electron density, between the two nuclei. The node contributes nothing to bonding but actually hinders it, hence the term antibonding orbital.

Now consider the combination of the p orbitals. Keep in mind that the p orbitals are directed in space, along the x, y, and z coordinate axes. The direction of their approach to form a molecular orbital is important. The z axis will be our reference axis, i.e., the p orbitals will be brought together along this line. The two p_z orbitals will then meet head-on [Fig. 4.12(a)]. Orbitals may overlap only if they have the proper symmetry

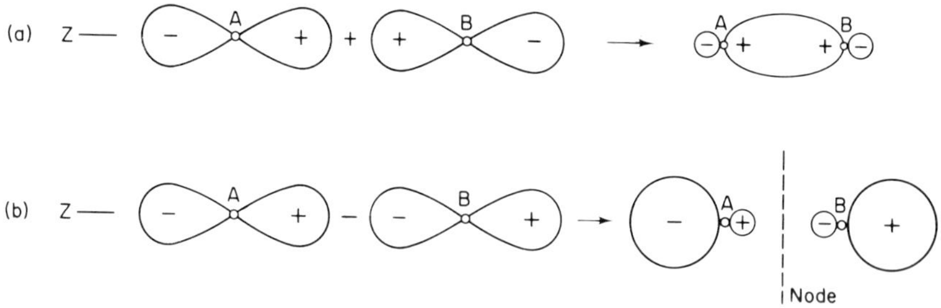

Figure 4.12. (a) Formation of σ_p bonding orbital; (b) formation of σ_p^* antibonding orbital.

(+ lobes overlap). In this case, it is apparent that a σ bond will form. If the two orbitals do not have the proper symmetry (+ and − lobes overlap) or the two functions are subtracted, an antibonding orbital will be formed [Fig. 4.12(b)]. The $2p_z$ orbitals therefore form a bonding sigma orbital and an antibonding sigma orbital. The equations representing these two linear combinations would be

$$\psi = N_3 (\phi_{PZA} + \phi_{PZB}) \quad [4.35]$$

$$\psi^* = N_4 (\phi_{PZA} + \phi_{PZB}) \quad [4.36]$$

There are two p orbitals remaining. As the atoms are brought together along the z axis, the orbitals will meet side-to-side (Fig. 4.13). Once again we point out that the orbitals must have the proper symmetry to have

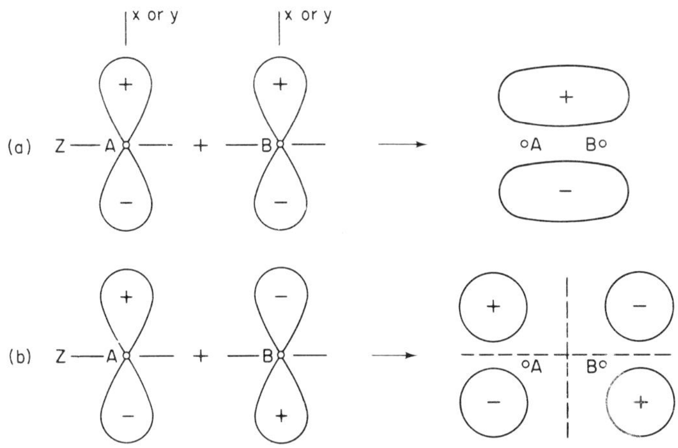

Figure 4.13. (a) Formation of π_p bonding orbital; (b) formation of π_p^* antibonding orbital.

effective overlap or bond formation. The orbitals pictured may represent either the p_x or p_y orbital if the z axis is taken to be the sigma bond axis. In Fig. 4.13(a), we see that the bond which is formed has its electron density above and below the internuclear axis, yet the electron density is in a volume concentrated between the two nuclei. This type of bond is called a π bond and is a bonding orbital. Note that in a π bond, there is a node containing the internuclear axis. If the two functions are subtracted or if the orbitals do not have the proper symmetry, a π antibonding orbital is formed. In this case, there are two nodal planes of zero electron density perpendicular to each other; thus, there is no contribution to bonding. The equations representing these combinations are

$$\psi(\pi) = N_5(\phi_{PXA} + \phi_{PXB}) \quad [4.37]$$

$$\psi^*(\pi) = N_6(\phi_{PXA} - \phi_{PXB}) \quad [4.38]$$

Similar bonds—pi bonding and pi antibonding orbitals—are formed by the p_y orbitals, and equations similar to [4.37] and [4.38] may be written to express their linear combination.

The final point that should be made in this aside concerns the energy relationships among these MO's. There are two facts that must be considered when determining the relative energy of the MO: (1) the energy of atomic orbitals combining to form the MO; and (2) the overlap between these atomic orbitals. The σ_{1s} and σ^*_{1s} are of much lower energy than the σ_{2s} and σ^*_{2s} molecular orbitals because of the relative energies of the 1*s* and the 2*s* atomic orbitals, which combine to form these MO's. Likewise, the σ_{2s} would be of lower energy than the σ_{2p}. One could arrive at the following order of MO's based on the relative energies of the atomic orbitals:

$$(\sigma_{1s})(\sigma^*_{1s})(\sigma_{2s})(\sigma^*_{2s})(\sigma_{2p})(\pi_{2px})(\pi_{2py})(\pi_{2px})(\pi_{2py})(\sigma^*_{2p}) \qquad [4.39]$$

$$\xrightarrow{\text{Energy increases}}$$

The molecular orbitals have the same properties as atomic orbitals relative to electrons. (1) They may contain a maximum of two electrons (when an MO contains two electrons, they must be paired). (2) Hund's rules must be obeyed when filling degenerate MO's (π_x) and (π_y) or (π^*_x) and (π^*_y).

Emission and absorption spectroscopy experiments lead to the electronic configuration of molecules. It has been found that in some cases of homonuclear diatomic molecules, the relative order of MO's in increasing energy is:

$$(\sigma_{1s})(\sigma^*_{1s})(\sigma_{2s})(\sigma^*_{2s})(\pi_{px})(\pi_{py})(\sigma_{2p})(\pi^*_{px})(\pi^*_{py})(\sigma^*_{2p}) \qquad [4.40]$$

This order may be rationalized by considering the σ_{2s} and σ_{2p} MO's. These orbitals occupy the same relative volumes in space, and if they were filled with electrons they would contain a sphere of great interelectronic repulsion. When the σ_{2s} MO is filled, the next electron prefers to occupy the π_x or π_y rather than the σ_{2p}. This interaction between the σ_{2s} and σ_{2p} is actually hybridization of the 2*s* and 2*p* atomic orbitals, whereby the atoms attain the largest charge separation. Hence, interelectronic repulsion is minimized.

These relative orders of energy are represented diagrammatically in Fig. 4.14. The interaction described previously increases in homonuclear diatomic molecules as the periodic table is traversed from left to right in the second row of the periodic table. In other words, this interaction decreases from Li_2 to F_2.

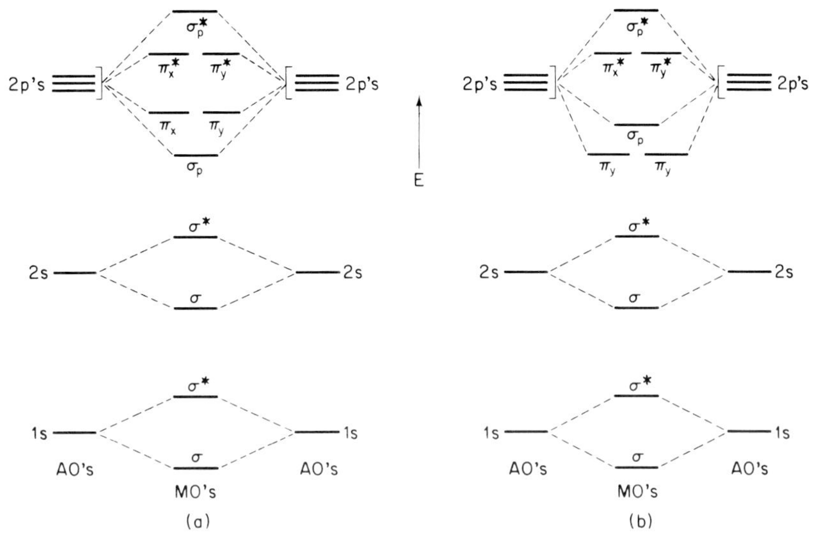

Figure 4.14. Energy level diagrams for homonuclear diatomic species. (a) Energy level diagram where there is no 2*s*—2*p* interaction and (b) where there is 2*s*—2*p* interaction.

At this point, we shall now return to our original discussion and determine the electronic configuration of homonuclear diatomic molecules. Use of Fig. 4.14 will help a great deal in this discussion. We shall assume that the energy relationship among MO's in Fig. 4.14(b) is valid up to and including N_2. In the species H_2^+, there are two orbitals available, σ_{1s} and σ_{1s}^*. The electron would be put into the σ_{1s} orbital to obtain the ground state configuration. The MO electronic configuration is $(\sigma_{1s})^1$—i.e., one electron is in the σ_{1s} bonding orbital. The next species would be H_2. Here we would place two electrons into the σ_{1s} MO and the electronic configuration would be $(\sigma_{1s})^2$. For He_2^+, the MO description of the electronic configuration is $(\sigma_{1s})^2$ $(\sigma_{1s}^*)^1$, whereas for He_2 it would be $(\sigma_{1s})^2$ $(\sigma_{1s}^*)^2$.

At this point, it would be worthwhile to introduce the concept of *bond order*, which simply means the number of bonds holding the two atoms together. A single bond is indicated by the bond order of 1. A bond order of 2 represents a double bond, and a bond order of 3 represents a triple bond. The bond order is found by subtracting the number of electrons in the antibonding orbitals from the number of electrons in the bonding orbitals and dividing the difference by two. The bond order of H_2 is 1 $[(2-0)/2=1]$. The bond order of He_2 is zero $[(2-2)/2=0]$. The bond order for H_2^+ and He_2^+ is $\frac{1}{2}$. He_2^+ has been found to exist, but He_2 has not been reported!

Table 4.2 summarizes the data on several homonuclear diatomic species derived from the second row of the periodic table. Several points should

TABLE 4.2 ***Ground State Configurations of Homonuclear Diatomic Species of the First and Second Rows of the Periodic Table***

Number of Electrons	Species	Electronic Configuration	Bond Order	Bond Dissociation Energy (kcal/mole)	Ground State Term	Bond Length (A)
1	H_2^+	$(\sigma_{1s})^1$	$\frac{1}{2}$	61	$^2\Sigma$	1.06
2	H_2	$(\sigma_{1s})^2$	1	103	$^1\Sigma$	0.74
3	He_2^+	$(\sigma_{1s})^2(\sigma^*_{1s})^1$	$\frac{1}{2}$	71	$^2\Sigma$	1.08
4	He_2	$(\sigma_{1s})^2(\sigma^*_{1s})^2$	0			
6	Li_2	$KK(\sigma_{2s})^2$	1	25	$^1\Sigma$	2.67
8	Be_2	$KK(\sigma_{2s})^2(\sigma^*_{2s})^2$	0			
10	B_2	$Be_2(\pi_p)^2$	1	69	$^3\Sigma$	1.58
12	C_2	$Be_2(\pi_p)^4$	2	150	$^1\Sigma$	1.31
13	N_2^+	$Be_2(\pi_p)^4(\sigma_p)^1$	$2\frac{1}{2}$	200	$^2\Sigma$	1.16
14	N_2	$Be_2(\pi_p)^4(\sigma_p)^2$	3	225	$^1\Sigma$	1.10
15	O_2^+	$Be_2(\pi_p)^4(\sigma_p)^2(\pi_p)^1$	$2\frac{1}{2}$	150	$^2\Pi$	1.12
16	O_2	$Be_2(\sigma_p)^2(\pi_p)^4(\pi_p)^2$	2	117	$^3\Sigma$	1.20
18	F_2	$Be_2(\sigma_p)^2(\pi_p)^4(\pi_p)^4$	1	36	$^1\Sigma$	1.41

become clear from a study of Table 4.2. First, when the bond order of a species is calculated, the electrons in the inner shells contribute essentially nothing to bonding. In the formulation of the electronic configuration of Li_2, the letters *KK* are sometimes used. This means that the σ_{1s} and σ^*_{1s} MO's are filled. The symbol Be_2 indicates that the inner core of electrons in the configuration has the arrangement of Be_2. It might also be pointed out that one of the major triumphs of MO theory is the production of the paramagnetism of O_2. The electronic configuration of oxygen shown in Table 4.2 indicates the two unpaired electrons.

It is worth mentioning at this point that beside the σ and π bonds, there is another type of bond that is rarely encountered—the delta bond, δ. The delta bond is formed by the combination of the four lobes of two *d* orbitals coming together as illustrated in Fig. 4.15. There would be four lobes of electron density resulting, defining two nodal planes mutually perpendicular to each other. The major use of this orbital is in the discussion of metal–metal bonds. If the metal–metal bond distance is less than twice the covalent radius of the metal, then bond shortening is attributed to the formation of δ bonds. In $(W_2Cl_9)^{3-}$, the metal–metal distance is

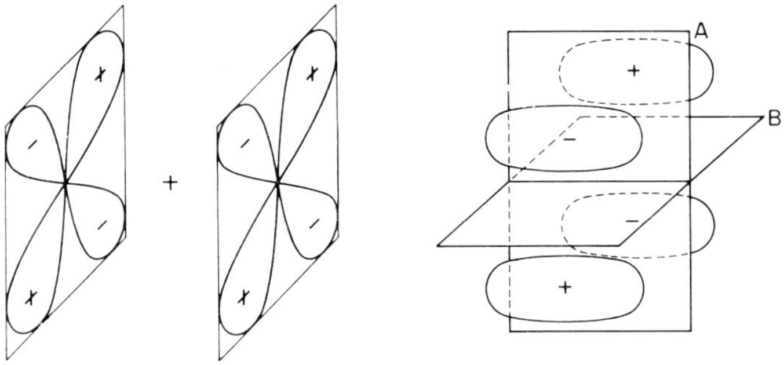

Figure 4.15. Formation of a delta bond by the proper orientation of two d orbitals, say the d_{xy} orbitals. The planes A and B are nodal planes.

2.46 A; however, twice the tungsten covalent radius is 2.74 A. The loss of expected paramagnetism in these species suggested electron pairing, and delta bonds were rationalized.

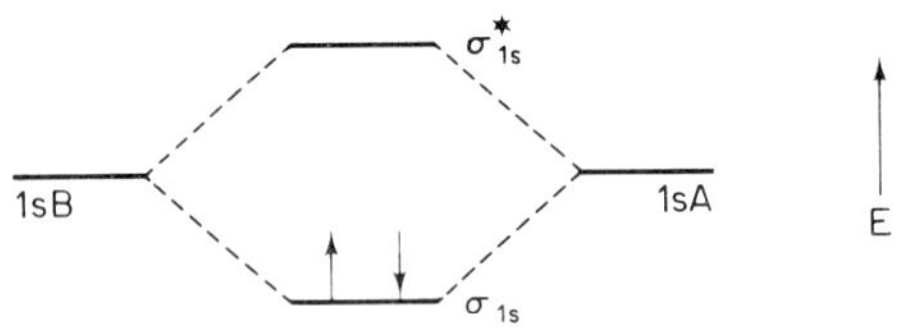

Figure 4.16. Molecular orbital energy level diagram for H_2.

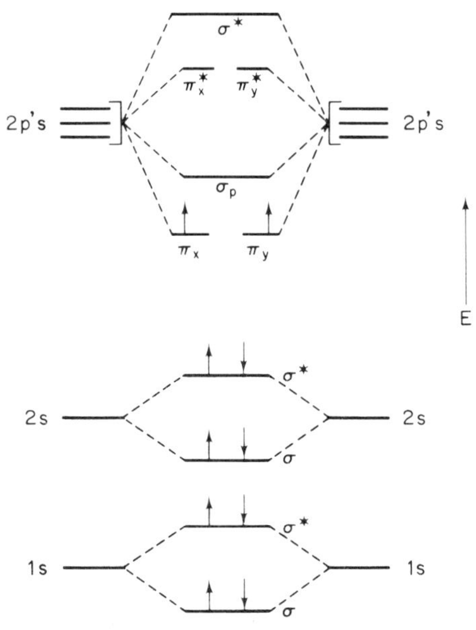

Figure 4.17. Molecular orbital energy level diagram for B_2.

It is possible to determine the ground state term symbols for homonuclear diatomic molecules by methods modelled after the Russell-Saunders method for atoms. Each molecular orbital type is given an angular momentum value, m_1. These are listed in Table 4.3.

TABLE 4.3 *Quantum Numbers for Molecular Orbitals in Some Linear Molecules*

Orbital	m_1
σ, sigma	0
π, pi	± 1
δ, delta	± 2

TABLE 4.4. *Symbols Representing Various Total Angular Momenta, M_L*

M_L	State Symbol
0	Σ
± 1	π
± 2	Δ
± 3	φ

The definition of a term is

$$^{2S+1}M_L$$

where S is the sum of the spin numbers and M_L is the sum of m_l's. The following examples will clarify the use of this system.

1. Calculate the ground state term for H_2^+.
   (a) Electronic configuration is $(\sigma_{1s})^1$.
   (b) $M_L = m_l = 0$.
   (c) $S = s = \frac{1}{2}$.
   (d) The term is $^2\Sigma$.

2. Calculate the ground state for H_2.
   (a) Electronic configuration is $(\sigma_{1s})^2$.
   (b) $M_L = m_l = 0 + 0 = 0$ (two electrons).
   (c) $S = s + s = (+\frac{1}{2}) + (-\frac{1}{2}) = 0$ (Pauli's principle).
   (d) The term is $^1\Sigma$.

3. Calculate the ground state term for C_2.
   (a) Electronic configuration is $(Be_2)(\pi_x)^2(\pi_y)^2$.
   (b) $M_L = 0$. The observant reader will see that a filled orbital will result in M_L being zero.

(c) $S = 0$. If the orbitals are filled, the sum of the electron spins is zero.

(d) The term is $^1\Sigma$.

4. Calculate the ground state term for O_2.

(a) The electronic configuration is $(Be_2)\,(\sigma_s)^2\,(\sigma_s^*)^2\,(\pi_x)^2\,(\pi_y)^2\,(\pi_x)^1\,(\pi_y)^1$.

(b) When calculating M_L and S, all you need to consider are the last two molecular orbitals, $(\pi_x)^1\,(\pi_y)^1$. All the other orbitals are filled. In the following representation, the dashes represent the two degenerate π antibonding orbitals and the numbers represent their respective m_l values. Four different possible arrangements of the electrons exist. They are, with their respective M_L and S values, the following:

(1) $\underset{+1}{\underline{\uparrow}}\ \underset{-1}{\underline{\uparrow}}$ $M_L = 0,\ S = 1$ (3) $\underset{+1}{\underline{\quad}}\ \underset{-1}{\underline{\uparrow\downarrow}}$ $M_L = 2,\ S = 0$

(2) $\underset{+1}{\underline{\uparrow\downarrow}}\ \underset{-1}{\underline{\quad}}$ $M_L = 2,\ S = 0$ (4) $\underset{+1}{\underline{\uparrow}}\ \underset{-1}{\underline{\downarrow}}$ $M_L = 0,\ S = 0$

Hund's rule of maximum multiplicity dictates that (1) is the correct ground state having a $^3\Sigma$ term, which agrees with experimental results.

HETERONUCLEAR DIATOMIC MOLECULES

An approach similar to that which was employed for determining the electronic configuration for homonuclear molecules may be used for heteronuclear diatomic molecules. The obvious difference lies in the fact that the energy level diagrams of the MO's are not symmetric. This is quite reasonable, because two different atoms, more than likely, will not have identical electronegativities. In comparing corresponding atomic orbitals, it will be found that the orbital in the more electronegative atom will be of lower energy because it will be nearer its nucleus. See the diagram in Fig. 4.18. The 3*s* orbital in *B* is of lower energy than that of *A*, because atom *B* is more electronegative than atom *A*. There will be a difference in the equations representing the linear combinations.

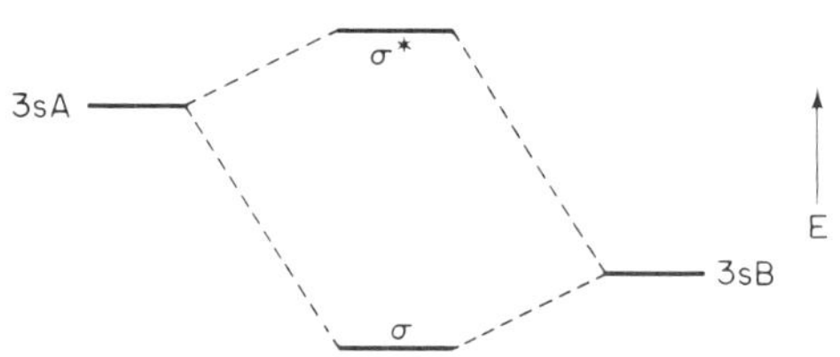

Figure 4.18. The molecular orbital energy level diagram of two corresponding atoms of different electronegativities. Atom *B* is more electronegative than atom *A*.

$$\psi = N_1(\phi_A + \lambda\phi_B) \qquad [4.41]$$

$$\psi^* = N_2(\phi_A - \lambda\phi_B) \qquad [4.42]$$

The λ appearing in equations [4.41] and [4.42] is called a *mixing coefficient*. It measures the relative contribution of a given atomic orbital to a given molecular orbital. The value of λ lies between zero and unity and will precede the atom of lesser electronegativity. An alternative method of writing equations [4.41] and [4.42] would be to use some constant, C, in the equation—namely,

$$\psi = C_1\phi_A + C_2\phi_B \qquad [4.43]$$

$$\psi^* = C_3\phi_A - C_4\phi_B \qquad [4.44]$$

The constants C contain the normalization factor and the mixing coefficient. If B is more electronegative than A, then $C_2 > C_1$ and $C_3 > C_4$. The rationalization of this fact is left to the reader.

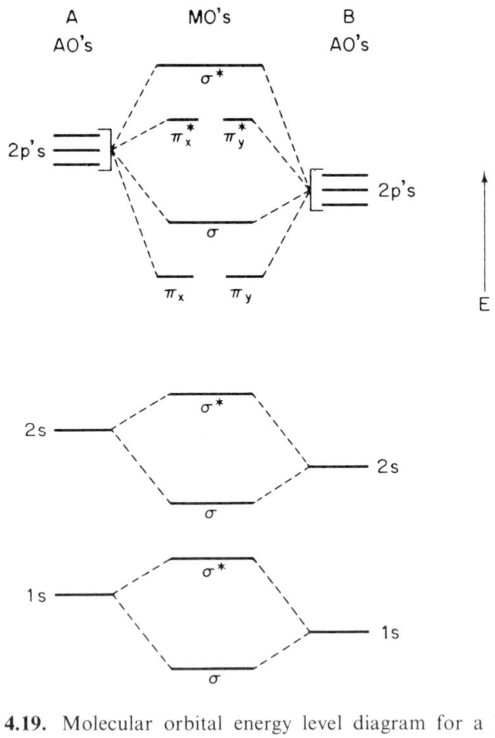

Figure 4.19. Molecular orbital energy level diagram for a heteronuclear diatomic species, atom B being more electronegative than atom A.

Figure 4.19 contains the energy diagram of the MO's for heteronuclear diatomic molecules. The electrons are placed in the MO's one-by-one, following Hund's rules and the Pauli principle. Table 4.5 contains the ground state electronic configuration of some heteronuclear diatomic species.

TABLE 4.5 ***The Ground State Electronic Configurations for Some Heteronuclear Diatomic Species***

Valence Electrons Only	Species	Electronic Configuration	Bond Order	Bond Dissociation Energy (kcal/mole)	Ground State Term	Bond Length (A)
8	BN	$KK(\sigma_{2s})^2(\sigma^*_{2s})^2(\pi_p)^4$*	2	92	$^1\Sigma$	1.28
9	BO	$KK(\sigma_{2s})^2(\sigma^*_{2s})^2(\pi_p)^4(\sigma_p)^1$	$2\frac{1}{2}$	185	$^2\Sigma$	1.20
	CN			188		1.17
	CO^+					1.11
10	CO	$KK(\sigma_{2s})^2(\sigma^*_{2s})^2(\pi_p)^4(\sigma_p)^2$	3	255	$^1\Sigma$	1.14
	NO^+					1.06
11	NO	$KK(\sigma_{2s})^2(\sigma^*_{2s})^2(\pi_p)^4(\sigma_p)^2(\pi^*_p)^1$	$2\frac{1}{2}$	162	$^2\Pi$	1.15

*Actually, $KK(\sigma_{2s})^2(\sigma^*_{2s})^2(\pi_p)^3(\sigma_p)^1$; term, $^3\Pi$

LINEAR TRIATOMIC MOLECULES

In determining MO's for linear triatomic molecules, it is necessary to position the molecules with respect to the coordinate axes and maintain that position throughout the argument. Two examples will be presented: beryllium hydride and carbon dioxide. Carbon dioxide offers itself as an example with π bonding, whereas there is no π bonding in beryllium hydride. In this discussion of beryllium hydride, the H—Be—H axis will be taken as the z axis.

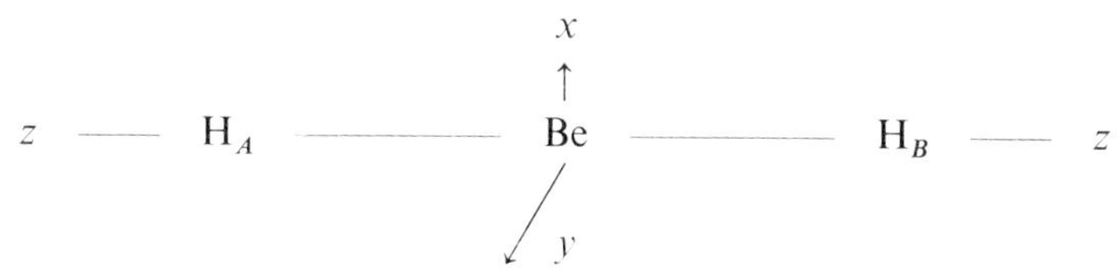

The beryllium atom will use its $2s$ and $2p_z$ orbitals to overlap with the $1s$ orbitals of the hydrogen atom (Fig. 4.20). First of all, when the $1s$ and $2s$

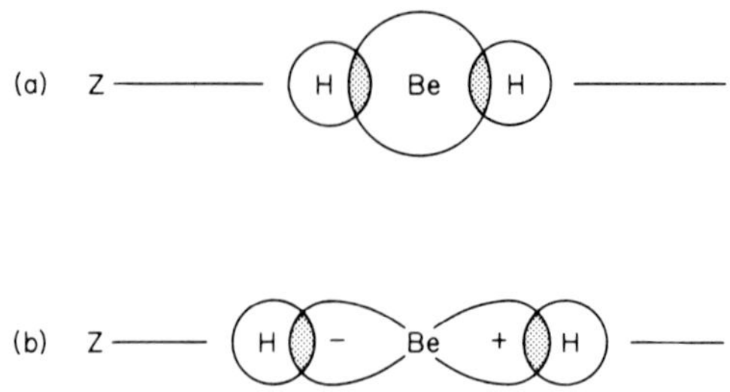

Figure 4.20. Diagrams of orbital overlap in BeH_2. (a) The $1s$ orbitals of the hydrogen atoms overlapping with the $2s$ orbital of the beryllium atom. (b) The $1s$ orbitals of the hydrogen atoms overlapping with the $2p_z$ orbital of the beryllium atom.

orbitals overlap, there is no problem because the s orbitals are totally symmetric, and the combination of atomic orbitals is

$$\psi_S = C_1 2s + C_2(1s_A + 1s_B) \qquad [4.45]$$

$$\psi_S^* = C_3 2s - C_4(1s_A + 1s_B) \qquad [4.46]$$

The subscripts A and B are used to distinguish between hydrogen atoms.

The overlapping of the $1s$ orbitals of the hydrogens with the $2p_z$ orbital is just a bit different because of the symmetry properties of the p orbital [Fig. 4.20(b)]. The equations representing the combination are

$$\psi_{PS} = C_5 2p_2 + C_6(1s_A - 1s_B) \qquad [4.47]$$

$$\psi_{PS}^* = C_7 2p_z - C_8(1s_A - 1s_B) \qquad [4.48]$$

The difference in symmetry of the lobes (+ and − signs on lobes) in the p orbital causes us to use the difference of $1s_a$ and $1s_b$ orbitals in the expression. Figure 4.21 contains the energy level diagram for the MO's of beryllium hydride.

Notice in Fig. 4.21 that the π orbitals contain no electrons. They are also at the same energy level in the MO diagram as they are in the isolated atom. They contribute essentially nothing to bonding. Orbitals of this type are called *nonbonding orbitals*.

The arrows represent the valence electrons of Be and the two H atoms. The reader should convince himself that the ground state term for this molecule is $^1\Sigma$.

The MO's for carbon dioxide are determined in the same manner. A frame of reference is chosen, the z axis being the principal axis. Several items must be carefully considered here. Carbon will use its s and p orbitals

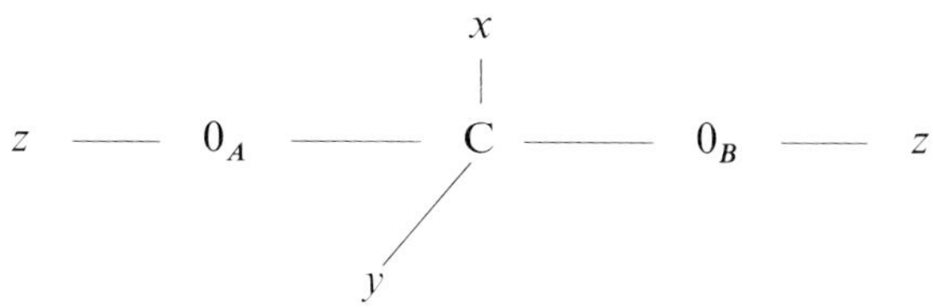

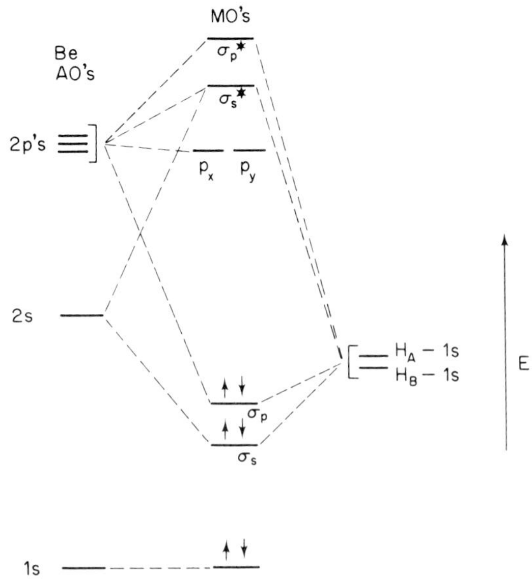

Figure 4.21. Molecular orbital energy level diagram for BeH_2.

for bonding but oxygen will use only its p orbitals for bonding. Once again linearity of the molecule is assumed, with the O—C—O axis as the z axis. Let us first look at the overlap of the $2p_z$ orbitals of the oxygen atoms with

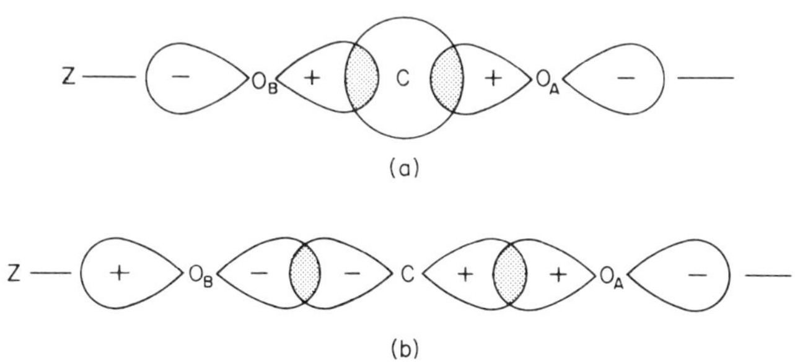

Figure 4.22. Diagram of the orbital overlap in CO_2. (a) The two $2p_z$ orbitals of oxygen overlapping with the $2s$ orbital of carbon. (b) The two $2p_z$ orbitals of oxygen overlapping with the $2p_z$ orbital of carbon.

the 2s orbital of carbon. The equations representing this combination are

$$\psi\sigma_S = C_1 2s + C_2(2p_{ZA} + 2p_{ZB}) \quad [4.49]$$

$$\psi\sigma^*_s = C_3 2s - C_4(2p_{ZA} + 2p_{ZB}) \quad [4.50]$$

Once again the subscripts A and B are arbitrarily used to distinguish between the two oxygen atoms. These bonds are sigma bonds because of the concentration of electron density along the internuclear axis. The same conditions will prevail when the $2p_z$ orbitals of the oxygen atoms will overlap with the $2p_z$ orbital of the carbon atom [Fig. 4.22(b)]. The equations representing this combination are

$$\psi\sigma p = C_5 2p_Z + C_6(2p_{zA} - 2p_{zB}) \quad [4.51]$$

$$\psi^*\sigma p = C_7 2p_z - C_8(2p_{zA} - 2p_{zB}) \quad [4.52]$$

However, the combination of the p_x and p_y orbitals presents a variation in the picture. These orbitals will meet side-to-side as the atoms are moved together. They must have the proper symmetry to overlap and have bond formation, i.e., π bond formation (Fig. 4.23). This proper symmetry is also

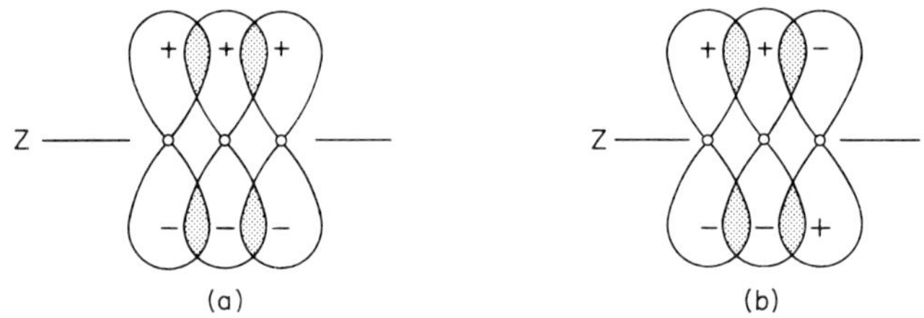

Figure 4.23. Diagram representing the possible overlap of the p_x and/or p_y orbitals as atoms are brought together along the z axis. (a) Representation of net overlap and bond formation, and (b) no effective bond formation due to no effective overlap.

necessary to write the correct equations expressing the combination of orbitals. The equations representing these combinations are

p_x orbitals: $\psi_\pi = C_9 2p_x + C_{10}(2p_{xA} + 2p_{xB})$ [4.53]

$\psi^*_\pi = C_{11} 2p_x - C_{12}(2p_{xA} + 2p_{xB})$ [4.54]

p_y orbitals $\psi_\pi = C_{13} 2p_y + C_{14}(2p_{yA} + 2p_{yB})$ [4.55]

$\psi^*_\pi = C_{15} 2p_y - C_{16}(2p_{yA} + 2p_{yB})$ [4.56]

The energy level diagram for CO_2 is given in Fig. 4.24.

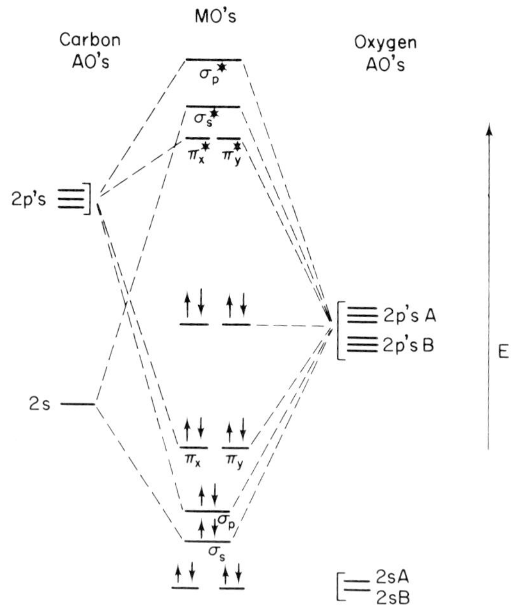

Figure 4.24. Molecular energy level diagram for CO_2.

The reader should confirm the fact that the ground state term for carbon dioxide is $^1\Sigma$. Note that there are four electrons in the nonbonding orbitals. These electrons do not participate in bonding, nor do they hinder bonding. The MO theory also predicts the existence of π bonding in carbon dioxide.

THE HEXACOORDINATED SPECIES

The MO's for an octahedral complex, ML_6, may be arrived at by our use of the atomic orbitals of the central metal alone and the orbitals of the ligands in a visual approach to the problem. The central metal atom will use an s orbital, three p orbitals, and the d_{z^2} and $d_{x^2-y^2}$ orbital to form an octahedral species. The VB method's designation is d^2sp^3 hybridization. The equations representing MO's, the σ bonds formed by the six ligands, will be derived in terms of Fig. 4.25. If the central atom is at the origin of the coordinate system, the ligands will approach along the coordinate axis. Thus, x_+ and x_- will represent the wave function of the orbital of the ligands approaching along the $+x$ and $-x$ axes. With the figures in Fig. 4.25, the equations for the σ bonding and antibonding orbitals will be determined.

The equations representing the bonding and antibonding orbitals

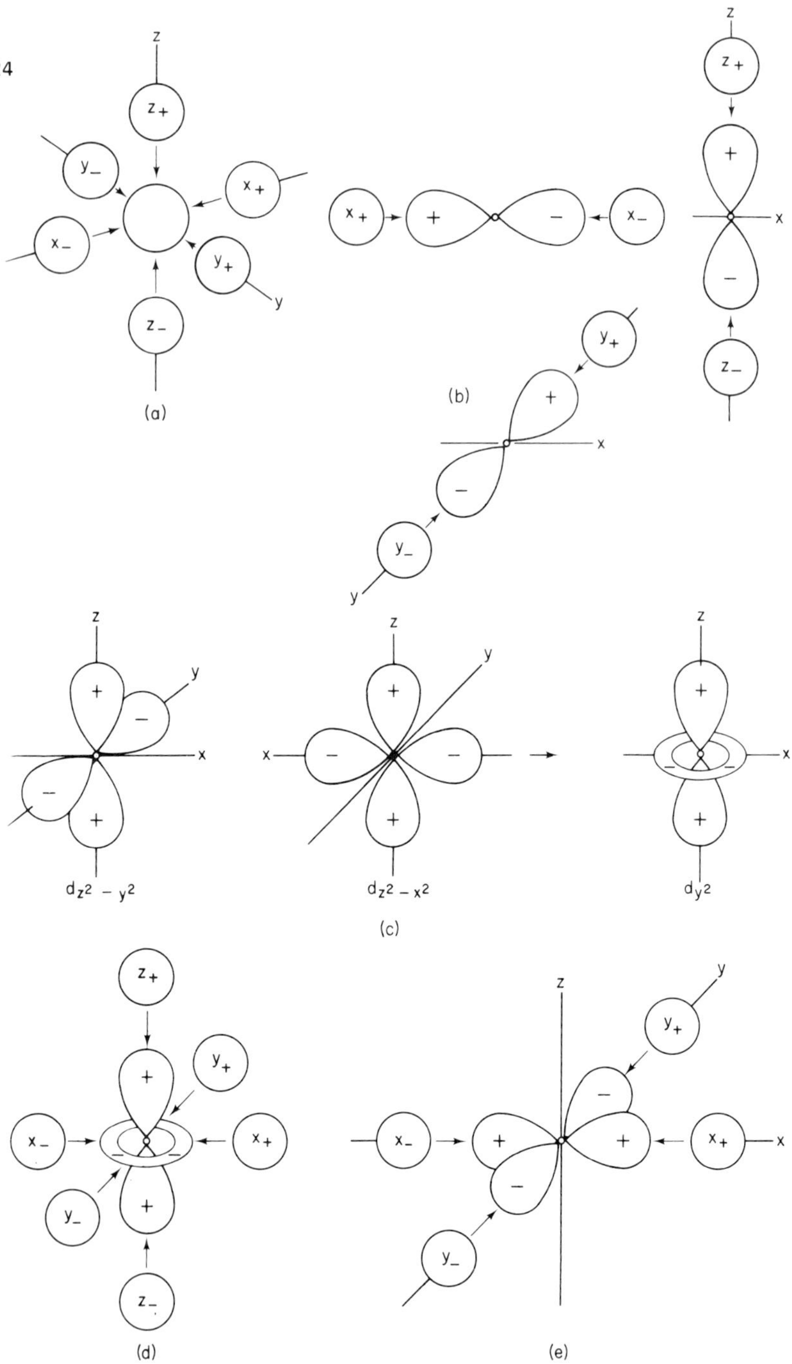

Figure 4.25. Pictorial representation of ligand orbitals approaching bonding orbitals in a hexacoordinated species. (a) Ligands approaching the s orbital; (b) ligands approaching the p orbitals; (c) the formation of the d_{z^2} orbital from the $d_{z^2-y^2}$ and $d_{z^2-x^2}$ orbitals; (d) ligands approaching the d_{z^2} orbital. (e) ligands approaching the $d_{x^2-y^2}$ orbital.

formed by the ligands and the s orbital of the central ion may be ascertained from Fig. 4.25(a).

They are

$$\psi\sigma_s = C_1 s + C_2(x_+ + x_- + y_+ + z_+ + z_-) \qquad [4.57]$$

$$\psi^*\sigma_s = C_3 s - C_4(x_+ + x_- + y_+ + y_- + z_+ + z_-) \qquad [4.58]$$

The energetically degenerate p orbitals form the following bonding and antibonding orbitals, represented by the following equations [Fig. 4.25(b)].

$$\psi\sigma_x = C_5 p_x + C_6(x_+ - x_-) \qquad [4.59]$$

$$\psi^*\sigma_x = C_7 p_x - C_8(x_+ - x_-) \qquad [4.60]$$

$$\psi\sigma_y = C_9 p_y + C_{10}(y_+ - y_-) \qquad [4.61]$$

$$\psi^*\sigma_y = C_{11} p_y - C_{12}(y_+ - y_-) \qquad [4.62]$$

$$\psi\sigma_z = C_{13} p_z + C_{14}(z_+ - z_-) \qquad [4.63]$$

$$\psi^*\sigma_z = C_{15} p_z - C_{16}(z_+ - z_-) \qquad [4.64]$$

The visual or pictorial approach fails when dealing with d_{z^2} orbital. The d_{z^2} orbital is actually constructed from two orbitals, the $d_{z^2-x^2}$ and the $d_{z^2-y^2}$ [Fig. 4.25(c)], so that probability along the z axis of the d_{z^2} orbital is twice that which is to be expected. Therefore, the equations representing the sigma bonding and antibonding orbitals formed by the ligands and central atom have the following forms [see Fig. 4.25(d) and 4.25(e)].

$$\psi\sigma d_{z^2} = C_{17} d_{z^2} + C_{18}(2_{z+} + 2_{z-} - x_+ - x_- - y_+ - y_-) \qquad [4.65]$$

$$\psi^*\sigma d_{z^2} = C_{19} d_{z^2} - C_{20}(2_{z+} + 2_{z-} - x_+ - x_- - y_+ - y_-) \qquad [4.66]$$

$$\psi\sigma d_{x^2-y^2} = C_{21} d_{x^2-y^2} + C_{22}(x_+ + x_- - y_+ - y_-) \qquad [4.67]$$

$$\psi^*\sigma d_{x^2-y^2} = C_{23} d_{x^2-y^2} - C_{24}(x_t + x_- - y_+ - y_-) \qquad [4.68]$$

The resultant energy level diagram is given in Fig. 4.26. It must be stressed that this energy level diagram represents only sigma bonding. It is also quite possible that π bonds form, but more will be said about this at another time.

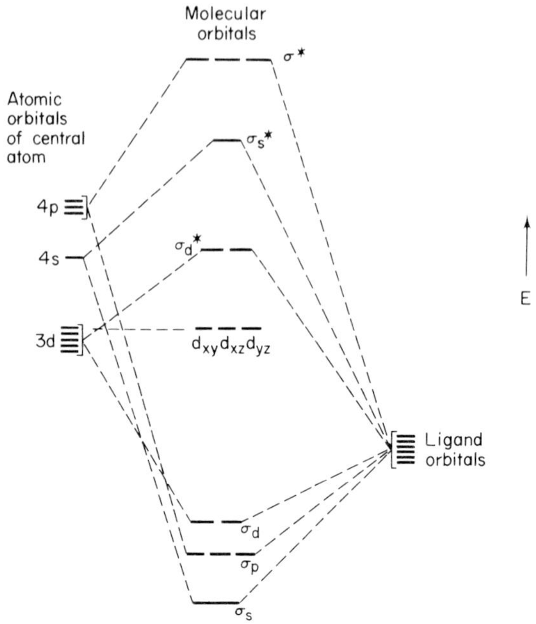

Figure 4.26. Molecular orbital energy diagram for a hexacoordinated species, ML_6.

HYBRIDIZATION

To many, the concept of hybridization may formally belong to the valence bond theory, but the concept may be used properly in molecular orbital theory as well. Often, it does not take a prohibitive amount of energy for an atom to hybridize a group of its orbitals. An atom may combine an *s* and a *p* orbital to form two *sp* hybrid orbitals. These hybrid orbitals are directed in space, usually at maximum separation. This maximum separation and direction in space offers opportunity for great overlap and consequently for the formation of strong bonds. The wave functions for such hybrids are

$$\phi_{Hy^1} = N_1(\phi_{2s} + \lambda\phi_{2p}) \quad [4.69]$$

$$\phi_{Hy^2} = N_2(\phi_{2s} - \lambda\phi_{2p}) \quad [4.70]$$

The mixing coefficient, λ, assumes values between zero and unity. When λ equals unity, the hybrid orbitals formed consist of 50% *s* character and 50% *p* character. There are two wave functions because there are two orbitals that must result. This is illustrated in Fig. 4.27. The + and − signs merely indicate that we turn the *p* orbital around when we combine it the

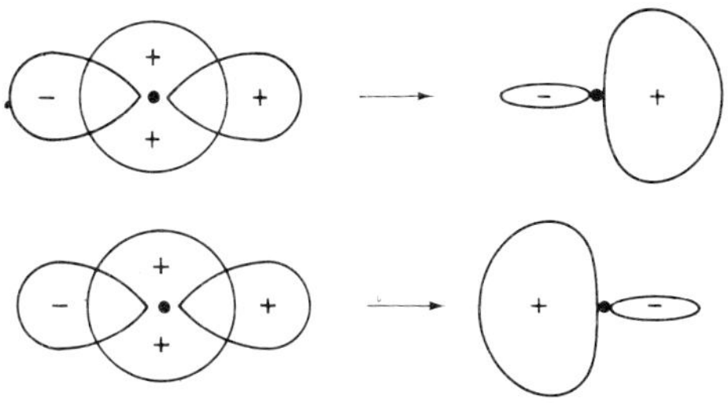

Figure 4.27. Diagrammatic scheme representing *sp* hybrid orbital formation.

second time. Figure 4.28 contains the energy diagram representing what has taken place in hybridization and the relative energies of the atomic orbitals

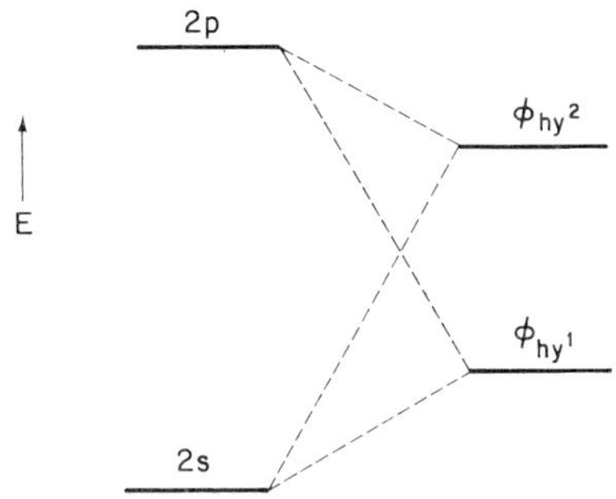

Figure 4.28. Molecular orbital energy level diagram for *sp* hybrid orbital formation.

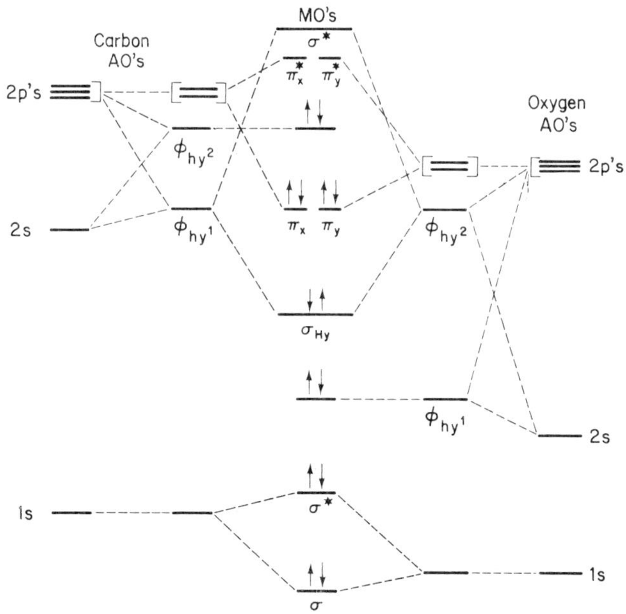

Figure 4.29. Molecular energy level diagram for carbon monoxide including the concept that both the oxygen atom and the carbon atom hybridize 2*s* and 2*p* orbitals prior to compound formation.

and the hybrid orbitals. The energy of the hybrid orbitals always lies between the energies of the atomic orbitals that are hybridized.

The application of the concept of hybridization to CO is as follows. It will be assumed that both the carbon atom and the oxygen atom form *sp* hybrid orbitals before overlapping and forming the molecular orbitals. Figure 4.29 contains the energy diagram of these operations and MO formation. Compare this diagram closely with Fig. 4.24.

After hybridization, one can imagine that the atoms move toward each other along the z axis. Assuming that the hybrid orbitals are *sp* hybrids, these orbitals will overlap to form a sigma bond. The lower energy *sp* hybrid orbital of carbon is of about the same energy as the higher energy *sp* hybrid orbital of oxygen. These two orbitals preferentially overlap. Therefore, both the carbon atom and the oxygen atom have a hybrid orbital containing a lone pair of electrons. The lone pair on the oxygen atom is in a hybrid orbital that contains a high percentage of s character—hence, is stable and relatively unreactive. The lone pair on the carbon is in a hybrid orbital with a greater percentage of p character—hence, is more loosely bound and of higher energy. This lone pair is more likely to react, which is consistent with the donor properties of carbon monoxide (see Chapter 13).

The p_x and p_y orbitals of both carbon and oxygen are involved in π bonding. These orbitals are not hybridized and meet side-to-side as the atoms are brought together. The other four electrons form these π bonds.

At this point, we should pause and observe some of the data in Tables 4.4 and 4.5. In general, bond energy increases with an increase in bond order and bond length decreases with an increase in bond order. Let the reader compare the isoelectronic systems CO and N_2. On ionization, these two species form CO^+ and N_2^+. The carbon-oxygen bond length in CO^+ is smaller than the carbon-oxygen bond length in carbon monoxide; however the nitrogen-nitrogen bond length in N_2^+ is larger than the nitrogen-nitrogen bond length in N_2. Molecular orbital theory seemingly has no rationalization to explain this anomalous behavior.

THE THREE-CENTERED BOND

It is a fitting conclusion to this brief, qualitative presentation to describe how MO theory is applied to the three-centered bond. This type of bond is postulated to explain the stability of electron-deficient molecules. Electron-deficient molecules are those where there are not sufficient bonding electrons to join all the atoms together by normal covalent bonds, electron pair bonds. The boron hydrides are prime examples of this situation (see Chapter 12).

The MO theory states that when two atomic orbitals overlap, two molecular orbitals are generated. This statement may be generalized to state that the number of MO's formed is equivalent to the number of atomic orbitals combining to form MO's. If three atomic orbitals overlap, three MO's will be formed.

Let us assume that atoms A, B, and C are going to combine to form a three-centered bond (Fig. 4.30). All these atomic orbitals will combine to

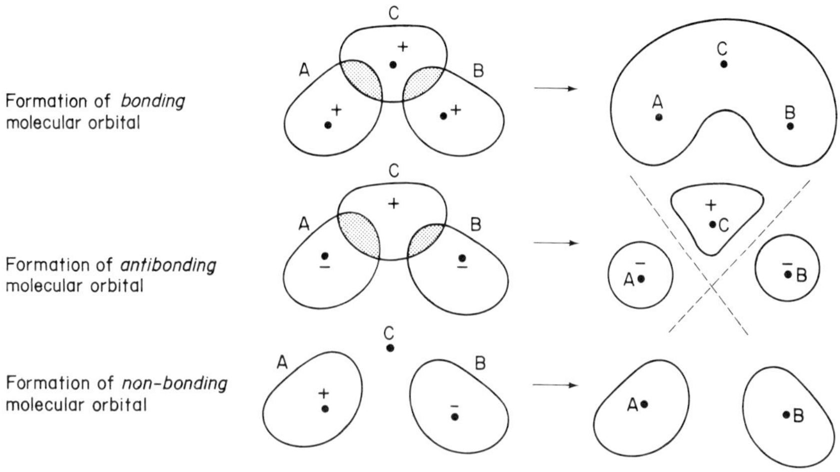

Figure 4.30. Diagrammatic representation of the orbitals formed in the formation of a three-centered bond.

form a bonding molecular orbital, represented by ψ_b. The second MO that is formed, represented by ψ^*, is an antibonding orbital, an orbital where there is a node of zero electron density between adjacent atoms. The third MO that is formed is formed by the nonadjacent atoms. There is negligible overlap, there is no appreciable energy change, and there is no aid or deterrent to bonding in the MO. It is a nonbonding MO, whose wave function is ψ_n. The energy level diagram for the three-centered bond is given in Fig. 4.31.

The equations representing these MO's are the following.

$$\psi_b = C_1\phi_A + C_2\phi_B + C_3\phi_C \qquad [4.71]$$

$$\psi^* \doteq -C_4\phi_A - C_5\phi_B + C_6\phi_C \qquad [4.72]$$

$$\psi_n = C_7\phi_A - C_8\phi_B \qquad [4.73]$$

Let us conclude this section by rationalizing the bonding in xenon difluoride. We shall assume that the $2p_z$ orbitals of each fluorine atom will overlap with the $5p_z$ of xenon. There are four electrons involved in bonding: two from xenon and one from each fluorine atom. Employing the energy level diagram of Fig. 4.31, assuming that A and B represent fluorine atoms,

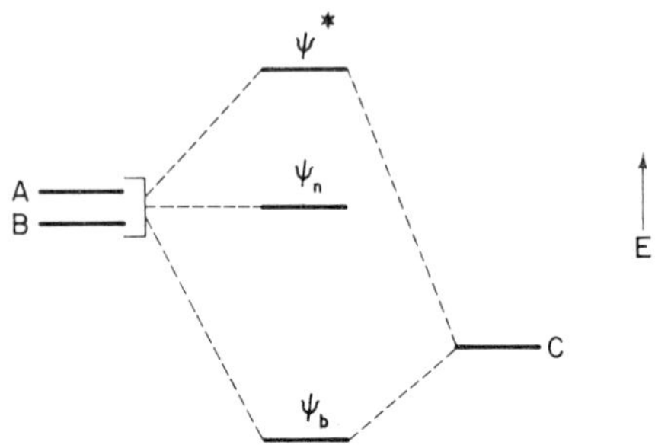

Figure 4.31. Molecular orbital energy level diagram showing the relative energies of the orbitals formed in a three-centered bond.

we see that the four electrons will be deployed in an energetically favorable situation for bond formation. Two electrons will be placed in the bonding orbital and two electrons will be placed in the nonbonding orbital. Therefore, the stability of xenon difluoride may be rationalized.

DOUBLE QUARTET THEORY OF LINNETT

An alternative method of approximating wave functions is proposed by Linnett. Where the VB method pairs electrons having opposite spins, Linnett approaches the problem of bonding by classifying valence electrons as belonging to two spin sets. The electrons possessing one type of spin belong to one spin set, whereas the electrons possessing the opposite spin belong to their own spin set. Therefore, the famous Lewis octet is now treated as a double quartet, according to Linnett. Herein lies the origin of the name, Linnett's double quartet theory. Linnett uses two concepts employed to some extent in both the VB and MO treatments in determining the electronic distribution about a given atom. The first of these is interelectronic repulsion, which will tend to separate the electrons. The second is the concept that two electrons can occupy the same region of space only if they have opposite spins. Electrons of like spin tend to keep apart, whereas those of unlike spin, but occupying different orbitals, tend to approach one another. The former effect is reinforced by interelectronic repulsion, but the latter is generally negated by it. Therefore, at a given atom, two electrons of the same spin tend to remain 180° apart. Three electrons of the same spin would tend to position themselves at the vertices of an equilateral triangle, and four electrons of the same spin tend to position themselves at the vertices of a tetrahedron, etc. Therefore, four pairs of electrons forming the octet now become two tetrahedra, which assume relative positions that keep electronic repulsion at a minimum. That is, the vertices of one tetrahedron are thus expected to be centered on the faces of the other (Fig. 4.32).

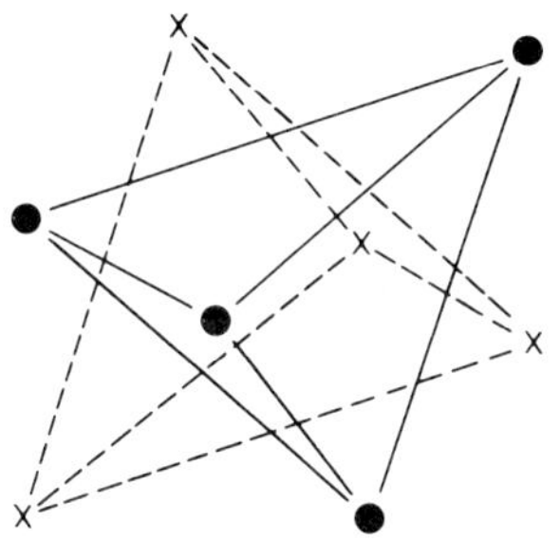

Figure 4.32. The double quartet of electrons illustrated by two interconnected tetrahedra.

The electron distribution in molecules is developed along similar lines; the preceding concepts are applied to each atom of the molecule. The octet rule and rules of formal charge—concepts used in valence bond theory—are also applied to each atom. Thus, Linnett's approach is, like the VB approach, a *localized* one—the situation about each particular atom is considered.

Because Linnett dispenses with the electron pair, he is able to handle many molecules having odd numbers of electrons without introducing resonance. In a number of cases, however, he does find resonance necessary. Which resonance structures are most important is decided by considering relative numbers of bonding electrons, formal charges, and the presence or absence of electron pairing in orbitals (a destabilizing effect, because of charge repulsion) in the various structures.

On the basis of these rules, approximate wave functions can be constructed for molecules; in many cases, these wave functions prove to be closer approximations to reality than ones derived using either VB or MO techniques. The superiority of Linnett's theory to the VB approximation is due to its reduced use of resonance; its superiority to the MO approximation is due to its allowance for charge repulsion effects. It is also superior to MO theory in its ability to predict the shapes of molecules, its extensibility to complex systems, and the fact that schematic chemical formulas can be drawn which represent the wave functions employed. That is, the rules used in developing the wave functions are also used to develop the formulas. As in the case of VB theory, these formulas can provide qualitative and semiquantitative predictions and explanations of molecular properties, without the use of calculations.

A few simple examples follow, which illustrate Linnett's approach to known compounds. He has, of course, successfully treated a very much larger number of systems than those presented here.

The fluorine molecule contains 14 valence electrons. If one uses crosses to represent electrons of one spin and circles to represent the electrons of opposite spin, the familiar Lewis (VB) formulation results:

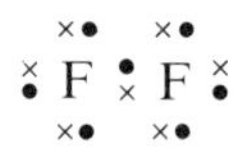

However, Linnett's explanation of the fluorine molecule is as follows. There are a total of 14 valence electrons in the molecule. For each atom to attain an octet, six of these electrons must be localized on each atom and a pair of electrons must be shared by the two atoms. So that the electrons will be accommodated in valence orbitals without violating the Pauli exclusion principle, seven electrons must have one spin and seven must have the other. Now the electrons in each spin set will tend to remain as far apart as possible. Compliance with the octet rule and Pauli principle results in seven electrons having one spin and seven electrons having the opposite spin. Therefore seven electrons of each spin set will be located at the vertices of two tetrahedra with one common apex. The fluorine nuclei will be the centers of the two tetrahedra with a bonding electron at a common apex.

The two spin sets will be superimposed so that the interelectron repulsion energy will be minimal. Therefore, the tetrahedra at each atom will be staggered relative to each other, and the only electrons in the entire molecule occupying the same region of space (spatial orbital) will be the bonding electrons. This is drawn in Fig. 4.33.

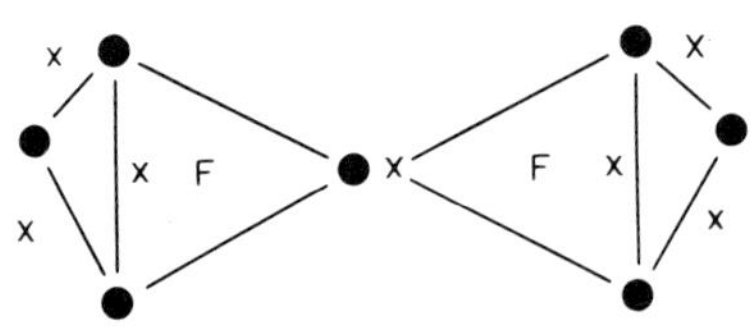

Figure 4.33. The double quartet structure of the fluorine molecule.

A schematic formula of the fluorine molecule may be represented in the following manner:

If, after Linnett, a light dash is used to represent a pair of electrons with opposite spins but not occupying the same spatial orbital, whereas a heavy dash is used to represent such a pair occupying the same spatial orbital, the formula for the fluorine molecules becomes

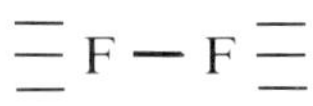

In the case of the nitrogen molecule and the isoelectronic acetylide ion, C_2^{2-}, the reverse electronic configuration is deduced. There are 10 valence electrons, and the spin states take on the form of staggered tetrahedra having a common base. This is drawn in Fig. 4.34. Both the Lewis structure and the Linnett structure predict a bond order of three, since both are

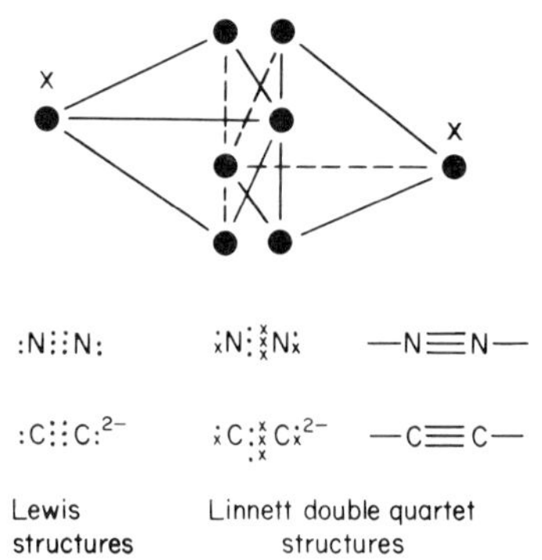

Figure 4.34. Double quartet and Lewis structures for the isoelectronic species N_2 and C_2^{2-}.

partially derived from the octet rule. The structures of carbon monoxide and cyanide ion are similar.

For the oxygen molecule, which contains 12 valence electrons, several Linnett structures can be drawn. In one of these, one spin state takes on the configuration of two tetrahedra with a common apex, and the other takes on the configuration of two tetrahedra with a common base, as shown:

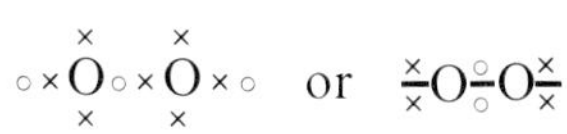

Note that the Pauli exclusion principle is fulfilled even though there are two more electrons of one spin than of the other.

An alternate Linnett electronic arrangement of the electrons in a molecule of oxygen is the following:

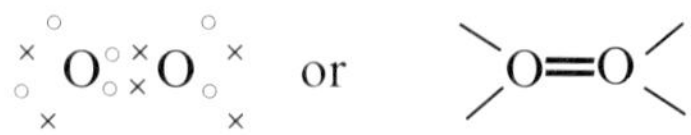

This spin set structure is that of two tetrahedra sharing a common edge. Since both spin sets have the same configuration in this structure, it does not separate the electrons so effectively as the first structure. Calculations have shown that this structure corresponds to the first excited state of the oxygen molecule, and that the first structure given corresponds to the ground state. Because of the differing numbers of electrons in the spin states of the first structure, the Linnett theory predicts the paramagnetism of oxygen which is likewise predicted by MO theory. Valence bond theory cannot account for this paramagnetism in any straightforward manner.

Data on nitric oxide, NO, which has 11 electrons, indicates that one spin set has one more electron than the other, that the atoms are joined by a 5-electron bond and that the molecule is quite stable. The electronic structure would be expected to be

which satisfies the octet rule and Pauli exclusion principle.

Since the nitric oxide molecule has an odd number of electrons, one might expect dimerization to occur, so that the odd electrons could be paired in a chemical bond. The electronic structure of the dimer would be expected to be

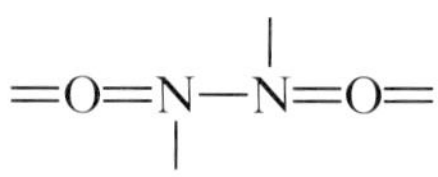

But this structure has no more bonding electrons than two molecules of the monomer, and all the electrons are present as pairs occupying the same spatial orbital. Thus this dimer is thermodynamically unstable relative to the monomer.

Nitrogen dioxide, NO_2, has 17 valence electrons and might well be expected to be a resonance hybrid of the following two structures, which are mirror images:

In both structures, the octet rule and the Pauli principle are obeyed. Comparison of the above structures with that for NO_2^+,

suggests that the ion should have slightly shorter NO bonds. This is indeed the case; the NO bond lengths in NO_2 and NO_2^+ are 1.19 and 1.15 A, respectively.

However, if the Linnett structures for nitrogen dioxide are redrawn as follows:

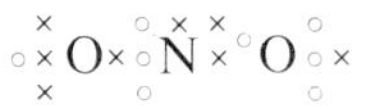 and 

we see that the spin set represented by the circles favors a linear structure, whereas the other spin set favors an ONO angle of approximately 120°. The ONO bond angle has been shown to be 134°. This value is understandable because the bending force constant for a bent molecule (O_3) is greater than that for a linear molecule containing a single and a triple bond (HCN).

Dimerization of nitrogen dioxide to dinitrogen tetroxide, N_2O_4, occurs readily. The Linnett structure for dinitrogen tetroxide is

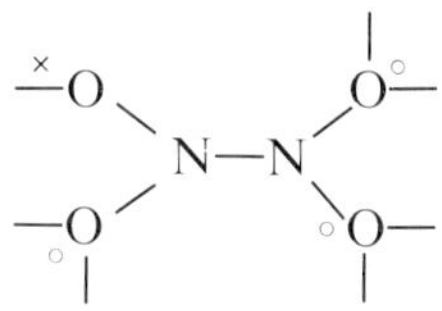

which has the same number of bonding electrons as two molecules of NO_2. This situation is similar to that encountered in the consideration of nitric oxide and its dimer. In dinitrogen tetroxide, both electron spin sets favor the same structure; therefore, the dimer is more stable than the monomer in this respect, which is shown as:

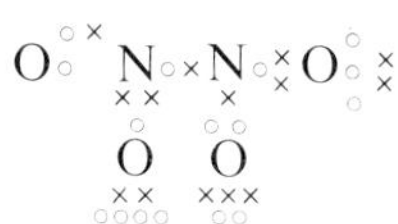

The heat of formation of N_2O_4 from NO_2 is about 12 kcal/mole, which is consistent with the view that the number of bonding electrons in the dimer molecule is the same as that in two molecules of the monomer.

Linnett explains the planarity of dinitrogen tetroxide by stating that each spin set of tetrahedra surrounding the nitrogen atoms is staggered relative to one another. This does not explain the molecule's planarity, but if one assumes some distortion of the tetrahedra due to the presence of oxygen atoms, there is a degree of plausibility in the rationalization.

For benzene, Linnett draws the following structure:

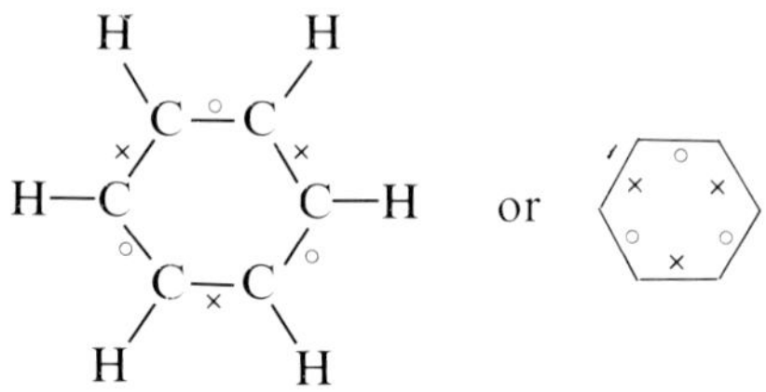

Linnett favors this structure over the Kekulé resonance structures because it reduces interelectron repulsion energy. Calculations using a wave function derived from the Linnett structure yield an electronic energy of 69 kcal/mole lower than that arrived at using a Hückle molecular orbital function without configuration interaction. Thus, this theory has merits that both the VB and MO descriptions do not possess.

The relationship between the Linnett and Kekulé structures for benzene is comparable to that between the ground state of the oxygen molecule and its second excited state. In the low energy structure of these compounds, some or all of the valence electrons are not paired to spatial orbitals, and the spin sets have different configurations; the opposite is true in the higher energy structures being considered. Now the energy difference between the ground state of O_2 and its second excited state is 38 kcal/mole; since spatial pairs are formed from 12 electrons, the molar pairing energy is about 6 kcal per electron pair. The corresponding difference for benzene is its so-called resonance energy, 36 kcal/mole; nine spatial pairs are present in the Kekulé structure which are absent from the Linnett structure, so the molar pairing energy here is about 4 kcal per electron pair. This is less than the figure obtained for oxygen, but it can be attributed to the difference in nuclear charge between carbon and oxygen. That is, the lower nuclear charge of carbon permits more diffuse orbitals than occur with oxygen, resulting in a lowering of the interelectron repulsion energy. Similar considerations can be applied to carbon dioxide; here, the molar pairing energy per electron pair is also about 4 kcal. All the valence electrons of the carbon dioxide molecule are affected, and one might object that oxygen orbitals are present here, although absent in benzene. Linnett admits the doubtful nature of these comparisons, but points out that the values obtained are at least approximately the same.

Suggested Readings

1. Ballhausen, C. J., and H. B. Gray, *Molecular Orbital Theory*, W. A. Benjamin, Inc., New York, 1964.
2. Barnard, A. K., *Theoretical Basis of Inorganic Chemistry*, McGraw-Hill Publishing Company, New York, 1965.
3. Coulson, C. A., *Valence*, Oxford University Press, Oxford, 1953.
4. Drago, R. S., *Physical Methods in Inorganic Chemistry*, Reinhold Publishing Corp., New York, 1965.
5. Firestone, R. A., *J. Org. Chem.*, **34,** 2621 (1969).
6. Gray, H. B., *Electrons and Chemical Bonding*, W. A. Benjamin, Inc., New York, 1964.
7. Linnett, J. W., *The Electronic Structure of Molecules*, John Wiley & Sons, New York, 1964; *J. Am. Chem. Soc.*, **83,** 2643 (1961).
8. Luder, W. F., *J. Chem. Educ.*, **44,** 206 (1967); **44,** 269 (1967).
9. Murrell, J. N., S. F. A. Kettle, and J. M. Tedder, *Valence Theory*, John Wiley & Sons, Inc., New York, 1966.
10. Orchin, M., and H. H. Jaffe, *Importance of Antibonding Orbitals*, Houghton Mifflin Company, Boston, 1967.
11. Pauling, L., *The Nature of the Chemical Bond*, 3rd ed., Cornell University Press, Ithaca, N.Y., 1960.
12. Pauling, L., *The Chemical Bond*, Cornell University Press, Ithaca, N.Y., 1967.

Study Questions

4.1. Nitrosyl fluoride has the formula FNO. (a) Draw the Lewis structure for this molecule. Indicate electrons from different atoms by different atoms by different symbols, e.g., ·, △, ×, etc. (b) Using an electron box type diagram, indicate what hybrid orbital is involved around the nitrogen atom. (c) Critically discuss the expected magnitude of the FNO bond angle.

4.2. From elementary molecular orbital considerations, would one expect the molecule FN to be capable of existence? If it should exist, would it be diamagnetic or paramagnetic?

4.3. What are the differences between σ and π bonds? (Diagrams would help!)

4.4. Given the following pieces of experimental evidence, draw covalent canonical forms of the cyanogen, $(CN)_2$, molecule.
(a) The normal C—C bond length is 1.54 A.
(b) The normal C=C bond length is 1.33 A.
(c) The normal C=N bond length is 1.18 A.
(d) The normal C≡N bond length is 1.15 A.
(e) The C to C bond length in $(CN)_2$ is 1.36 A.

(f) The C to N bond length in $(CN)_2$ is 1.16 A.
(g) The cyanogen molecule is linear.
(h) The cyanogen molecule shows no magnetic properties.

4.5. Discuss briefly and draw fully labelled orbital diagrams to illustrate (a) intermolecular π bonding and (b) intramolecular π bonding. Give examples to show how intramolecular π bonding has been used to account for unexpected bond angles, structures, dipole moments, and base strengths of certain compounds.

4.6. Give two different possible hybridization states of atom A in the regular tetragonal pyramidal structure, AB_5. State clearly what p, d, etc., orbitals are involved.

4.7. The concept of $(p \rightarrow d)\pi$ bonding is used to explain some of the properties of the SiF bond in H_3SiF. Use a clearly labeled diagram to show the overlap of appropriate atomic orbitals consistent with this hypothesis. Include signs of wave functions.

4.8. In the construction of sp hybrids from an s orbital and a p_x orbital, the new wave functions are in the form

$$\psi = N(s + kp_x)$$
$$\psi^* = N(s - kp_x)$$

where N is the normalization constant, k is the mixing coefficient, and s and p_x are the wave functions for the s and p orbitals that are being mixed to form the hybrid orbitals. What is the value of k? Evaluate the normalization constant, N, in terms of the equations. Indicate the shape of the resulting hybrid orbital.

4.9. Indicate the ground state spectroscopic term symbol for NO^-. Using the reasoning based on the double quartet theory as advanced by Linnett, indicate the double quartet drawing of NO^-.

4.10. Determine the molecular orbital description of the water molecule—equations and energy level diagram.

4.11. (a) Draw the homonuclear diatomic molecular orbital diagram for N_2 and fill in the electrons.
(b) Sketch the six top molecular orbitals (including sign). Label them. (One drawing, with a one-sentence explanation, will suffice for any pair of orbitals.)
(c) Sketch the valence bond approach to N_2 (orbital not Lewis diagram) and again include any signs.
(d) Use both valence bond and molecular orbital theories to predict what will happen to N_2 on ionization to N_2^+.

4.12. Using the Linnett double quartet approach *alone*, show *clearly* whether the odd electron molecules NO and CN would be expected to be monomeric or dimeric.

4.13. Draw and explain in detail the molecular orbital diagram for molecules with prior hybridization (such as NO). Compare the simple valence bond treatment with the molecular orbital treatment. What is the molecular term symbol for NO?

4.14. Write the molecular orbital designation for the superoxide ion, O_2^-. Calculate the bond order and indicate whether the ion should be paramagnetic or diamagnetic. Express the term symbol of this ion.

4.15. Compare the molecular orbital formulation for NO, NO^+, and NO^-, and give the ground state symbol for each.

CHAPTER 5

Stereochemistry of the Isolated Molecule

This chapter will consist in three parts. The first section concerns itself with the valence shell electron pair repulsion (VSEPR) theory as a method of determining the stereochemistry of molecular or aggregate species. The second section presents the concept of symmetry and its relationship to structure. The third and last section presents the idealized geometries of the various ML_n species and their classification according to their symmetry elements.

THE VALENCE SHELL ELECTRON PAIR REPULSION (VSEPR) THEORY

The concept basic to this theory is that pairs of electrons binding ligands to a central atom, as well as lone pairs of electrons, are considered to be point charges in space. The presentation of this theory had begun with Sidgwick and Powell as early as 1940. Gillespie and Nyholm had developed the theory to the extent that in 1957 they concluded that electron pairs surrounding the central atom actually determined the structure of a given species.

It would be well to begin by defining terms. The general representation ML_nE_r will be used extensively. The letter M represents the central atom to which the ligands, L, are bound, and n represents the number of ligands. The letter E represents a lone pair of electrons associated with the central atom, M; r indicates the number of lone pairs of electrons. The total number and kind of pairs of electrons $(n + r)$ determines the structure of the species under consideration. The water molecule is represented as OH_2E_2, according to this presentation of the VSEPR theory, indicating that oxygen is the central atom, two hydrogen atoms are attached to the oxygen (ligands), and oxygen also has two lone pairs of electrons associated with it. Note that $OF_2(OF_2E_2)$ would have the same formulation. However, the lone pairs of electrons on the fluorine are presumed not to affect the structure of the species, since they are relatively distant from the central atom.

Calculations of repulsive forces on the surface of a sphere as a function of distance have resulted in the following arrangements of lowest energy for the number of point charges in space:

Number of Point Charges	*Geometry*
2	Linear
3	Trigonal plane
4	Tetrahedron
5	Trigonal bipyramid
6	Octahedron
7	Pentagonal bipyramid (one possibility)

Figure 5.1 contains drawings of the preceding geometries, and we shall, for the moment, restrict our discussion to the representative elements.

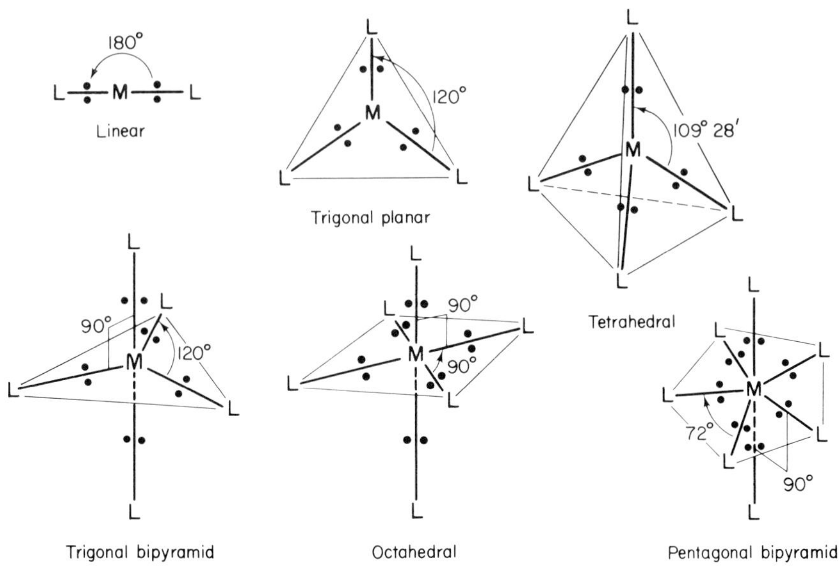

Figure 5.1. Idealized structures of the VSEPR theory, considering electron pairs as point charges in space.

Application of these concepts to an isolated molecule of BeH_2 leads to this train of reasoning. (1) Beryllium is in Group II of the periodic table. (2) Hydrogen pairs its electron with one of the electrons on beryllium. (3) Beryllium is surrounded by two pairs of electrons, i.e., two points of negative charge. (4) *Conclusion*: BeH_2 is a linear molecule.

The application of VSEPR theory concepts to BF_3 is similar. (1) Boron is in Group III of the periodic table, hence three valence electrons. (2) The fluorine atoms share their unpaired electrons with the three electrons of boron. (3) Boron is surrounded by three electron pairs. (4) *Conclusion*: BF_3 is a trigonal planar molecule.

The application of these ideas to NH_3 differs mainly in convention. (1) The central atom, nitrogen, is in Group V of the periodic table and has five valence electrons. (2) The three hydrogen atoms in ammonia share three of the five valence electrons, leaving a lone pair on the nitrogen atom. (3) The general formula for ammonia in VSEPR theory is NH_3E. (4) *Conclusion*: The four point charges arrange themselves tetrahedrally around the nitrogen. However, the ammonia molecule is described as being pyramidal; the lone pair of electrons is not usually included in the structural description.

The last application of these fundamental principles is to water. (1) The central atom is oxygen, in Group VI of the periodic table, having six valence electrons. (2) Hydrogen atoms share two of these six valence electrons in bond formation, leaving four electrons or two lone pairs. (3) The general VSEPR formula for water is OH_2E_2, the oxygen atom being surrounded by four point charges. (4) *Conclusion*: The four point charges arrange themselves tetrahedrally around the oxygen. However, since there are three atoms (nuclei) in the molecule and three points define a plane, the isolated water molecule is generally described as being a planar molecule that is *bent*. Compare H_2O and BeH_2.

At this point, we return to Fig. 5.1 to rationalize the angles associated with the idealized geometries of the various structures. The angles associated with the linear, trigonal planar, and tetrahedral structures are self-explanatory and the student should be readily familiar with them. Note the following relationships in the following structures:

Trigonal bipyramid	Axial ligands subtend an angle of 180° An axial and equatorial ligand, 90° Equatorial ligands subtend an angle of 120°
Octahedron	Axial ligands subtend an angle of 180° An axial and equatorial ligand, 90° Equatorial ligands subtend an angle of 90°

Pentagonal bipyramid	Axial ligands subtend an angle of 180°
	An axial and equatorial ligand, 90°
	Equatorial ligands subtend an angle of 72°

TABLE 5.1 ***L—M—L Angles of Some Molecular Hydrides***

Compound	Angle	Compound	Angle
NH_3	106°47′	H_2O	104°27′
PH_3	93°37′	H_2S	92°16′
AsH_3	92°	H_2Se	91°
SbH_3	91°	H_2Te	90°

The student should now peruse Table 5.1 with the idealized geometries (shapes and angles) in mind. The ideal tetrahedral angle is 109° 28′. Both methane and carbon tetrafluoride are experimentally known to possess this ideal tetrahedral angle. As is evident from Table 5.1, both ammonia and water do not possess this ideal tetrahedral angle. But the VSEPR theory analysis predicted a basic tetrahedral structure for the four point charges surrounding both the nitrogen and oxygen atoms, considering the lone pairs of electrons as ligands. Another argument may be presented from the aspect of the directional character of the p orbitals. These are directed along the coordinate axes and theoretically should subtend an angle of 90°. Calculations based on ligand repulsion as a cause for the observed $L—M—L$ angles in water and ammonia would only widen the angle a few degrees. The real angles are much closer to the tetrahedral angle than they are to 90°. The distortion from idealized geometries must then be based on electron pair repulsions. The fact that CF_4 and CH_4 have idealized tetrahedral geometries leads to two conclusions: (1) bonding electron pairs have relatively weak repulsions relative to lone pairs, since the bonding electron pairs are under the control of two nuclei; (2) in comparison, the lone pairs are under the control of only one nucleus, and the volume that they occupy is much larger and their repulsive forces more effective. The nitrogen atom in ammonia has one lone pair, the oxygen atom in water has two lone pairs; the H—N—H bond angle is smaller than the tetrahedral angle, and the H—O—H bond angle is smaller yet. The comparative size of the H—O—H bond angle may be attributed to the fact that oxygen has two lone pairs of electrons in this molecule. From the preceding discussion and the data in Table 5.1, we can arrive at an order of electrostatic repulsion among electron pairs:

Lone pair − lone pair > lone pair − bonding pair > bonding pair − bonding pair

——— Electrostatic repulsion decreases ———→

Possibly a not-so-obvious point should be mentioned here. The $L—M—L$ angles of the hydrides below NH_3 and H_2O are very close to 90°, the idealized angle of the undisturbed p orbitals, and quite removed from 109° 28′, the idealized tetrahedral angle. The student must realize that as the size of the central atom increases (descends the group), there is less and less need for hybridization!

We now return our attention to the concept of electron pair repulsion and the rationalization of the structure of ClF_3. The chlorine atom is in Group VII of the periodic table, and therefore has seven valence electrons. The fluorine atoms share three of these electrons. Chlorine then has five pairs of electrons surrounding it, three bonding pairs and two lone pairs. Figure 5.2 shows three possible structures for this molecule. The correct

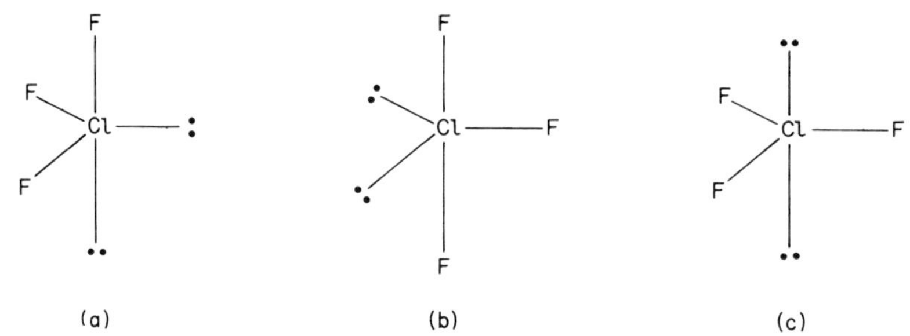

Figure 5.2. Possible structures of ClF_3.

structure is the T-shaped molecule in (b). The positioning of the lone pairs of electrons is essential in this consideration. In Fig. 5.2(c), the lone pairs are 180° apart and the lone pairs-bond pair interactions are 90° apart. In (b), the lone pairs are 120° apart, and these are 120° from a bonding pair of electrons; there are four lone pairs-bond pair interactions at 90°. The structure illustrated in (b) has minimal repulsion and is the preferred structure. The determination of the structure of ClF_3 illustrates that the concepts of VSEPR theory are not always easy and clear cut.

The molecule sulfur tetrafluoride presents the problem of where to place the lone pair of electrons. Sulfur, which is in Group VI-A of the periodic table, has six valence electrons, four of which are used to bond the fluorine atoms. Therefore, sulfur is surrounded by five electron pairs: four bonding pairs and one lone pair. Figure 5.3 illustrates the two possible structures. In Fig. 5.3(a), the lone pair is 90° from three bonding pairs, whereas in Fig. 5.3(b), it is 90° from two bonding pairs and 120° from two other bonding pairs. Repulsion between electron pairs falls off very rapidly as

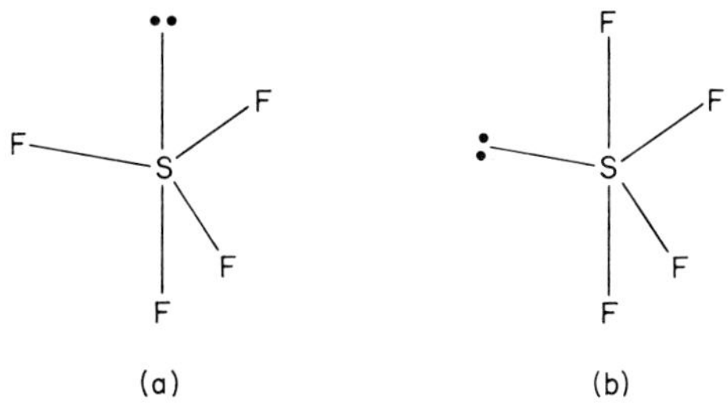

Figure 5.3. Two possible structures for SF_4.

the angle between them increases. Hence, (b), which is a slightly distorted tetrahedron, would be the preferred structure.

As a general rule, we might state that lone pairs of electrons will tend to occupy those positions in a molecule in which their repulsive action will be at a minimum.

The concepts of this theory may be applied readily to species containing multiple bonds. If we use the same representation, ML_nE_r, it will be seen that it really is not necessary to be aware of the type of bonding and that the general posture of the molecule is simply determined by $n + r$. One simply considers that a double bond is formed by a single orbital capable of containing four electrons, and that a triple bond is formed by a single orbital capable of containing six electrons! In simple language, both a double and a triple bond are considered to be *single point charges in space*.

The sulfur oxides—SO_2, SO_3, and SO_4^{2-}—offer themselves as pertinent examples. Sulfur has six valence electrons, as does oxygen. Since oxygen is more electronegative than sulfur, it may be said that sulfur donates a pair of electrons to oxygen to complete its octet; and, an oxide ion may donate a pair of electrons to sulfur to complete its octet. See the following:

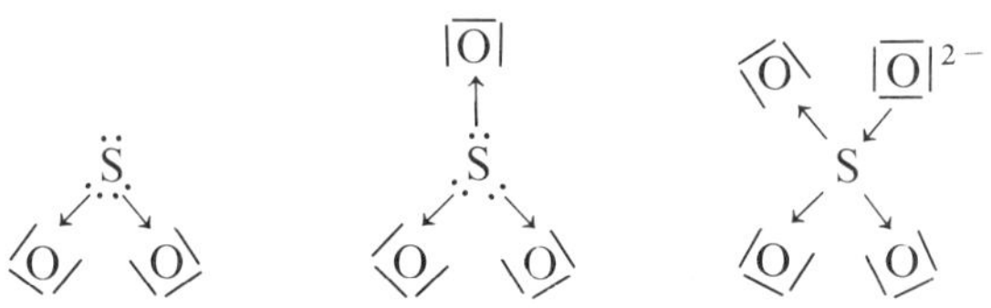

SO_2 SO_3 SO_4^{2-}

Now the electrons on the oxygen atoms may *back bond* to the d orbitals of sulfur to form π bonds, giving double character to the bonds, but these bonds are still considered point charges in space. Then the structures of the preceding species are SO_2, bent; SO_3, trigonal planar; and SO_4^{2-}, tetrahedral. (The double charge on oxygen is used for illustrative purposes: in reality, the charge is distributed over the entire ion.)

Even the structures of the hydrocarbons may be rationalized. The alkanes are really no problem, with their basic tetrahedral carbon atom. The alkenes may also be approached in the manner just used. As an example, take ethylene, $H_2C{=}CH_2$. The general formula representing this molecule is ML_3, considering CH_2 as a ligand. Therefore, each carbon atom will be surrounded by three point groups, and they will assume trigonal planar geometry:

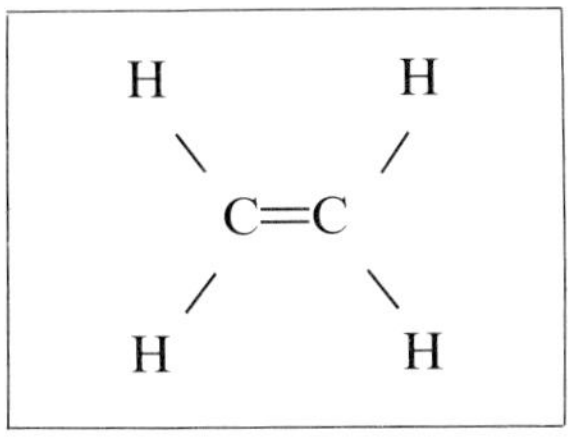

A molecule such as allene contains a subtlety. The formula for allene is $H_2C_x{=}C_y{=}C_zH_2$. The subscripts are used merely to distinguish the carbon atoms. The center carbon atom, C_y, is surrounded by two point charges, and the two $-CH_2$ species will be considered as ligands; the formula for this species is ML_2 and the C=C=C chain is linear. Both C_x and C_z are surrounded by three point charges, and they give a trigonal planar configuration to the two hydrogen atoms and the double bond. Here it must be remembered that the central atom is using the p_z orbitals for sigma bonding. Now if it uses the p_x orbital for π bonding with C_x, it must use p_y orbitals for bonding with the C_z atom. This results in the fact that the plane containing the two hydrogen atoms on the C_x is perpendicular to the plane containing the two hydrogen atoms on the C_z. See Fig. 5.4.

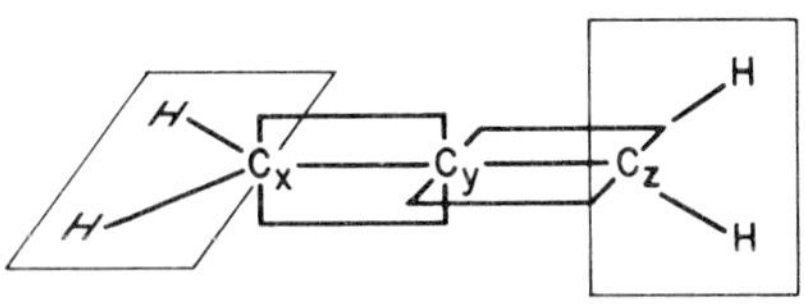

Figure 5.4. The perpendicular planes of allene.

In conclusion we might state that in species where multiple bonds occur, only the σ bonds are important for structural considerations. And the structure may be rationalized using the formula ML_nE_r, by taking $n + r$ as the number of point charges in space. In reference to angle deviation from the ideal or in terms of repulsive forces, are there any trends? It would seem quite obvious that a triple bond exerts a stronger repulsive force than a double bond; and a double bond exerts more repulsive force than a single bond. This trend will aid in rationalizing structural deviations, such as angle deviation from the ideal.

At this point, a few remarks about the electronegativity of the ligand and its effects on structure are in order. As a first approximation, it might

TABLE 5.2 ***Bond Angles (X-M-X) for Group V Halides***

NF_3 102.1°	PF_3 104°	AsF_3 102°	
	PCl_3 100°	$AsCl_3$ 98.4°	$SbCl_3$ 99.5°
	PBr_3 101.5°	$AsBr_3$ 100.5°	$SbBr_3$ 97°
	PI_3 102°	AsI_3 101°	SbI_3 99°

seem that as the electronegativity of the ligand increases, the bonding electron pairs will be pulled away from the central atom. Therefore, the $X—M—X$ angle will decrease. Generalizing, as the electronegativity decreases in a series of similarly formulated compounds, the $X—M—X$ bond angle will decrease. A study of Table 5.2 will reveal that this trend is true for all the compounds below the dotted line—i.e., with the exception of the fluorides.

The reason that the MF_3 angle is larger than expected from electronegativity concepts is that back bonding exists. VSEPR theory postulates that when an atom with a filled valence shell possesses one or more lone pairs of electrons and is bonded to an atom with an incomplete valence shell, there is a tendency for the lone pairs of electrons to be transferred from the filled shell to the incomplete shell. This back bonding gives the original bond some double bond character and shortens the $M—X$ bond length. This effect automatically increases electron density near the central atom and the bond angle must open to accommodate this increase. This is the rationalization as to why the F—P—F and F—As—F bond angles are the largest in the PX_3 and AsX_3 series.

SYMMETRY AND STRUCTURE

Symmetry is difficult to define immediately. It may be better described in terms of its operations and elements. There are two distinct parts to symmetry: (1) *the symmetry operation*; and (2) *the symmetry element*. The *symmetry operation* is an operation performed on a molecule that results in a new orientation of the molecule which is indistinguishable from and superimposable on the original orientation. For example, consider the water molecule,

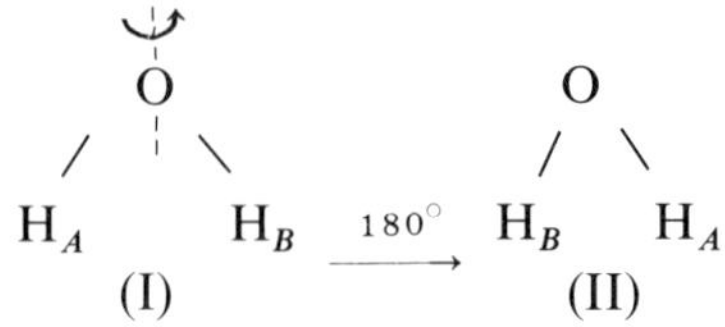

(I) (II)

If one imagines an axis bisecting the H—O—H angle in the plane of the molecule, and then labels the hydrogen atoms A and B, one may rotate the molecule through an angle of 180°. This new orientation of the molecule is superimposable on the first. The *symmetry operation* is *rotation* of the molecule 180°, and the *symmetry element* is the *axis* around which the molecule is rotated. The symmetry operations and elements are distinct yet intimately related. Symmetry may now be defined as the property of a molecule in which parts of the molecule may be interchanged without altering its appearance.

The symmetry of a molecule is inherent in its structure. There are only five basic symmetry elements, but they may combine to give a variety of compound symmetry elements. It is possible for a molecule to possess a great number of symmetry elements—one then says it has a high symmetry —or, it may have very few or no symmetry elements—one then says the molecule possesses low symmetry. It can be shown that all the elements of symmetry constitute a mathematical group. Consequently, mathematical group theory may be applied to symmetry elements and hence to molecules. The applications of group theory are many and varied, especially in the relationship between structure and spectroscopy. This powerful tool has given the modern chemist a key to unlock the doors to many insights.

It was said that there are only five types of symmetry operations. They are these:

(1). *Rotation about some symmetry axis.* The symbol is C_n when n is defined as $2\pi/n$ or $360°/r$. The letter r represents the number of degrees necessary to rotate a molecule about a rotational axis (symmetry element) to obtain a superimposable representation. The illustration of rotating the water molecule through an angle of 180° exemplifies r and the axis would be a C_2 axis.

We should now stress, to maintain uniformity and understanding, that the Cartesian axes are used in determining the rotational axes of molecules. The z axis is the reference axis, the axis of highest symmetry, and it is usually determined in the following manner. Place the center of gravity of the molecule at the origin of the Cartesian coordinate system. Then if:

(a) there is only one rotational axis, make it the z axis.

(b) the molecule has several rotational axes, the one of highest symmetry is taken as the z axis.

(c) there are several axes of highest order, the axis that passes through the greatest number of atoms is taken as the z axis.

It now remains to assign the x axis. Two cases need be considered: planar molecules and nonplanar molecules. With regard to planar molecules, these two points are made: (1) if the molecule is planar and the z axis lies in this plane, the x axis must then be perpendicular to this plane; (2) if the z axis is perpendicular to the plane of the molecule, the x axis is in the plane of the molecule, passing through the greatest number of atoms. Finally, in the case of nonplanar molecules, a plane of atoms is chosen that contains the greatest number of atoms. This plane is then treated as the molecular plane.

(2). *Reflection through a plane of symmetry*. The symbol for reflection is sigma, σ. Reflection may be defined as moving a point from x to $-x$ through the xy plane on a path perpendicular to the xy plane. For this operation to be a symmetry operation, there must be a point at $-x$ that moves to x simultaneously as the first point moves from x to $-x$. The water molecule possesses two planes of symmetry (Fig. 5.5), one containing

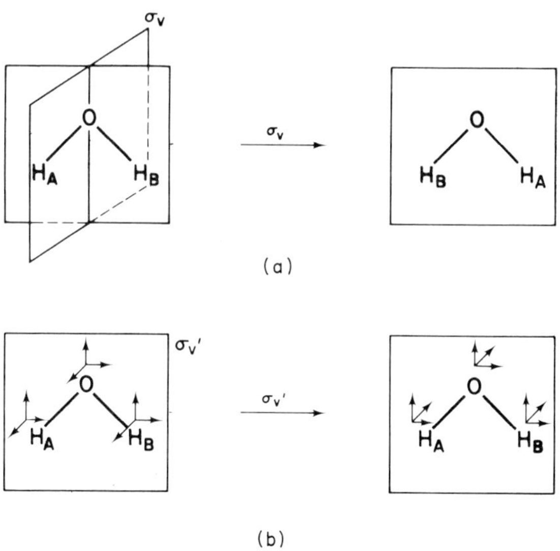

Figure 5.5. Reflection planes in the water molecule and the results of the symmetry operation. (a) Reflection through a plane perpendicular to the plane of the molecule and (b) reflection through a plane containing the molecule.

the entire water molecule and the other perpendicular to this plane, bisecting the water molecule. The plane containing the water molecule is a real plane of symmetry, even though the atoms do not move upon reflection. The halves of each atom change sides of the plane. Each atom may possess three vectors; the one perpendicular to the plane changes sign. Figure 5.5(b) illustrates this concept.

(3). *Inversion at· a center of symmetry.* The.symbol for a center of inversion is i. If one draws a straight line from an atom through the center of a molecule and encounters an identical atom, and this is true for all the atoms in the molecule, the molecule is said to possess a center of inversion (see Fig. 5.6). In the staggered form of ethane, hydrogen atoms A and B, C and D, and E and F may be interchanged with each other by inversion to give an identical, superimposable configuration. It is important to realize that the inversion process requires that every point (x, y, z) is converted to the position $(-x, -y, -z)$.

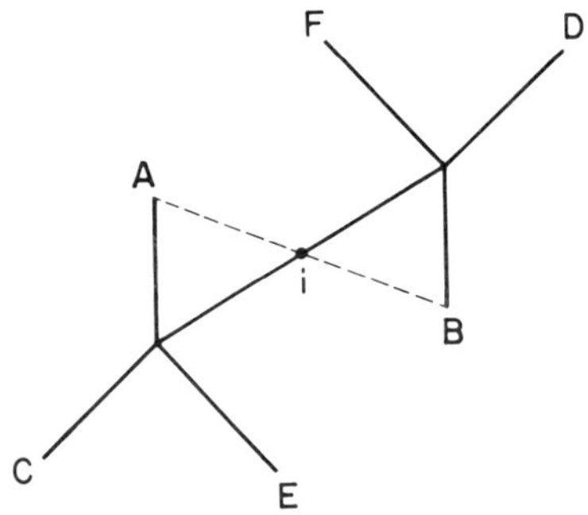

Figure 5.6. The center of inversion i in the staggered form of ethane. On inversion, A goes to B, C goes to D, and E goes to F.

(4). *Rotation about an axis, followed by reflection through a plane perpendicular to this axis.* The symbol for this combined operation is S_n, where $n = 360°/r$ and r is the angle of rotation necessary to obtain a superimposable configuration. Figure 5.7 contains the drawing of a

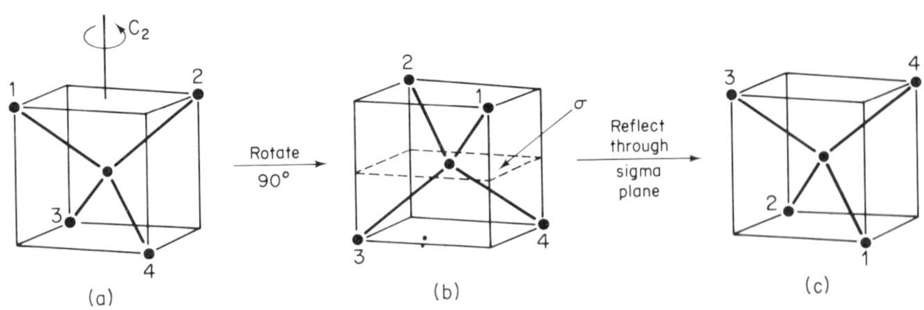

Figure 5.7. The S_4, improper rotation axis (rotation reflection) in a tetrahedral molecule.

tetrahedral molecule—say, methane—relative to a cube. From the figure, we see that if we rotate the molecule 90°, it is not a symmetry operation, but if we then reflect through the σ plane perpendicular to the C_2 axis, we do obtain a superimposable configuration. The reader should note that neither operation, rotation by 90° (*A* to *B*) or reflection (*B* to *C*), is a symmetry operation in itself, but combined they are. The reader should prove to himself that these operations commute, i.e., you could reflect first and then rotate to obtain a superimposable orientation. The axis around which we rotated the molecule is referred to as an S_4 axis because the rotation is 90°. Does the staggered form of ethane have an S_n axis? Where does this axis lie and what is the value of n?

In the illustration of the S_4 axis using a tetrahedral molecule, we pointed out that the rotation and reflection were not symmetry operations in themselves, but this is not always the case; they may be real symmetry operations. This criterion is not essential for the rotation-reflection axis. These rotation-reflection axes are also called *improper rotational axes* to distinguish them from *proper rotational axes*, which are simple rotational axes, C_n.

(5). *Identity operation.* The symbol for the identity operation is commonly E, and on rare occasions I is used. The identity operation leaves the molecule unchanged. It causes the molecule to go into itself (Fig. 5.8).

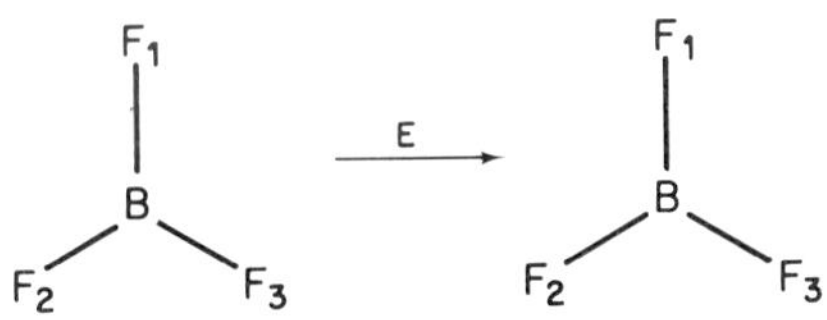

Figure 5.8. The identity operation E on BF_3.

This operation may seem trivial, but it is a result of the requirements of group theory. One of the fundamental postulates of group theory states that at least one member of the group must commute with all the other members of the group and leave them unchanged. In symmetry considerations, this member is the identity operation. Successive operations may

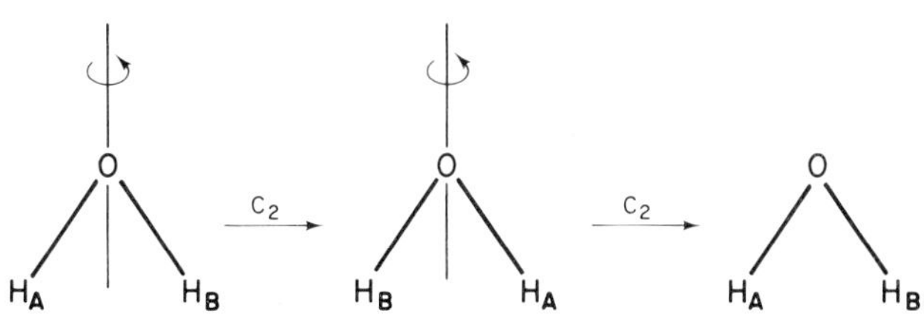

Figure 5.9. Illustration of how two successive symmetry operations may be equivalent to the identity operation ($C_2 \cdot C_2 = E$).

result in the identity operation. If we rotate the water molecule twice around the C_2 axis (Fig. 5.9), symbolized as C_2^2, the result is an identical configuration. Therefore, we may say that $C_2^2 = E$.

Recognition of symmetry elements in a molecule or a proposed structure of a molecule is not always an easy assignment. Look at a rather simple molecule, such as BF_3, and account for the symmetry elements present (Fig. 5.10). The molecule is planar; therefore, it contains a horizontal plane

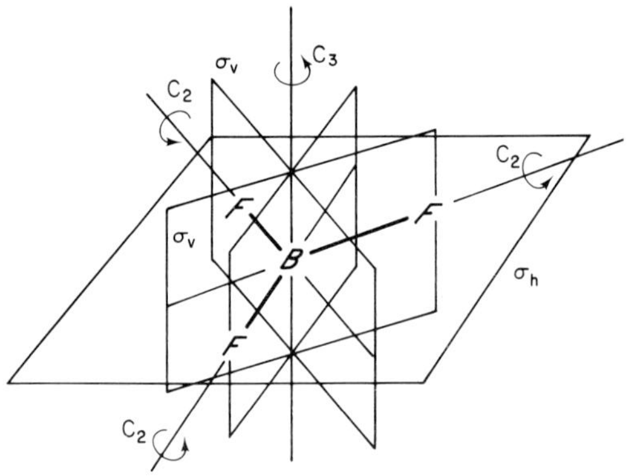

Figure 5.10. The symmetry elements in the BF_3 molecule (D_{3h}).

of symmetry. Perpendicular to this plane is the reference axis, a C_3 axis. There are three vertical planes of symmetry, each containing one of the B—F bonds and the C_3 axis. Then there are three C_2 axes perpendicular to the C_3 axis coincident with each of the B—F bonds. This assemblage of symmetry elements classifies the BF_3 molecule in the D_{3h} point group. The symbol D_{3h} represents a point group. The point group label immediately tells the initiated person the number of symmetry elements in a given molecule. Point groups are tags that are placed on molecules to indicate their structure. This tag may be determined experimentally by X-ray analysis, NMR, microwave spectroscopy, and infrared spectroscopy. We may divide the point groups into three types: (1) C; (2) D; and (3) special groups. Each of these types will be examined.

The C group generally indicates that there is a rotational axis in the molecule. The following are the exceptions. In a case where there is no symmetry element other than E, the molecule is given the label C_1, i.e., FHSiClBr. There is a case where there is only a plane of symmetry in the molecule. The symbol normally used for this case is C_s, for example, HSD. Finally, there is the possibility that a molecule may possess only an inversion center, i; the point group representing this case is C_i. This case is rare.

The groups that now follow really possess a rotational axis, C_n. The molecule hydrogen peroxide exemplifies this classification. In Fig. 5.11, the H_2O_2 molecule is pictured. The O—O bond and the C_2 axis lie in a plane.

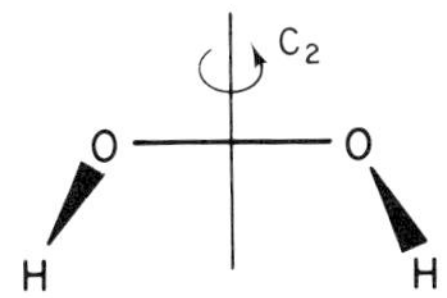

Figure 5.11. The hydrogen peroxide molecule H_2O_2 in the staggered form. In this configuration, the molecule belongs to the C_2 group.

One hydrogen atom is above and the other is below the plane. The configuration of H_2O_2 has only one symmetry element, the C_2 axis. The point group of H_2O_2 in the configuration is C_2.

When a molecule possesses not only a proper axis, C_n, but also a set of n vertical planes of symmetry, it belongs to the point group C_{nv}. The water molecule contains a C_2 axis and two vertical planes; its point group is C_{2v}. A vertical plane is a plane that contains the C_n axis. Convince yourself that ammonia belongs to the C_{3v} point group.

If a molecule possesses a proper axis, C_n, and a plane of symmetry perpendicular to the C_n axis, it belongs to the C_{nh} point group. The molecule *trans*-ClHC=CHCl is classified as being in the C_{2h} point group, which does not preclude the possibility of other symmetry elements being present.

The D classifications are next. A molecule belonging to the D_n point group has a C_n axis and nC_2 axes perpendicular to the C_n axis. Molecules possessing this point group are rare. The (Co en$_3$)$^{3+}$ ion possesses this symmetry. The *en* symbolizes ethylenediamine; it is a bidentate ligand spanning two adjacent sites in an octahedral structure. The student should realize that an octahedron is a trigonal antiprism in which the bidentate ligand will span vertices of the upper triangle (solid line) and the lower

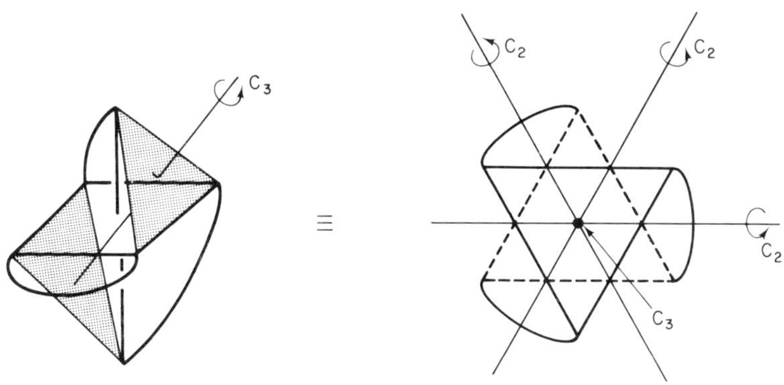

Figure 5.12. The rotational axes in $Co(en)_3^{3+}$. This ion belongs to the D_3 group.

triangle (dashed line), which is illustrated in Fig. 5.12. The C_3 axis is perpendicular to the plane of the paper and has three C_2 axes perpendicular to the C_3 axis as drawn. The symmetry point group of this ion is D_3.

The D_{nh} point group is exemplified by a molecule that has a C_n axis, nC_2 axes perpendicular to the C_n axis, and a horizontal plane of symmetry (i.e., a plane of symmetry perpendicular to the C_n axis). Benzene, C_6H_6, belongs to this classification. Benzene has a C_6 axis perpendicular to the plane of the molecule. The plane of the molecule is a horizontal plane of symmetry, and there are six C_2 axes in this plane perpendicular to the C_6 axis. Hence, its point group classification is D_{6h}. There are also vertical planes of symmetry in this molecule, but remember that the horizontal plane is an element essential to this point group and it receives the preferential recognition.

When, in addition to C_n and the set of nC_2 axes perpendicular to the C_n axis, there exist vertical planes that bisect the angles made by the C_2 axes, the classification of the point group is D_{nd}. The d is representative of dihedral planes, which are represented by σ_d. Ferrocene is $Fe(C_5H_5)_2$, which consists of an Fe^{2+} ion sandwiched between two cyclopentadienyl anions that have a staggered configuration with respect to one another. (This is purely a qualitative description of the bonding in ferrocene; it is much more complicated.) This is illustrated in Fig. 5.13(a). The C_5 axis is

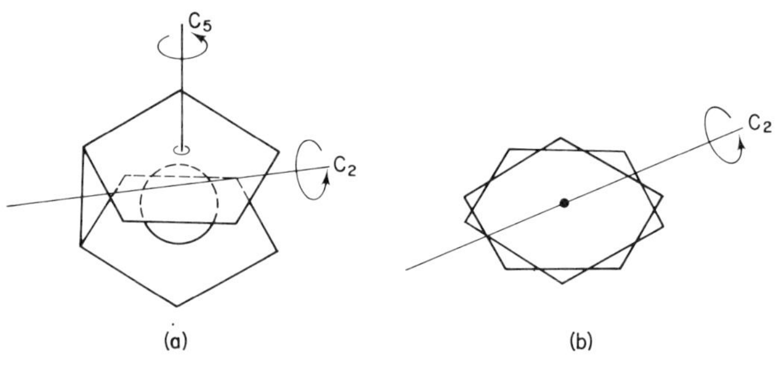

Figure 5.13. Two views of the rotational axes in ferrocene, $Fe(C_5H_5)_2$. This molecule belongs to the D_{5d} point group.

perpendicular to the plane of the cyclopentadienyl ions. There are five C_2 axes perpendicular to the C_5 axis. The C_2 axes pass through the Fe^{2+} ion and bisect the line joining a carbon atom on the upper ring with the nearest carbon atom on the lower ring [Fig. 5.13(b)]. The planes of symmetry are perpendicular to the page [Fig. 5.13(b)], bisecting the angles formed by the C_2 axes. This molecule belongs to the D_{5d} point group. The reader should compare this with ruthenocene, $Ru(C_5H_5)_2$, which is a pentagonal prism. What is its symmetry point group?

Finally, there are the special groups. All linear molecules fall into two classes: $C_{\infty v}$ or $D_{\infty h}$. The compounds CO, HCN, and ClC≡CH possess molecules that have a C_∞ axis coincident with the internuclear axis and also an infinite number of vertical planes. These molecules belong to the $C_{\infty v}$ point group. On the other hand, there is a group of more symmetrical molecules, such as Cl_2, CO_2, H_2 and HC≡CH, which have a C_∞ axis coincident with the internuclear axis and ∞C_2 axes perpendicular to the C_∞ axis. There is a plane of symmetry also perpendicular to the C_∞ axis. These symmetrical linear molecules belong to the $D_{\infty h}$ group.

The final special groups classify molecules that are highly symmetrical. The tetrahedron belongs to the T_d point group. The cube and octahedron belong to the O_h point group. The regular pentagonal dodecahedron and the icosahedron belong to the I_h point group. Group T_d is illustrated by SiF_4, O_h is illustrated by SF_6, and I_h is illustrated by the $B_{12}H_{12}^{2-}$ ion. Table 5.3 contains the essential symmetry elements of each point group.

TABLE 5.3 ***Essential Elements of the Point Groups***

	Point Group	*Essential Symmetry Elements*
C Groups	C_1	E
	C_s	Plane of symmetry
	C_i	Center of inversion
	C_n	Axis of rotation of order n
	C_{nv}	Axis of rotation of order n; n vertical planes of symmetry
	C_{nh}	Axis of rotation of order n; plane of symmetry perpendicular to C_n axis
	$C_{\infty v}$	Applicable to unsymmetrical linear molecules; C_∞ axis coincident with the internuclear axis; vertical planes
D Groups	D_n	C_n axis; nC_2 axes perpendicular to the C_n axis
	D_{nh}	C_n axis; nC_2 axes perpendicular to the C_n axis; a plane of symmetry perpendicular to the C_n axis
	D_{nd}	C_n axis; nC_2 axes perpendicular to the C_n axis; dihedral planes of symmetry
	$D_{\infty h}$	Applicable to symmetrical molecules; C_∞ axis; $_\infty C_2$ axes perpendicular to the C_∞ axis; plane of symmetry perpendicular to the C_∞ axis
Special Groups	T_d	A set of four C_3 axes; a set of three S_4 axes; a set of six dihedral planes of symmetry
	O_h	A set of three C_4 axes; a set of four C_3 axes; a set of six planes each containing one C_4 axis and bisecting two of the angles formed by the other C_4 axes
	I_h	Contains all the elements of O_h; six C_5 axes

The symbols used in this presentation are called the *Schonflies symbols*, after their originator. Crystallographers use different symbols. These are presented in Chapter 6.

Finally, the student will quickly realize that assigning a molecule to a point group with any confidence comes only with practice and three-dimensional thinking. He should become proficient in this area, because the present system of describing the structure of molecular species is done by listing the point group.

STEREOCHEMISTRY: IDEALIZED GEOMETRIES AND STRUCTURAL RIGIDITY

Stereochemistry deals with the structures of chemical moieties. It is necessary to include such topics as rigidity in this discussion. Not all molecules are rigid and many of them have a dynamic character. For the most part, this aspect of isolated molecules is usually ignored.

If ammonia undergoes a chemical reaction so that the species $NRR'R''$ is produced, where the R's represent three different ligands, it should be theoretically possible to separate optical isomers because of the theoretical existence of nonsuperimposable mirror image molecules [Fig. 5.14(a)]. Such derivatives have been made, but the optical isomers have not been isolated. This fact has two possible explanations: (1) very rapid exchange of ligands among molecules (intermolecular exchange); or (2) very rapid molecular rearrangement (intramolecular exchange). This is what is meant by a dynamic process in a given molecule.

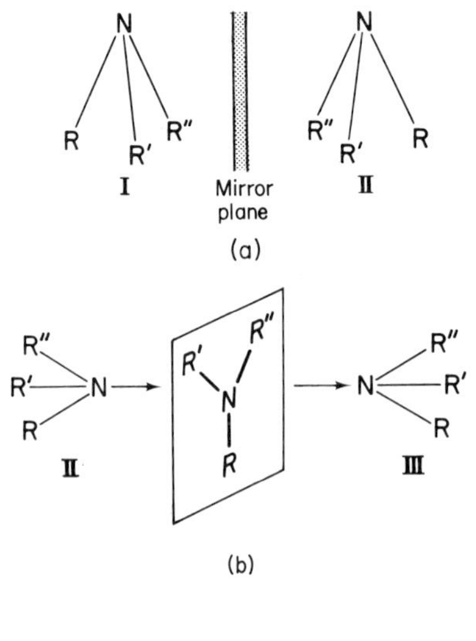

Figure 5.14. (a) Possible mirror images of the species $NRR'R''$. (b) Mechanism of the inversion process in this molecule.

In Fig. 5.14(b), isomer II is pictured undergoing a molecular rearrangement. This rearrangement is an operation similar to that of an umbrella being opened and then turning inside out. Isomer III is now superimposable on isomer I. If this process were diagrammed with potential energy, E, plotted against some molecular structural parameter, Q_i, it would take on the shape of a double humped curve, as is shown in Fig. 5.15. For isomer II

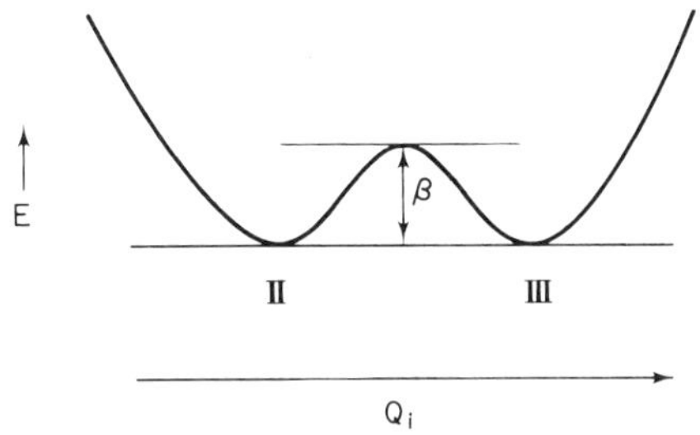

Figure 5.15. Diagram relating potential energy E to some molecular parameter Q_1.

to convert to isomer III, this isomer has to pass a potential energy barrier, β; this term, β, would be the potential energy barrier to inversion. The *lifetime* of isomers I and II would depend upon the height of this energy barrier. Some ground state lifetimes of nonrigid molecules are given in Table 5.4.

TABLE 5.4 ***Ground State Lifetimes for Non rigid Molecules****

Molecule or Ion	Lifetime (sec)	Origin of Data
NH_3	2.5×10^{-11}	calculated
ND_3	2.7×10^{-10}	estimated
AsH_3	10	estimated
CH_4	10^{15}	estimated
PF_5	10^{-5}	estimated
IF_7	10^{-3} to 10^{-12}	NMR, IR
ReH_9^{2-}	10^{-3} to 10^{-12}	NMR, IR
$(CH_3)_2NPF_4$	10^{-4}	NMR

*This table is reproduced in part from Muetterties, E. L., *Inorg. Chem.*, **4**, 769 (1965).

Any structural determination must take into account the possibility of a dynamic molecular system. This immediately brings two points to mind. First, the physical state of the species being studied is important, because the structure of a given molecule, ion, or aggregate may not be the same in the solid, liquid, solution, or vapor states. In the solid state, packing factors may alter the geometry of a species; in the solution state, solvation may be an important factor in ordering a particular geometric form. The gaseous state would be ideal for obtaining a molecule in an unperturbed form, but this is not always realistic, nor are the instruments capable of operating dependably at elevated temperatures. Second, the probe that we are using must have the correct time scale of response necessary to study a dynamic molecule in a given state. For example, the lifetime of ammonia in the ground state is approximately 10^{-11} sec. If the time scale of the probe we are using is greater than 10^{-11} sec—say, 10^{-6} sec—then the structural picture we obtain with this probe is the average picture of the states ammonia may assume. If the time scale of the probe is less than 10^{-11} sec—say 10^{-13} sec—then one has a measure of confidence that the picture obtained by the probe is a reliable picture in a given state. Table 5.5 contains the approximate time scale for different structural probes.

TABLE 5.5 ***Time Scale for Structural Techniques****

Technique	Approximate Time Scale (sec)
Electron diffraction	10^{-20}
X-ray diffraction	10^{-18}
Ultraviolet	10^{-15}
Visible	10^{-14}
Infrared-Raman	10^{-13}
Electron spin resonance	$10^{-4} - 10^{-8}$
Nuclear magnetic resonance	$10^{-1} - 10^{-9}$
Quadruple resonance	$10^{-1} - 10^{-8}$
Mössbauer (iron)	10^{-7}
Molecular beam	10^{-6}
Experimental separation of isomers	10^{2}

*This table is reproduced in part from Muetterties, E. L., *Inorg. Chem.*, **4**, 769 (1965).

In the remainder of this section, we shall be speaking of different types of compounds. Therefore, it is necessary to make the following distinctions. First, we shall be speaking of monomeric coordination compounds. A coordination compound is one in which the ligands (L) are attached to one central atom (M) by two-electron bonds. If the central metal atom possesses one or more lone pairs of electrons, these will also be considered as ligands because they will noticeably affect the molecular geometry. In general, these structures are relatively nonrigid because of the absence of bonding interactions between adjacent ligands. Second, consider *molecular polyhedra* that possess no central atom. The boron hydrides exemplify this type of polyhedron; even $Pb_4(OH)_4^{4+}$ is considered to be a polyhedron (Fig. 5.16).

The third type of structure to be considered is the *metal cluster* system, in which there is definite metal-metal interaction. In this relatively rigid system, the metal atoms (ions) describe a polyhedron, and the ligands also describe a polyhedron. A classic example is $Mo_6Cl_8^{4+}$ (Fig. 5.17).

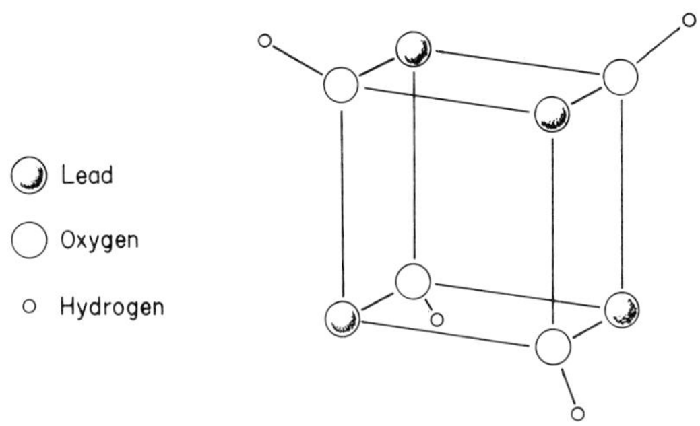

Figure 5.16. Polyhedral $Pb(OH)_4^{4+}$.

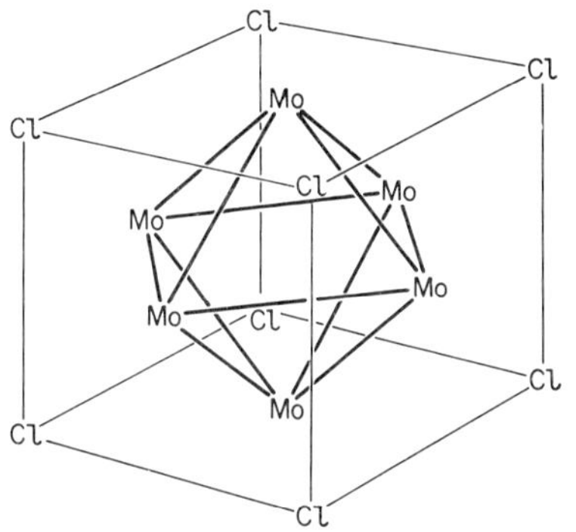

Figure 5.17. The $Mo_6Cl_8^{4+}$ unit in the $Mo_6Cl_{14}^{2-}$ anion. Each molybdenum atom has a terminal chlorine within bonding distance.

To remove the idea that these structures are absolutely rigid, we must state that they are not. A change of structure (sometimes referred to as scrambling) may be achieved by heating. For example, it has been shown that the 1, 6 isomer of B_8H_8 $(CCH_3)_2$ is converted to the 1, 10 isomer by heating for a 22-hour period at 350°.

In summation, the least rigid of the systems are the coordination compounds. The polyhedra and metal clusters are more rigid. It has been observed that polyhedra and metal clusters do have bonding interactions that the coordination compounds do not. The ligands define the vertices of the structure in a coordination compound, but there is no interaction between ligands. The polyhedra and metal clusters differ because there are multicentered bonds, and the atoms themselves, bound to each other, describe the structure. The unique feature of the polyhedra and metal clusters is that their structures are triangulated. That is, the structures described by these species may be constructed from triangles. Possibly their comparative rigidity may be due to the Euclidian observation that the triangle is the only polygon whose shape cannot be changed without changing the length of one of its sides.

It now remains to point out the geometrical structures that coordination compounds, polyhedra, and metal clusters may assume, and to give their symmetry. The classification may not always be strict. For example, a coordination compound of the type ML_4 will be considered under a four-atom system because the four ligands determine the geometry.

Four-Atom Systems

Most four-atom systems (i.e., ML_4) belong to the T_d point group. Square planar species exist, especially among transition metal d^8 systems (Ni^{2+},

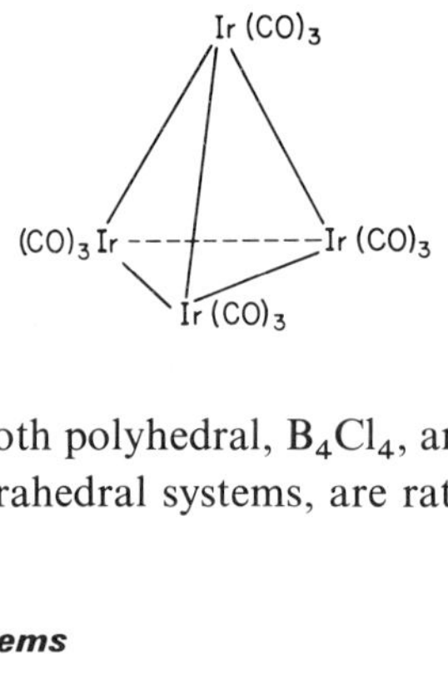

Figure 5.18. The tetrahedral metal cluster $Ir_4(CO)_{12}$.

Pd^{2+}, Pt^{2+}). Both polyhedral, B_4Cl_4, and metal cluster systems, $Ir_4(CO)_{12}$, which form tetrahedral systems, are rather rigid.

Five-Atom Systems

The idealized geometries, illustrated in Fig. 5.19, for the five-atom system are the trigonal bipyramid (D_{3h}) and the tetragonal pyramid (C_{4v}). These geometries are illustrated in Fig. 5.19(a). If one could look down the C_2 axis of I and the C_4 axis of II, these two geometries would appear to be astonishingly similar. As a matter of fact, the compound $(CH_3)_2NPF_4$ undergoes rapid interconversion at room temperature. The energy barrier to inversion is quite low, and from Fig. 5.19(b) one can see that this interchange between the two symmetries may be achieved by a simple bending of the molecule. Polyhedra such as $B_3C_2H_5$ and metal clusters such as $Pt_3Sn_2Cl_x$ are rather rigid (Fig. 5.20). The $Ni(CN)_5^{3-}$ ion appears in both C_{4v} and D_{3h} symmetries in the same crystal lattice!

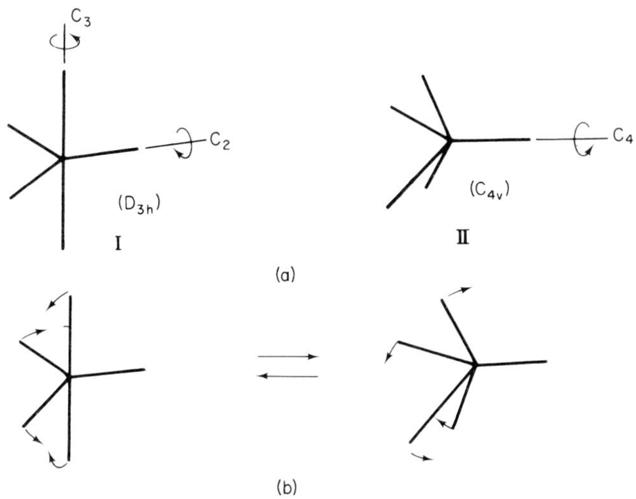

Figure 5.19. Rotational axes of the pentacoordinated systems: trigonal bipyramid (D_{3h}) and square pyramid (C_{4v}) in (a). (b) Bending modes for interconversion of D_{3h} to C_{4v}, commonly referred to as the Berry mechanism.

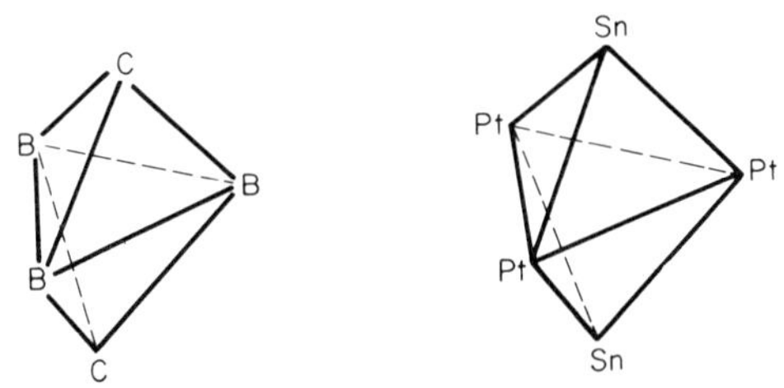

Figure 5.20. Trigonal bipyramidal polyhedra and metal cluster. (a) $B_3C_2H_5$ with the extrapolyhedral hydrogen atoms omitted. (b) $Pt_3Sn_2Cl_x$ with the extrapolyhedral chlorine atoms omitted.

Six-Atom Systems

The coordination compounds of the type ML_6 that are commonly formed by the transition metals are invariably octahedral in structure (O_h). In some cases, certain chelating ligands, such as

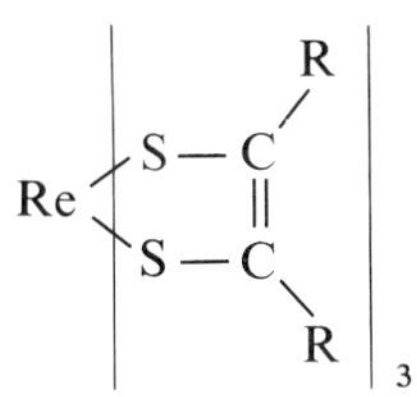

give structures that are based on a trigonal prism (D_{3h}). The apparent cause for this change in structure is that there is significant bonding action between the adjacent sulfur atoms. Actually, there is not a great deal of difference between the two geometries. If one remembers that an octahedron looked at along one of its faces is a trigonal antiprism, a simple twist of this face will result in a trigonal prism (see Fig. 5.21). This twisting

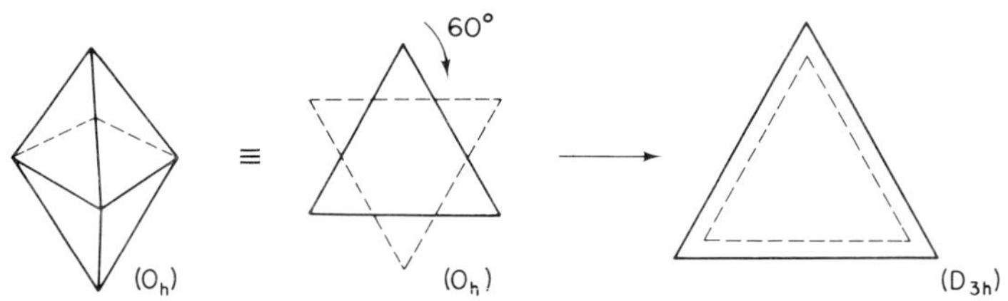

Figure 5.21. The "Bailar twist" in an octahedral configuration.

illustrated in Fig. 5.21 is called the *Bailar twist.* Both the polyhedra ($B_6H_6^{2-}$) and metal clusters ($Mo_6Cl_8^{4+}$) (Fig. 5.17) are rather rigid.

Seven-Atom Systems

There are three idealized symmetries for heptacoordinated structures: the pentagonal bipyramid (D_{5h}); the capped trigonal prism (C_{2v}); and the monocapped octahedron. There is also another structure, a square-base triangular-cap polyhedron (C_s) that has been established for metal clusters and polymeric oxides. A monocapped octahedron has been reported for the structure of ZrF_7^{3-} ion (C_{3v}), and there has also been a theoretical configuration not observed in real systems, based on a sphere, that belongs to the C_2 point group. These models, with examples, are given in Fig. 5.22. Actually, there is very little energy difference between these configurations. Heptacoordinated coordination compounds are basically nonrigid.

Eight-Atom Systems

There are several idealized geometries for eight-atom systems. The most symmetrical and also the configuration of highest energy is the cube (O_h). Solid crystal lattices of CaF_2, CsCl, and $\alpha - UO_3$ assume this geometry.

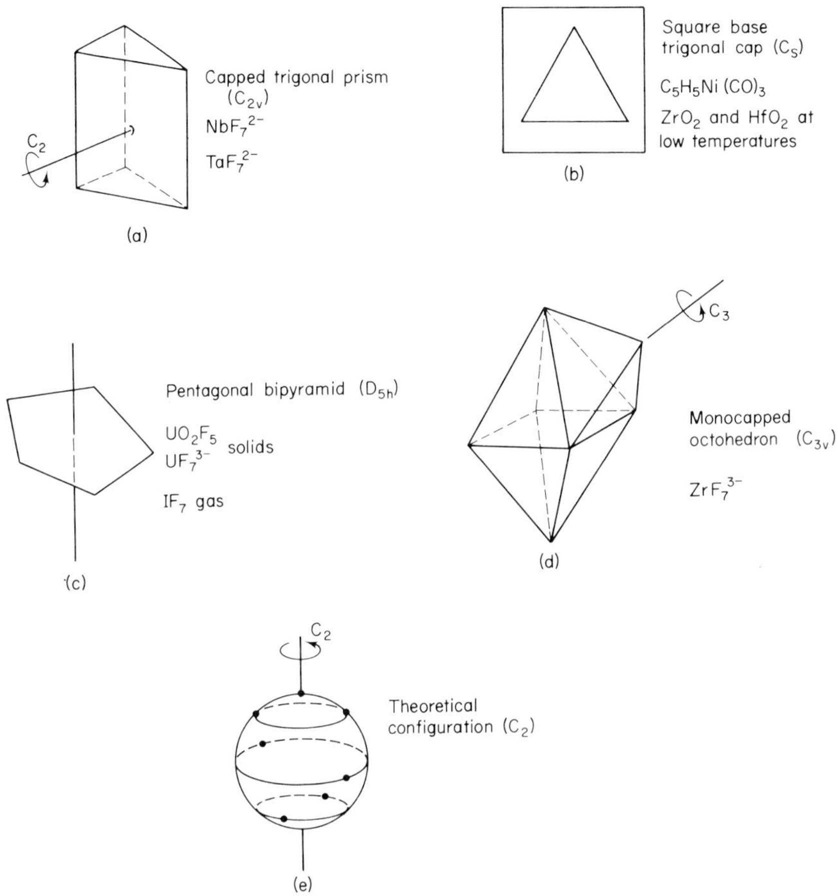

Figure 5.22. Idealized heptacoordinated structures.

If one of the faces of a cube is rotated 45°, a square antiprism results (D_{4d}). A distortion of the cube in this manner reduces ligand-ligand repulsion. The cube may also be distorted (Fig. 5.23) to give a dodecahedron (D_{2d}). This dodecahedron is easily visualized to give its point group by considering it as two perpendicular trapezoids. Another idealized geometry belonging to this group is the hexagonal bipyramid (D_{6h}). The ring of the hexagonal bipyramid may be puckered to give a symmetrically bicapped trigonal antiprism (D_{3d}). Some lanthanide and actinide halides have been observed with lattices that are hybrids of D_{2d} and D_{4d} symmetries by forming a hendecahedron, which is a trigonal prism with two of the square faces

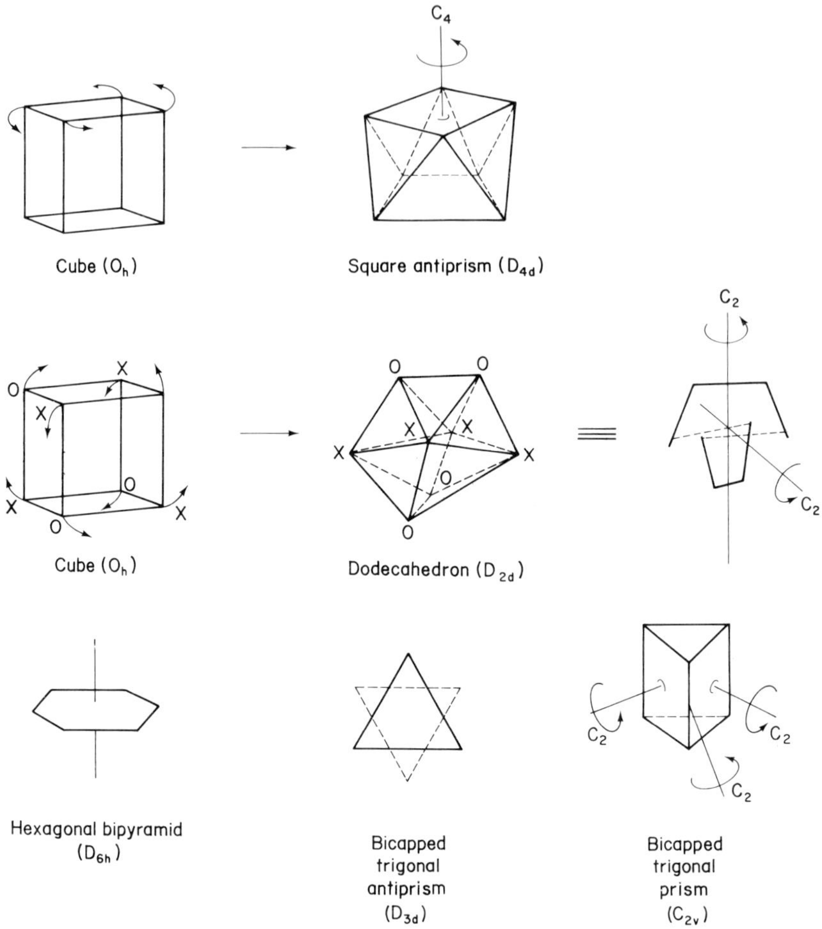

Figure 5.23. Idealized structures for octacoordinated species with the first two rows of illustrations indicating methods of distorting the cube.

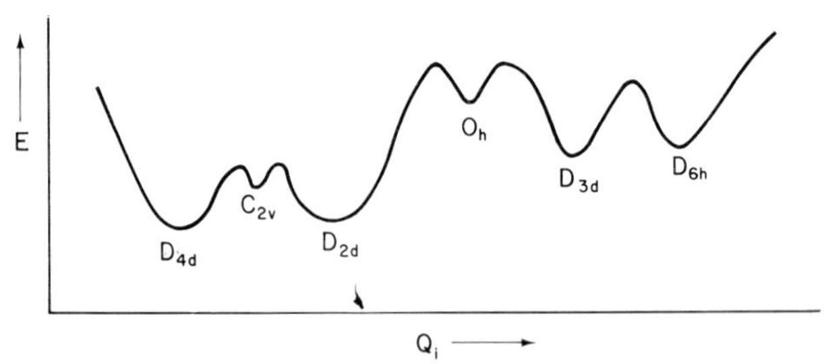

Figure 5.24. Energy vs. structural parameter plot giving qualitative energy relationships among octacoordinated geometries.

capped (C_{2v}). Figure 5.24 illustrates the qualitative energetic relationships among these geometries by showing a plot of energy E vs. some shape parameter, Q_i.

The following polyhedra, $B_8H_8^{2-}$, $B_6C_2H_8$, and B_8Cl_8, all have D_{2d} symmetries and are rather rigid. A very rigid metal cluster is Pt_6Cl_{12}. Here, the platinum atoms form an octahedron and the chlorine atoms describe a cube around this octahedron (Fig. 5.25). Each platinum atom is coplanar with four chlorine atoms, and each is bonded to four other platinum atoms, so that the coordination number of each platinum atom is eight.

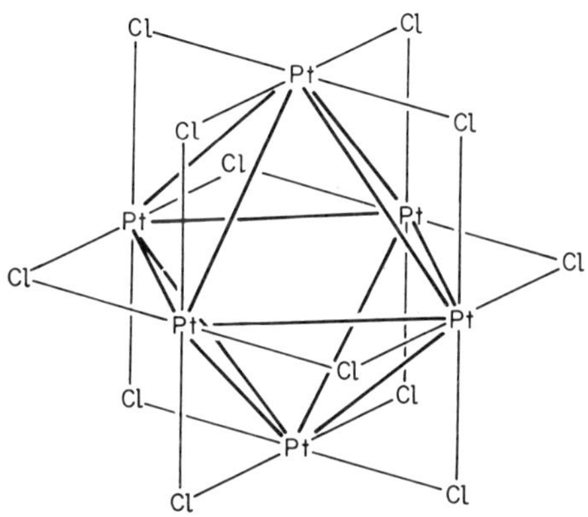

Figure 5.25. Structure of Pt_6Cl_{12}.

Nine-Atom Systems

Nonacoordinated species appear to be fairly common in molecular complexes and in ionic lattices of the larger lanthanides and actinides. One

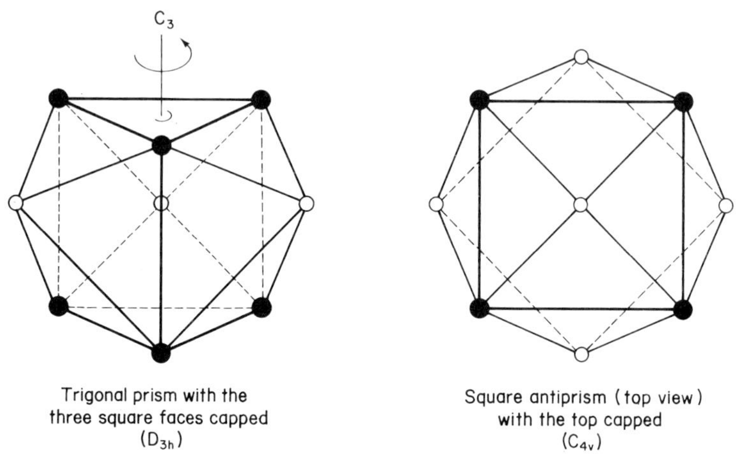

Figure 5.26. Two idealized structures for noncoordinated species.

geometric form has been reported for the nine-coordinated complex, i.e., the symmetrically tricapped trigonal prism (D_{3h}). The species is ReH_9^{2-}. Some metal clusters exhibit the monocapped square antiprism structure (C_{4v}). Structural studies (^{11}B NMR) on $B_9H_9^{2-}$ rule out the C_{4v} geometry and support the D_{3h} configuration.

Systems with More Than Nine Systems Atoms

The idealized geometry for deacoordination is the doubly capped square antiprism, the caps being on the square faces (D_{4d}). The polyhedralboranes, $B_{10}H_{10}^{2-}$, $B_9CH_{10}^-$, and $B_8C_2H_{10}$, appear to conform to this symmetry. It is quite possible to convert the 1, 6 isomer of $B_8C_2H_{10}$ to the 1, 10 isomer by heating it to 300°. No metal clusters have been reported for this group.

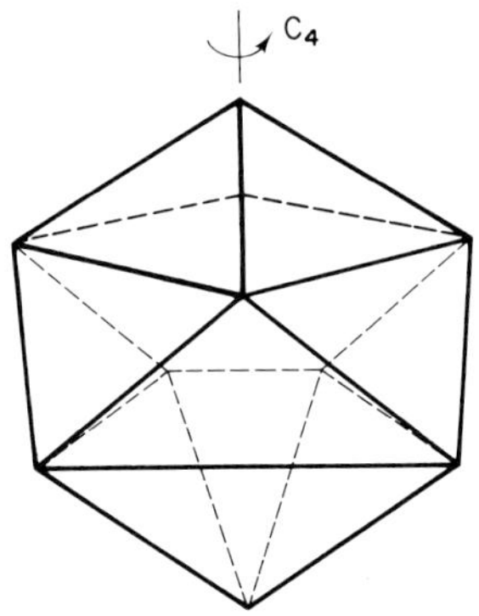

Figure 5.27. Bicapped Square antiprism (D_{4d}).

Undecacoordination is rare in inorganic chemistry. In $Th(NO_3)_4 \cdot 5H_2O$, thorium is seemingly within bonding distance of 11 oxygen atoms—3 from water and 8 from nitrate ions. The polyhedron described is far from being regular. The polyhedron described by the boron atoms in $B_{11}H_{11}^{2-}$ is a monocapped pentagonal antiprism (C_{5v}). This structure is rigid up to 400°. At this temperature, the cesium salt is converted to equimolar mixtures of $B_{10}H_{10}^{2-}$ and $B_{12}H_{12}^{2-}$.

The dodecacoordinated species are relatively common. These species are formed by tightly packing ligands such as SO_4^{2-} and NO_3^-. In $Ce_2Mg_3(NO_3)_{12}{\cdot}24H_2O$, a unit $Ce(NO_3)_6^{3-}$ exists. The Ce ion is surrounded by 12 oxygen atoms from the nitrate ions. In this case, the nitrate ions are bidentate ligands, and they form a distorted octahedron with handles on it. The figure is called a patohedron after the Argentine pato ball and has the special point group T_h. Other idealized geometries for the 12-atom system are the cube octahedron (O_h) and the truncated tetrahedron (T_d) (Fig. 5.28).

The polyhedral borane species, such as $B_{12}H_{12}^{2-}$, have icosahedral (I_h) form, which is actually a bicapped pentagonal antiprism.

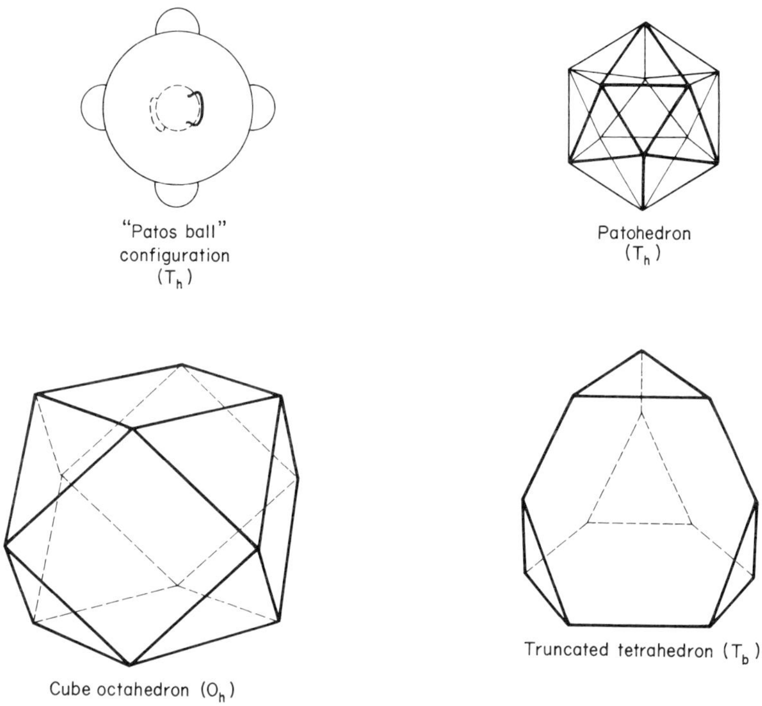

Figure 5.28. Idealized geometries for dodecacoordinated systems. The icosahedron (bicapped pentagonal antiprism) is not shown.

This section, concerning the idealized geometries of multiple atomic systems, has intended to make the reader aware of several points. First, it is very difficult to make a structural study really definitive. That is to say, not all instruments and methods always give unambiguous results. Three-dimensional X-ray analysis and neutron diffraction seem to have the greatest reliability as far as solids are concerned. No probe at present can measure a reliable parameter that will give sound information concerning the structure of materials in the liquid and solution state. Where structural studies are concerned, it is always important to eliminate the possibility that a dynamic process gives a hazy picture of the real structure of a given species. This is the case with PF_5 and some of its derivatives. Don't be surprised that species with the same empirical formula have different structures, because different geometries for given systems of atoms are not that energetically diverse. Finally, some of the higher coordinated species are frequently questioned as to whether such species actually exist as stated. This will depend upon the concept of bonding distance.

Suggested Readings

1. Barrow, G. M., *Introduction to Molecular Spectroscopy*, McGraw-Hill Book Company, New York, 1962.
2. Cotton, F. A., *Chemical Applications of Group Theory*, John Wiley & Sons, Inc., New York, 1963.
3. Cotton, F. A., and G. Wilkinson, *Advanced Inorganic Chemistry*, 2nd ed., John Wiley & Sons, Inc., New York, 1966.
4. Gillespie, P. J., and R. S. Nyholm, *Quart. Rev. London,* **11,** 339 (1957).
5. Jaffe, H. H., and M. Orchin, *Symmetry in Chemistry*, John Wiley & Sons, Inc., New York, 1965.
6. Jaffe, H. H., and M. Orchin, *J. Chem. Educ.,* **47,** 246 (1970); **47,** 373 (1970).
7. Muetterties, E. L., *Inorg. Chem.,* **4,** 769 (1965).
8. Muetterties, E. L., series of lectures given under the auspices of the W. P. Philips Fund, Haverford College, Haverford, Pa., 1968.
9. Muetterties, E. L., and W. H. Knoth, *Polyhedral Borones,* Marcel Dekker, Inc., New York, 1968.
10. Muetterties, E. L., and R. A. Schun, *Quart. Rev. London,* **22,** 245 (1966).
11. Muetterties, E. L., and C. M. Wright, *Quart. Rev. London,* **21,** 109 (1967).
12. Zeldin, M., *J. Chem. Educ.,* **43,** 17 (1966).

Study Questions

5.1. The ^{19}F NMR ($I_F = \frac{1}{2}$) of ClF_3, at $-60°C$, consists of five components. Draw the spectrum and clearly indicate the relative intensities expected for these components; indicate how these data are related to the structure of the molecule.

5.2. Explain why XeF_2 is a linear molecule, whereas OF_2 is bent.

5.3. The dipole moment of NF_3 (0.234 *D*) is much lower than the dipole moment of NH_3 (1.47 *D*), even though the N—F bond is more polar than the N—H bond.

5.4. Using electrostatic concepts of stereochemistry, predict and draw the structure of:

(a) ClO_2^-	(d) IF_5	(g) SF_4	(j) BrO_2
(b) ICl_4^-	(e) $SOCl_2$	(h) H_3O^+	(k) BrF_5
(c) XeF_4	(f) I_3^-	(i) $POCl_3$	

5.5. Give a brief description of structure, including geometry and bonding orbitals, for each of the following: acetone, the hydronium ion in dilute aqueous solution, benzene, ferocene, acetic acid dimer.

5.6. Explain why $TeCl_4$ has an appreciable dipole moment, whereas $SiCl_4$ has no dipole moment.

5.7. Draw the Lewis structure for HNO_3. What type of hybridization would you expect about the nitrogen atom of HNO_3? Explain. The N—O—H angle in this molecule has not been completely evaluated. Would you expect this part of the molecule to be bent or linear? Explain. Would you expect *all* the O—N—O bond angles in this compound to be the same? Explain in detail, commenting on the magnitude of the O—N—O bond angle(s). Would you expect all the N—O bond lengths in HNO_3 to be the same? Explain.

5.8. Predict the structure, show that your answer is not a guess, and give an indication of any distortions of bond angles: SbF_5; $UO_2F_5^{3-}$ ($U = 5f^36d^17s^2$); ICl_2^-; $IO_2F_2^-$; $(CH_3)_6Al_2$.

5.9. (a) Give the stereochemistry of BrF_4^-, BrF_3, SPF_3, $PbCl_6^{2-}$.
(b) In what point group is BrF_4^-?

5.10. Give the detailed stereochemistry of XeO_2F_2 and explain how you arrive at it. To what symmetry point group does it belong?

5.11. Give and explain the detailed stereochemistry of the following: SO_3^{2-}, $TlCl_3$, NO_2^-, ICl_3, SbF_6^-.

5.12. Carbon tetrafluoride belongs to the T_d point group. List *all* the symmetry elements in a T_d point group. Suppose one of the fluorine atoms is replaced by a chlorine atom (Cl—CF_3). To what symmetry point group does $ClCF_3$ belong? How many symmetry operations does this new point group possess? How many symmetry operations were lost?

5.13. The difluorochlorate(I) anion, ClF_2^-, has successfully been prepared in the form of its nitrosyl salt [Criste and Guertin, *Inorg. Chem.*, **4**, 905 (1965)]. Nitrosyl fluoride, NOF, reacts with ClF in a 1:1 molar ratio, forming a white solid at $-78°$, which is completely dissociated in the gas phase at 25°. Discuss the structure of the ClF_2^- anion using the Gillespie-Nyholm VSEPR approach. Compare the structure of the ClF_2^- anion with that of ClF_2^+ cation.

CHAPTER 6

The Inorganic Solid State

Inorganic solids are classified according to a variety of criteria. One can describe a solid as being amorphous or crystalline. The distinction between these two types of solid rests on the degree of regularity of environment existing throughout the solid. The immediate environment of many of the atoms of an element or compound in the solid state can be the same, whether the substance is amphorphous or crystalline. This similarity of immediate environment is often referred to as short-range order. If a

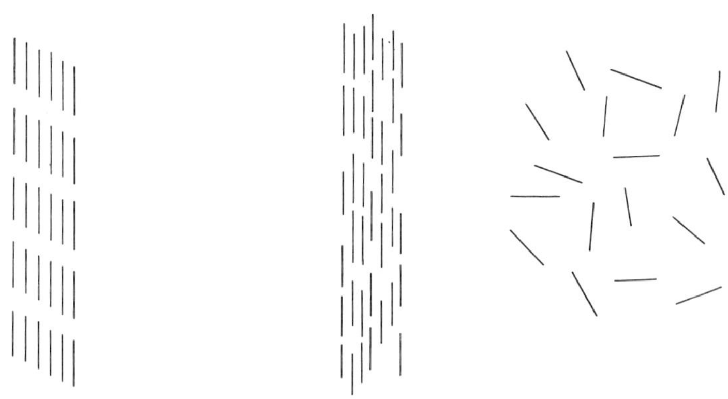

(a) (b) (c)

Figure 6.1. Schematic illustration of three states of matter. (a) A solid with long-range and short-range order; (b) a paracrystalline substance with a partial long-range and short-range order; and (c) an amorphous substance with only short-range order present in the unit parts.

regular atomic arrangement persists over distances very great compared to interatomic distances then the solid is said to exhibit long-range order. When a solid exhibits both short-range and long-range order, it is found to have properties attributed to crystals and is said to be crystalline. If only short-range order is found, then the solid is amorphous. Short-range order is also exhibited in liquids. When certain inorganic liquids, e.g., silicates, are cooled rapidly, the stable solid state does not form but an amorphous solid is produced.

Substances may be further described as belonging to a paracrystalline state. Such substances are referred to as liquid crystals. This state usually arises when a solid containing long rigid molecules begins to melt. This disorder introduced by melting is often only partial, and groups of well-oriented molecules remain, imparting properties of both the liquid and solid state to the material (Fig. 6.1). Characterized liquid crystal systems have been limited to organic substances.

CRYSTAL GEOMETRY

Crystals can be described as solid bodies bounded by naturally formed plane faces. These crystal faces are the outcome of the natural process of crystal growth, even under controlled laboratory conditions. During crystal growth, single crystals or polycrystalline material may be obtained, depending on the conditions of formation. Although many substances are often not looked upon as crystalline because of their complexity, most solids are crystalline. From rocks to sand and soil, from concrete to bones and teeth, from most powdered, inorganic chemical reagents on the laboratory shelf to cleaning powders, careful examination reveals that complex and seemingly amorphous substances are aggregates of microcrystals that have grown together, each crystal possessing its own peculiar faces and properties. We shall follow the procedure of the solid state inorganic chemist, emphasizing the properties of single crystals for purposes of simplicity. Moreover, we stress the morphological characteristics (external appearances) before we investigate the arrangements of ions, atoms, and molecules in the crystal. Keep in mind that an intimate relation of external crystal geometry and internal atomic arrangements does exist, and remember the effect of these structural relationships on the properties of inorganic solids.

The shape of crystals of a particular substance is not a fixed characteristic. In fact, depending on the conditions in which the crystal formed, the relative sizes of a particular face of the crystal as well as the number of faces may vary considerably. To minimize surface energy, the faces of crystals are planar. The characteristic form or combination of forms which crystals of a particular substance can take is called the *crystal habit* of the substance. For example, in a stirred solution, crystals of sodium chloride grow as cubes (Fig. 6.2), but if they grow near the bottom of a beaker, they grow as square tablets. Moreover, if a small amount of urea is added to the solution, the salt crystallizes in octahedra. Thus, the bottom of the beaker prevents growth downward, and the absorption of the organic molecule on preferred faces alters the growth rate of specific faces, forcing a change in the crystal habit of the sodium chloride crystal. This sensitivity of crystal habit is common, and, since the internal arrangement in crystals was not definitely known prior to the discovery of X-ray diffraction, the full meaning of crystals and crystal faces was difficult to understand. For example, Fig. 6.2(d) illustrates stages of growth which can occur in formation of cubic crystals of fluorite, CaF_2. The original octahedral crystal develops truncated faces because of slower growth at its corners and is gradually converted into a cube.

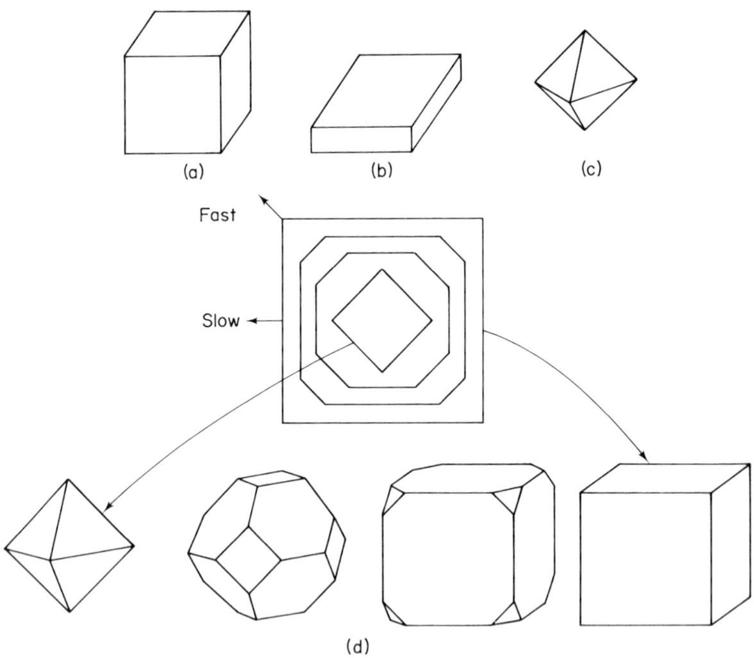

Figure 6.2. Variation of habit of sodium chloride crystal with conditions of growth (a) in stirred solution, (b) at bottom of beaker, (c) from a solution containing urea; (d) growth of cubic fluorite crystal from an octahedron seed.

In 1669, Steno, in studying quartz crystals (SiO_2), made the significant observation that the angles between corresponding faces of crystals have a constant value. This observation, the first law of crystallography, has been found to apply to crystals of all substances. Corresponding faces may not be developed to the same extent in different crystals [Fig. 6.3(a), (c), (d)] of the same substance, but their interfacial angles do not change. The interfacial angles were measured by a simple contact goniometer [Fig. 6.3(e)].

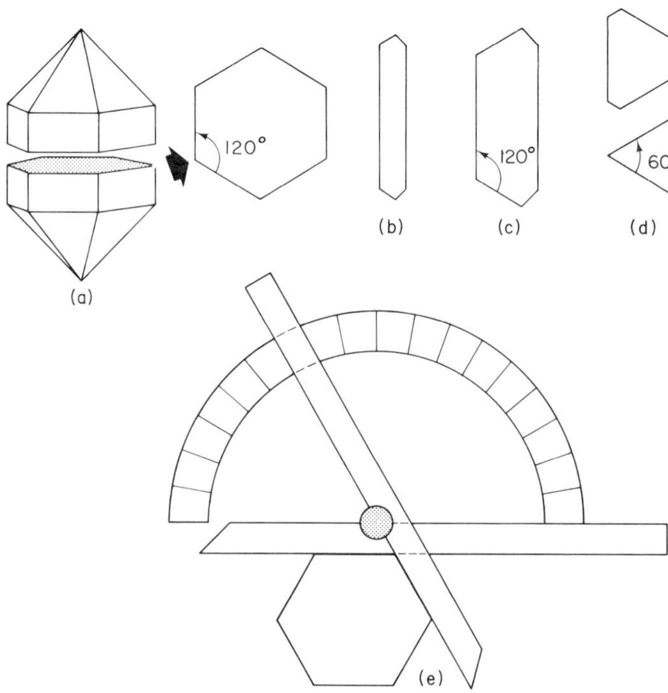

Figure 6.3. Diagram of quartz crystal and cross sections of perfect and imperfect crystals with a goniometer, showing law of constancy of interfacial angles.

Figure 6.3 presents several cross sections from quartz crystals, similar to those studied by Steno. The cross section of a perfect hexagon at the left has a contact goniometer on it [Fig. 6.3(e)], showing the simple method of measuring the interfacial angles still used by crystallographers. The cross sections on the right are distortions [Fig. 6.3(b), (c), (d)], but note, as Steno did, the 120° angle between adjoining faces. In the case of the triangular cross section, this particular crystal developed alternate faces in growing at the expense of the others; therefore, the interfacial angle is 60°.

The external regularity of structure in crystals led to speculation regarding probable internal structural arrangement. In 1784, Hauy, who found that despite numerous deceptive external features, crystals of calcite, $CaCO_3$, split along definite planes no matter how small the crystals. Hauy also found that the same angle of 74° 55′ existed between two faces of the cleavage plane and the crystals take the same shape. After many experiments on other substances, Hauy proposed that continued reduction of the size of the crystals of any substance would lead to a smallest possible unit. He further showed how regular stacking of these fundamental units, or *unit cells*, of identical size and shape would produce the particular faces and slopes found on crystals of the particular substances. It is common to find crystals of the same substance with different faces (Fig. 6.2), and Hauy showed how these different faces and their respective interfacial angles could be produced from the unit cells of the substance.

Solid state inorganic chemists are familiar with the chunky, cubic crystals of the mineral galena, PbS, an important source of lead. Figure 6.4 shows how cubes of galena could form from unit cell building blocks and how the cubes could be modified by octahedral faces, as Hauy proposed. In fact, octahedral faces occur in many crystals of galena.

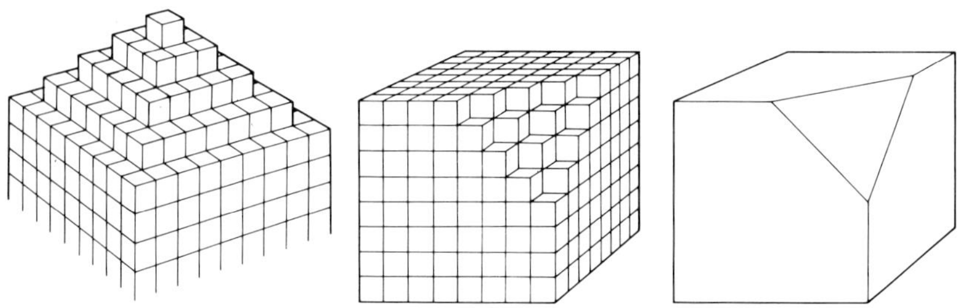

Figure 6.4. (a) Cube built of small cubes; (b) cube modified by octahedral face.

Crystals of inorganic substances can be considered as nothing more than enlarged versions of these fundamental parallelpipeds, complicated by development of numerous faces which these fundamental units permit. The relation of these fundamental structural units to crystallographic unit cells will be discussed. When possible shapes for these fundamental units are classified systematically, it is found that there are only seven different types of building blocks that can be packed together snugly without gaps. These seven crystal systems will be listed and shown in Fig. 6.8.

We can show why there are only seven crystal systems and why the faces have fixed angles. For example, Fig. 6.5 shows the stacking of cubic cells in vertical to horizontal ratios of 1:1, 1:2, and 1:3, respectively. Interfacial angles of 45, 26.5, and 18.5° are measured, respectively, and are

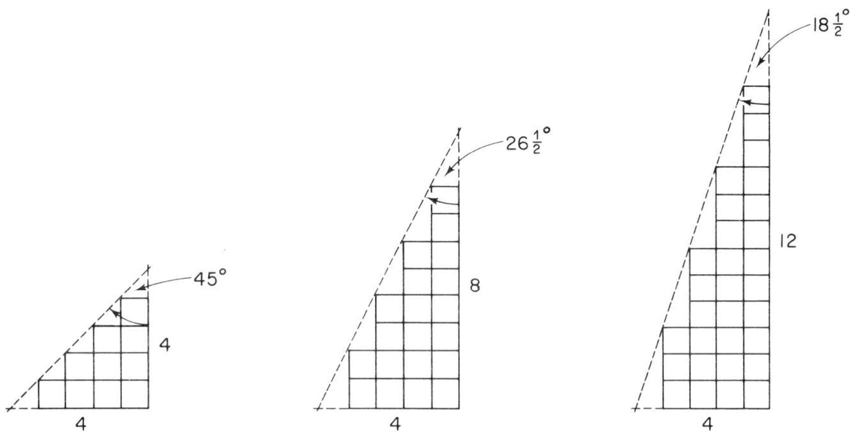

Figure 6.5. The relation of angles between faces to axial ratios in cubic crystals.

anticipated for crystals of the cubic class. In fact, the 1 : 1 ratio, and consequently the 45° interfacial angle, exist in octahedral crystals of sodium chloride crystals grown in 10% urea solutions. The 26.5° interfacial angle (1 : 2 ratio) and the 18.5° interfacial angle (1 : 3 ratio) exist in the dodecahedron and tetrahexahedron faces, respectively, in the cubic system. Thus, by measuring interfacial angles, crystallographers are actually predicting the ratio of the lengths of the faces in fundamental units. Hauy and those after him attributed the smoothness of the crystal face to the small size of the unit cell. With growing interest in atomic theory, attention was transferred to packing of spheres as a model of crystal structure. Thus, if each fundamental unit of an infinite extension of cubic building blocks is replaced by a representative point, such as its center of gravity, and adjacent points are joined, a new unit is outlined, as illustrated in Fig. 6.6. Note that the points of this structural unit do not necessarily represent ions or atoms. Each unit contains a complete representation of the structure of the crystal.

Repetition of these units by parallel translations produces an arrangement of points in three dimensions such that a straight line drawn through any two adjacent points of the network and continued will intersect at equal

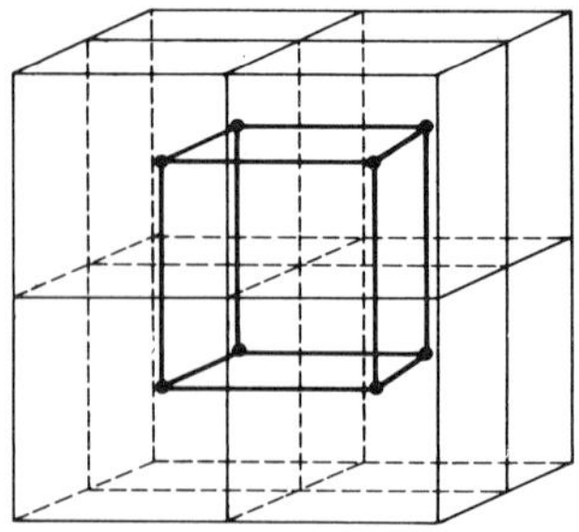

Figure 6.6. Development of a new structural unit from fundamental units of Haüy.

intervals the same succession of points. This arrangement, in which all points have identical environments, is called a *space lattice*. The structural unit is called a *unit cell*. Figure 6.7 illustrates a space lattice of points with some possible unit cells.

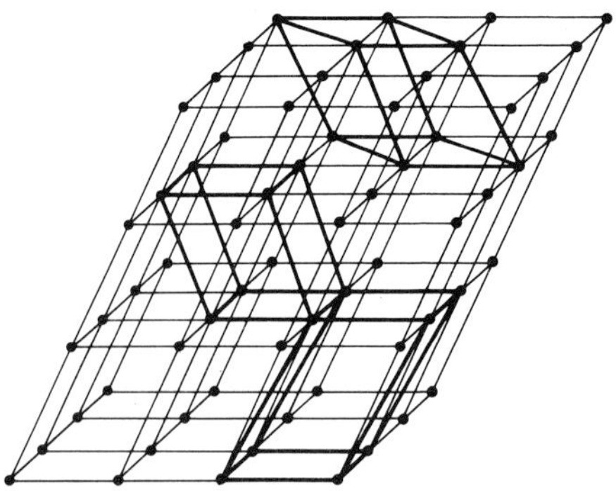

Figure 6.7. A space lattice of points with some possible unit cells heavily outlined.

In 1848, Bravais showed that there can only be 14 independent lattices distributed among the seven crystal systems, in which all points have identical environments. These Bravais space lattices are illustrated in Fig. 6.8. Notice that in addition to the simple cube, called a unit cell in a primitive (simple cube) cubic space lattice and represented by the letter *P*, there can also be nonprimitive space lattices, represented by the letters *I*, *C*, and *F*. In the *P* space lattices, every point has six nearest neighbors and each point is shared among eight unit cells, making one equivalent point per unit cell. The *I* lattices have a body-centered unit, every point has eight nearest neighbors, and there are two equivalent points per unit cell. The *C* lattices have a so-called side-centered unit and can also be referred to as *A* or *B* lattices; every point has four nearest neighbors and there are two equivalent points per unit cell. The *F* lattices have face-centered units; every point has 12 nearest neighbors and there are four equivalent points per unit cell.

In contrast to the crystal systems, the type of Bravais lattice cannot be recognized from the external form of the crystal. Moreover, since lattice points represent a geometrical repeating unit, they are not necessarily occupied by ions, atoms, or molecules.

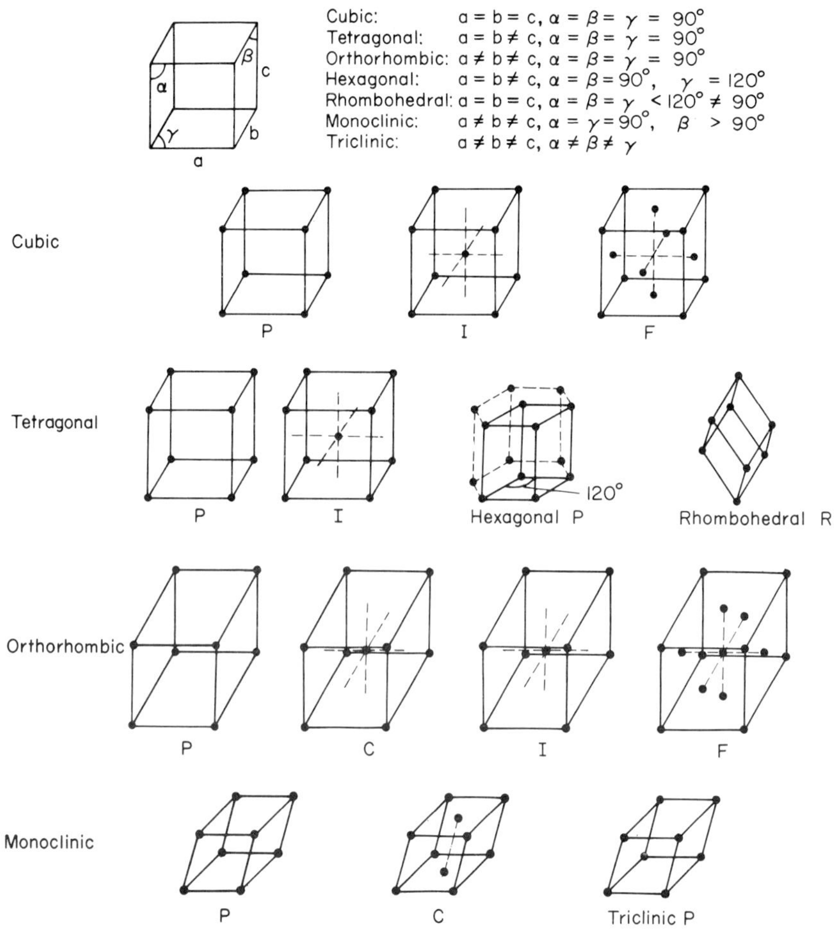

Figure 6.8. The fourteen Bravais lattices distributed among the seven crystal systems. The primitive lattice P has one equivalent point per unit cell at (000); the body-centered lattice I has two equivalent points per unit cell at (000, $\frac{1}{2}\frac{1}{2}\frac{1}{2}$); the side-centered lattice C has two equivalent points per unit cell at (000, $\frac{1}{2}\frac{1}{2}0$); and the face-centered lattice has four equivalent points per unit cell at (000; $0\frac{1}{2}\frac{1}{2}$; $\frac{1}{2}0\frac{1}{2}$; $\frac{1}{2}\frac{1}{2}0$). The numerals in parentheses are Miller indices. See the text for a discussion of their meaning.

DESIGNATION OF PLANES IN CRYSTALLINE SOLIDS

Following his suggestion that crystals are built up by stacking of fundamental units. Hauy proposed that arbitrarily placed external plane faces of crystalline solids do not occur. Only those faces exist which bear a rational relationship to one another. This statement is the *law of rational indices*. It means that all the plane faces on a crystal may be described with reference to three axes by three rational numbers. The standard notation to describe these crystalline faces is *Miller indices*. The intercepts of the face on the three axes are measured by a goniometer. Conventionally, crystallographic axes are chosen so that they are parallel to the principal edges of an ideal crystal of the system to which the particular crystal belongs. As an example, consider an orthorhombic crystal, such as that illustrated in Fig. 6.9. The question is, what are the Miller indices of faces, *a*, *b*, *c*, *d*,

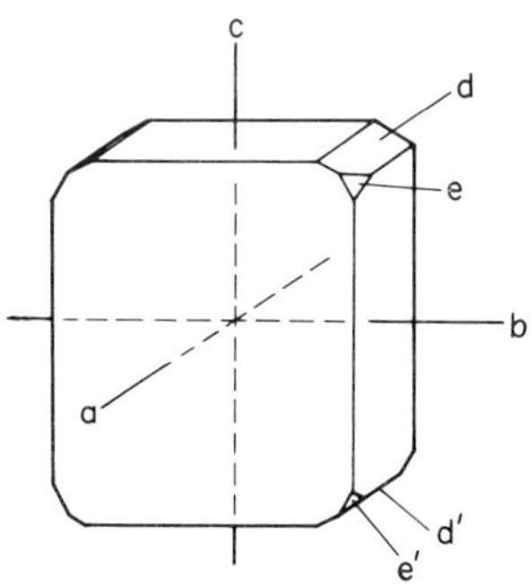

Figure 6.9. Orthorhombic crystal model for the determination of the Miller indices.

and *e*? In the orthorhombic system, the faces subtend angles of 90° (see Fig. 6.8); hence, the reference axes on *a*, *b*, and *c* are perpendicular to each other. It remains to determine where the following faces intercept the axes. These are listed in Table 6.1.

TABLE 6.1 ***Measured Intercepts of the Orthorhombic System Where Face Intercepts Axes***

	Where Face Intercepts Axes		
Face	*a*	*b*	*c*
a	1.04	∞	∞
b	∞	1.12	∞
c	∞	∞	0.97
d	∞	3.36	1.94
e	2.08	2.24	3.88

The actual units of length of the intercepts are unimportant, but only their relative values have importance. According to the Miller scheme, a face can be expressed in terms of three quotients

$$\frac{a \text{ parameter}}{a \text{ intercept}}, \quad \frac{b \text{ parameter}}{b \text{ intercept}}, \quad \frac{c \text{ parameter}}{c \text{ intercept}} \qquad [6.1]$$

where the parameter value is the relative length on that axis of some arbitrarily chosen unit plane (parametral plane), which cuts all three axes. Choosing face e as our parametral face, we find that the relative lengths of the three intercepts on the a, b, and c faces as plane e is extended in space are 2.08, 2.24, and 3.88, respectively. Usually the b parameter is simplified to equal unity; then the intercepts of the parametral plane become 0.92, 1, and 1.73, respectively, where a/b and c/b are called the axial ratios. One can then adjust the relative values of all the intercepts in terms of the same ratio as shown in Table 6.2.

TABLE 6.2 ***Table of Reduced Intercepts***

Face	Reduced Intercepts a	b	c
a	0.46	∞	∞
b	∞	0.50	∞
c	∞	∞	0.43
d	∞	1.50	0.87
e	0.92	1.00	1.73

Miller indices may now be obtained for each face:

Face e: $$\frac{0.92}{0.92} : \frac{1.00}{1.00} : \frac{1.73}{1.73} = 1:1:1 \quad \text{or} \quad (111)$$

Face a: $$\frac{0.92}{0.46} : \frac{1.00}{\infty} : \frac{1.73}{\infty} = 2:0:0$$

It is conventional to index planes with the smallest whole numbers possible. Therefore, the Miller indices of a face whose ratios come out 2:0:0 are written (100), since a (200) plane is parallel to a (100) plane.

Face b: $$\frac{0.92}{\infty} : \frac{1.00}{0.50} : \frac{1.73}{\infty} = 0:2:0 = (010)$$

Face c: $$\frac{0.92}{\infty} : \frac{1.00}{\infty} : \frac{1.73}{0.43} = 0:0:4 = (001)$$

Face d: $$\frac{0.92}{\infty}:\frac{1.00}{1.50}:\frac{1.73}{0.87}=0:\tfrac{2}{3}:2=(013)$$

Face d': $$\frac{0.93}{\infty}:\frac{1.00}{1.50}:\frac{1.73}{-0.87}=0:\tfrac{2}{3}:-2=(01\bar{3})$$

The bar over a digit of a Miller index indicates that the plane crosses the reference axis at a negative intercept. For any face, f, if a, b, and c are the intercepts of the parametral plane, if x, y, and z are the intercepts on the three axes, and if h, k, and l are the Miller indices, then

$$h:k:l=\frac{a}{x}:\frac{b}{y}:\frac{c}{z} \tag{6.2}$$

In solid state chemistry, the word "habit" denotes external shapes and forms in which a compound crystallizes. The word "form" itself has a restricted sense. A crystal's symmetry usually demands the presence of several identical faces. For example, the (111) plane in the orthorhombic crystal in Fig. 6.9, face e, requires the presence of seven other faces; namely, face e', which is $(11\bar{1})$, also faces $(1\bar{1}1)$ $(\bar{1}\bar{1}1)$ $(\bar{1}11)$ $(1\bar{1}\bar{1})$ $(\bar{1}\bar{1}\bar{1})$ $(\bar{1}1\bar{1})$. The total of eight faces constitutes a form. This form can be abbreviated by the enclosure in braces of any face belonging to the form, such as $\{111\}$.

We can now describe the faces in our orthorhombic crystal in Fig. 6.9 on a space lattice. For the sake of simplicity, we assume the unit cell is P orthorhombic. Figure 6.10 illustrates the axial ratios of the unit cell $a:b:c=0.92:1.00:1.73$ and planes a, b, c, d, and e, respectively.

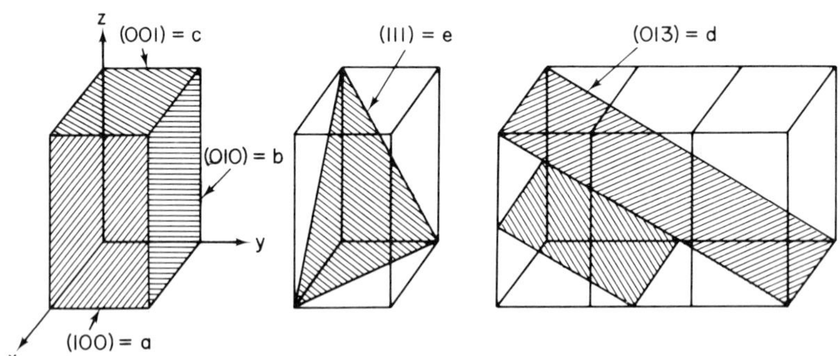

Figure 6.10. Miller indices for faces of the crystal in Fig. 6.9.

At this stage, it might be asked which faces on particular crystals are most likely to form. The mechanisms involved in crystal growth are still not entirely understood and are under intensive study because of the technological importance of crystals. Sometimes the shapes of a crystalline product seemingly bear little resemblance to shapes of other crystals of the same substance, and it has been found, as was illustrated in the formation of NaCl crystals in Fig. 6.2, that slight changes in conditions can produce significant changes in the appearance of crystals of a substance. The usual methods of identification used by solid state chemists, therefore, are diffraction and microscopy techniques, or identification of the symmetry class to which the particular crystal belongs. The following analysis of planes through an orthorhombic lattice similar to that which would represent the crystal in Fig. 6.9, however, will lead to a significant conclusion. These planes are illustrated in Fig. 6.11.

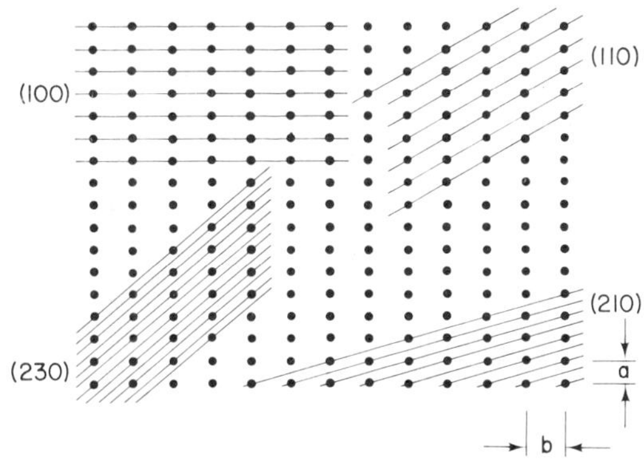

Figure 6.11. Families of planes through an orthorhombic lattice as seen along the c axis of a crystal.

It has been found from accumulated data taken from thousands of crystals that the most prominently developed faces are most thickly populated with lattice points. Considering faces (Fig. 6.11) parallel to the c axis, faces (100) have the greatest density of lattice points. The (110), (210), and (230) planes are less densely packed. Crystals with more complex forms than {100}, for example, have faces having greater distances between neighboring lattice points. However, these plane faces are often small and poorly developed. Thus, for example, in Fig. 6.9 it might be expected that the {013} form might not have prominent faces; in fact, these faces might be missing from many of the crystals.

SYMMETRY IN CRYSTALS

Even a brief study of the distribution of faces on crystals often reveals a sense of geometrical proportion and balance. This external symmetry of crystals has not only pleased the aesthetic sense of people of every age but is important to the chemist in identifying substances and studying their properties in the solid state. From examination of thousands of crystals, it can be demonstrated that the elements necessary to classify all crystals are few in number. These same symmetry elements were later applied to describe molecular systems with equal success; they were discussed in Chapter 5. The symmetry operations, the act that brings an object into a position indistinguishable from its former position, and the symmetry elements, the representation of these acts, are the same for crystal systems as for isolated molecules. The symmetry elements and operations for crystals are the same as those for isolated molecules, except for one small distinction related to the improper rotation (S_n). For crystal symmetry operations, this improper rotation is called the rotation-inversion operation. It is simply a distinction of nomenclature.

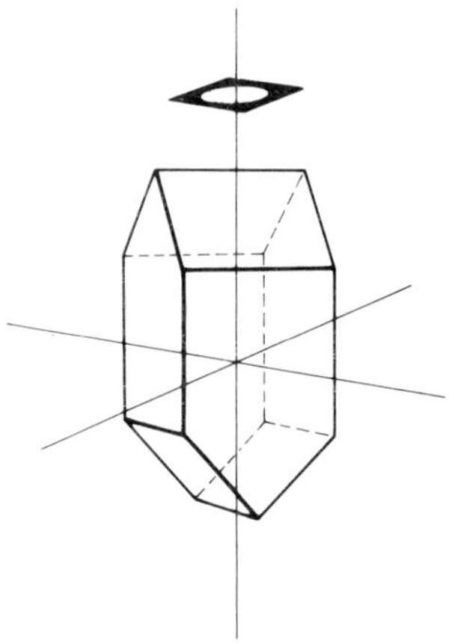

Figure 6.12. Fourfold axis of rotation-reflection and rotation-inversion manifest the same symmetry in the solid. The rotation-inversion is denoted by the symbol ◘.

To classify solids, we can apply symmetry operations to crystals in the same way as to isolated molecules. For example, Fig. 6.13 shows 23 symmetry elements consisting of nine planes, three fourfold axes, four threefold axes, six twofold axes, and one center of symmetry. This defines the cube and therefore its point group. The determination of these symmetry elements in crystals, particularly if they are well formed, may reveal not only the shape of the unit cell but also, to some extent, the symmetries of the atomic arrangement.

Unfortunately, solid state scientists use the Hermann-Mauguin symbols to describe point groups and not the Schoenflies symbols, described in Chapter 5. However, the Hermann-Mauguin symbols, sometimes called

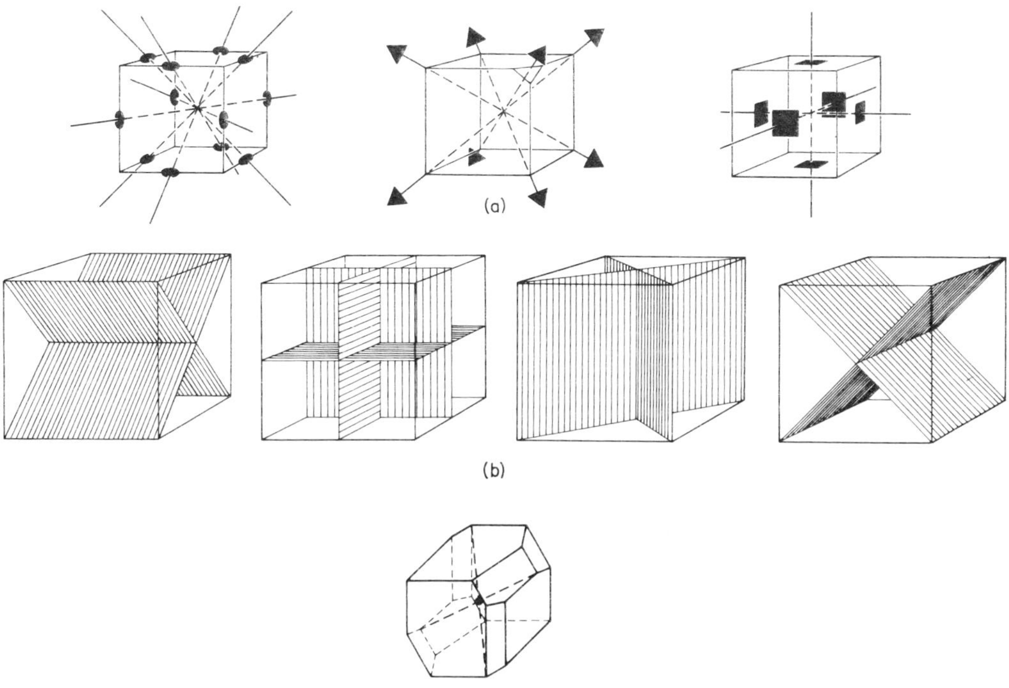

Figure 6.13. (a) Six twofold axes, denoted by ⬬; four threefold axes, denoted by ▲; three fourfold axes, denoted by ■; (b) the nine symmetry planes; (c) the center of symmetry is present when a straight line passing through the center of the solid intersect similar points, lines, or faces on opposite sides of the solid.

International Crystallographic Symbols, are quite concise and self-explanatory. Two-, three-, four-, and sixfold rotation axes are represented by the numbers 2, 3, 4, and 6, respectively, whereas two-, three-, four-, and sixfold rotation-inversion axes are represented by $\bar{2}$, $\bar{3}$, $\bar{4}$, and $\bar{6}$, respectively. The identity element is represented by 1 (only one repetition in a 360 degree rotation), and a center of symmetry (inversion through a point) is represented by $\bar{1}$. A plane of symmetry is represented by the letter *m* (a mirror plane).

To express the point group of a crystal, we conventionally give the symmetry of the principal axis first; for example, 4 means there is a fourfold axis of symmetry along the *c* axis,[1] the principal axis. If a plane of symmetry exists which is perpendicular to the principal axis, then the two symbols

[1] Whereas spectroscopists often refer to *x*, *y*, and *z* axes, solid state chemists as often refer to these axes as the *a*, *b*, and *c* axes.

are associated as $4/m$ (read "four over m"). Then follow the symbols for secondary axes, if any, and other symmetry planes. There is no need to mention more than one secondary axis if they exist in sets. For example, if one finds two secondary twofold axes perpendicular to a principal fourfold

TABLE 6.3 ***The 32 Crystallographic Point Groups***

System	Hermann-Mauguin Point Group Symbol	Schoenflies Point Group Symbol	*Symmetry Elements Present* m	2	3	4	6	$\bar{1}$	$\bar{2}$	$\bar{3}$	$\bar{4}$	$\bar{6}$
Triclinic	1	C_1										
	$\bar{1}$	C_i						1				
Monoclinic	2	C_2		1								
	m	C_s	1						1			
	$2/m$	C_{2h}	1	1				1	1			
Orthorhombic	222	D_2		3								
	$mm2$	C_{2v}	2	1					2			
	mmm	D_{2h}	3	3				1	3			
Tetragonal (every point group has one 4 or $\bar{4}$ axis)	4	C_4				1						
	$\bar{4}$	S_4									1	
	$4/m$	C_{4h}	1			1		1	1		1	
	422	D_4		4		1						
	$4mm$	C_{4v}	4			1			4			
	$\bar{4}2m$	D_{2d}	2	2					2		1	
	$4/mmm$	D_{4h}	5	4		1		1	5		1	
Rhombohedral (trigonal) (every point group has one threefold axis)	3	C_3			1							
	$\bar{3}$	C_{3i}								1		
	32	D_3		3	1							
	$3m$	C_{3v}	3		1				3			
	$\bar{3}m$	D_{3d}	3	3					3	1		
Hexagonal (every point group has one 6 or $\bar{6}$ axis)	6	C_6					1					
	$\bar{6}$	C_{3h}	1									1
	$6/m$	C_{6h}	1				1	1	1	1		1
	622	D_6		6			1					
	$6mm$	C_{6v}	6				1		6			
	$\bar{6}m2$	D_{3h}	4	3					4			1
	$6/mmm$	D_{6h}	7	6			1	1	7	1		1
Cubic (every point group has four threefold axes)	23	T		3	4							
	$m3$	T_h	3	3					3	4		
	432	O		6	4	3						
	$\bar{4}3m$	T_d	6		4				6		3	
	$m3m$	O_h	9	6		3			9	4	3	

axis, there is no need to mention more than one of the twofold axes. The action of the fourfold axis requires a second twofold axis at right angles to the first axis. Since the symbols for a particular point group are conventional, it can happen that abbreviated forms may be used at times in the literature. Table 6.3 lists the appropriate symmetry elements for each point group. Similar keys to the nomenclature are available in advanced texts.

As an example of expressing a point group, let us refer to the urea crystal in Fig. 6.12. Figure 6.14 illustrates the symmetry axes and symmetry planes of this crystal, which belongs to the tetragonal system. The principal axis is a fourfold rotation-inversion axis, which also appears morphologically as a twofold rotation axis. However, the higher symmetry element of fourfold rotation-inversion prevails. The $a = b$ crystallographic axes are twofold rotation axes. At right angles to these axes are two vertical dihedral

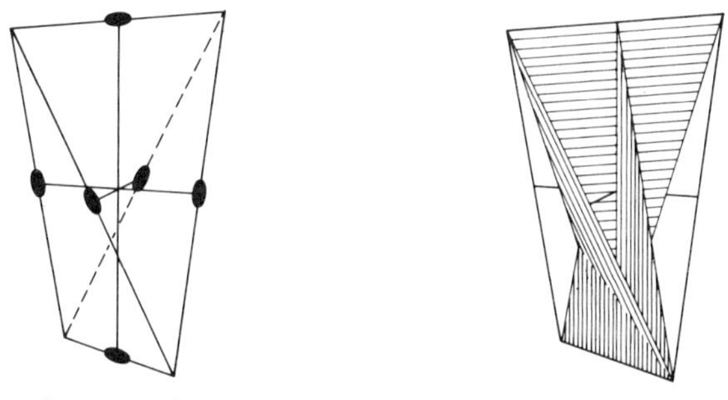

Figure 6.14. Symmetry axes and symmetry planes for point group $\bar{4}2m$.

planes. Therefore, crystals of the mineral chalcopyrite and urea can be represented as $\bar{4}2m$. It can be shown that the Schoenflies notation for this point group is D_{2d}. It is important to realize that the symmetry of a crystal of a compound is not necessarily the symmetry of atomic arrangement within an individual molecule. As noted earlier in this chapter, crystals consist of groups of ions, atoms, or molecules repeated regularly in space. The crystal structure may be imagined as built up from the assemblage of this group and then repetition of the same grouping in the same orientation at regular intervals in space. Each such group is construed as a unit and may be regarded as associated with a lattice point, i.e., each unit is imagined as replaced by a symbolic point. The possible arrangements of lattice points are the 14 Bravais lattices. The arrangement of atoms, ions, or molecules around each lattice point might not exhibit the same symmetry. The urea molecule in Fig. 6.15(a) has two vertical planes, one passing through the two nitrogen atoms and one intersecting them. These planes intersect in a twofold axis. Thus, the molecule has the symmetry found in crystals belonging to the orthorhombic system; the point group is $mm2(C_{2v})$.

By means of X-ray diffraction, optical microscopy, and density measurements combined with chemical intuition, the structure of urea is readily

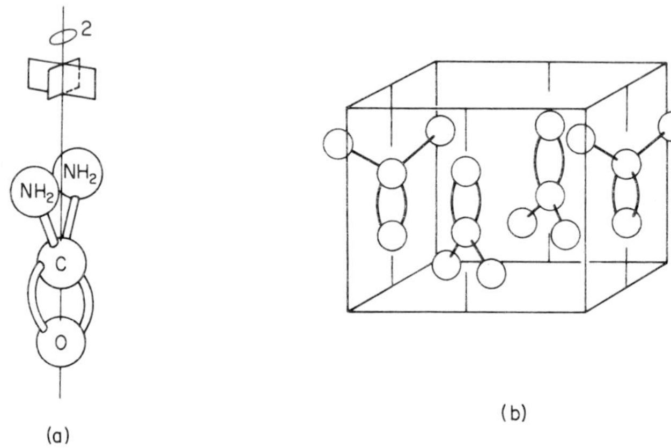

Figure 6.15. (a) The symmetry of the urea molecule; (b) the structure of urea.

deduced [Fig. 6.15(b)]. The student should recognize in the unit cell the $\bar{4}2m$ symmetry that leads to $\bar{4}2m$ morphological symmetry.

How many point groups exist? The number of possible symmetries of isolated groups of atoms is unlimited; however, in crystals, these symmetries must exist in repeating space patterns which permit packing. The 10 symmetry elements may be combined to give 32 crystallographic point groups distributed among the seven crystal systems. Table 6.3 lists these point groups by symbols, along with the symmetry elements of each group. The appropriate Schoenflies symbols are also tabulated for comparison; however, these symbols usually refer to the isolated molecules, as discussed in Chapter 5. In fact, the vast majority of molecules possesses one of the 32 crystallographically allowed point groups.

SPACE GROUPS

When one considers point group symmetry, one is concerned with symmetry operations associated with some lattice point. Space groups are obtained by the application of point group symmetry to infinite lattices, with the additional possibility of translational symmetry. The symmetry elements involved are the glide plane, which involves simultaneous reflection (a mirror plane) and translation, and the screw axis, which involves simultaneous rotation and translation. By continued repetition of these symmetry elements one does not arrive back at the atom from which he started; rather he reaches a corresponding atom associated with the next lattice

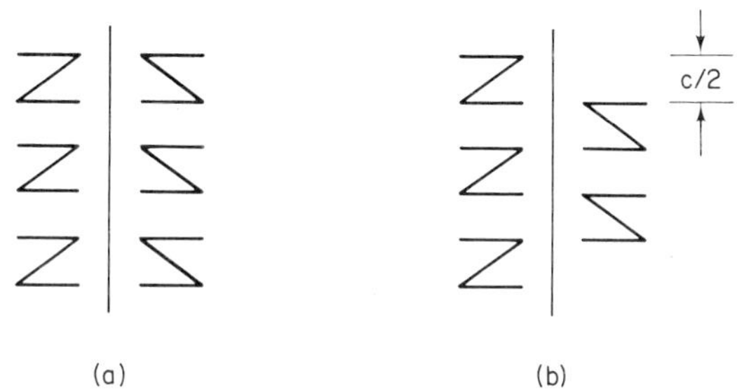

Figure 6.16. (a) Mirror plane; (b) glide plane.

point. Figure 6.16(a) and (b) illustrate a reflection plane and glide reflection plane, respectively, where the figure represents some structural unit. If one assumes in Fig. 6.16(b) that the plane is on the *c* axis and the unit distance along the axis is *c*, then translation of the figure along half the unit length combined with a reflection gives the corresponding structural unit associated with the next lattice point. The structural arrangements in Fig. 6.16(a) and (b) are closely related; in fact, both give the same external crystal planes arranged about the *c* axis.

An example of a threefold screw axis is shown in Fig. 6.17. Rotation is counterclockwise, as seen from the top. Each rotation through 120° is

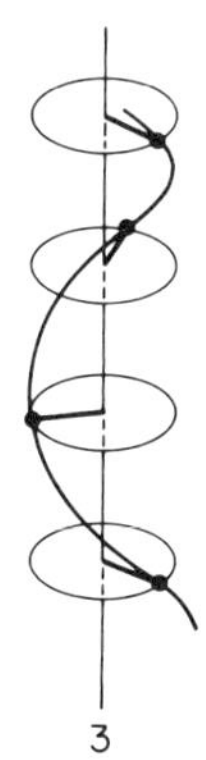

Figure 6.17. A threefold screw axis.

accompanied by a translation downward. In a particular structure, certain types of glide planes and screw axes are possible because their operations must yield the corresponding structural unit (ions, atoms, or molecules) when extended to an adjacent unit cell.

Federov (1890), Schoenflies (1891), and Barlow (1895) independently discovered that the possible combinations of symmetry elements involving the 32 crystallographic point groups with permitted translations involving the Bravais lattices, glide planes, and screw axes, totaled 230. It was not possible to exemplify all 230 space groups until diffraction techniques were utilized to work out the internal symmetry in crystals. With reflection planes and glide planes, morphological symmetry reveals that a crystal has an axis of symmetry, but it does not tell whether it is a simple rotation axis or a screw axis. However, systematic absences of certain types of reflections in diffraction data reveal the presence of structural units related by translation.

CLOSE-PACKED STRUCTURES

Our attention is now shifted from the morphology of solids and its relation to internal symmetry to the actual positions of ions, atoms, and molecules in crystals. Thirty years before the resolution of the sodium chloride structure by X-ray diffraction, the English physicist Barlow proposed a model to explain why many metals crystallize with hexagonal or cubic symmetry, and, in the former case, why many of the observed values of axial ratios lie near the ideal value of 1.633 for hexagonal closest packing. He showed that there are only two ways of attaining closest packing of equivalent spheres—one with cubic and one with hexagonal symmetry. By *close packing* is meant an arrangement of equidimensional objects in space so that unused space is minimized. It is known that the bonding character, be it ionic, covalent, or van der Waals, will have a significant effect on structures. However, the Barlow's model is a sound basis for discussion of inorganic solids. After a discussion of close packing of equidimensional spheres and its use in characterizing some structural trends, especially for the metals, we can investigate voids created by the close packing of spheres.

Any student of pool knows that the most efficient spatial arrangement consists in an array of spheres where every sphere has six touching neighbors. In two dimensions, the spheres become circles, and such a closest-packed layer about the center of each sphere has hexagonal symmetry, $6mm$. The symmetry of the layer about each void is $3m$ (Fig. 6.18).

There are two sets of triangular voids in the layer—one with the vertices of the triangles pointing down, O, and one set with the vertices of the triangles pointing up, T. A second closest-fitting layer can be packed on layer A by either placing all the spheres on O sites or all on T sites. Placing

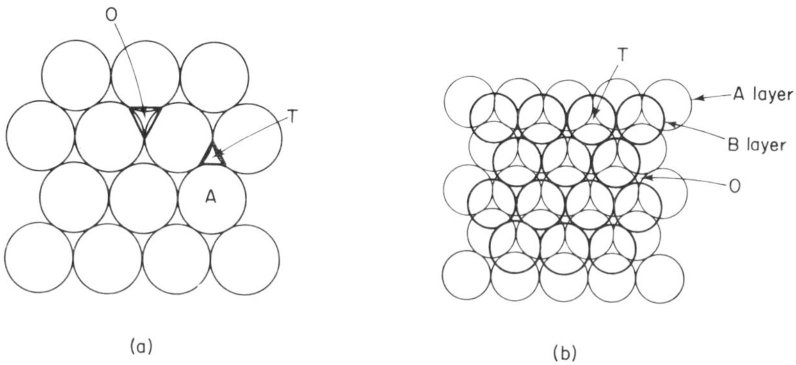

Figure 6.18. (a) Spheres arranged in a closest-packed layer; (b) closest-packed *AB* layers.

the spheres on O sites does not give a different structure from that obtained by placing the spheres on T sites, since the structure can be converted by a simple change of coordinate system. Suppose the second layer is a B layer. The third layer can now be either a C layer or another A layer. These possibilities can be seen in Fig. 6.18(b). In any case, each sphere in these closest packings touches 12 other spheres, 6 in its own layer and 3 from each of the layers above and below. Indeed, it is possible to have less than 12 spheres touching each sphere, but then the structure is not closest packed. If the packing sequence is *ABAB*..., it is called hexagonal closest packing (Fig. 6.19). Notice how spheres in the two A layers occupy the sites of the hexagonal Bravais lattice. The threefold axis through the void in layer B can be considered as limiting the symmetry of the hexagonal structure. However, higher symmetry can exist in the lattice by application of the two space group symmetry operations. Referring to Fig. 6.18(b), we note that the voids marked O are continuous through the structure to form a channel.

An alternative form for the closest packing of equidimensional spheres is the sequence of layers *ABCABC*..., (Fig. 6.20), which is called cubic closest packing. In layer C, the spheres cover the voids in layer B that were also voids in layer A, i.e., the O voids. A second layer of A spheres is then added to renew the sequence. The sequence of closest packed layers are

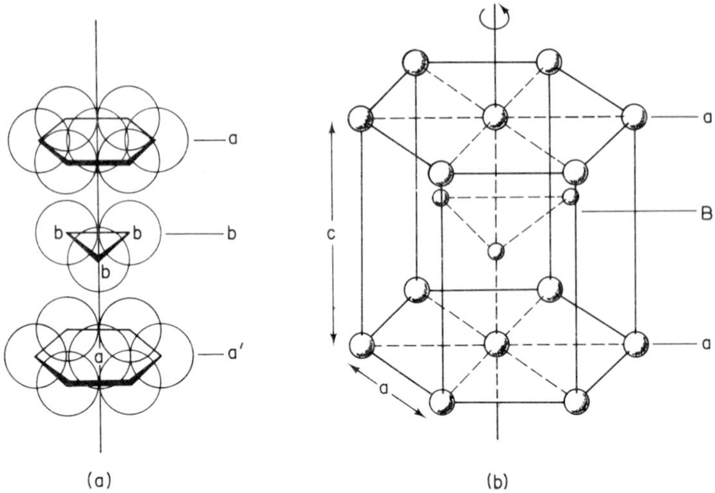

Figure 6.19. Hexagonal closest-packing structure. (a) Stacking sequence of spheres; (b) same structure with spheres shrunken, showing threefold axis through void in layer *B*. The closest-packed layers are 001 planes.

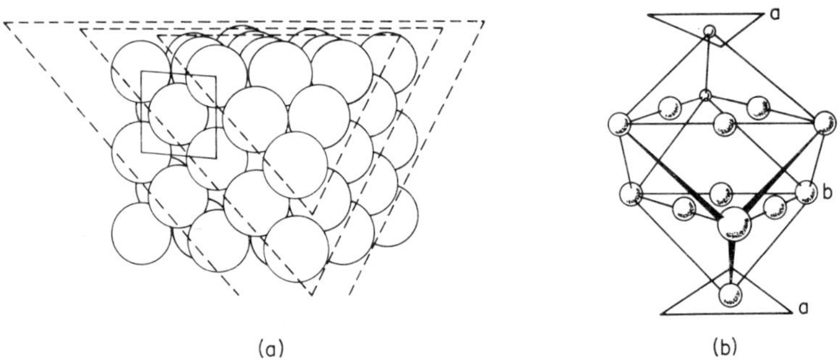

Figure 6.20. Cubic closest-packing structure. (a) Stacking sequence of spheres; (b) same structure with spheres shrunken.

(111) planes of a face-centered cubic structure, illustrated in Fig. 6.20(b), that is normal to the direction of the threefold axis of rotation of a cube (Fig. 6.13). Notice how the spheres in the *ABC* layers occupy the sites of the *F* cubic lattice. Examples of other sequences of closest packings such as *ABACABAC* . . . are known but are less common than those described. Some will be discussed later. All closest packings have an efficiency for filling 74% of the space.

Other common forms of cubic structure are the body-centered cube (spheres are at points in the *I* cubic lattice of Fig. 6.8) and the simple cube (spheres are at points in P cubic lattice in Fig. 6.8). The former has a packing efficiency of 68% and the latter a packing efficiency of 52%. As we see in Fig. 6.8, the body-centered cube has 8 nearest neighbors and 6 next nearest neighbors, as opposed to 12 nearest neighbors in closest packing.

With the packing efficiencies of equidimensional spheres in mind, one can rationalize predictions about thermodynamically stable structures of many solids that are closest packed, providing one assumes they are made

TABLE 6.4 ***Crystal Structures of the Metallic Elements*** *

I-A	II-A	III-A	IV-A	V-A	VI-A	VII-A	VIII-A			I-B	II-B	III-B	IV-B	V-B	
Li A_2 A_1 A_3	Be A_2 A_3 X			A							B				
Na A_2 A_3	Mg A_3											Al A_1			
K A_2	Ca A_1 A_3 X	Sc A_1 A_3	Ti A_3 A_2	V A_2	Cr A_2 A_3 X	Mn X A_1 A_2	Fe A_2 A_1	Co A_3 A_1	Ni A_1 A_3	Cu A_1	Zn X A_3	Ga X			
Rb A_2	Sr A_1 A_3 A_2	Y A_2	Zr A_3 A_2	Nb A_2	Mo A_2 A_3	Tc A_3	Ru A_3	Rh A_1	Pd A_1	Ag A_1	Cd X A_3	In X	Sn X	Sb X	
Cs A_2	Ba A_2	La A_1 A_3	Hf A_3 A_2	Ta A_2	W A_2	Re A_3	Os A_3	Ir A_1	Pt A_1	Au A_1	Hg X	Tl A_3 A_2	Pb A_1	Bi X	Po X
Fr —	Ra —	Ac A_1													
Ce A_1 A_3	Pr A_1 X	Nd X	Pm —	Sm X	Eu A_2	Gd A_3	Tb A_3	Dy A_3	Ho A_3	Er A_3	Tm A_3	Yb A_1	Lu A_3		
Th A_1 A_2	Pa X	U X A_2	Np X	Pu X A_1 A_2	Am —	Cm —	Bk —	Cf —	E —	Fm —	Mv —	No —			

* A_1, cubic closest packing; A_2, cubic body-centered structure; A_3 hexagonal closest packing; X, other more complex structures.

up of incompressible spheres. Cubic closest-packed, body-centered cubic, and hexagonal closest-packed arrangements are conventionally represented by the symbols A_1 or c.c.p., A_2 or b.c.c., and A_3 or h.c.p., respectively. Table 6.4 indicates which of these packings the metallic elements will assume. The symbol X indicates that one or more other complex structures also exist for that element.

Table 6.4 indicates that, in general, metallic elements pack as if they are equidimensional spheres and, therefore, give close-packed structures. Thus, determination of the unit cell dimensions of these structures is a means of deriving the atomic radii of the metals. Metals that crystallize in A_1 and A_3 arrangements exhibit equality of interatomic distances, showing the general validity of the Barlow model of packing incompressible spheres. The A_3 structures can show variations from the axial ratio, c/a, of 1.633, which is predicted for the arrangement of equidimensional spheres in hexagonal closest packing. This is particularly noticeable in metals that lie near the metal-nonmetal borderline. Thus, beryllium, zinc, cadmium, and thallium are distorted in their A_3 structures, having six atoms in one layer at equal distances and six atoms in the adjoining layers, three in each layer, at slightly different distances. Gallium, indium, tin, antimony, bismuth, and polonium continue this trend, and are hence quite distorted in terms of having near neighbors at close-packed equal distances. This structural trend may be explained in terms of band theory, described in Chapter 7. For present purposes and as a brief introduction, one might visualize a crystal as a single giant molecule. As the atoms are brought together, corresponding outer orbitals of the atoms overlap, giving bands of electron

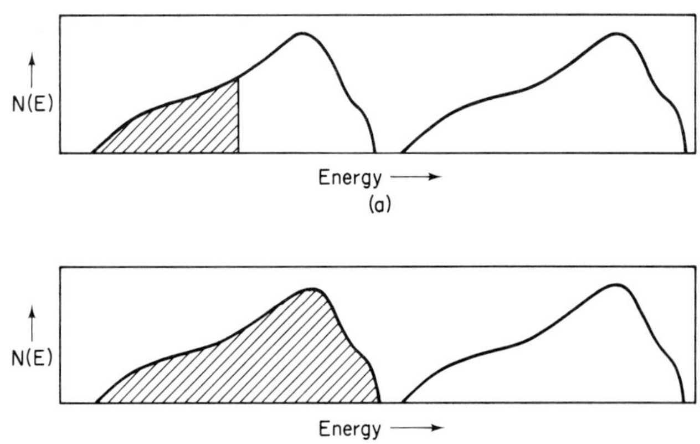

Figure 6.21. The distribution of energy states for (a) a metal and (b) a nonmetal. The shaded areas are occupied states.

energy states instead of discrete energy states of the atoms. Moreover, overlap can occur between *ns*, *np*, and $(n-1)d$ bands at favorable internuclear distances. Figure 6.21 shows hypothetical energy bands for a metal and a nonmetal and distributions of energy states within the bands. The value of $N(E)$ represents the number of available energy states with energies in the range E to $(E + dE)$ and is plotted against E.

According to band theory, the population of electrons in the energy states, such as in Fig. 6.21, determines how metallic the crystalline substance will be. Metals have bands incompletely filled with electrons [Fig. 6.21(a)], whereas nonmetals have bands either completely filled or completely empty [Fig. 6.21(b)]. Lower parts of the bands are quite insensitive to the directional bonding influences associated with atomic orbital bonding, whereas the upper parts of the bands are quite sensitive. Hence, we find the *A* metals (Table 6.4) able to form a continuous solid of closest-packed atoms. On the other hand, *B* metals (Table 6.4) have $(n-1)d$ orbitals filled with electrons and are not able to pack ideally, being sensitive to directional influences of bonding associated with molecular units.

Table 6.4 indicates that many of the metals are polymorphous, i.e., they have two or more structures. The thermodynamic stability of these structures is, of course, dependent on temperature and pressure conditions, but we would expect from the band theory model that transitions between close-packing structures could take place fairly readily, due to the relative insensibility of the lower parts of bands to directional influences. This also accounts for the relative ductility of these metals. The *B* metals are also polymorphic, but the trend is toward complex structures that are interpreted in terms of the directional influences associated with atomic orbitals. Reasons for the preference of Group A metals for particular close-packed structures are also related to directional influences.

We noted previously that closest-packing sequences other than *ABAB*... and *ABCABC*... are known. For example, *ABACABAC*... and sequences with longer periods have been found, notably among lanthanides and actinides. This phenomenon has been attributed to possible long-range directional bonding overlaps due to the presence of electrons in *f* orbitals of these elements. Another type of polymorph that has been studied intensively is the polymorph of silicon carbide, known as carborundum. A number of complex stacking sequences of closest-packed layers have been found to be responsible for variations in properties of industrial interest. However, structurally they are intriguing because the complex sequences are maintained over distances of several thousand angstroms. The ordering forces effective over such large distances are difficult to explain if they are regarded as conventional crystals. The best explanation at present is in terms of the mechanism of growth of these crystals permitting such structures, but as yet this is not fully understood.

Mechanisms during crystal growth or heat treatment can permit "mistakes" in stacking sequences. For example, hexagonal close-packed cobalt has been found to have sequences such as *ABABABCBCBAB*..... Such arrangements are called stacking faults and lead to defect structures. The effect of these stacking faults on such physical properties of metals as tensile strength is technologically important.

Nondirectional bonds link molecules or atoms that are essentially spherical in shape or become spherical by rotation or random orientation in closest-packing structures. Hence, all noble gases crystallize in either A_1 or A_3 structures. Since the electron cloud in H_2 is nearly spherical in shape, crystals of solid hydrogen can pack in an A_1 arrangement. The hydrogen halides have the relatively large spherical halogen atom and are only slightly distorted by the small hydrogen atom; hence, they pack in the A_3 arrangement. By means of closest packing, these atoms and molecules can achieve a maximum number of nearest neighbors, thereby maximizing the van der Waals forces in the crystal.

RADIUS RATIO

A large proportion of all known inorganic compounds exhibit structures that can be related to close-packing models. So far, we have emphasized the packing of atoms. Before going into the ionic bond itself, we shall extend the model of close packing to voids generated by the stacking of spheres. In this way, the model will permit classification of the structures of many ionic compounds.

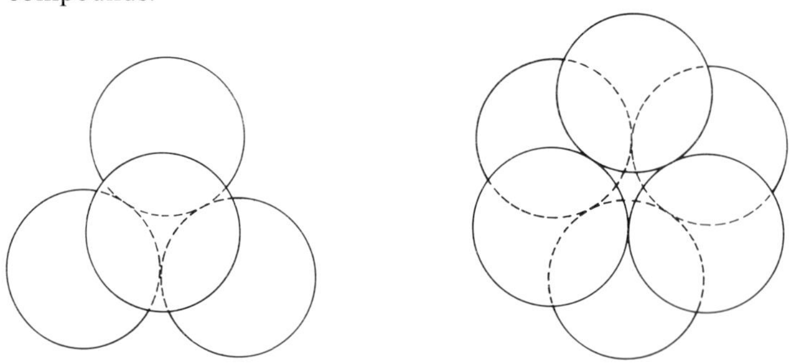

(a) (b)

Figure 6.22. Holes formed in closest-packed structures. (a) Tetrahedral hole; (b) octahedral hole.

Figure 6.18(a) indicated that there are six triangular voids around each sphere in a closest-packed layer. If a triangular void is covered with a sphere, then the void has four spheres around it, each sphere at the corner of a tetrahedron. This is illustrated in Fig. 6.22(a). On the other hand, in closest packings, a triangular void pointing down, which was labeled as an *O* in Fig. 6.18(b), can also be covered by a triangular void pointing up. In this case, a void surrounded by six spheres exists, each sphere sited at the corner of an octahedron [Fig. 6.22(b)]. These voids are called tetrahedral and octahedral holes. Examination shows that they are the only voids formed by closest packing of equidimensional spheres, no matter what the sequence of packing. The position of these holes in the A_1 and A_3 structures is illustrated in Fig. 6.23.

What is the number of tetrahedral or octahedral holes to spheres in closest-packed structures? In a hexagonal closest-packed structure, Fig. 6.18(b) indicates that three tetrahedral holes, which have been labeled as

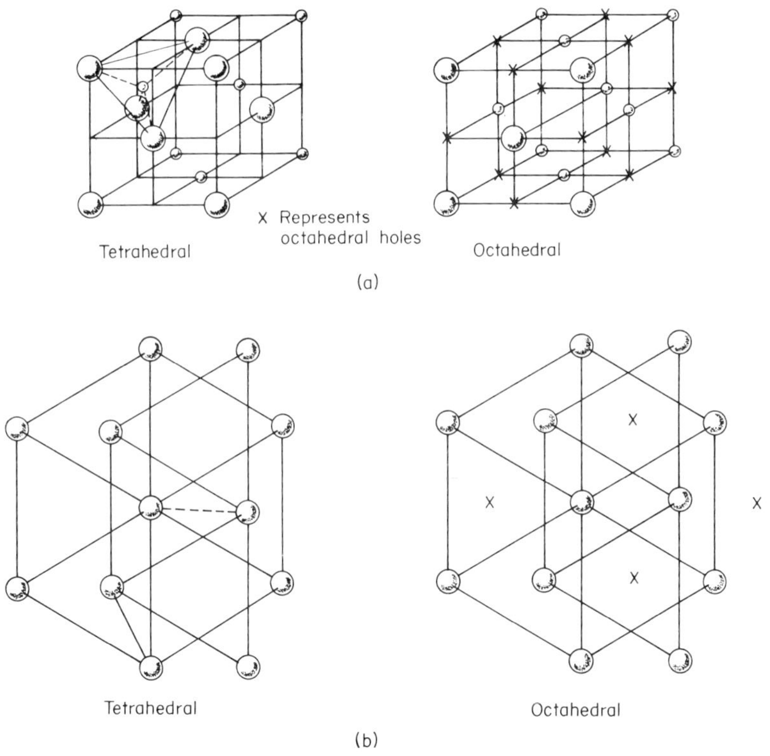

Figure 6.23. Tetrahedral and octahedral holes in (a) cubic closest-packing structure; (b) hexagonal closest-packing structure.

T holes, and three octahedral holes, which have been labeled as *O* holes, surround each sphere. In the same way, three tetrahedral and three octahedral holes are generated around each sphere from packing spheres under layer *A*. This gives a total of six tetrahedral and six octahedral holes per sphere. However, the top of each sphere in the *A* layer is surrounded by spheres from the *B* layer, causing a triangular void above each sphere in the *A* layer in Fig. 6.18(b). When a new layer is placed above the *B* layer to renew the *AB* sequence, each sphere in the *A* layer will, therefore, have one tetrahedral void above it and one below it. Hence, there are a total of eight tetrahedral holes surrounding each sphere. Now, since, as discussed previously, there are four spheres surrounding each tetrahedral hole, there are a total of $\frac{8}{4}$ or 2 tetrahedral holes for each sphere. Similarly, since, as discussed previously, there are six spheres surrounding each octahedral hole, there are a total of $\frac{6}{6}$ or 1 octahedral hole for each sphere. This analysis can be applied also to a cubic closest-packed structure and the same values are found. Models can assist the reader greatly.

In view of the fact that many ionic crystals, including metallic halides and oxides, can be thought of as relatively large anions in closest packing with smaller cations occupying the holes, it is helpful to calculate the limiting size of the cations that might fit into these holes efficiently. The model should also help to predict the solubility of atoms in metals having close-packing structures.

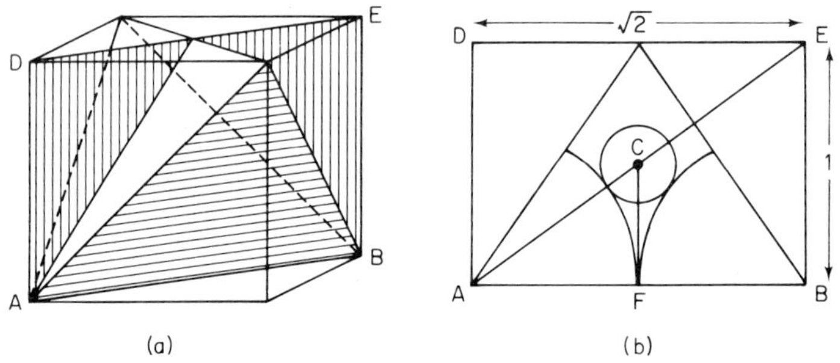

Figure 6.24. (a) Tetrahedron within a cube with (110) plane passing through the tetrahedron; (b) front view of the (110) plane.

Figure 6.24(a) illustrates a tetrahedron within a cube with sides equal to unity. The (110) plane of the cube, *ABDE*, cuts through the center of the tetrahedron. Figure 6.24(b) shows the front view of the plane with the outline of the packing spheres and a sphere whose center is at *C* and which could be most efficiently packed inside the tetrahedron so that it touches all four surrounding spheres. Now the sides of the right triangles in the plane are proportional:

$$\frac{AF}{AC} = \frac{AB}{AE} = \frac{\sqrt{2}}{[(\sqrt{2})^2 + (1)^2]^{1/2}} \qquad [6.3]$$

Let r_c be the radius of the small sphere in the center of the tetrahedron and r_a the radius of the spheres in closest packing. Then,

$$\frac{AF}{AC} = \frac{r_a}{r_a + r_c} = \frac{\sqrt{2}}{\sqrt{3}} = \frac{r_c}{r_a} \qquad [6.4]$$

The ratio of radii, r_c/r_a, is, therefore, 0.225. This is the ratio of radii that permits maximum efficiency of packing around and inside the tetrahedron. It is called the radius ratio, and, in ionic compounds, r_c and r_a refer to cation and anion radii, respectively. The number of nearest neighbors surrounding an ion is called its coordination number, C.N. Using the same type of reasoning, we can calculate limiting radius ratios for triangular holes, octahedral holes, and holes with other coordination numbers. The application of radius ratios will be made when ionic structures are discussed.

The body-centered cubic structure is also close packed. However, although slightly more open, it does not contain holes as large as those in the two closest-packed structures. One potential hole in the A_2 structure is tetrahedral. It occurs in the face at the upper edge of the cube and is illustrated in Fig. 6.25(a). The limiting value of the radius ratio for the largest sphere that can fit into this void is only 0.291. A second hole is octahedral. It occurs at the center of the cube faces and forms a compressed octahedron. It is illustrated in Fig. 6.25(b). The radius ratio in the two compressed directions is 0.154 and is 0.633 in the four elongated directions.

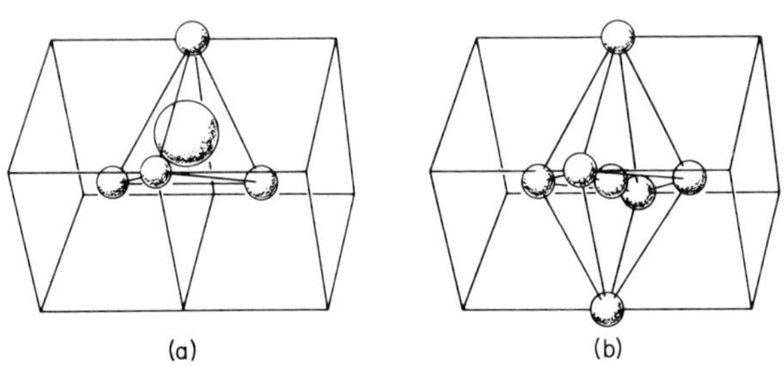

Figure 6.25. Holes in body-centered cubic structure. (a) Tetrahedral holes; (b) octahedral holes.

In terms of the low radius ratios for A_2 structures, it would be expected that far fewer holes can be filled in this structure than in A_1 and A_3 structures. Thus, only 0.025 weight percent carbon can fill the octahedral holes in alpha iron, known as ferrite, which has an A_2 structure. On the other hand, up to 2.0 weight percent carbon can occupy the octahedral holes in gamma iron, known as austenite, which has an A_1 structure. The properties of these phases are important in the steel industry.

THE IONIC MODEL

Since many inorganic solids are described as ionic in character, familiarity with the ionic model is important when these solids are classified structurally. Moreover, as was seen in Chapter 3, models based on the covalent bond can require elaborate mathematical calculations to obtain approximate bond energy values that agree with experimental values. Even then, close agreement is obtained only for simpler chemical systems. On the other hand, relatively simple calculations based on an ionic model can often afford good agreement with experimental binding energies. This is especially true for salts composed of pretransition metals and elements with high electronegativity.

As with the treatment of closest packing, a simple pure ionic model will be adopted for the sake of simplicity. It is presumed that ionic species assume the most efficient packing possible—closest packing of the larger anions, usually. Necessary modifications of the model, due to covalent character in the bonds, are then more clearly understood. The following presuppositions are posited:

(1) Ions are charged, incompressible, nonpolarizable spheres.
(2) Arrangements of ions of one charge around an ion of the opposite charge are stable only if the central ion is in contact with each of its neighbors. Consequently, cations, which are usually smaller than anions, are limited in their coordination numbers by a radius ratio.
(3) Ions tend to have as large coordination numbers as possible, subject to the restrictions of chemical composition in the compound.
(4) Coordinating groups around an ion so arrange themselves as to limit electrostatic repulsion.

It follows from (4) that MX_2 species should be linear to minimize repulsion. The MX_3, MX_4, and MX_6 species should be symmetrically arranged in triangular, tetrahedral, and octahedral configurations, respectively—the same arrangements found in closest packing (see Chapter

5). If MX_8 species were arranged according to (4), then many inorganic structures would have a square antiprism structure and not the more common body-centered cubic arrangement. Figure 6.26(a) shows how electrostatic repulsion would be less in square antiprisms than if the eight coordinating ions were at the corners of a cube, as they exist in Fig. 6.26(b) in the cubic CsCl structure. Ions, such as TaF_8^{3-}, do, in fact, take the square antiprism arrangement; however, it is not common, due to the difficulty of building an extended structure in space in which every ion is surrounded by a square antiprism of ions of opposite charge. Thus the greater stability of long-range order dictates body-centered cubic structures for most compounds with a coordination number of eight. The existence of MX_5 trigonal bipyramids [Fig. 6.26(c)] in ionic compounds can be considered in light of conditions (2) and (3). The angle between the axial and equatorial bonds is 90° and the M-X distance is not smaller than the M-X distance in an octahedron. Hence, (3) would lead us to expect more octahedra than trigonal bipyramids in ionic structures. Favorable arrangements for MX_7 and MX_9 species are illustrated in Chapter 5.

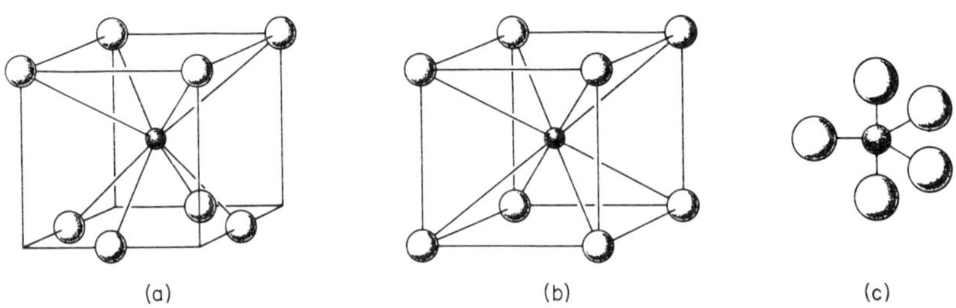

Figure 6.26. (a) Square antiprism for eight coordination; (b) cubic (CsCl) for eight coordination; (c) trigonal bipyramid for five coordination.

It is difficult to arrange these groups in extended three-dimensional space, and, therefore, there are fewer solids found with these arrangements than with the more symmetrical ones. Distortion of structure is common among these arrangements to provide for most efficient packing and orbital overlap, especially where partial covalent character exists.

Table 6.5 lists the limiting values of some radius ratios, r_c/r_a, which can be obtained by the same principles used to derive the value of 0.225 for the tetrahedral arrangement. The more common structures are underlined.

TABLE 6.5 ***Limiting Radius Ratios for Various Structures***

Structure	Coordination Number	Radius Ratio
Triangular	3	0.155
Tetrahedral	4	0.225
Trigonal biprism	5	0.414
Octahedral	6	0.414
Square antiprism	8	0.645
Cubic	8	0.732

To test the pure ionic model, Table 6.6 lists radius ratio values of MX compounds, which are generally considered to be among the most ionic in the periodic table. The radii values used are those of Pauling. The structure of each compound at standard temperature and pressure is also listed.

TABLE 6.6 ***Radius Ratio Values and Structures of Some Ionic Compounds***

Compound	r_c/r_a	Structure	Compound	r_c/r_a	Structure
LiF	0.44	*R**	BeO	0.22	*W*
LiCl	0.33	*R*	BeS	0.17	*Z*
LiBr	0.31	*R*	BeSe	0.16	*Z*
LiI	0.28	*R*	BeTe	0.14	*Z*
NaF	0.70	*R*	MgO	0.46	*R*
NaCl	0.52	*R*	MgS	0.35	*R*
NaBr	0.49	*R*	MgSe	0.33	*R*
NaI	0.44	*R*	MgTe	0.29	*W*
KF	0.98	*R*	CaO	0.71	*R*
KCl	0.73	*R*	CaS	0.54	*R*
KBr	0.68	*R*	CaSe	0.57	*R*
KI	0.62	*R*	CaTe	0.45	*R*
RbF	1.09	*R*	SrO	0.81	*R*
RbCl	0.82	*R*	SrS	0.63	*R*
RbBr	0.76	*R*	SrSe	0.57	*R*
RbI	0.69	*R*	SrTe	0.51	*R*
CsF	1.24	*R*	BaO	0.96	*R*
CsCl	0.93	CsCl	BaS	0.75	*R*
CsBr	0.87	CsCl	BaSe	0.69	*R*
CsI	0.78	CsCl	BaTe	0.62	*R*

*The letter *R* refers to the rocksalt, NaCl, structure and is illustrated in Fig. 6.27. The letters *W* and *Z* refer to the wurtzite and zincblende structures, respectively, which have tetrahedral coordination around the metal ion. These structures are also illustrated in Fig. 6.27.

According to condition (2) of the ionic model, a radius ratio less than 0.414 should not form an octahedral arrangement because the anions are so large that the cation cannot touch each neighbor. On the other hand,

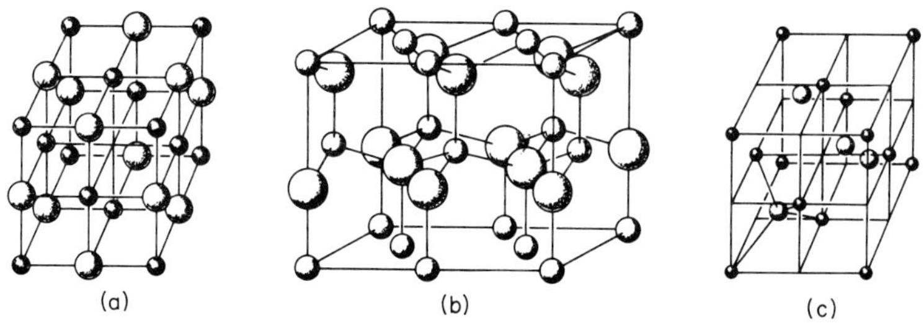

Figure 6.27. (a) Rocksalt structure for six coordination; (b) wurtzite structure for four coordination; (c) zincblende structure for four coordination.

according to condition (3), a radius ratio greater than 0.732 should not permit an octahedral arrangement because it is then possible to fit eight ions around the central ion. A discussion of the alkali halides follows.

In the lithium chloride, bromide, and iodide lattices, the lithium ions are sometimes described as *rattling* inside the crystal. It can be shown by electrostatic calculation that there can be an increase in lattice energy if a small cation such as Li^+ can move from the center of symmetry of its hole in closest-packing arrangement of the anions and become situated closer to a corner. It would then have three nearest neighbors and three next-nearest neighbors, or, a distorted octahedral structure instead of the expected tetrahedral structure. Since distorted octahedral structures are also found among high valence transition metal oxides, this concept of *rattling* has merit. However, it has neither been verified nor rejected, since the small size of the lithium ion does not permit its exact location from X-ray data.

Although KF and RbF have an r_c/r_a near 1.00, they cannot adopt a closest-packed arrangement of equal spheres because it is geometrically impossible to have equal numbers of oppositely charged ions packed so that each ion is surrounded by 12 ions of opposite charge. Rubidium chloride, bromide, and iodide adopt the cesium chloride structure at higher pressures; however, theoretical and quantitative affirmation has not yet been produced to show why all these compounds adopt the rocksalt structure at standard conditions instead of the cesium chloride structure.

Outside of beryllium and magnesium compounds, the Groups II-VI, *MX*-type compounds, generally, have radius ratios within the rocksalt region and exhibit the structure predicted by the simple ionic model. The oxides of strontium and barium are exceptions, probably for the same unknown reason as that which causes the anomalies of structure of rubidium chloride and other alkali halides with r_c/r_a values greater than 0.732.

Magnesium compounds often show a close similarity of behavior to compounds of lithium following the diagonal relationship mentioned in Chapter 13. Hence, it is noteworthy that, like the lithium halides, the sulfide and selenide of magnesium tend to prefer octahedral coordination instead of the predicted tetrahedral coordination.

One structural trend in the periodic table can be seen from *MX* compounds. The rocksalt structure is assumed by M^{2+} oxides of titanium, vanadium, manganese, iron, cobalt, and nickel, although some of these oxides can be deficient in the amount of metal. However, outside of manganese compounds, none of the sulfides, selenides, or tellurides of these metals, despite radius ratios being in the rocksalt range, form the rocksalt structure. The importance of the d^5 electronic configuration in Mn^{2+} compounds will be discussed. The predominant structures for these *MX* compounds are variants of the nickel arsenide structure, and their properties become more characteristic of intermetallic compounds than of ionic compounds. For example, they are opaque, they have metallic luster, and they have variable composition. The nickel arsenide structure is illustrated in Fig. 6.28(a). This change to the nickel arsenid structure introduces a contraction in the *M*—*M* distance from that expected in more ionic structures. This can be explained by the decrease in electronegativity from oxygen to the other Group VI-B elements, which permits lower positive charge on the cations. This decreases electrostatic repulsion between cations. The structure of nickel arsenide is close-packed hexagonal, belonging to the $6/mmm$ point group. The characteristic axial ratio, c/a, of 1.633 for hexagonal closest packing decreases as r_c decreases for a given r_a, e.g., from Ti^{2+} to Ni^{2+}. Thus, greater distortion of the structure is necessary to permit the *M*—*M* bonding. The expanded radial distribution due to the availability of an orbital overlap makes this trend toward intermetallic compounds even more possible with lower series transition metal ions. Hence, there is a gradual change in the structure of *MX* compounds when *X* represents a Group VI-A element. When *M* is an alkali or alkaline earth metal, the compound is ionic and stoichiometric; when *M* is a transition metal, *MX* compounds may be intermetallic. Usually the oxides are insulators, and less highly colored, the sulfides are semiconductors or metalliclike and often not stoichiometric, and the selenides and tellurides of the transition metals are often alloys which are conductors and intensely colored, usually black. This type of intermetallic structure with similar properties is also characteristic of compounds of Group IVB and VB elements with transition metals, as might be expected from this trend.

Table 6.6 indicates that *MX* compounds of beryllium show tetrahedral coordination. Since there are twice as many tetrahedral holes as spheres in closest packings, it would be expected that an ionic MX compound in the radius ratio range of from 0.225 to 0.414 would have cations surrounded by four anions. In fact, if the packing of anions is h.c.p., then the wurtzite structure occurs; on the other hand, if the packing is c.c.p., then it is the zincblende structure. Wurtzite and zincblende are two forms of ZnS and are illustrated in Fig. 6.27. In both structures, the distance to nearest neighbors is the same; however, there is less possibility of interaction between atoms of the same kind in wurtzite than in zincblende. This is manifested by the greater ease of preparation of wurtzite and by a slight difference in the Madelung constants (to be discussed shortly)—1.641 for wurtzite and 1.638 for zincblende. Both structures are also taken by MX compounds made up of Group II-B metals, such as zinc, with Group VI-A elements. Table 6.7 lists some substances showing these structures, along with their respective radius ratios.

TABLE 6.7 ***Some MX Crystals with Tetrahedral Coordination***

Wurtzite Structures (h.c.p.):				
ZnO (0.53)	CdS (0.53)	AlN (0.29)	GaN (0.36)	InN (0.47)
ZnS (0.40)	CdSe (0.49)			
ZnTe (0.34)				
Zincblende Structures (c.c.p.):				
CuF (0.71)	ZnO (0.53)	CdS (0.53)	HgS (0.60)	
CuCl (0.53)	ZnS (0.40)	CdSe (0.49)	HgSe (0.56)	
CuBr (0.49)	ZnSe (0.37)		HgTe (0.50)	
CuI (0.44)	ZnTe (0.34)			
AlP (0.24)	GaP (0.29)	InP (0.38)		
AlAs (0.23)	GaAs (0.28)	InAs (0.36)		
AlSb (0.20)	GaSb (0.25)	InSb (0.33)		

It is clear from Table 6.7 that the pure ionic model and the radius ratio fail to account satisfactorily for the structure of these crystals. Many of the Group B metal ions fill tetrahedral holes instead of octahedral holes, which the radius ratio predicts. This tetrahedral selectivity by these metals is attributed to the importance of covalent bonding, and specifically of sp^3 hybrid orbitals in the overall bonding in these crystals. Thus, MX structures containing beryllium behave more like MX aluminium structures (diagonal relationship) and zinc structures than like MX structures containing other Group III-A elements. This also is due to the importance of covalent bonding in these beryllium compounds. Hence, we see from Table 6.7 that many Groups II–VI and III–V MX-type compounds take these two structures as do beryllium compounds. In all these crystals, it is probable that the bonding is best described as covalent with some ionic character. In fact, if both the zinc and sulfur atoms in the zincblende structure are replaced by carbon atoms then the diamond structure results. This arrangement is also shown by the Group IV-A elements, silicon, germanium, and tin with tetrahedra formed which are characteristic of their sp^3 covalent bonding. It is also noteworthy that manganese(II) sulfide and manganese(II) selenide with radius ratios of 0.43 and 0.40, respectively, exist in the wurtzite and zincblende structures. Thus Mn^{2+} with a d^5 electronic configuration tends to crystallize in structures like the $d^{10}b$ elements and not in the nickel arsenide structure which is common to sulfides and selenides of the transition metals. The MX compounds showing wurtzite and zincblende structures are stoichiometric. Hence the relatively wide ranges of composition stability and other properties which are found to exist in MX compounds of the transition metals with the nickel arsenide structure are attributed to d orbital interactions which are still under intensive study.

Ionic solids of composition MX_2 generally follow predictions made from the pure ionic model. The two principal structural types of MX_2 ionic crystals, fluorite and rutile, are illustrated in Fig. 6.28. The cation M

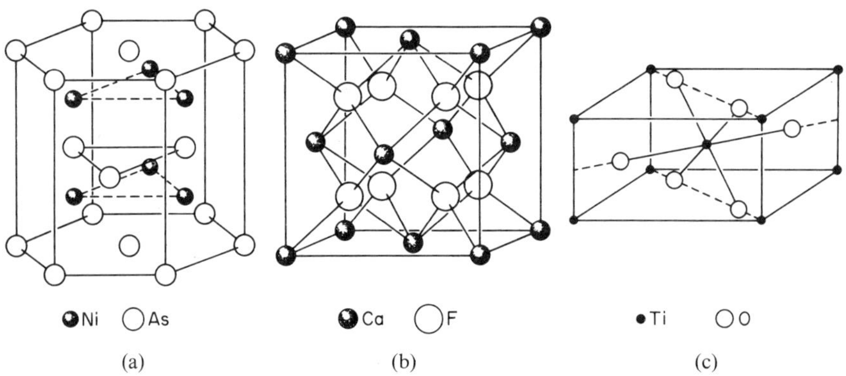

Figure 6.28. (a) Nickel arsenide structure; (b) fluorite structure; (c) rutile structure.

in the fluorite, CaF_2, structure has a face-centered cubic structure. Each cation is surrounded by eight anions in a cubic arrangement and the anion X is surrounded by four cations in a tetrahedral arrangement. In terms of closest packing, this is an A_1 (c.c.p.) structure with all the tetrahedral holes filled. When an MX_2 compound crystallizes with all its metal ions taking the F^- sites and all its X^- ions taking the Ca^{2+} sites of the fluorite structure then the MX_2 compound has an antifluorite structure. This structure is adopted by most alkali metal oxides, sulfides, selenides, and tellurides. The fluorite and antifluorite structures should be stable for radius ratio values greater than 0.732.

Ionic MX_2 compounds with radius ratio values less than the limiting value of 0.732 show the six coordinated rutile arrangement. Rutile is a form of titanium(IV) oxide. As Fig. 6.28(b) indicates, the structure is tetragonal and the symmetry permits deviations from a perfectly octahedral environment around the cations. In terms of closest packing, rutile is a slightly distorted h.c.p. lattice of oxygen atoms having one-half of the octahedral holes filled with titanium atoms. Table 6.8 lists some MX_2 compounds, where X is fluoride, which crystallize with fluorite and rutile structures. The radius ratio values are in parentheses and F indicates fluorite and T indicates rutile structures, respectively.

TABLE 6.8 ***Radius Ratio Values and Structures of Some MX_2 Compounds***

MgF_2 (0.23) T	MnF_2 (0.59) T	ZnF_2 (0.54) T	SnF_2 (0.82) F
CaF_2 (0.73) F	FeF_2 (0.56) T	CdF_2 (0.71) F	PbF_2 (0.88) F
SrF_2 (0.83) F	CoF_2 (0.54) T	HgF_2 (0.81) F	
BaF_2 (0.99) F	NiF_2 (0.53) T		

The ionic and semi-ionic MX_2 compounds, where X is oxygen, also take these two structures and generally follow the radius ratio rule quite faithfully. The MX_2 compounds with radius ratio values less than the lower limiting value of 0.414 for octahedra tend to take structures in which the M atom is surrounded by four X atoms. These are characteristic of such compounds as equilibrium forms of silicon dioxide, of germanium dioxide, and beryllium bifluoride; however, the linking and arrangement of these various tetrahedra will be discussed under silicate structures.

Distortions from perfect eightfold cubic arrangement and from perfect octahedra are common in MO_2 structures, where M is a transition metal. It is believed that these distortions have their origin in the breakdown of the assumptions of the pure ionic model—namely, that ions are incompressible and that a stable arrangement of ions of one charge about an ion of opposite charge requires simultaneous contact between the central ion and all its neighbors. If these restrictions are relaxed, it is found that, although the closest packing model is retained, there is not necessarily a sharp transition from cubic to octahedral and from octahedral to tetrahedral coordination at the respective limiting radius ratios. Rather, there is a gradually developing instability of structure leading to a change from a cubic to octahedral and from an octahedral to tetrahedral structure as the size of the central ion decreases. This leads successively to off-center displacements of the central ion and to the same phenomenon of rattling that was suggested to occur among some of the alkali halides. Thus, for example, with the titanium(IV) oxide radius ratio of 0.49 being near the critical size for the titanium atom to have an octahedral environment, oxygen-oxygen repulsions can come into play, the octahedron cannot contract any further, and a slight distortion occurs. The Ti^{4+} ion then must have some deviation from octahedral symmetry surrounding it.

It has also been found that metal ions such as titanium with ionic radii close to this critical size that allows distortion are found in ferroelectric compounds. For example, from 1460°C to the melting point of 1612°C, barium titanate ($BaTiO_3$) has a hexagonal structure. From 120° to 1460°C, it is cubic with the perovskite structure, the mineral $CaTiO_3$, which is illustrated in Fig. 6.29. Between 5 and 120°C, barium titanate is tetragonal

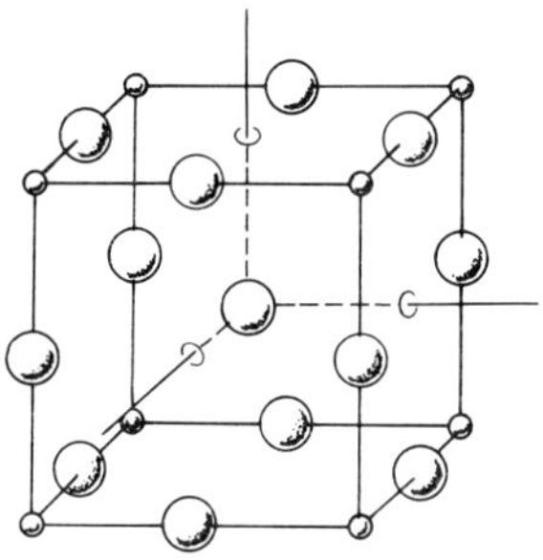

Figure 6.29. The cubic $BaTiO_3$ (perovskite) structure. The small Ti^{4+} ions are at the corners and octahedrally surrounded by O^{2-} ions. The large Ba^{2+} ion is in cubic c.c.p. packing with twelve oxygen neighbors.

with displacements of the Ti^{4+} ion occurring along the fourfold axis of the octahedron, the Ti—O distances being 2.00 (four), 1.87, and 2.17 A. At lower temperatures, orthorhombic and rhombohedral distorted structures occur with further displacements of the Ti^{4+} ion. The tetragonal, orthorhombic, and rhombohedral distorted structures of barium titanate are ferroelectric. A large number of XYO_3 compounds with metal ions that permit these distortions are known, and their ferroelectric and ferromagnetic properties are of interest.

The nickel arsenide structure discussed previously can be characterized as being an h.c.p. array of X atoms with all its octahedral holes filled with M atoms. When only half the octahedral holes are filled in the h.c.p. array, the cadmium iodide structure occurs. A similar structure is taken by cadmium chloride, in which half the octahedral holes in a c.c.p. of anions are filled by M atoms. Some substances exhibiting these structures are listed in Table 6.9, and the cadmium iodide structure is illustrated in Fig. 6.30.

The cadmium iodide structure is similar to the nickel arsenide structure, except that alternate layers of M atoms are missing. Therefore, the X atoms

TABLE 6.9 *Some Crystals with Cadmium Iodide and Cadmium Chloride Structures*

$CdCl_2$ Structure:
$MgCl_2$, $MnCl_2$, $FeCl_2$, $CoCl_2$, $NiCl_2$, $CdCl_2$
$ZnBr_2$
NiI_2

CdI_2 Structures:
$TiCl_2$, VCl_2
$MgBr_2$, $FeBr_2$, $CoBr_2$, $NiBr_2$, $CdBr_2$
MgI_2, CaI_2, TiI_2, MnI_2, FeI_2, CoI_2, ZnI_2, CdI_2, PbI_2
$Mg(OH)_2$, $Ca(OH)_2$, $Mn(OH)_2$, $Fe(OH)_2$, $Co(OH)_2$, $Ni(OH)_2$, $Cd(OH)_2$
TiS_2, ZrS_2, PtS_2, SnS_2
$TiSe_2$, VSe_2
$TiTe_2$, VTe_2, $CoTe_2$ (at high temperature), $NiTe_2$

are shielded from each other by cations on one side only. This arrangement is incompatible with a simple ionic structure because of the repulsion forces involved between anions. Van der Waals forces of attraction between these large polarizable anions whose charge has been greatly reduced by covalent bonding account for the stability of the structure. However, perfect cleavage of the crystals along these layers where weaker Van der Waals forces exist would be anticipated.

As might be expected, compounds showing the cadmium iodide structure tend to show ranges of composition like some compounds showing the nickel arsenide structure. This phenomenon is more pronounced among the sulfides, selenides, and tellurides. For example, Table 6.9 lists the compound $CoTe_2$ as stable in the cadmium iodide structure at

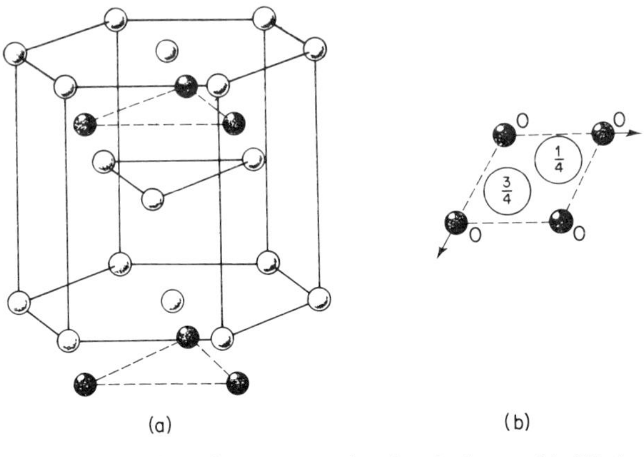

Figure 6.30. (a) The CdI_2 structure, showing the h.c.p. of iodide ions; (b) plan of the unit cell of the CdI_2 structure projected on a plane perpendicular to the *c* axis.

high temperature. This structure with 66.7 atomic percent tellurium can be retained by quenching the sample. However, CoTe is stable at lower temperatures and it shows the nickel arsenide structure. In the range of from 50 to 66.7 atomic percent tellurium, a continuous change from CoTe to $CoTe_2$ occurs and there is a slight gradual change in the dimensions of the unit cell. This is explained in terms of closest packing by realizing that the tellurium atoms (ionic radius 2.21 A) are arranged in a hexagonal closest-packed structure. The cobalt atoms (ionic radius 0.74 A) occupy all the octahedral holes at the composition CoTe. As the proportion of cobalt decreases, some of these octahedral holes become vacant, until at composition $CoTe_2$ only half these holes are occupied. Both structures, therefore, can be explained in terms of the packing model, but the preference of cobalt for octahedral holes shows that the strict ionic model is unsatisfactory, as was discussed for the nickel arsenide structure. Thus, radius ratio rules are a useful guide to predictions of structure, but the crudeness of the ionic model and the uncertainty of values of ionic radii, especially as directional bonding increases, make the radius ratio at most a semiquantitative tool for predicting structure.

The closest-packing model itself can be quite helpful in classifying structures that are preferred by many inorganic substances. Table 6.10 lists some of the more familiar inorganic structures.

TABLE 6.10 ***Some Inorganic Structures***

Holes Filled: Tetrahedral	Holes Filled: Octahedral	Closest Packing	Formula	Type Structure	Examples
Cubic Closest Packing (c.c.p., A_1):					
		All M	M	c.c.p.	Ni, Cu, Ag, etc.
	All M	All X	MX	Rocksalt	MgO, chlorides, sulfides
All X		All M	MX_2	Fluorite	CaF_2, etc.
All M		All X	M_2X	Antifluorite	Most oxides, sulfides, selenides, and tellurides of alkali metals
$\frac{1}{2}M$		All X	MX	Zincblende	Many II-VI, III-V compounds
$\frac{1}{8}M$	$\frac{1}{2}M$	All X	MM_2O_4	Spinel	$MgAl_2O_4$, etc.
Hexagonal Closest Packing (h.c.p., A_3):					
		All M	M	h.c.p.	Zn, Cd, Mg, etc.
	All$M$	All X	MX	NiAs	CrS, FeS, etc.
	$\frac{2}{3}M$	All X	M_2X_3	Corundum	Al_2O_3, La_2O_3, etc.
$\frac{1}{2}M$		All X	MX	Wurtzite	ZnO, CdS, etc.

More complex substances can be discussed in terms of closest packing. For example, the crystallographic unit cell for the mineral spinel, $MgAl_2O_4$, may be considered to have 32 oxide ions in cubic closest packing. This arrangement makes available 32 octahedral holes and 64 tetrahedral holes in the unit cell. A large number of compounds having a variety of atoms present in different ratios crystallize in the spinel structure. Hence, the distribution of cations among tetrahedral and octahedral holes has been studied carefully. If, for the sake of simplicity, one assumes predominently ionic bonding, then the compositional formula AB_2O_4 requires that the positive charge must equal $+8$. Combinations such as $M(\text{II})M(\text{III})_2O_4$, $M(\text{IV})M(\text{II})_2O_4$, and $M(\text{VI})M(\text{I})_2O_4$ are possible. In so-called normal spinels, AB_2O_4, B ions occupy octahedral sites and A ions occupy tetrahedral sites. It is also possible for half of the B ions to occupy tetrahedral sites, the remaining half together with the A ions being distributed over the octahedral sites, this is the so-called *inverse* spinels, $B[AB]O_4$. The brackets are sometimes used to enclose the ions in octahedral sites. Thus, $MgAl_2O_4$, in which Mg(II) and Al(III) are in tetrahedral and octahedral sites, respectively, is a normal spinel. On the other hand, $Fe[CuFe]O_4$ has an inverse structure, in which Fe(III) occupies tetrahedral sites and both Cu(II) and Fe(III) are in octahedral sites. Table 6.11 lists some typical $M(\text{II})M(\text{III})_2O_4$ spinels of known structure. All known $M(\text{IV})M(\text{II})_2O_4$ spinels, e.g., $SnCo_2O_4$ and $TiZn_2O_4$, are inverse.

TABLE 6.11 ***Some Crystals with the Spinel Structure***

Normal Structure	Inverse Structure
$ZnAl_2O_4$	$CuFe_2O_4$
$FeAl_2O_4$	$NiFe_2O_4$
$ZnAl_2S_4$	$CoFe_2O_4$
$CoCr_2O_4$	$FeFe_2O_4$
$MgCr_2O_4$	$NiMn_2O_4$
$HgCr_2S_4$	$MgGa_2O_4$
$MnMn_2O_4$	$FeIn_2O_4$
$CoCo_2O_4$	$CoIn_2O_4$
$FeCr_2O_4$	$NiIn_2O_4$

Radius ratio rules predict that tetrahedral sites would be occupied by the smaller cations. However, the A cations generally have radii between 0.6 and 0.8 A, whereas the B cations generally have radii between 0.5 and 0.7 A. A major factor which can explain why the radius ratio rules are not followed in some spinels is the increased stability in octahedrally vs. tetrahedrally coordinated central metal ions as a result of the number of d electrons in the particular metal ion. The resulting energy factor, the crystal field stabilization energy, is discussed in Chapter 9, where the structures of the minerals Fe_3O_4, Co_3O_4, and Mn_3O_4 are considered. Spinels ideally have cubic symmetry, as expected from the closest packing model. However, some compounds show distortion along one axis in the octahedral sites, causing tetragonal symmetry in the crystals. These distortions from cubic symmetry are explained by Jahn-Teller stabilizations, also discussed in Chapter 9.

If the tetrahedral holes containing Zn(II) in the zincblende structure are replaced by alternate Cu(II) and Fe(II), the structure is that of chalcopyrite (Fig. 6.12). As a result of these substitutions, the atoms at the corners of the zincblende unit cell are not all the same kind, so that the repeat unit is doubled in one direction. Thus, chalcopyrite belongs to the tetragonal crystal system, whereas zincblende belongs to the cubic system, and crystals of each would exhibit their own respective external symmetry.

The spinel and zincblende crystals described are two examples of possible combinations of cations occupying sites in close-packed inorganic systems. However, for oxides and sulfides, conditions for substitution of cations differ. In spinels and other oxide crystals, where ionic character often predominates substitution depends primarily on similar ionic radius of the cations and maintenance of electric neutrality in the crystal. Chemical properties of the ions are relatively unimportant. Thus, one might predict some substitution of Mg^{2+} (0.65 A) for Zn^{2+} (0.74 A) in sites in an oxide structure. However, since zinc is considerably more electronegative than magnesium, directional bonds become important in zinc sulfides, and substitution of magnesium for zinc in zincblende or wurtzite might not be expected. This type of chemical reasoning was first used by Goldschmidt to explain the relative abundance of elements in the earth's crust in oxide and sulfide minerals.

ESTIMATION OF LATTICE ENERGIES BASED ON AN IONIC MODEL

Since a thermodynamic consideration of reactions and properties of solids often involves crystals being pulled apart into their constituents at some stage of the consideration, knowledge of the forces holding crystals together and crystal lattice energies is significant. Known methods that satisfactorily measure directly the lattice energy of predominantly ionic substances are difficult to measure. Moreover, crystal lattice energies can not be obtained for some compounds from a Born-Haber treatment because of the difficulty in measuring electron affinity values for the anions. However, by taking advantage of the simple ionic model, these lattice energies can often be calculated with accuracy.

If R is the distance between the center of a sodium ion and the center of the six nearest chloride ions in Fig. 6.27(a), then the next nearest neighbors surrounding the sodium ion are 12 other sodium ions at a distance of $R\sqrt{2}$. Beyond are eight more chloride ions at $R\sqrt{3}$ and then six sodium ions at $2R$. This geometrical analysis thus generates an infinite series in terms of R for all ionic substances with the rocksalt structure. The electrostatic (coulombic) interaction energy, E_{coul}, of the sodium ion with each of these surrounding ions will be equal to the product of the charge on each ion, Z^+ and Z^-, divided by the respective distances. Hence, the total electrostatic energy for a sodium ion will be

$$E_{\text{coul}} = \frac{6e^2}{R}(Z^+Z^-) + \frac{12e^2}{R\sqrt{2}}(Z^+Z^+) + \frac{8e^2}{R\sqrt{3}}(Z^+Z^-) + \frac{6e^2}{2R}(Z^+Z^+) \ \ldots \quad [6.5]$$

This equation can be simplified algebraically to

$$E_{\text{coul}} = \frac{e^2(Z^+Z^-)}{R}\left[6 - \frac{12}{\sqrt{2}} + \frac{8}{\sqrt{3}} - \frac{6}{2} + \frac{24}{\sqrt{5}} - \frac{24}{\sqrt{6}} \cdots\right] \quad [6.6]$$

Equation [6.8] represents the electrostatic potential energy for the process

$$\text{Na}^+\,(\text{g}) + \text{Cl}^-\,(\text{g}) = MX(\text{s}) \quad [6.7]$$

and if multiplied by Avogadro's number, N, it is the energy per mole of NaCl, since the same geometric analysis is applicable in the rocksalt structure to both cations and anions. The expression inside the brackets in equation [6.6], commonly denoted A, is thus the sum of an infinite series depending only on the location of the ions in the lattice. The sum converges toward the value 1.748 and is called the *Madelung constant* for a rocksalt structure. Madelung constants have been evaluated for many common structures, and some are given in Table 6.12.

TABLE 6.12 ***Madelung Constants for Some Structures***

Structure		Madelung Constant, A
Rocksalt	M^+X^-	1.74756
Cesium chloride	M^+X^-	1.76267
Zincblende	$M^{2+}X^{2-}$	1.63806
Wurtzite	$M^{2+}X^{2-}$	1.64132
Fluorite	$M^{2+}X_2^-$	5.03878
Cadmium iodide	$M^{2+}X_2^-$	4.71
Corundum	$M_2^{3+}X_3^{2-}$	25.0312

In a simple ionic model, the approach distance between adjacent ions due to electrostatic interactions is limited by repulsions occurring due to overlap of electron clouds of these ions. Born assumed that this interaction between ions is a function of interionic distance, which gives rise to a repulsion energy of the form

$$E_{\text{rep}} = \frac{B}{R^n} \qquad [6.8]$$

where B and n are adjustable parameters related to the ions and type of lattice involved and R is the distance between nearest unlike ions. Moreover, Born assumed that the total crystal energy per mole, U, is the sum of the electrostatic and repulsion terms

$$U = E_{\text{coul}} + E_{\text{rep}} = \frac{NAe^2(Z^+Z^-)}{R} + \frac{NB}{R^n} \qquad [6.9]$$

At the equilibrium bond length, R_0, the electrostatic attraction must equal the repulsion term. Hence, the value of U is a minimum at this equilibrium

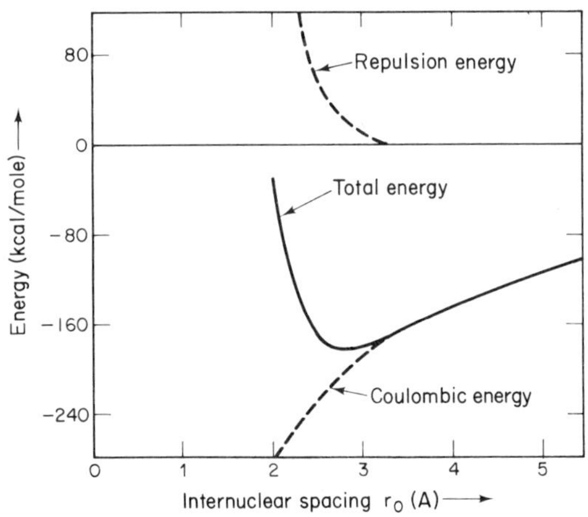

Figure 6.31. Variation of crystal energy U with internuclear distance R_0 in NaCl.

distance, as indicated in Fig. 6.31. This observation may be used to eliminate the parameter B,

$$\left(\frac{dU}{dR}\right)_{R=R_0} = 0 = -\frac{NAe^2(Z^+Z^-)}{R_0^2} - \frac{nNB}{R_0^{n+1}} \qquad [6.10]$$

which yields

$$B = -\frac{Ae^2(Z^+Z^-)}{n}R_0^{n-1} \qquad [6.11]$$

Substituting the value of B back into equation [6.5], we obtain

$$U = \frac{NAe^2(Z^+Z^-)}{R_0}\left(1 - \frac{1}{n}\right) \qquad [6.12]$$

The numerical value of n was shown by Born to be related to the compressibility of the crystal. Table 6.13 contains values of n that Slater derived from his own compressibility measurements and Table 6.14 contains theoretical values of n. Values for closed-shell electronic configurations in Table 6.12 were estimated theoretically by Pauling and show reasonable correlation with experimental data. It is clear from equation [6.12] that a change of n by 1.0 affects the value of U by only 1 to 2%. Average values of n, for example in ZnO $n = 8.0$, may be used to estimate lattice energies if experimental compressibilities of substances are not available. However, accuracy decreases as covalent bonding character increases between ions.

TABLE 6.13 ***Values of n Derived from Compressibility Measurements***

Structure	Repulsion Term, n
LiF	5.9
LiCl	8.0
LiBr	8.7
NaCl	9.1
NaBr	9.5

TABLE 6.14 ***Values of n Estimated Theoretically***

Ion Type	Example	Repulsion Term, n
He	Li^+	5
Ne	Na^+, Mg^{2+}	7
Ar, Cu^+	K^+, Zn^{2+}	9
Kr, Ag^+	Br^-, Se^{2-}	10
Ze, Au^+	Cs^+, I^-	12

Quantum mechanical calculations indicate that the energy associated with interionic repulsions is better related to interionic distance in an exponential form

$$E_{\text{rep}} = be^{-R/\rho} \qquad [6.13]$$

where b and ρ are constants. Using the same procedure as was used to solve the Born approximation, we find the lattice energy equation

$$U = \frac{NAe^2(Z^+Z^-)}{R_0}\left(1 - \frac{\rho}{R_0}\right) \qquad [6.14]$$

The value of ρ, calculated directly from measurement of the bulk compressibility of the crystal, is found to be 0.311 A for many salts. Hence, equation [6.14] is more useful than equation [6.12] when more accurate calculations are made. Figure 6.31 shows the variation of the energy terms with internuclear separation for sodium chloride using equation [6.14].

Consideration of van der Waals forces and the zero point energy of a crystal were also included in more exact calculations of the lattice energy by Born and Mayer. The two added energy terms are relatively small but can be considerably important in determining the relative stability of two different structures for a substance, since they are comparable with the difference in energy of the potential modifications. The van der Waals energy term is related to the attraction that adjacent atoms or ions have for each other due to a synchronization of instantaneous oscillating dipoles existing in them. The attractive energy is approximately proportional to $1/R^6$ and can be calculated from polarizability and ionization potential values of the ions involved. The zero point energy term is related to the vibrations of ions persisting in a crystal even at absolute zero. This zero point energy is therefore subtracted from the attraction terms in determining the total lattice energy. Table 6.15 gives calculated values of the

TABLE 6.15 ***Relative Contributions of Energy Terms to Lattice Energies (in eV)***

Substance	Madelung	Repulsion	Van der Waals	Zero Point
LiF	−12.4	1.93	−0.17	0.17
NaF	−10.8	1.54	−0.20	0.13
NaCl	−8.8	1.02	−0.23	0.07
NaBr	−8.4	0.90	−0.24	0.06
NaI	−7.7	0.74	−0.27	0.05
CsI	−6.4	0.63	−0.48	0.03

energy terms of some alkali halides to indicate the relative contributions to the total lattice energy. Note that attraction energies have inherently negative values.

IONIC RADII

The useful presupposition for the simple ionic model—that ions are incompressible, nonpolarizable spheres—breaks down when ionic radii are considered. Since electronic wave functions approach zero asymptotically, it is not possible to attach a precise radius to an atom or ion. When atoms have covalent bonds, radii may be estimated from X-ray diffraction measurements, a conventional method for determination of interatomic distances in solids. However, in ionic solids, the radii can not be calculated by simply halving interatomic distances. There is a further complication that the radius of an ion will vary, depending on the type and number of its neighbors in the crystal lattice.

A number of semiempirical methods have been proposed to estimate ionic radii. One common set of values of ionic radii used by chemists was proposed by Pauling. The derivation of his numerical values begins with the generalization that in an isoelectronic series, ions with the greatest nuclear charge are smallest. For example, the radii of ions with a neon configuration would be $S^{2-} > Cl^- > K^+$. Using the alkali halide crystals as standards, one assumes that the relative sizes of a pair of isoelectronic ions—e.g., K^+ and Cl^-—are inversely proportional to the effective nuclear charge operating on the outermost electron shell. This relationship can be found in Bohr's expression for the radius of an electron in an orbit around its nucleus,

$$r = \frac{n^2h^2}{Z4\pi^2\mu e^2} \qquad [6.15]$$

where all the letters connote the usual representations. The *most* probable radius may be calculated from equation [1.9]. This relation of the radius of an ion to the effective nuclear charge is assumed to be

$$r = \frac{C}{Z-S} \qquad [6.16]$$

where Z is the nuclear charge and S is Slater's screening constant (see Chapter 2), and C is a constant determined by the principal quantum number of the outer electrons on the ion.

Calculation of the value of the constant C for the argon configuration is possible if the KCl internuclear distance, known from X-ray measurements, is introduced:

$$R_{KCl} = 3.14\,\text{A} = \frac{C}{7.75} + \frac{C}{5.75} \quad \text{and} \quad C = 10.38 \qquad [6.17]$$

The semiempirical values for the radii of K^+ and Cl^- are then found:

$$r_{K+} = \frac{10.38}{7.75} = 1.33\ \text{A}, \quad r_{Cl-} = \frac{10.38}{5.75} = 1.81\ \text{A} \qquad [6.18]$$

Following this procedure with appropriate refinements, we find radii for the other isoelectronic alkali halides: sodium fluoride, rubidium bromide, and cesium iodide. A radius of 0.60 A is given to lithium(I) ion on the basis that the observed Li—O distance in lithium oxide is 2.00 A and the best empirical value for the oxide ion is 1.40 A. Interionic distances for all alkali halides derived from these radii show good correlation with measured values, except for lithium chloride, lithium bromide, and lithium iodide, whose radius ratios fall below 0.414. As would be expected, anion-anion repulsions in these crystals cause larger observed interionic distances than those predicted.

Use of the values of S and C in equation [6.16] for the appropriate noble gas configuration and the ionic radii of multiply charged ions, e.g., Ca^{2+} and S^{2-}, obtained from the same procedure, give interionic distances greater than measured values. Apparently increased electrostatic attraction draws the multiply charged ions closer together. The radii values are actually for multivalent ions, which retain their electron configurations but enter into electrostatic interaction as if they were univalent, like the alkali halides. More realistic values are obtained if this multivalency is taken into account, which we do by solving for the internuclear distance, R_0, in equation (6.10) and calling it R_z,

$$R_z = \left(\frac{Bn}{Ae^2Z^2}\right)^{1/n-1} \qquad [6.19]$$

Assuming the values of B, n, and A remain unchanged, the ratio of R_z and the univalent interionic distance R_1 (where $Z = 1$), and therefore of the respective radii, becomes

$$\frac{R_z}{R_1} = Z^{-2/n-1} \qquad [6.20]$$

New crystal radii, r_z, are found to reproduce satisfactorily measured interionic distances.

Pauling adopted a set of ionic radii values based on his own data and estimates of other crystallographers. Table 6.16 lists these radii for some important ions, using the rocksalt structure as standard, where there are six nearest neighbors of opposite charge.

TABLE 6.16 ***Crystal Radii of Ions (A)***

Ion	Radius	Ion	Radius	Ion	Radius	Ion	Radius
Li^+	0.60	Ca^{2+}	0.99	Al^{3+}	0.50	Pb^{4+}	0.84
Na^+	0.95	Sr^{2+}	1.13	Ga^{3+}	0.62	Ti^{4+}	0.68
K^+	1.33	Ba^{2+}	1.35	In^{3+}	0.81		
Rb^+	1.48	Zn^{2+}	0.74	Tl^{3+}	0.95	O^{2-}	1.40
Cs^+	1.69	Cd^{2+}	0.97	La^{3+}	1.15	S^{2-}	1.84
Cu^+	0.96	Hg^{2+}	1.10	Cr^{3+}	0.69	Se^{2-}	1.98
Ag^+	1.26	Sn^{2+}	1.12	Mn^{3+}	0.66	Te^{2-}	2.21
Au^+	1.37	Pb^{2+}	1.20	Fe^{3+}	0.64		
Tl^+	1.44	Cr^{2+}	0.84	Co^{3+}	0.63	H^-	2.08
		Mn^{2+}	0.80			F^-	1.36
Be^{2+}	0.31	Fe^{2+}	0.75	Si^{4+}	0.41	Cl^-	1.81
Mg^{2+}	0.65	Co^{2+}	0.72	Ge^{4+}	0.53	Br^-	1.95
		Ni^{2+}	0.69	Sn^{4+}	0.71	I^-	2.16

Interionic distances, and therefore ionic radii, in structures other than rocksalt may be estimated by taking a ratio of these values for each structure. This ratio may be derived from equation (6.10).

$$\frac{R_1}{R_2} = \left(\frac{A_2 B_1}{A_1 B_2}\right)^{1/n-1} \qquad [6.21]$$

If the value of the repulsion term B is assumed to be proportional to the coordination numbers in the respective structures—i.e., the number of contacts between positive and negative ions—approximate agreement with measured interionic distances in the new structure is found. However, for crystals in which the charge of the anion and cation are unequal—e.g., crystals with fluorite and rutile structures (Table 6.8—this structural correction is not necessary and the direct sum of the crystal radii in Table 6.16 is sufficient.

We emphasize again that the ionic model used to develop the set of crystal radii in Table 6.16 is only a model. Real cations in solids contain a positively charged nucleus embedded in a cloud of electrons insufficient to shield the nucleus from interacting with electrons of neighboring ions. Moreover, these cations possess low-energy outer shell orbitals capable of accepting charge. Hence, neighboring anions with positively charged nuclei embedded in a surplus of electrons are capable of donating electron pairs so that a sharing of their charge with the cations is possible. Electron density maps of solids capable of plotting the positions of nuclei and relative amounts of negative charge around these nuclei verify that there is some sharing of negative charge even in so-called ionic substances such as potassium chloride. The relative amount of sharing of negative charge by nuclei corresponding to a free energy minimum varies from substance to substance; therefore, the effective radius of a particular ion depends to some degree on its environment.

Suggested Readings

1. Budnikov, P. P., and A. M. Ginstling; translated and edited by Kenneth Shaw, *Principles of Solid State Chemistry*, Gordon and Breach Science Publishers, New York, 1968.
2. Bunn, C. W., *Chemical Crystallography*, Oxford University Press, London, 1961.
3. Dunitz, J. D., and L. E. Orgel, "Stereochemistry of Inorganic Solids," in *Advances in Inorganic Chemistry and Radiochemistry*, Vol. II, edited by Emeleus, H. J., and A. G. Sharpe, Academic Press, Inc., New York, 1959.
4. Hannay, N. B., *Solid-state Chemistry*, Prentice-Hall, Inc., Englewood Cliffs, 1967.
5. Kittel, C., *Elementary Solid State Physics*, John Wiley & Sons, Inc., New York, 1962.
6. Phillips, F. C., *An Introduction to Crystallography*, John Wiley & Sons Inc., New York, 1963.
7. Wells, A. F., *Structural Inorganic Chemistry*, Oxford University Press, London, 1962.

Study Questions

6.1. One crystal is a square tablet and a second crystal is a cube. Each crystal contains 1 mole of KBr. Would one crystal dissolve more rapidly in water than the other? Why?

6.2. A crystal belonging to the hexoctrahedral class of the cubic system possesses the highest possible crystalline symmetry, $4/m\ 3\ 2/m$, often written as $m3m$. Sketch a crystal belonging to this system containing {001} and {111} forms.

6.3. The X-ray diffraction pattern of a newly synthesized inorganic compound is unable to show readily to which crystal system the new compound belongs. It is noticed, however, that some crystals of the compound show two cleavages of equal quality at right angles and a cleavage whose surface shows poorer quality at right angles to the first two. Discuss the crystal systems that might describe this.

6.4. Give the Miller indices of the faces belonging to the CaF_2 crystals in Fig. 6.2(d).

6.5. Give the Miller indices of the faces belonging to forms (a) cubic {001} and (b) tetragonal {110}.

6.6. The mineral barite, $BaSO_4$, the chief source of barium, is often found as a tabular crystal belonging to the orthorhombic system with symmetry mmm. Some crystals show principally {001, 010, 100} forms with {011} and {101} forms developed to a slight extent. Sketch such a crystal.

6.7. What are the symmetry elements of:

(a) a shoe,
(b) the human body, disregarding the insides,
(c) a square pyramid,
(d) the four crystals in Fig. 6.2,
(e) the quartz crystal in Fig. 6.3(a),
(f) the orthorhombic crystal in Fig. 6.9.

6.8. Compare the packing efficiency of spheres of the same size in two dimensions (therefore as circles) in closest packing with an arrangement of circles on the points of a square lattice.

6.9. Show that the triangular and octahedral holes in a closest packed arrangement can accommodate spheres with radius ratios of 0.155 and 0.414, respectively.

6.10. Show that the packing efficiency of spheres is 74% in c.c.p. and h.c.p. and 68% in b.c.c.

6.11. Discuss the trends to non-stoichiometric solids among inorganic compounds.

6.12. For the following ionic crystals compare calculated values of lattice energies with the values obtained from a Born-Haber calculation which are in parentheses (kcal $mole^{-1}$):

$$LiCl(198),\ LiI(181),\ \text{c.c.p. } ZnO(972),\ ZnSe(845).$$

6.13. Consider the hypothetical compound KrBr. Assuming it would have the same structure as RbBr, estimate the interionic distance Kr^+-Br^-. If the first ionization potential of krypton is 14.0 electron volts, show whether the formation of this compound is to be expected. Neglect the entropy factor.

6.14. Table 6.6 indicates that CaS takes a rocksalt structure. Using Pauling radii, what is the interionic distance in this salt? What would be the interionic distance if univalent radii were used for the Ca^{2+} and S^{2-} ions?

6.15. The interionic distance in solid CsCl measured from X-ray data is found to be 3.47 A when it has a rocksalt structure and 3.56 A when it has a cesium chloride structure. Calculate the crystal radii of Cs^+ and Cl^- in each case. Which structure would probably exist at high pressure?

6.16. Crystals of the natural mineral chrysoberyl, $BeAl_2O_4$, are very hard and have a high melting point. Would you predict these properties?

6.17. Draw a rocksalt structure and prove that the values of the second and third terms of the Madelung constant for this structure contain $12/R\sqrt{2}$ and $8/R\sqrt{3}$ terms, respectively.

6.18. Aluminum is known to have four- and sixfold coordination in solids. If one wished to synthesize a feldspar (AB_4O_8 in which A is Na, K, Ca and Ba, and B is Si and Al) the most common of all minerals, in which as many aluminum atoms as possible substitute for fourfold coordinated silicon atoms, would one use high or low temperature and pressure controls? Assume substitution of Al^{3+} for Si^{4+} can be compensated for by appropriate replacement of A^+ ions by A^{2+} ions.

CHAPTER 7

Bonding in Metals

Both the valence bond and molecular orbital treatments have had a measure of success in their rationalization of the metallic bond. Valence bond approximations better explain the structural properties, whereas the molecular orbital treatment seems to be more suitable in explaining physical properties. Brief, qualitative presentations of both theories will be presented.

The valence bond treatment considers all possible interactions between nearest neighbor atoms in the metallic crystal. The number of nearest neighbor atoms in a crystal is usually 8 or 12. Metals, however, do not have sufficient valence electrons to form covalent bonds with this many nearest neighbors. Consequently, a large number of resonance structures are invoked to account for all the interactions. Implied in this conclusion is that bonds holding the metallic atoms together are not localized but highly delocalized. This is illustrated in Fig. 7.1.

The lithium atoms form a body-centered cubic unit cell in the crystal lattice; therefore, each lithium atom has eight nearest neighbors. If all eight nearest neighbors are bound covalently to the central metal atom, 16 electrons are required but only 9 valence electrons are available. Therefore, the synchronous [Fig. 7.1(a)] and unsynchronous [Fig. 7.1(b)] resonance forms account for the bonding strength holding a metallic crystal together.

This simple picture model offers a qualitative explanation as to why metals are conductors of electricity and nonmetals are nonconductors (insulators). In general, the nonmetals that obey the $8 - N$ rule (they form $8 - N$ covalent bonds, where N is the periodic group number of the nonmetal) have employed all their available low energy orbitals in the formation of these covalent bonds. Now they have no low energy orbitals available for the acceptance of extra electrons and cannot form the ionic resonance structures shown in Fig. 7.1(b). Lithium may use one of its $2p$

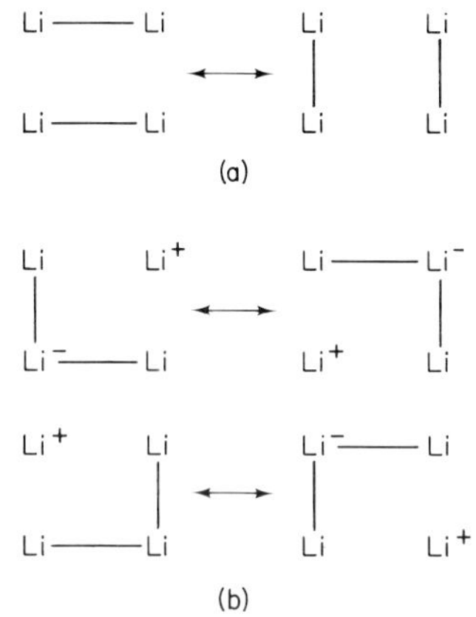

Figure 7.1. (a) The synchronized resonance forms of lithium metal crystals; (b) unsynchronized resonance forms resulting in ionic species in the metallic crystal.

orbitals, which does not differ too much in energy from the $2s$ orbital to form the ionic resonance forms. Still available to lithium are two more vacant $2p$ orbitals that may be used for more resonance forms or for bonding.

Pauling has employed the terms *metallic valence* and *metallic orbital* in his application of the valence bond theory to metals. Basing his concept of metallic valence on physical properties of hardness, density, melting point, etc., Pauling assigned a metallic valence of 1 for potassium, 2 for calcium, 3 for scandium, 4 for titanium, and 5 for vanadium. Chromium, manganese, iron, cobalt, and nickel are said to have metallic valences of 6, but copper has a metallic valence of 5 and zinc 4. Consider copper with its ground state configuration of the outer electrons being $3d^{10}4s^1$. Copper is said to have a total of nine orbitals available for bonding: five $3d$, one $4s$, and three $4p$ orbitals. One stable orbital must remain available to serve as the metallic orbital. Eight orbitals are now available for placing electrons, and the resulting configuration is

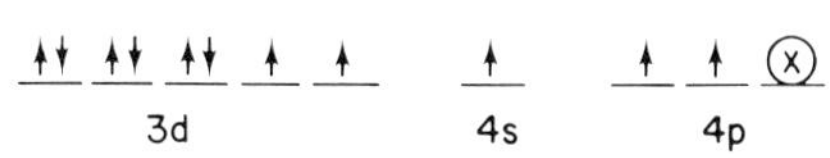

The (X) indicates the metallic orbital, which is available to provide mobility to the electrons from the other atoms. The unpaired electrons are available for bonding to the nearest neighbors describing a metallic valence of 5. Since copper is diamagnetic, the unpaired electrons are not unpaired for long.

Iron has a metallic valence of 6. Consideration of the metallic orbital would yield the following configuration:

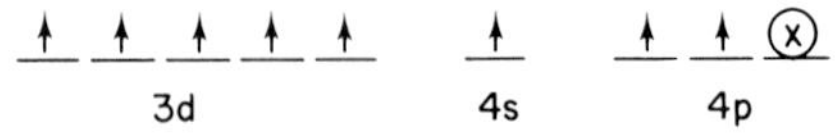

The remaining two unpaired electrons account for the observed paramagnetism of iron.

Tin also serves as an interesting example, illustrating that structural differences can cause differences in properties. Tin exists in two modifications—white and gray. Gray tin appears to be tetravalent and is an insulator. For the tetravalent gray tin, all the low energy orbitals are used in covalent bonding. Thus, there are no low energy orbitals available for the resonance structures; consequently, metallic conduction is not possible. On the other hand, white tin occurs in a crystalline form in which the tin is divalent; this modification is a metallic conductor. Since this form of tin is divalent, there are still two low energy orbitals remaining to be used for ionic resonance forms; thus, metallic conduction occurs.

The molecular orbital (MO) treatment of bonding in metals begins by placing nuclei at their equilibrium positions in the crystals, building molecular orbitals through the system, and then placing the electrons in the molecular orbitals.

The MO description is based on the following experimental facts. When an isolated atom absorbs electromagnetic radiation, an electron may be excited from a filled low-lying orbital to an empty orbital of higher energy. The transition that will occur depends on the energy of the radiation absorbed. Most weakly held electrons will absorb less than 10 eV for their lowest energy transition. This absorption will occur in the visible or ultraviolet region. A series of such transitions produces an absorption spectrum. In like manner, when the excited electron returns to its ground state, a quantum of radiation, $h\nu$, is emitted. This behavior would give rise to an emission spectrum.

When X-ray radiation is used, electrons from even lower energy levels in the atom may be excited to empty orbitals of higher energy. An electron from one of the filled higher orbitals will then fall back to a vacancy created by the excited electron and an X ray is emitted. The transition $3p \rightarrow 1s$ gives rise to what is called a K emission spectrum (because an electron is falling back to the K level). The study of these absorption and emission spectra gives information relative to the electron population in materials under study.

The K emission spectrum of gaseous aluminum atoms, for example, is found to consist of sharp lines. The same absorption in solid materials containing aluminum (the pure metal or aluminum oxide) is observed to be much broader. It is reasonable to suggest that the lowest atomic orbitals of an element would not be perturbed very much by other atoms when the element forms a solid compound. It can be said that the lower atomic orbitals in a solid may be considered to be of the same energy as those of an isolated free atom. On the other hand, the higher energy atomic orbitals ($3s$ and $3p$) are on the periphery of the atom and would be easily perturbed by approaching neighbors. Therefore, these orbitals would have different energies in comparison with the isolated atoms, depending on the degree of perturbation. The K emission spectrum would not be sharp because the frequencies of radiation emitted would not all be identical. They would be close but not the same. In the solid form of aluminum, the varying energies of the $3s$ and $3p$ orbitals are said to form an *energy band* rather than discrete levels.

Comparison of the $3p \rightarrow 2s$ emission spectrum of gaseous aluminum atoms and aluminum metal also shows the broader line in the case of the metal. However, comparison of the $2p \rightarrow 1s$ transition in the gaseous free atoms and the solid metal shows that they are almost identical, indicating that there is not much change in these atomic orbitals on proceeding from the gaseous free atom to the solid metal.

Thus, X-ray spectroscopy suggests that when metallic atoms combine to form a crystal, the lower atomic orbitals do not undergo much perturbation, whereas the higher atomic orbitals appear to form bands of energy. These ideas are illustrated in Fig. 7.2.

The comparison of both the absorption and emission spectra of aluminum metal and aluminum oxide furnishes interesting insights. Figure 7.3 contains the plots of the intensity of the K emission (and the K absorption) vs. the energy of the given emission. The K emission spectrum of the pure metal falls off rather sharply at its upper energy limit, which seems to indicate that there is a definite upper energy limit on the electrons in the metal. The absorption spectrum rises rather sharply at its lower limit, indicating a sharply defined lower limit for absorption. Probably the most important feature on which to focus our attention is the relative positioning of the absorption and emission spectra. These spectra are essentially adjacent, absorption beginning at the point at which emission ceases. The inflection point (or discontinuity) of the emission is found to be within 0.1 eV of the absorption edge of the metal. What is suggested is that there is essentially no energy difference between the highest energy level occupied by electrons and the lowest unoccupied energy level in the metal. Stated in a more simple fashion, there is a continuous band of available energy levels the number of which is greater than the number of valence electrons (Fig. 7.2).

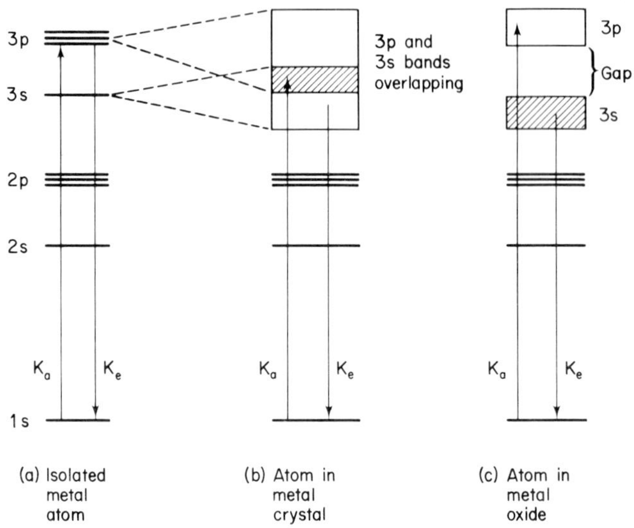

Figure 7.2. Mechanism of X-ray emission and absorption spectra of an isolated aluminum atom (a), aluminum metal (b), and aluminum oxide (c). $K_A = K$ absorption and $K_E = K$ emission.

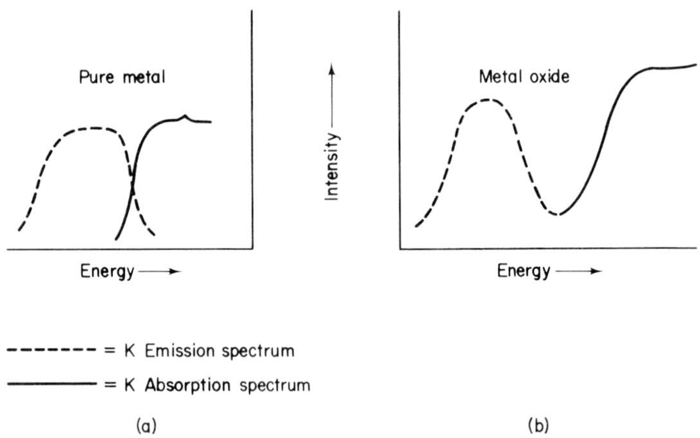

Figure 7.3. X-ray (K) emission and absorption spectra of aluminum metal and aluminum oxide. The intensity of K emission or K absorption is a function of the probability of a given electronic transition; i.e., an $s \rightarrow s$ transition is unlikely while an $s \rightarrow p$ transition is highly probable.

The ideas in the preceding paragraph lead to very interesting observations. The so-called electron deficient bonds in species such as the boron hydrides (see Chapter 12) can be considered as intermediate between ordinary covalent bonds and metallic bonds. In the *normal* covalent bond, there are just enough electrons to completely fill all the valence orbitals. In the boron hydrides, there are not enough electrons to fill all valence orbitals. The deficiency is relatively small and will usually amount to a few electrons per molecule. In the case of metallic crystals, in general, the deficiency is relatively extensive. Thus in these three classes, there is a gradual change from the exact number of electrons necessary (ordinary covalent bond) to a large deficiency (metallic bond), with the boron hydrides in between.

The spectra observed for aluminum oxide in Fig. 7.3(b) differ from those of the pure metal in that the K emission spectrum has no relatively sharp edge at the upper limit; instead, a tailing off is observed. This fact points out that no sharply defined limit exists from which electrons may come. Besides the transitions between the M—L shells of aluminum, electron transitions between the orbitals of oxygen ($2p$) and aluminum ($1s$) may be observed. Correspondingly, the absorption spectrum does not exhibit a sharp edge at its lower limit. The other noticeable difference between the two spectra in Fig. 7.3 is that the emission and absorption spectra are no longer adjacent, but a gap exists between them. All indications point to the fact that there is an energy gap between the highest occupied orbital and the lowest unoccupied orbital [see Fig. 7.2(c)].

The X-ray data for solids indicate that the valence electrons occupy bands of energy levels rather than discrete atomic orbitals. These bands are described as shown in Fig. 7.4, with the lowest energy bands being filled first as expected. These graphs are a plot of the number of available energy levels per increment of energy vs. the energy of the available energy levels.

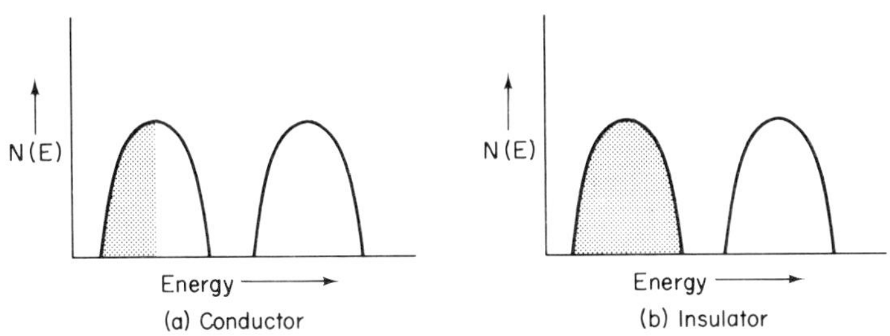

Figure 7.4. Energy bands for a metallic conductor and an insulator. Filled bands or filled portions are indicated by cross hatching.

All factors that would complicate the diagrams have been omitted, and therefore the diagrams are highly simplified. The complicating features are considerations of the probability of X-ray transitions, because this is the source of data. In addition, certain geometric areas (Brillouin zones) around the lattice may cause a distortion of the spherical Fermi surface (uppermost electron levels) near these zones. These factors will be ignored to emphasize the general picture rather than details. What the diagram tells us is that in a conductor, the electrons have mobility, whereas in an insulator, the electron mobility is hindered by the absence of empty orbitals.

X-ray spectral evidence supports the existence of bands. The molecular orbital theory seems to have the best rationalization of these data, and it appears to be most satisfactory for metals involving *d* orbitals in the bands, which would include the majority of metals.

In the molecular orbital approach (Chapter 4), one may recall that as two hydrogen atoms approach each other to form a covalent bond, one bonding MO and one antibonding MO are formed (see Fig. 4.10). The energetics of this operation are plotted in Fig. 4.9. Now, if the separation between the two atoms were continually decreased, a point would be reached where the nuclei would coalesce. The new atom, the *united atom*, would have two new atomic orbitals, which would be derived from the two original atomic orbitals. Now aluminum has an emission line near 1550 eV, which corresponds to a $3p \rightarrow 1s$ transition. The first ionization potential of aluminum is 6 eV, which is a $3p \rightarrow \infty$ transition. Therefore, one can say that the difference between the highest and lowest orbitals in this atom would be 1556 eV or 3.6×10^4 kcal (1eV = 23 kcal). This figure may serve as an estimation for the *maximum* energy difference between atomic energy levels of a *united atom* formed from two equivalent aluminum atoms. The actual energy difference should be much lower, since atomic orbitals generally differ from one another only by a few hundred kilocalories.

A simple two-atom system has been described, but remember that real systems are more complicated. In a body-centered cubic system, each atom is surrounded by eight nearest neighbors, and there are six atoms just slightly farther away than are the nearest neighbors. Remember that the number of atomic orbitals is conserved in the molecular orbital approach. If two atomic orbitals overlap to form a bond, two molecular orbitals are formed—one bonding and one antibonding orbital. When three atoms unite to form a unit, three molecular orbitals are found—one bonding, one antibonding, and one nonbonding orbital (see Figs. 4.30 and 4.31). Therefore, for N (Avogadro's number) atomic orbitals uniting to form a united species, there are N molecular orbitals formed. If this united species

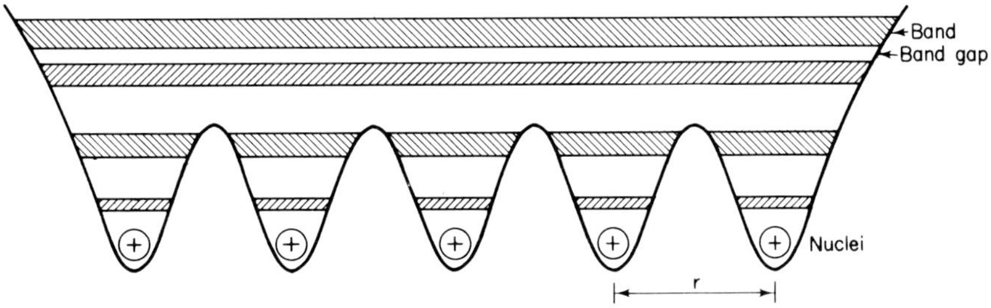

Figure 7.5. Band and band gaps at internuclear distance r.

is considered to be a united atom—say, of aluminum—then the difference between the highest energy level and the lowest energy level may be approximated to be 3.6×10^4 kcal. Now the energy difference between each of the N molecular orbitals is approximately only 10^{-18} kcal. Since this energy difference is obviously much smaller than that accounted for by ordinary thermal energy, this group of molecular orbitals may be best described as a *band* of energy levels. This example is crude and qualitative, but it serves well in describing what is meant by a band.

An overall view of the bands and band gaps throughout the metal may be obtained from Fig. 7.5. In this diagram, bands are denoted by cross-hatched areas and gaps by open areas. Note that the width of band widens as its distance from the nucleus increases, and the width of the gap shrinks as its distance from the nucleus increases. The explanation of this phenomenon lies in the extent of overlap that occurs in each case. The nearer the atomic orbitals are to the nucleus, the smaller is the overlap with atomic orbitals on neighboring nuclei. Thus, less perturbation occurs and essentially atomic orbitals remain. The farther away an orbital is from the nucleus, the greater is the amount of overlap with atomic orbitals of neighboring atoms; thus, perturbation of the atomic orbitals is larger and the size of the band gap is smaller.

Figure 7.6 contains the same species as Fig. 7.5, but the internuclear

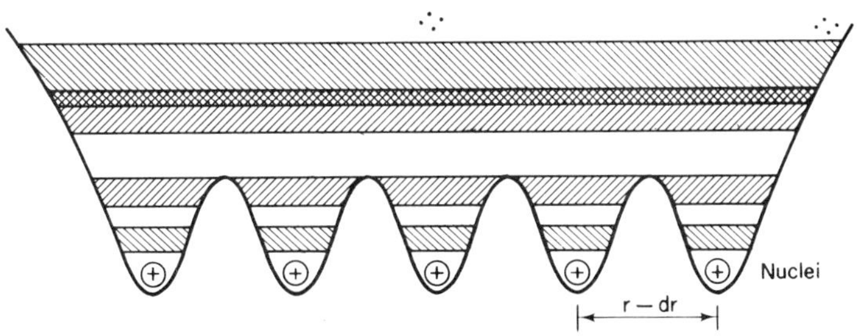

Figure 7.6. Band and band gaps at internuclear separation $r - dr$.

distance is now $r-dr$. It is now expected that overlap of orbitals would be greater and more perturbation would result. This is shown to be so because the bands are widened and the gaps are decreased, even to the extent that some gaps disappear and bands overlap.

These concepts may be employed to explain the metallic properties of sodium and magnesium. The metallic bands of sodium are formed by the overlap of the 3*s* orbitals of the sodium atoms. If *N* atoms exist in a piece of sodium, then *N* molecular orbitals are formed. There are *N* valence electrons in these *N* orbitals. The *N* orbitals are only half-filled, since the capacity of these *N* orbitals is 2*N* electrons. Thus, mobility is available to sodium electrons, and hence its metallic properties. However, *N* atoms of magnesium contain 2*N* valence electrons, and the 3*s* band of a piece of magnesium would be filled. The 3*s* and 3*p* bands are known to overlap to give a whole, new, and very broad band to account for metallic behavior.

These concepts may also be employed to illustrate why metallic behavior increases as one descends a group of the periodic table. As one descends a group, the orbitals of the atoms containing the valence electrons extend farther into space and have more extensive overlap with the atomic orbitals of the neighboring atom. Thus, bands are more readily obtained, and the overlap of filled bands with empty bands produces the partially filled bands that are characteristic of metals. The decrease in metallic character on traversing the periodic table from left to right is explained by the fact that the bands are becoming more and more filled with electrons, and consequently electron mobility decreases.

Some interesting properties of the nonmetals may be explained in terms

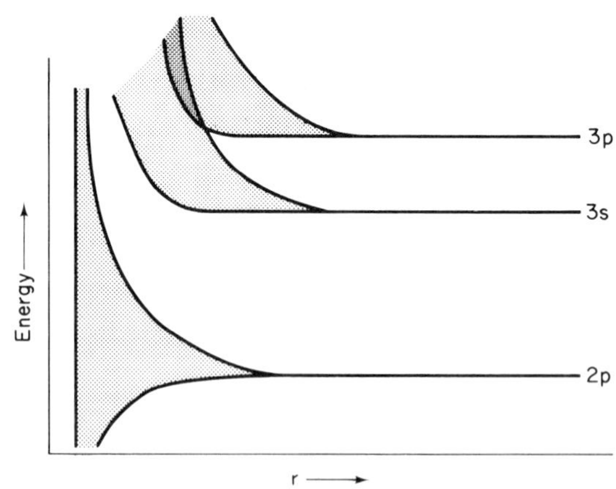

Figure 7.7. Diagram illustrating band formation as a function of internuclear distance.

of variation of band width and band gap width with internuclear distance. Figure 7.7 shows that as the internuclear distance decreases, more and more overlapping of the bands occurs. This is the case for all substances. Eventually a point is reached at which it appears that all bands have completely overlapped and one continuous band is formed. This concept leads one to conclude that one may induce metallic properties into any substance by decreasing the internuclear distance. The most simple method of decreasing the internuclear distance in nonmetals is by an increase in pressure (generally shock techniques). It has been observed that when phosphorus and carbon are subjected to pressures of 100,000 and 600,000 atmospheres (atm), respectively, they display metallic properties.

The band theory is used in a very satisfying way to explain the physical properties of metals. It was mentioned previously that the X-ray spectra showed rather sharp lower and upper limits for absorption and emission spectra, respectively. These spectra, however, give evidence of tailing, illustrated in Fig. 7.8. The electron density in these bands is considered

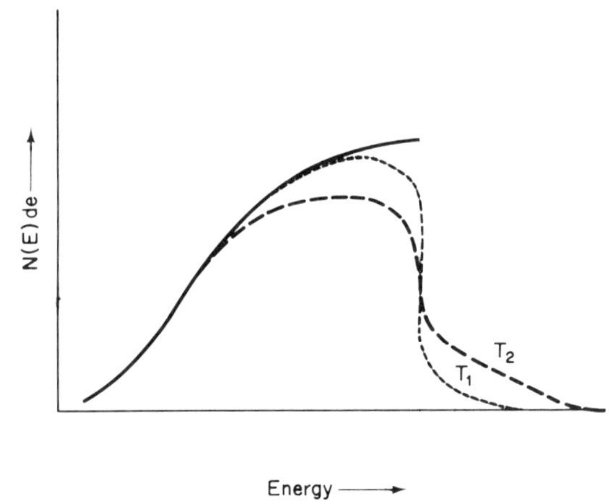

Figure 7.8. The effect of temperature on band population.

analogous to a sea and is called the Fermi sea. The surface of the electron density is called a Fermi surface. At absolute zero (0° K), the Fermi surface is perfectly flat [Fig. 7.8(a)]; consequently there is a very sharp line of demarcation between the filled and empty portions of the band. At slightly higher temperatures, some of the electrons are excited from the surface just as the wind causes waves on a real sea. Figure 7.8(b) represents this irregularity. At a higher temperature, T_2, even more electrons are excited from the surface. Since X-ray spectra are not recorded at absolute zero, it is reasonable that there will not be a sharp separation between the filled and empty portions of the band, and tailing is to be anticipated.

Since the filled and empty portions of a metallic band are so close in energy, the electrons are able to absorb light very easily and emit this light as they return to their original energy level. Essentially, metals reflect almost totally all incident light, which gives them luster.

One of the most successful applications of the band theory of metals is the ease with which it explains the conduction of metals and non-conduction of insulators. Referring to Fig. 7.4(b), the insulator, note that all the low energy bands are filled with electrons. Electrons must be excited to empty bands where they gain mobility in order that the insulator can conduct electricity or heat. The amount of energy required to excite these electrons into the conduction band depends on the band gap width; thus, band gap width is closely related to the insulating properties of the non-metal. Increasing the potential or the temperature across the nonmetal causes more electrons to be excited into the conduction band, and the material becomes a conductor. Theoretically, a point can be reached where all substances become conductors. Therefore, all materials can be regarded as being intrinsic semiconductors. When an insulator (or intrinsic semi-conductor with a large band gap) is caused to become a conductor by applying an electrical potential to it, the material is considered to have experienced *dielectric breakdown.* Heat also promotes electrons into the conduction band. The conductivity of germanium is increased 100% by increasing its temperature from room temperature to 150°C.

In the case of a metal, the electrical conductivity decreases with an increase in temperature. Nonmetals behave in just the opposite fashion. The qualitative explanation is this: In an isolated metallic crystal, the electron waves travel in a completely random fashion. If a potential is applied to a metal, a wave traveling in the direction of the potential is aided by the applied potential, whereas a wave traveling in the opposite direction is hindered by the applied potential. The net flow of current that can then be observed can be related to the amount of interference with these electron waves. If a metallic crystal is heated, the heat causes the metallic lattice to vibrate, thus interfering with the passage of the waves and decreasing the conductivity of the metal. The resistance of the metal at room temperature is approximately doubled at the melting point of the metal.

The conduction of heat by metals is dependent on the electrons at the Fermi surface; thus, the transition elements, which generally have large Fermi surfaces due to the presence of *d* orbitals in their band structure, will conduct heat readily.

The semiconductor mentioned previously was labeled as an intrinsic or band gap semiconductor. There is another type of semiconductor known as an extrinsic or doped semiconductor. These are of two general types. The so called *p*-type semiconductors can be formed by adding a small amount of some substance, such as gallium or another Group III metal that can occupy a lattice position with little or no distortion, to a Group IV semiconductor, such as silicon or germanium. The doping substance (gallium) has fewer valence electrons per atom than does silicon or germanium. The gallium has an empty conduction band lying just above the filled conduction band of silicon or germanium. Electrons from the silicon are excited to this empty band, and, in so moving, they are said to leave positive holes behind. As electrons move into these holes, the holes appear to migrate. Thus, the conduction may be due to the movement of the *positive* holes—hence the name *p*-type semiconductor. The second type of semiconductor is called an *n*-type. In this instance, the silicon or germanium is doped with a small amount of a substance, such as arsenic or another Group V metal, which has more valence electrons per atom than does the germanium or silicon. The arsenic has a filled conduction band just beneath the empty conduction band of the pure species. These electrons can be excited easily into the conduction band of the silicon or germanium. The conduction occurs because of the negative electrons in the conduction band, and it is spoken of as an *n*-type semiconductor.

For both the *p*- and *n*-type of semiconductor, the amount of conduction depends on the amount of material used in the doping and varies little with temperature, since all impurity holes are filled or all impurity electrons are excited at ordinary temperatures. The effect of temperature change on intrinsic semiconductors has already been noted. These materials can be

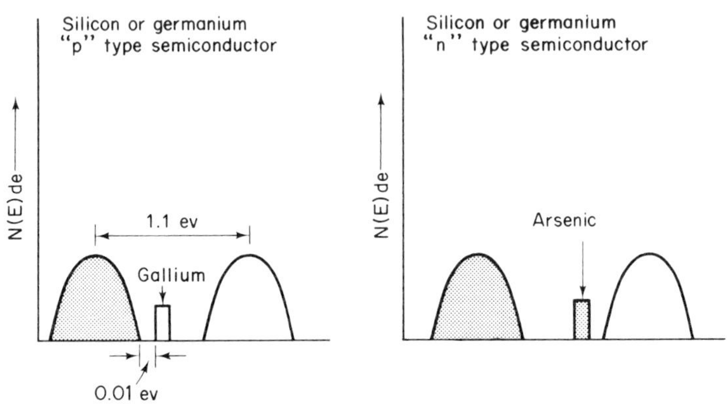

Figure 7.9. Doped semiconductors.

joined end-to-end or even on opposite ends of the same material, etc., to give rise to the whole series of solid state materials that are delighting the electronics industry. These concepts are summarized in Fig. 7.9.

It is appropriate to conclude the chapter with a brief mention of alloys, because alloys are essentially solid solutions of one metal in another. We shall be brief, because a discussion of alloys leads into metallurgy, which is itself a vast science.

Generally speaking, alloys may be classified as substitutional or interstitial alloys. Substitutional alloys arise when one atom occupies some of the lattice sites normally used by another metal atom. Several factors affect the formation of this type of alloy. The first of these to be considered is atomic size and crystal structure. The amount of substitution of one atom in the lattice of another is observed in general to be small if the atomic sizes differ by more than about 15%. This is not really too surprising, since large differences in sizes of the atoms would lead to distortions of the lattice, and the energy factors involved in causing this distortion would be expected to decrease the solubility. A good example of the effect of atomic size is given by a consideration of the formation of alloys among the members of Group IA. The important percentage increases in size are Li → Na, 22%; Na → K, 24%; K → Rb, 6%; K → Cs, 14%; and Rb → Cs, 7%. A study of these figures should lead to the prediction that lithium and sodium do not form any alloys with other Group IA members. On the other hand, potassium and rubidium would be expected to form alloys with any element below them (not, however, potassium with francium), since the largest percentage increase is less than 15%. Experimentally, it is observed that lithium is completely immiscible with any other element in the group, and that K-Rb, K-Cs, and Rb-Cs mixtures form a continuous series of solid solutions. This also implies that both atoms prefer the same crystal structure, or large distortions would result and thus solubility would decrease.

The second factor to be considered in substitutional alloys is the electronegativity difference between the metals. The larger is the difference in electronegativity between the two elements, the greater is the tendency toward compound formation rather than alloy formation, since the electrons will tend to be more highly localized on one of the elements. An interesting example is evident in Table 7.1. The fact that there is no band gap associated with Tl-Bi indicates that this species should be an alloy, as is observed experimentally.

In some alloy systems, the structure is found to be dependent on the relative ratio of atoms. One such class is called *electron compounds*, and the factor used to explain it is called the electron/atom ratio effect. This is

TABLE 7.1 ***Electronegativity and Band Gap of Compounds of III-IV Descending the Table***

Difference in electronegativity*	1.0	0.6	0.4	0.2	0.0
Band gap (eV)	6.0	3.0	1.4	0.3	0.0
Elements	B-N	Al-P	Ga-As	In-Sb	Tl-Bi

*Pauling's method.

sometimes called the *electron concentration*, and is often observed to be important in cases in which the electronegativity and size factors are unimportant. Stated in a very qualitative fashion, these concepts indicate that the alloy takes up the crystal structure that will accommodate the valence electrons with the lowest free energy. Hume-Rothery has performed extensive studies on this subject.

Brass is observed to have three distinct structures, which are labeled β, γ, and ε brass. The formulas corresponding to these three phases are CuZn, Cu_5Zn_8, and $CuZn_3$. A range of composition is possible for each of these structures; however, the preceding may be set up as approximate concentrations. A ratio is set up between the number of available electrons and the total number of atoms. Each atom contributes the number of electrons equal to its periodic group number, except for Group VIII, which is considered to contribute no electrons. Thus, CuZn has an electron/atom ratio of $\frac{3}{2}$, Cu_5Zn_8 has a ratio of $\frac{21}{13}$, and $CuZn_3$ has a ratio of $\frac{7}{4}$. Many other intermetallic compounds will then have the same structure as β, γ, and ε brass at the same electron/atom ratio. There are many exceptions and modifications to these rules, but limited time and space do not allow their discussion (see Table 7.2).

TABLE 7.2 ***Structures of Alloys***

β Brass($\frac{3}{2}$)	γ Brass($\frac{21}{13}$)	ε Brass($\frac{7}{4}$)
CuZn	Cu_5Zn^8	$CuZn_3$
AuZn	$Na_{31}Pb_8$	Cu_3Si
FeAl	Pt_5Zn_{21}	Ag_5Al_3

The other interesting type of alloy is the interstitial alloy. These materials are formed by the occupation of interstitial holes in the metallic lattice by a small substituent. Carbides and nitrides of the transition metals generally fall into this classification, and, as will be seen later, can possess some very interesting properties.

The interstitial compounds are generally regarded as being formed in the following manner. If one looks at the method of packing of metal atoms, one notices voids in the lattice, as Fig. 7.10 indicates. These voids, or interstices, can be occupied by atoms of the appropriate size. The manner in which the voids are filled will give rise to various structures. All these voids need not be filled, and thus several species can be formed, such as Fe_3C and Fe_4C. Calculations show that metal lattices are only slightly expanded; therefore, there is little effect on M—M bonding, and the holes are just about right for the interstitial atom. Thus, metal–nonmetal bonding and metal–metal bonding are important for properties of such compounds.

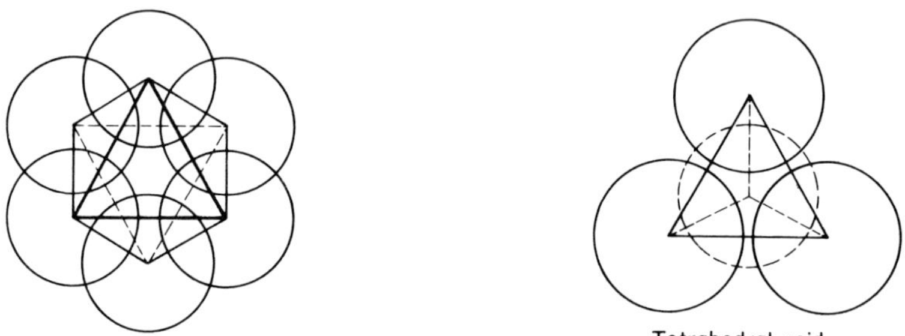

Figure 7.10. The packing of metal atoms and the voids formed.

Due to some of its very interesting properties, further consideration will be given to one of the carbide structures. If all the octahedral holes in a c.c.p. structure of a metal are occupied, a sodium chloride type of lattice is obtained. For the formation of this type of species, a nonmetal/metal size ratio of approximately 0.59/1 appears to be desirable. If the radius ratio exceeds this value, then simple interstitial alloys are not formed; instead, a much more complex type of structure is observed. Among the many simple carbides of this type are TiC, WC, and UC. These carbides possess some rather unusual properties. They have a high electrical conductivity that decreases with increasing temperature, high thermal conductivity, and metallic luster, just as in the pure metals. They are brittle, however, rather than ductile, as pure metals would be. In addition, they are somewhat inert chemically, but this appears to be due to kinetic rather than thermodynamic factors. They have high melting points and are extremely hard. In fact, the highest melting point recorded appears to be the 3940° C reported for the mixed carbide 4TaC + ZrC. It is also intriguing that one of these species, NbC, becomes superconducting at 10.1° K, which is a higher temperature than that observed for the superconduction of any pure metal.

Suggested Readings

1. Coulson, C. A., *Valence*, 2nd ed., Oxford University Press, Oxford, 1961.
2. Douglass, B. E., and D. H. McDaniel, *Concepts and Models of Inorganic Chemistry*, Blaisdell Publishing Co., New York, 1965.
3. Hume-Rothery, W., *Electrons, Atoms, Metals and Alloys*, Dover Publications, Inc., New York, 1963.
4. Hume-Rothery, W., and G. V. Raynor, *The Structure of Metals and Alloys,* The Institute of Metals, London, 1962.
5. Phillips, C. S. G., and R. J. P. Williams, *Inorganic Chemistry*, vols. 1–2, Oxford University Press, Oxford, 1966.
6. Sebera, D. K., *Electronic Structure and Chemical Bonding,* Blaisdell Publishing Co., New York, 1964.

Study Questions

7.1. Use the molecular orbital treatment to discuss the bonding in metals. Show how this model can be used to explain some physical properties of metals.

7.2. Give the X-ray emission and absorption diagram expected for Ga_2O_3. Explain in detail the information that can be obtained from it. What could be done to the gallium trioxide to make the diagram resemble that of gallium metal?

7.3. Compare the stabilities of Be_2 vs. F_2 and solid beryllium vs. solid fluorine.

7.4. Explain the difference between an intrinsic and an extrinsic semiconductor.

7.5. Classify the following as having the β, γ, or ε brass structures: AgZn, $Cu_{31}Sn_8$, Rh_5Zn_{21}, CoAl, Cu_3Sn.

7.6. Show which binary alloys of Group I-B are allowed by relative size differences.

7.7. Would silicon doped at one end with indium be a *p*- or *n-type* semiconductor? If the other end were doped with arsenic, in which direction would you expect electrons to flow most readily (potentials up to 1000 V cannot make the current flow in the opposite direction)?

7.8. Use band theory to explain any differences in electrical properties between calcium and phosphorus.

7.9. A Nernst glower is a metal oxide whose resistance is inversely proportional to temperature. It can be kept hot by the passage of a large current and the energy it emits is used as a source of infrared radiation in infrared spectroscopy. Compare in detail, with the use of diagrams, the property of electrical resistance in the oxide and the pure metal.

PART III

The Behavior of Atoms and Molecules

A. Some Principles of Behavior

CHAPTER 8

Acid-Base Systems and Non-aqueous Solvents

Acid-base theory is the result of an attempt to classify chemical reactions. Three acid-base classifications have proven useful in relating the structure, properties, and reactions of the molecules included in these systems. These systems are those of Arrhenius, Bronsted-Lowry, and G. N. Lewis. The Arrhenius system considers only protons and hydroxyl groups in aqueous solutions. It is the most limited of the acid-base theories, and we shall not discuss it further.

BRONSTED-LOWRY SYSTEM

The Bronsted system classifies acids and bases according to their proton-donating or proton-accepting ability. Thus, an acid is a substance that has a tendency to lose one or more protons, and a base is a substance that has a tendency to gain one or more protons. This can be illustrated by equation (8.1), which indicates an acid giving up a proton (reaction proceeds to the right) and a base accepting a proton (reaction proceeds to the left).

$$\text{Acid}_1 \leftrightharpoons H^+ + \text{base}_1 \qquad [8.1]$$

A common example would be the reaction between HCl (H_3O^+) and NaOH (OH^-). This can be shown in the simplified equation [8.2]. The first equation, [8.1]

$$\underset{\text{acid}_2}{H_3O^+} + \underset{\text{base}_1}{OH^-} \leftrightharpoons \underset{\text{acid}_1}{H_2O} + \underset{\text{base}_2}{H_2O} \qquad [8.2]$$

however, is very useful because it points out what is referred to as the conjugate acid-base system. Thus, base$_1$ is spoken of as being the conjugate base of acid$_1$, or vice-versa. This arises because each time an acid gives up a proton in acting as an acid, a species is left which is capable of accepting a proton to form the original acid. Thus, it is the conjugate base of the original acid. The same situation obviously holds for the base-conjugate acid system. A look at the conjugate acid-base equation also indicates the obvious fact that the stronger is the acid the weaker is the conjugate base, and vice-versa. If the reaction proceeds very far to the right, indicating a strong acid, the base cannot be very strong, because it is not accepting the proton.

Considering HCl in water, the reaction can be written as a competition,

$$\underset{\text{acid}_2}{HCl} + \underset{\text{base}_1}{H_2O} \leftrightharpoons \underset{\text{acid}_1}{H_3O^+} + \underset{\text{base}_2}{Cl^-} \qquad [8.3]$$

between the two bases, H_2O and Cl^-, for the proton [equation (8.3)]. Water is then seen to be the stronger base. There is also a competition between HCl and H_3O^+ to donate a proton. Since HCl donated the proton, it must be the stronger acid. In general, acid-base reactions tend to form the weakest acid and the weakest base. It follows, then, that the strength of an acid or a base depends on competition with other acids or bases. Now it is possible to construct a table that would list acids or bases in their relative order or strengths. Table 8.1.

TABLE 8.1 ***List of Some Conjugate Acids and Bases According to Their Relative Strengths***

	Conjugate Acid	*Conjugate Base*	
Decreasing strength ↓	$HClO_4$	ClO_4^-	Decreasing strength ↑
	HCl	Cl^-	
	HNO_3	NO_3^-	
	H_2SO_4	HSO_4^-	
	H_3O^+	H_2O	
	H_3PO_4	$H_2PO_4^-$	
	HF	F^-	
	H_2S	HS^-	
	NH_4^+	NH_3	
	OH^-	O^{2-}	

One important point to note is that in water, all those acids above H_3O^+ in Table 8.1 appear to have the same strength. This effect is called the *leveling effect* of water. All acids stronger than H_3O^+ are able to donate all their protons to the water; therefore, the strongest acid available in water is the H_3O^+ ion. For the case of $HClO_4$, the competition is between ClO_4^- and H_2O for the proton. The H_2O is a stronger base and thus accepts the proton,

$$\underset{\text{acid}_1}{HClO_4} + \underset{\text{base}_2}{H_2O} \leftrightharpoons \underset{\text{acid}_2}{H_3O^+} + \underset{\text{base}_1}{ClO_4^-} \qquad [8.4]$$

which leaves H_3O^+ as the only acid remaining in the solution.

If the acid is a weaker acid than H_3O^+, as is HF, two acids would result—a small amount of H_3O^+ and a large amount of HF.

$$HF + H_2O \leftrightharpoons H_3O^+ + F^- \qquad [8.5]$$

In such a situation, the leveling effect is not in operation, and the acid strength observed is that of HF. An alternative view is to consider that the conjugate base, fluoride ion, is a stronger base than water. The acid exists as essentially undissociated, and the strength observed is that of the acid itself rather than H_3O^+. The same concepts are applicable to bases. All bases that are stronger than OH^- in water are leveled to the strength of the hydroxyl ion in aqueous solution, i.e.,

$$Na_2O + H_2O \rightarrow 2NaOH \qquad [8.6a]$$

or

$$O^{2-} + H_2O \rightarrow 2OH^- \qquad [8.6b]$$

This leveling effect is noted in other solvents. One can have differentiating and leveling solvents, and, by proper choice of the solvent, the apparent acid or base properties of a substance can be increased or decreased. To make acids appear to be stronger, we choose a more basic solvent. In liquid ammonia, hydrogen fluoride and hydrogen bromide appear to be of equal strength. To decrease the apparent acid strength, a strong acid, such as nitric acid, is used as a solvent. Very few acids would be able to donate a proton to an acid of the strength of nitric acid. Perchloric acid is a strong enough acid to do so.

The use of differentiating solvents is the method by which the relative order of acid strengths, such as that given in Table 8.1, is obtained. It might have been expected that relative strengths might also be studied in aprotic solvents (those which neither donate nor accept a proton). The data obtained in this type of solvent are difficult to interpret, however, because of the possibility of the ions associating to form ion pairs and even higher order associations. This is not too surprising because aprotic solvents will have a low dielectric constant, which generally leads to the formation of ion pairs. Therefore, the method is not widely used. In addition, differentiating solvents sometimes react with solutes and thereby cause problems.

Not all Bronsted acids present themselves as immediately obvious as those mentioned so far. Metal ions in water are always solvated. The solvation energy may be very large and the water very tightly bound to the metal ion. It is possible that a proton can be released from one of solvating water molecules. In the case of Al^{3+}, the following could arise:

$$[Al(H_2O)_6]^{3+} \rightleftharpoons [Al(H_2O)_5(OH)]^{2+} + H^+(aq) \qquad [8.7]$$

Here, the hydrolysis of a metal ion produces an acidic solution. The $[Al(H_2O)_6]^{3+}$ species is a slightly weaker acid than acetic acid: $K_1[Al(H_2O)_6]^{3+} = 1.1 \times 10^{-5}$ and $K_a(CH_3COOH) = 1.8 \times 10^{-5}$. The degree of hydrolysis and its effect on the pH of the solution can be related to the base strength of the metal hydroxide. The stronger is the base, hydrolysis will occur to a lesser extent ($K_aK_b = K_w$ for a conjugate acid-base pair).

For cations with a noble gas type of electron configuration, the interaction between the metal ion and water can be considered solely from the electrostatic point of view; therefore, the charge and radius can be used to help decide the extent to which hydrolysis will occur. In Group I-A, lithium has the smallest cation and therefore might be expected to be hydrolyzed to the greatest extent of its group ($pK_b = 0.18$; therefore, $pK_a = 13.82$). The sodium atom is a stronger base ($pK_b = 0.7$) than the lithium ion and therefore hydrolyzes less. Hydrolysis is essentially nonexistent for the potassium, rubidium, and cesium ions, as these ions are progressively larger. Compared with Group I-A cations, Group II-A cations would be expected to be hydrolyzed to a greater extent, since the cations are smaller and the charge is greater. For a noble gas ion, this may be nicely expressed in terms of ionic potentials, i.e.,

$$\text{Ionic potential} = \frac{\text{ionic charge}}{\text{radius}} \qquad [8.8]$$

A low ionic potential generally indicates less hydrolysis of the ion. For non-noble gas ions, polarization is extensive and makes this type of calculation unreliable. In fact, the beryllium ion, Be^{2+}, which is expected to be most hydrolyzed in this group, is so extensively hydrolyzed that polynuclear species are formed in solution, and at pH = 4 the most likely species is

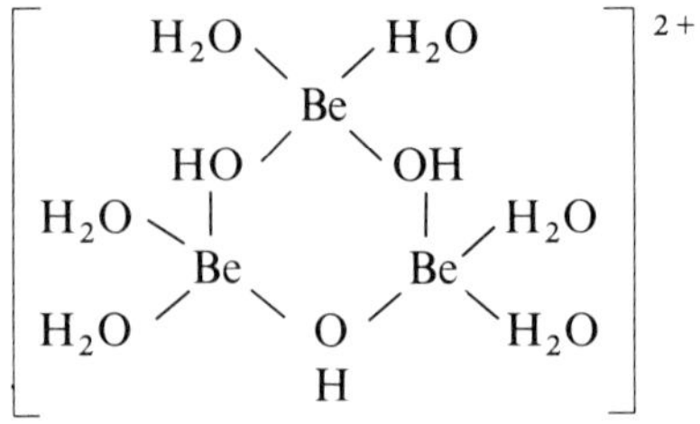

It appears, then, that Group III-A ions in their highest oxidation states will be more highly hydrolyzed still. The extent to which they may be hydrolyzed may be noted from the *pK* value for Al^{3+} (aq), or from its high ionic potential (3/0.5 = 6).

In the following series, the size of the ion increases and the ionic potential number in parentheses decreases: Mg^{2+} (3.1), Cu^{2+} (2.9), Zn^{2+} (2.7), Cd^{2+} (2.1), and Hg^{2+} (1.8); a decrease in hydrolysis is expected. Experimentally we note that hydrolysis increases in this series and that solutions of mercuric ion are most acidic. The series contains the trend that the electronegativity of the ions increases from Mg^{2+} to Hg^{2+}, and their polarizing power increases in the same fashion. This would indicate an increase in covalent character in an *M*—O bond on going from Mg^{2+} to Hg^{2+} in the preceding series. Inductive effects might remove electron density from the hydrogen and make it more susceptible to proton loss.

In general, successive *K* values for hydrolysis reactions differ by much less than successive *K* values for stepwise dissociations of ordinary hydrogen acids (H_3PO_4, etc.). This appears to indicate that the effect of loss of a proton is not transferred to the central metal ion in the same manner as for ordinary acids. A change in oxidation state of the metal, however, has a very large effect on the acidity of the species, as inductive effects would predict.

Since Bronsted acids are proton donors, a relationship between a base's position in the periodic table vs. its base strength might be established. A Born-Haber approach using enthalpy terms is valid, because enthalpy is a thermodynamic state function, i.e., a change in enthalpy depends only on the initial and final states of the system and is independent of the path. Consequently, if the total energy of a given process is known along with all the energy steps except one, the energy of this final step may be calculated very easily. The concept of proton affinity may now be defined in conjunction with the analysis of the energy steps involved in proton addition (see Fig. 8.1).

Proton affinity is the energy released when a gaseous atom, ion, or molecule combines with a gaseous hydrogen ion to form a protonated gaseous species. The larger is the proton affinity, the stronger is the attraction of the base for a proton and the weaker is the acid strength of the

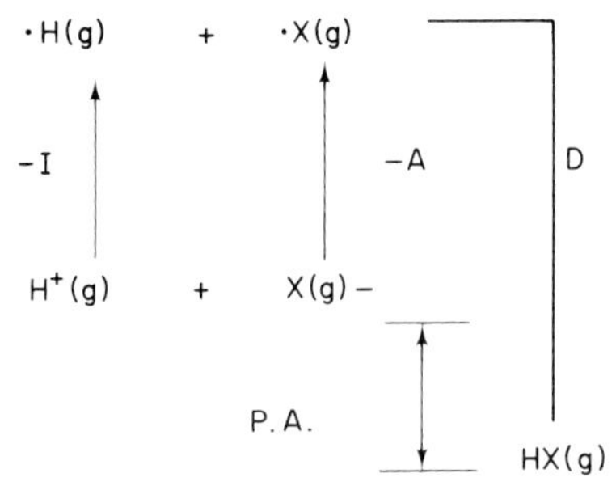

Figure 8.1. A Born-Haber approach to proton affinity, where *I* is the ionization potential, *A* is the electron affinity, *D* is the bond energy, and P.A. is the proton affinity.

protonated base. For HCl, the following values have been reported: for hydrogen, the ionization potential is 313 kcal/mole; the electron affinity of chlorine is − 85 kcal/mole, and the hydrogen-chlorine bond energy is 103 kcal/mole. From Fig. 8.1,

$$P.A. = (-I) + (-E) - (D) \qquad [8.9]$$

For Cl^-,

$$P.A. = (-313) + (85) - 103$$
$$= -331 \text{ kcal/mole}$$

TABLE 8.2 ***Proton Affinities of Various Species (—kcal/mole)****

H^-	402	CH_3^-	405	NH_2^-	395	OH^-	375	F^-	367
				PH_2^-	392	SH^-	346	Cl^-	331
				AsH_2^-	374	SeH^-	340	Br^-	321
CH_4	161	NH_3	209	OH_2	82	HF	91		

*Reprinted by permission of the publisher, from Bodie E. Douglas and Darl H. McDaniel: *Concepts and Models of Inorganic Chemistry*, Xerox Corporation, through Xerox College Publishing, a Xerox Company.

Reprinted from R. P. Bell: *The Proton in Chemistry*. C 1959 by Cornell University. Used by permission of Cornell University Press.

Waddington, T. C., "Lattice Energies" in *Advances in Inorganic and Radiochemistry*, Vol. 1, H. J. Emeleus and A. G. Sharpe, editors, Academic Press, Inc., New York, 1959.

A study of Table 8.2 will reveal that the variation in proton affinity almost parallels the variation of bond energies. As a group of the periodic table is descended, the proton affinity decreases with increasing radius. This trend is expected on the basis of ionic potential [equation (8.8)]. The decrease in proton affinity in the series $CH_3^- \rightarrow F^-$ appears to reflect the greater repulsion between nuclear protons and external protons as compared with repulsion between external protons), due to the distance between nuclear protons and external protons being smaller than between external protons. Thus, in going across the series, the number of nuclear protons increases, causing greater repulsive forces toward the external proton and a decrease in proton affinity. Calculations using Slater's rules support this argument. The screening constant for neon is 4.5; therefore, the effective nuclear charges for the following species are $CH_3^- = 6 - 4.5 = 1.5$; $F^- = 9 - 4.5 = 4.5$. This greater effective nuclear charge has a greater repulsive effect on external protons, resulting in HF being a stronger acid than CH_4.

It is interesting that the methane molecule, CH_4, a molecule with no free electron pairs, has a higher proton affinity than HF, in which there are several free electron pairs. Methane forms the electron deficient species CH_5^+, which very frequently appears in mass spectra of organic molecules. Thus it appears that in the gas phase the only thing a proton will not cling to is a species that is already positive.

From the foregoing discussion, we may make the following generalizations concerning hydrogen acids:

(1) Acid strength of binary hydrides increases on going from left to right across a period of the periodic table. Note the pK_a values in Table 8.3.

TABLE 8.3 ***Some selected pK Values of Some Binary Hydrides***

CH_4	NH_3	H_2O	HF
52 (1.5)*	35 (2.5)	15.74 (3.5)	3.17 (4.5)
	PH_3	H_2S	HCl
	27 (3.4)	7 (4.4)	−7.4 (5.4)
		H_2Se†	HBr
		3.8 (4.7)	−9 (5.7)
		H_2TE†	HI
		2.64 (4.7)	−9.5 (5.7)

*The values in parentheses are effective nuclear charges based on calculations using Slater's rules as described in Chapter 2. These are provided for purpose of comparison.

†pK_1 values listed.

The smaller or more negative is the pK value ($pK = -\log K$), the stronger is the acid. The pK values follow the proton affinity values very nicely. If one wished, one could also explain this phenomenon on the basis of electronegativity and size. Size does not change greatly across a period. Carbon is only 0.02 A larger than fluorine. Electronegativity values, however, show a rather large change ($C = 2.5$; $F = 4.0$). Therefore, electronegativity is seemingly the more important factor. The electronegativity of the nonmetal increases as we go to the right of the periodic table and the compounds are generally more ionic, thereby allowing the donation of hydrogen ion more easily.

(2) Acid strength of a binary hydride increases as a group of the periodic table is descended; hydrogen fluoride is a weaker acid than hydrogen iodide. This fact may be rationalized on the basis of both size and electronegativity. The electronegativity difference between fluorine and iodine is 1.5, identical to that between carbon and fluorine. However, the size of the halogen nearly doubles: fluorine, 0.64 A; iodine, 1.33 A. The variation in ionic potential for the halide ions also indicates this general trend. The negative ion with the highest ionic potential should have the greatest attraction for the proton: F^- (0.75); Cl^- (0.55); Br^- (0.51); I^- (0.46). This picture is not complete because the covalent contribution to the bond is completely neglected.

Equation (8.10) represents the gas phase ionic dissociation reaction, which is the reverse of the equation used in the definition of proton affinity.

$$HX\,(g) \rightarrow H^+\,(g) + X^-\,(g) \qquad [8.10]$$

The reverse process may also be described by two hypothetical steps, so that in a relative study, the only factors that will vary in the proton affinity cycle as the halogens are varied are the H—X bond energy and the electron affinity of the halogen [see equation (8.11)].

$$HX\,(g) \overset{(1)}{\rightarrow} H\cdot\,(g) + X\cdot\,(g) \overset{(2)}{\rightarrow} H^+\,(g) + X^-\,(g) \qquad [8.11]$$

In the case of hydrogen fluoride, the bond energy is 135 kcal/mole and the electron affinity of fluorine is 79 kcal/mole; the difference is 56 kcal/mole. For hydrogen iodide, however, the bond energy is 71 kcal/mole and the electron affinity of iodine is 70 kcal/mole. In this case, there is only a 1 kcal difference. The conclusion is that gaseous hydrogen iodide should dissociate more easily into gaseous hydrogen ion and gaseous iodide ion than gaseous hydrogen fluoride. In water, hydration effects on the anions tend to reverse this trend. The fluoride ion should have a higher hydration energy than the iodide ion. However, the larger degree of hydration of the fluoride ion will decrease the entropy of the solution and thereby cancel out part of the effect of the enthalpy of hydration. The overall effect still remains the same, viz., that the acid strength of hydrogen acids generally increases as a group of the periodic table is descended. The solvation effects will cause two acids, such as hydrochloric and hydrofluoric to be closer in strength in the liquid phase than in the gas phase. The hydration energies of the neutral molecules are relatively small and do not have much effect on the overall picture.

In general, the proton affinities follow the same order in solution as in the gas phase; however, there are some inversions. In the gas phase, SH^- has a smaller proton affinity than the fluoride ion; consequently, hydrogen sulfide is a stronger acid than hydrogen fluoride. In solution, however, hydrofluoric acid has a lower pK_a and is a stronger acid than hydrosulfuric acid. Obviously, solvation energies are going to affect the order of acid strength in solution. For this type of acid, the solution of the anion is the only important factor, for this is the most important species that is changing.

As was mentioned earlier, the solvation of the neutral molecules does not contribute much to the overall effect. The energy of solvation will depend on the size of the anion and its ability to interact with the solvent (hydrogen bonding). The smaller is the anion, the greater is the solvation energy. Thus, the higher members of a group will have higher energies of solvation, which will tend to decrease the difference in acid strength of the neutral members of the group in solution, as compared with the gas phase. It is not too surprising, then, that the solvation energy of the fluoride ion is much larger than that of HS^-, and that this energy is large enough to make hydrofluoric acid a stronger acid than hydrosulfuric acid. In this situation, the fluoride ion and HS^- were closer in proton affinity than were fluoride and the iodide ions. Thus, the solvation effects were large enough to cause the reversal in one case but not in the other.

The solvation effects influence the cationic acids in a different manner. Solvation will stabilize the undissociated cation acid in contrast to the neutral acid. The pK_a's for ammonia and phosphine are 35 and 27, respectively, making phosphine the stronger acid ($\Delta pK = 8$). The cationic acids of ammonia and phosphine, NH_4^+ and PH_4^+ have pK_a's of 9.5 and -12, PH_4^+ being the stronger acid ($\Delta pK = 21.5$). The marked difference in the change in pK in the case of the *onium* salts may be explained by the fact that NH_4^+ is a smaller ion than PH_4^+, and the solvation energy has a greater stabilization effect on the cation in this case and makes NH_4^+ a weaker acid than PH_4^+.

Enthalpy and entropy values for these reactions can be obtained by use of cyclic thermodynamic processes again. Substitution into the two equations, $\Delta G^\circ = \Delta H^\circ - T\Delta S^\circ$ and $\Delta G^\circ = -RT \ln K$, allows the pK values to be calculated. For the series HF, HCl, HBr, HI, observed pK's are 3, −7, −9, −10; predicted by calculation, they are 2, −7, −9, −10.

Some generalizations may also be made concerning the strengths of inorganic oxyacids. It is found that acid strength increases in the order HOI < HOBr < HOCl. This order is the reverse of that for the binary hydrohalogen acids. Size and electronegativity effects may be employed to explain this order of acid strengths. Iodine is larger in size but its electronegativity is less than chlorine. In the species H—O—*X*, the more covalent is the O—*X* bond, the more ionic (acidic) is the H—O bond. Although iodine is larger and more easily polarized, chlorine's electronegativity causes electron density to be pulled from the oxygen to the chlorine. The chlorine–oxygen bond is more covalent than the iodine-oxygen bond. Hence hypochlorous acid is stronger than hypoiodus acid.

As the number of nonhydrogenated oxygen atoms increases around the central atom in an oxyacid, the acid strength increases; sulfuric acid is stronger than sulfurous acid. The greater is the number of oxygen atoms around the central atom, the greater is the charge dispersal and the weaker will be the anionic part of the acid as a base. A weak base has a strong conjugate acid. Therefore, the strength of an oxyacid increases with the number of nonhydrogenated oxygen atoms in the anion.

A combination of the preceding ideas presents an a priori method of determining the relative strengths of acids. Perchloric acid, $HClO_4$, is stronger than H_2SO_4, because chlorine is smaller and in a higher oxidation state than sulfur. Consequently, the chlorine-oxygen bond is more covalent than the sulfur–oxygen bond. The hydrogen–oxygen bond in $HClO_4$ is more ionic than in H_2SO_4. Sulfuric acid is stronger than sulfurous acid for the same reasons.

Pauling formulated an empirical rule to aid in determining the pK's of oxyacids that are uncharged. If the formula of the acid is written as $(HO)_y MO_x$, one may approximate the pK's of an acid depending on the values of x. The successive pK values of a given oxyacid usually differ by five pK units. If $x = 0$, then the pK of the acid is 7; if $x = 1$, $pK = 2$; if $x = 2$, $pK = -3$; and if $x = 3$, $pK = -8$. It is interesting to note that the number of OH groups appears to have little or no effect on the acid strength of the species. The actual effect of the hydroxyl groups appears to be a weakening of acid strength. The solvation of the OH groups will push electrons toward the central element and lessen its inductive ability to make the hydrogen more ionic. This is nicely shown by the change in pK value as we replace the hydrogens in H_3PO_4 by methyl groups. The methyl group does not solvate, and thus the acid is strengthened:

$$\begin{array}{rl} (OH)_3PO & pK_1 = 2.16 \\ (CH_3O)(OH)_2PO & pK_1 = 1.52 \\ (CH_3O)_2(OH)PO & pK_1 = 0.76 \end{array}$$

The same type of explanation may be used to rationalize the strength of the following series of acids:

$$\begin{array}{rl} (OH)_3PO \text{ or } H_3PO_4 & pK_1 = 2.16 \\ (OH)_2HPO_2 \text{ or } H_3PO_3 & pK_1 = 2.00 \\ (OH)H_2PO \text{ or } H_3PO_2 & pK\ = 1.0 \end{array}$$

Since structure is indicative of acid strength, it is not unreasonable that acid strength should be an indication of the type of structure possessed by the acid. Consider H_3PO_3, $pK_1 = 2.0$, and H_3AsO_3, $pK_1 = 9.2$. The pK's of these acids differ by a factor of 10^6, and this great difference in pK's suggests a difference in structure. In reality, the structures are significantly different:

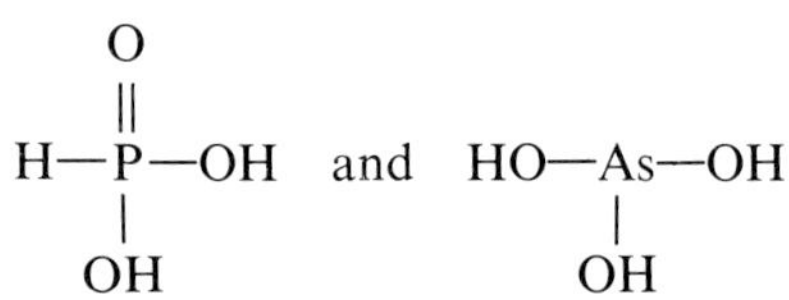

These structures and pK's are in fair agreement with Pauling's concepts.

For pyrophosphoric acid, $H_4P_2O_7$, $pK_1 = 1.52$, $pK_2 = 2.36$, $pK_3 = 6.60$, and $pK_4 = 9.25$. It may be observed that pK_1 and pK_2 are rather similar; pK_3 and pK_4 are also similar. But pK_1 and pK_2 differ significantly from pK_3 and pK_4. The order of dissociation might be expected to be as follows:

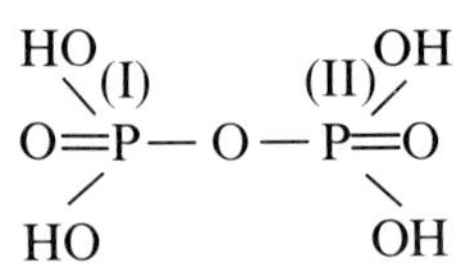

A hydrogen ion dissociates from an oxygen attached to P(I), and then one dissociates from an oxygen on P(II); the same process is followed in the dissociation of the two remaining acidic hydrogens.

An empirical relationship for the calculation of the pK of an acid has been developed by Ricci. The relationship is

$$pK = 8.0 - m(9.0) + n(4.0)$$

where m = formal charge on the central atom (see Chapter 4 on valence bond theory for calculation of formal charge) and n = number of non-hydrogenated oxygen atoms. Thus, for HSO_4^-, $m = 2$ and $n = 3$. According to Ricci's relationship, $pK = 2$ (experimental value is 1.9). This type of calculation gives adequate results in many cases. It tells us nothing, however, about the processes involved, and thus its significance is greatly decreased. An indication of the results as compared with experimental values is given in Table 8.4.

TABLE 8.4 ***Some pK Values Calculated by the Ricci Formula***

Acid	m	n	pK (exp)	pK (calc)
H_3AsO_3	0	0	9.2	8
H_3PO_4	1	1	2.1	3
H_2SeO_4	2	2	−2.0	−2

To this point, only those species which have definite acid or basic properties have been considered. Some species are intermediate in character, and their behavior depends upon the nature of the solution in which they are placed. For example, when zinc oxide is put into solution, it becomes zinc hydroxide, which may behave in either of two ways, depending on the pH of the solution. It may form Zn^{2+} and H_2O on reacting with an acidic solution or it may form $Zn(OH)_4^{2-}$ in basic solution. Such a species is said to be *amphoteric*. To define a substance precisely as amphoteric is very difficult, due to the fact that there are no definite criteria for such a state. Just how much acid property and how much base property must a species have to be considered amphoteric? Amphoterism is probably best described in a practical sense rather than a theoretical sense, and one such classification refers to the oxides of the elements. In general, metallic oxides are basic anhydrides and nonmetallic oxides are acidic anhydrides, with some exceptions (see Chapter 13). A gray area does exist between the two types of oxides, and no sharp line of distinction may be drawn between metallic and nonmetallic oxides. The oxides of Be^{2+}, Al^{3+}, Ga^{3+}, In^{3+}, and Pb^{4+} oxides are amphoteric. The oxides of Pb^{2+}, Sn^{2+}, Ge^{2+}, As^{3+}, and Sb^{3+} also are amphoteric. Among the transition metals, Ti^{4+}, V^{4+}, Cr^{4+}, Mn^{4+}, Fe^{4+}, Co^{3+}, Ni^{3+}, Cu^{2+}, and Zn^{2+} oxides have amphoteric properties. One would expect amphoterism to occur among the elements in that part of the periodic table in which elements undergo the transition from strictly basic oxides to strictly acidic oxides. In the transition metal series, the metals toward the end of the series are strongly polarizing, and hence are more acidic. This brings amphoterism to a lower oxidation state. The situation is similar in the other two transition metal series, except that amphoterism is generally found at a somewhat higher oxidation state: Mo^{6+} and W^{3+}; Au^{3+}. In general, a working rule may be that the smaller and more highly charged an ion, the more acidic it is. Various ions of manganese bear this out:

$$\underset{\text{acidic}}{Mn^{6+}} > Mn^{5+} \gg \underset{\text{amphoteric}}{} Mn^{4+} > \underset{\text{basic}}{Mn^{2+}}$$

THE LEWIS SYSTEM

There are many instances in which a proton is not involved in reactions that are evidently acid-base reactions. As one may no doubt recall, in this type of situation the Lewis system is generally used, and in this system an electron pair donor is a *base* and an electron pair acceptor is an *acid.*

The Lewis system is probably the most widely used at the present time because of its breadth. This system also makes the distinction between primary and secondary acids. A *primary acid* is one whose electronic structure indicates its acid nature (BF_3), whereas a *secondary acid* is one whose electronic structure does not immediately indicate its acid nature (CO_2).

The Lewis system has several inherent disadvantages, which should be discussed briefly before we consider the practical applications of the theory. First, the Lewis system is rather weak in its treatment of acid and base properties of conventional acids and bases; their treatment is indirect. Consider the reaction of HCl (g) and NH_3 (g):

$$HCl\,(g) + NH_3\,(g) \rightarrow Cl - H \ldots NH_3 \rightarrow Cl^-NH_4^+ \qquad [8.12]$$

It is believed that a hydrogen bond is formed in an intermediate, which then breaks down into its final form. Second, it has been stated that a Lewis acid should have catalytic properties. Two points must be considered. In some cases, it has been found that the apparent properties of Lewis acids have been due to the formation of the hydrogen ion. This suggests that the Lewis acid itself is not the catalyst. In addition, some reactions catalyzed by Lewis acids are not catalyzed by ordinary protonic acids. This could possibly mean that protonic acids are not Lewis acids. Finally, a rapid chemical reaction between a Lewis acid and base seems to be essential in a Lewis acid-base reaction. In aqueous solution, the following reactions occur:

$$Cr^{3+} + 6NH_3 \rightleftarrows Cr(NH_3)_6^{3+} \qquad [8.13]$$

$$Ag^+ + 2NH_3 \leftrightharpoons Ag(NH_3)_2^+ \qquad [8.14]$$

The reaction of the silver(I) ion is rapid and is considered a Lewis acid-base reaction, whereas the reaction of the chromium(III) ion is slow and is not considered a Lewis acid-base reaction. This seems to exemplify hair-splitting!

Probably the greatest disadvantage of the Lewis theory is that it is much more difficult to get a quantitative determination of the relative order of acid or base strength. The strength of an acid or a base depends upon its particular environment and changes as the environment changes. Two examples follow to clarify this point. (1) Steric effects may affect the strength of an acid or a base. To a given base, $(CH_3)_3B$ could appear to be a stronger acid than $(t-C_4H_9)_3B$, due to the steric hindrance of the boron by the much larger *t*-butyl group. But remember that a small base may not be hindered by the size of the *R* group, and may appear equally strong toward both acids. (2) In some coordination compounds, situations such as the following are noted. It is found that BeF_4^{2-} is much more stable than CuF_4^{2-}. Therefore, Be^{2+} is a stronger acid than Cu^{2+}. On the other hand, it has been observed that $Cu(NH_3)_4^{2+}$ is more stable than $Be(NH_3)_4^{2+}$, which indicates that Cu^{2+} is more acidic than Be^{2+}. The conclusion of this entire discussion is that one must be very careful in making comparisons of systems in terms of the Lewis theory.

Various methods have been used in studying the strengths of Lewis acids. They have included such diverse measurements as the stability of complexes, heats of reaction, and competition between acids for the same base. Again, comparison of the strengths obtained from one type of measurement with those from another type of measurement can be dangerous, and caution must be used in any such comparisons.

In general, the strengths of Lewis bases will parallel their basicity toward a proton if only σ bonds are formed in adduct formation and if steric factors are not important. Thus, $(CH_3)_3N$ is seen to be a stronger base than $(CH_3)_3Sb$. The same trend is observed when HCl is used as the reference acid.

$$(CH_3)_3N + (CH_3)_3B \rightarrow (CH_3)_3NB(CH_3)_3 \qquad \Delta H = -17.6 \text{ kcal/mole} \qquad [8.15]$$

$$(CH_3)_3P + (CH_3)_3B \rightarrow (CH_3)_3PB(CH_3)_3 \qquad \Delta H = -16 \text{ kcal/mole} \qquad [8.16]$$

$$(CH_3)_3As + (CH_3)_3B \rightarrow (CH_3)_3AsB(CH_3)_3 \qquad \text{exists only at } -80^\circ \qquad [8.17]$$

$$(CH_3)_3Sb + (CH_3)_3B \qquad \text{(no adduct)} \qquad [8.18]$$

Thus steric effects and π bonding are of little importance in these cases. The fact that $(CH_3)_3N$ is a stronger base than $(CH_3)_3Sb$ may be rationalized by the use of size relationships. Antimony is much larger than nitrogen. The electron density of the lone pair should be much more diffuse in the case of the antimony. The percentage overlap with the empty orbital of the acid would be much less in this adduct than with the nitrogen. In general, then, one might expect Lewis base strength to decrease as a group of the periodic table is descended. Similarly, base strength is found to decrease in the series (8.19); $(CH_3)_3N$ is found to displace both $(CH_3)_2O$

$$(CH_3)_3N > (CH_3)_2O > CH_3F \qquad [8.19]$$

and CH_3F, whereas $(CH_3)_2O$ will displace only CH_3F from $(CH_3)_3B$ complexes. The reasons for this order have been mentioned earlier in the discussion on proton affinity. With Lewis bases, then, there does appear to be a reasonable correlation between strength and the periodic table. Still, extensive caution must be employed in making generalizations and suggesting relationships.

Substituents can play a large role in determining acid-base properties of molecules for several reasons; inductive effects, back bonding, and steric effects are among the factors that are used extensively in discussions of this nature. For example, in the series of nitrogen bases, the base strength is $(CH_3)_3N > NH_3 > NF_3$. This relationship can be explained in terms of the electron withdrawing effects of the substituents. In the series of boron acids, $(CH_3)_3B < BH_3 < BF_3$ (the BH_3 has been corrected for the fact that this does not actually exist as such, but as B_2H_6), the order observed is again related to the electron withdrawing effects of the substituents.

In the case of the series of boron halides, however, the observed order is $BF_3 < BCl_3 < BBr_3$. This is not the order expected from inductive effects. These effects alone would make boron trifluoride the strongest acid. Steric reasons, if important in these molecules, would again lead to the opposite order of strength from that observed. In this series $(p \rightarrow p)$ π back bonding, from lone pair electron density on the halogen to the empty *p* orbital on boron, is felt to play a dominant role. Back bonding is dependent on the sizes of the orbitals involved and should be largest in the case of the fluorine, which has the most compact free electron pairs. In general, this type of bonding is believed to decrease as a group of the periodic table is descended. The back bonding energy that has to be replaced in the Lewis acid-base interaction is: boron trifluoride, 48 kcal/mole; boron trichloride, 30 kcal/mole; boron tribromide, 26 kcal/mole. Back bonding would place more electron density on the boron and reduce the acid properties of the molecule. Then boron trifluoride, which back bonds to the greatest extent, would assume the position of the weakest acid in the series. It also has been suggested that this order of acid strength follows the order of nonbonding repulsions. Fluorine, with the most compact electron density and short B—F bond length may offer the most repulsion to the incoming base, and thus make boron trifluoride the weakest acid.

The preceding order will depend again, however, on the reference base used. Thus, it has been reported that phosphorus and sulfur were stronger in base strength toward boron trichloride than were nitrogen and oxygen, respectively (P > N and S > O). However, toward boron trifluoride, the reverse was true. Thus, the following experimental information was obtained: for boron trifluoride as reference acid, $(CH_3)_3N > (CH_3)_3P$ and $(CH_3)_2O > (CH_3)_2S$. However, for boron trichloride and diborane as reference acids, the order is $(CH_3)_3P > (CH_3)_3N$ and $(CH_3)_2S > (CH_3)_2O$.

Steric effects are seemingly unimportant in discussing strengths of simple protonic acids and bases, but they are important in more complicated systems. For cases in which steric effects do play a role, displacement reactions are used in studying them. Thus, trimethylamine is a stronger base toward trimethylboron than is ammonia.

$$(CH_3)_3N + H_3NB(CH_3)_3 \rightarrow NH_3 + (CH_3)_3NB(CH_3)_3 \qquad [8.20]$$

However, if $(t-C_4H_9)_3B$ is used as the reference acid, the inverse order is obtained experimentally. This results from a strain caused by steric interference of atoms or groups, which are attached to the different central atoms. This strain results in a force that tends to separate the two central atoms; it is sometimes termed the *F* strain.

THE LUX-FLOOD AND FRANKLIN SYSTEMS

Various oxides are used many times as solvents in high-temperature chemistry. The Lux-Flood acid-base system was devised for work under such conditions. In liquid oxide systems, Lux proposed the following:

$$\text{Base} \rightleftarrows \text{O}^{2-} + \text{acid} \qquad [8.21]$$

A base is an oxide ion donor and an acid is an oxide ion acceptor. In reaction [8.22], calcium oxide is a base and silicon dioxide is an acid:

$$\text{CaO} + \text{SiO}_2 \rightarrow \text{CaSiO}_3 \qquad [8.22]$$

In a somewhat analogous fashion, but placing more stress on the solvent involved, Franklin developed the solvent system theory of acids and bases. In liquid NH_3, a proton would be transferred in the autoionization of the solvent:

$$2\text{NH}_3 \rightarrow \underset{\text{acid}}{\text{NH}_4^+} + \underset{\text{base}}{\text{NH}_2^-} \qquad [8.23]$$

In view of this, Cady and Elsey further defined an acid as a substance that, either by direct dissociation or by reaction with the solvent, gives the cation characteristic of the solvent, and a base as a substance that, either by direct dissociation or by reaction with the solvent, gives the anion characteristic of the solvent. Neutralization is thus a combination of an anion and a cation to form the neutral solvent. For example, in SO_2 the ionization might be postulated as

$$2\text{SO}_2 \rightleftarrows \text{SO}^{2+} + \text{SO}_3^{2-} \qquad [8.24]$$

Thus, $SOCl_2$ would presumably be an acid (assuming the ionization was ($SOCl_2 \rightleftarrows SO^{2+} + 2Cl^-$), and $[(CH_3)_4N]_2SO_3$ would be a base. Neutralization would then proceed,

$$\underset{\text{acid}}{\text{SOCl}_2} + \underset{\text{base}}{[(\text{CH}_3)_4\text{N}]_2\text{SO}_3} \rightarrow \underset{\text{salt}}{2(\text{CH}_3)_4\text{NCl}} + \underset{\text{solvent}}{2\text{SO}_2} \qquad [8.25]$$

It should be stressed that such a theory is no replacement for the Lewis or Bronsted-Lowry theories. Essentially, it is just a restatement of the Arrhenius system for each solvent used. Therefore, it is subject to all the disadvantages of the Arrhenius system. Its major usefulness is its ability to allow us to make some simple predictions concerning the course of many reactions in these solvents. At this point, then, it might prove worthwhile to take a look at some nonaqueous solvent systems.

NONAQUEOUS SOLVENTS

A solvent has two important functions in chemical reactions. The primary purpose of the solvent is to allow rapid reaction at a relatively low temperature by bringing together substances in a finely divided state that would otherwise mix poorly and thus react slowly. Thus, for the reaction

$$BaBr_2 + 2AgNO_3 \rightarrow 2AgBr + Ba(NO_3)_2 \qquad [8.26]$$

the thermodynamics are favorable, but the reaction proceeds only very slowly in the solid state. However, in aqueous solution, the reactions form the products very rapidly. Therefore, the solvent is used to accelerate the rate of the reaction. In addition, it is quite possible for the solvent to alter the course of the reaction. In water, barium chloride and silver nitrate react to form a precipitate of silver chloride. The reaction in liquid ammonia is completely reversed [equation (8.27)]:

$$2AgCl + Ba(NO_3)_2 \xrightarrow{NH_3(l)} BaCl_2 + 2AgNO_3 \qquad [8.27]$$

The solvent also affects the rate of reaction. Reaction (8.28),

$$C_2H_5I + (C_2H_5)_3N \rightarrow (C_2H_5)_4N^+I^- \qquad [8.28]$$

occurs at a given rate in hexane; if the reaction is carried out in benzene, the rate increases by a factor of 80; in acetone, by a factor of 500; and in nitrobenzene, by a factor of 2800.

The second major function of a solvent is to allow rapid separation and purification of the products. The precipitation of the chlorides in the preceding reactions allows for an easy separation of the products. Another example is the preparation of anhydrous copper sulfate. This is impossible to prepare in water, but the desired product precipitates out nicely in the solvent $HgBr_2$:

$$CuBr_2 + HgSO_4 \xrightarrow{HgBr_2(l)} HgBr_2 + CuSO_4\downarrow \qquad [8.29]$$

A few of the many solvents used with some of their properties are given in Table 8.5. The solvent to be used in a reaction is generally chosen with regard to some specific properties that may help to carry out the reaction and separate the products more conveniently. Some of the more important properties of solvents follow.

Melting Point (M.P.) and Boiling Point (B.P.)

These determine the temperature range within which the solvent is useful. Thus, a product that is unstable at room temperature might better be prepared in liquid ammonia than in water. The ease of separation of the product from the solvent is also determined by these factors. A volatile substance that boils near 100°C would be difficult to separate from water.

TABLE 8.5 ***Physical Properties of Some Solvents***

Solvent	Dielectric Constant (Temperature, °C)	Postulated Ionization	M.P. (°C)	B.P. (°C)
HCN	123 (16)	H_2CN^+ and CN^-	−13.4	25
HF	83 (0)	H_2F^+ and F^-	−85	19.5
H_2O	81 (18)	H_3O^+ and OH^-	0	100
H_2NNH_2	51 (25)	$N_2H_5^+$ and $N_2H_3^-$	2	113.5
IF_5	36 (25)	IF_4^+ and IF_6^-	9.6	98
NH_3	22 (−34)	NH_4^+ and NH_2^-	−78	−33.5
$HgBr_2$	9.8 (240)	$HgBr^+$ and $HgBr_3^-$	238	320
SO_2Cl_2	9.2 (20)	SO_2Cl^+ and Cl^-	−54	70

Acid-Base Properties

The range of acid-base properties is set by the solvent. Consideration of protonic solvents in terms of the Bronsted-Lowry system leads to the conclusion that the acid-base range is set by the ability of the solvent to act as a proton donor and acceptor. Recalling the leveling effect mentioned earlier, we see that in water, we cannot have a stronger acid than H_3O^+ or a base stronger than OH^-. If a stronger acid is desired, glacial acetic acid might be used as a solvent. This, however, lowers the strength of the base avilable in solution. If a stronger base is desired, liquid ammonia may be used as the solvent. It, too, limits the strength of the acid available (NH_4^+ is the strongest acid in liquid ammonia).

A good example of the effect of the acid-base properties of the solvent occurs in the preparation of the sodium salt of urea. In water, urea acts as a base, and if the anion were prepared, it would take H^+ from the H_2O and produce the weaker base, OH^-.

$$^{-}HNCONH_2 + H_2O \rightarrow H_2NCONH_2 + OH^- \quad [8.30]$$

However, the sodium salt can be prepared in liquid ammonia:

$$NaNH_2 + H_2NCONH_2 \rightarrow NH_3 + NaNHCONH_2 \quad [8.31]$$

Dielectric Constant

Since this property is a measure of the polar nature of the solvent, it may be used to estimate the solubility of polar or nonpolar materials in this solvent. The dielectric constant is the ratio of the field between two charged plates when there is a vacuum between them to the field between the two charged plates with a given substance (the dielectric) between them. Coulomb's law for the force between two charged bodies is given by

$$F = \frac{e_1 e_2}{Er^2} \quad [8.32]$$

where E is the dielectric constant of the medium, e_1 and e_2 represent the electrostatic unit of charge on each of the bodies, and r is the distance between the two bodies. Thus, in solvents of high dielectric constant, the force of attraction between the two charged particles is less, and greater solubility should result.

With a decrease in dielectric constant, there is a marked increase in the formation of ion pairs and undissociated electrolyte molecules, and thus a marked decrease in the solubility of salts of the more highly charged ions. For example, $E_{H_2O} = 80$ and $E_{NH_3} = 22$. Few, if any, salts of polyvalent anions such as O^{2-}, S^{2-}, SO_4^{2-}, PO_4^{3-}, and CO_3^{2-} are soluble in liquid ammonia. The general order of halide solubility in liquid ammonia is $I^- > Br^- > Cl^- > F^-$ (I^- forms the more covalent salts due to its greater polarizability). Note that the order of solubility in water is reversed: $F^- \gg Cl^- > Br^- > I^-$.

In general, a solvent is considered to have a high dielectric constant if the value lies between 50 and 80, a medium dielectric constant if from 20 to 50, and a low dielectric constant if from 2 to 20. Based on the dielectric constant alone, hydrogen fluoride, $E_{HF} = 83$, and hydrogen cyanide, $E_{HCN} = 123$, would not dissolve organic molecules too readily. Contrary to this expectation, these acids have been found to be good solvents for organic molecules. Therefore, consideration of other factors, such as the degree of association, is important. Many organic molecules that dissolve in liquid hydrogen fluoride will do so because hydrogen fluoride acts as a base toward them. However, acetone dissolves in hydrogen fluoride by solvolytic dissociation:

$$(CH_3)_2C{=}O + HF \rightarrow (CH_3)_2C{=}OH^+ + F^- \qquad [8.33]$$

Therefore, in itself, the dielectric constant is of minimal value in predicting solubilities.

Factors, other than dielectric constant, that might be important in predicting solubility are dipole-dipole interactions, hydrogen bonding, and π complexing reactions. Thus, silver salts are soluble in aromatic and olefinic solvents due to π bonding effects, such as those discussed with coordination compounds. Similarly, sulfur dioxide is a solvent with a high electron-accepting power, and is, therefore, important in the petroleum industry for separating paraffins from aromatics and olefins.

Redox Characteristics

Just as the acid-base characteristics of the solvent indicate the strength of acid or base that might be used, so also the range of redox potentials is limited by the ease with which the solvent is oxidized or reduced. Both kinetic considerations and thermodynamic considerations are important. Some species that are thermodynamically capable of oxidizing water to produce oxygen may do so only slowly, and thus can be used in aqueous solution (in acid, for MnO_4^-/Mn^{2+}, $E^\circ = 1.53$ V, and for O_2/H_2O, $E^\circ = 1.23$ V). If a stronger redox agent is desired, a different solvent may be used. Liquid ammonia is a solvent in which we can have strong reducing agents. In comparison to water, ammonia is less easily reduced. However, ammonia is more easily oxidized than H_2O, and therefore only weaker oxidizing agents may be used in it. Of particular interest in regard to the stability of liquid ammonia toward reduction is the fact that alkali metals may be dissolved in it with the apparent production of a solvated electron. This action is represented by the following equation:

$$M \xrightarrow{NH_3(l)} M^+ \text{(am)} + e^- \text{(am)} \qquad [8.34]$$

Reductions in which the free electron is the reducing agent may be performed. The following is typical of such reductions:

$$Ni(CN)_4^{2-} + e^- \rightarrow Ni(CN)_3^{2-} + CN^- + e^- \rightarrow Ni(CN)_4^{4-} \qquad [8.35]$$

In this case, a zero valent complex of nickel is formed. This topic will be discussed when liquid ammonia is studied as a solvent.

The types of reactions that may take place in a solvent can be categorized. There is no need to repeat those already described. Precipitation reactions have been mentioned and it has been pointed out that a judicious use of solvent may influence a course of a reaction by the precipitation of a particular product. Acid-base reactions may be useful, as was seen from examples of solvent system theory, since such a concept allows a simple prediction of reactions that might occur and of the products of such reactions. For example, one might easily predict that the cation and anion characteristic of the solvent would react in a neutralization reaction.

The one remaining kind of reaction that is of importance is the solvolytic reaction. In these reactions, the dissolved solute reacts with the solvent in a manner such that the normal cation and anion concentrations of the solvent are changed. The dissolution of $POCl_3$ in water results in the following reaction:

$$POCl_3 + 3H_2O \rightarrow (HO)_3PO + 3HCl \qquad [8.36]$$

The solution is decidedly acidic. This solvolytic reaction is distinguished from simple solvation, which is an attachment of solvent molecules to solute particles, i.e.,

$$MgCl_2\,(s) \xrightarrow{H_2O} Mg(H_2O)_x^{2+} + 2Cl(H_2O)_y^- \qquad [8.37]$$

There is no change in cation and anion concentrations themselves, in contrast to reaction [8.37]. The utility of the solvolytic concept lies in the fact that it allows comparisons among reactions in various solvents. It might be expected that RCOOH would behave in water as $RCONH_2$ would behave in liquid ammonia.

LIQUID AMMONIA

As another illustration of some of the ideas that have been presented, we now look at some particular examples of nonaqueous solvents. Probably the most widely studied system is liquid ammonia. Liquid ammonia could be classified as a basic solvent. The first class of reactions to look at is the *acid-base* reaction. It has already been pointed out that KNH_2 would be basic and NH_4Cl would be acidic in liquid ammonia. Neutralization would then proceed as

$$NH_4NO_3 + KNH_2 \longrightarrow KNO_3 + 2NH_3 \quad [8.38]$$

In this solvent, there are bases other than KNH_2 available. For example, the following bases react with NH_4I:

$$2NH_4I + PbNH \longrightarrow PbI_2 + 3NH_3 \quad [8.39]$$

$$3NH_4I + BiN \longrightarrow BiI_3 + 4NH_3 \quad [8.40]$$

Thus, in the liquid ammonia system, there are three primary bases: NH_2^-, NH^{2-}, and N^{3-}; whereas in the water system, there are only two: O^{2-} and OH^-. A number of comparable acids and bases in the two systems are *isosteric*. Isosteres are species that contain the same number of electrons and have the same nuclear charge. Some isosteres are

$$NH_4^+, NH_3, NH_2^-, NH^{2-}, N^{3-} \quad [8.41]$$

$$H_3O^+, H_2O, OH^-, O^{2-} \quad [8.42]$$

Note that N^{3-} has no isostere in the water system.

The preparation of bases in the liquid ammonia system should be analogous to the preparation of the bases in the water system. The following reactions have been found to occur:

$$PbI_2 + 2KNH_2 \longrightarrow 2KI + PbNH + NH_3 \quad [8.43]$$

$$BiI_3 + 3KNH_2 \longrightarrow 3KI + BiN + 2NH_3 \quad [8.44]$$

Note that some of these materials are very explosive; care must be taken in their preparation. This is the only method of preparation for some of the metal nitrides.

As in aqueous solutions, phenolphthalein is magenta in liquid ammonia when the basic characteristic predominates. That is, phenolphthalein will be magenta in a liquid ammonia solution of KNH_2 and colorless in a liquid ammonia solution of NH_4I. The ammonium ion, NH_4^+, is the acid characteristic of the liquid ammonia system, and the behavior of ammonium ion in liquid ammonia is analogous to the behavior of the hydronium ion in aqueous solutions. Note the following reactions:

$$2Na + 2H_3O^+ \xrightarrow{H_2O} H_2 + 2Na^+ + 2H_2O \qquad [8.45]$$

$$2Na + 2NH_4^+ \xrightarrow{NH_3(l)} H_2 + 2Na^+ + 2NH_3 \qquad [8.46]$$

In addition, ammonium salts in liquid ammonia dissolve the oxides or hydroxides of some metals, even though these are bases in terms of another solvent system.

We now note another interesting relationship between the oxygen and nitrogen system in the case of amphoteric reactions. We can write the following series of reactions:

$$Zn(NO_3)_2 + 2KOH \xrightarrow{H_2O} 2KNO_3 + Zn(OH)_2 \qquad [8.47a]$$

$$Zn(NO_3)_2 + 2KNH_2 \xrightarrow{NH_3(l)} 2KNO_3 + Zn(NH_2)_2 \qquad [8.47b]$$

$$Zn(OH)_2 + 2KOH \xrightarrow{H_2O} K_2Zn(OH)_4 \qquad [8.48a]$$

$$Zn(NH_2) + 2KNH_2 \xrightarrow{NH_3(l)} K_2Zn(NH_2)_4 \qquad [8.48b]$$

$$Zn(OH)_2 + 2HCl \xrightarrow{H_2O} ZnCl_2 + 2H_2O \qquad [8.49a]$$

$$Zn(NH_2)_2 + NH_4I \xrightarrow{NH_3(l)} ZnI_2 + 4NH_3 \qquad [8.49b]$$

Another interesting aspect of liquid ammonia chemistry is the study of redox reactions that occur in this solvent. It appears that the reductant in metal-ammonia solutions is actually a free electron because of the high reducing capabilities of these solutions. If elemental sulfur is added to a sodium-liquid ammonia solution, sodium sulfide is precipitated immediately.

$$S + 2e^- \rightarrow S^{2-} \qquad [8.50a]$$

$$S^{2-} + 2Na^+ \rightarrow Na_2S \qquad [8.50b]$$

Examples of this type are very well known.

The possibility of the existence of a free electron in solution merits a look at the structure of alkali metal-liquid ammonia solutions. Apparently all dilute solutions of alkali metals in liquid ammonia have nearly identical infrared spectra, indicating that the same species is probably present in each case. At low concentrations of sodium, it appears that equation [8.53] represents the dissolution process

$$M \xrightarrow{NH_3(l)} M^+ \text{(am)} + e^- \text{(am)} \qquad [8.51]$$

According to the *cavitation model* used to explain these phenomena, the deep blue, dilute solutions consist of sodium ions and free electrons nestled in their separate *cavities* formed by clusters of ammonia molecules. Now as the concentration of the metal is increased, the solutions become bronze colored (up to the limits: Li, 16 *M*; Na, 10 *M*; K, 12 *M*). The dilute solutions are paramagnetic, but the paramagnetism decreases with increasing concentration of the metal. Conductance studies indicate that the concentrated solutions of alkali metals in liquid ammonia have metallic-like conduction, with evidence pointing to the fact that these metal solutions are structurally analogous to molten metals. This, basically, is the evidence for the concept of the structure of metal-liquid ammonia solutions suggested by equation (8.51). Now, as the solution becomes more concentrated, the electrons pair [equation (8.52)], or *ion pairs* are formed in solution [equation (8.53)]

$$e^-(\text{am}) + e^-(\text{am}) \longrightarrow e_2^{2-}(\text{am}) \qquad [8.52]$$

$$M^+(\text{am}) + e^-(\text{am}) \longrightarrow M^+e^-(\text{am}) \qquad [8.53]$$

These equations account for the decrease in paramagnetic, color, and conduction properties of concentrated solutions.

Solutions of alkali metals in liquid ammonia are metastable, and the smallest amount of impurity may act as a catalyst to initiate the production of an amide. A rusty nail will serve to catalyze the formation of sodium amide.

$$2\text{Na} + 2\text{NH}_3\,(\text{l}) \xrightarrow{\text{Fe}_2\text{O}_3} 2\text{NaNH}_2 + \text{H}_2 \qquad [8.54]$$

However, if both the sodium and ammonia are pure, the metal may be dissolved in liquid ammonia and then recovered unreacted on evaporation of the solvent.

In addition to the alkali metals, other metals, such as barium, calcium, magnesium, strontium, aluminum, and silver, dissolve in liquid ammonia. When solutions of the alkaline earth metals are evaporated, the free metals are not obtained. Instead, ammoniates of the type $M(\text{NH}_3)_6$ result. When these ammoniates are placed under a vacuum to remove the ammonia, some of the metal is recovered as the amide.

The conclusion of this section on liquid ammonia is a discussion of ammonolysis (compare with hydrolysis in aqueous systems). Nearly all covalent inorganic halides will undergo ammonolysis. A familiar example would be the following:

$$\text{BCl}_3 + \text{NH}_3 \longrightarrow \text{Cl}_3\text{BNH}_3 \xrightarrow{\text{NH}_3(\text{l})} \text{Cl}_2\text{BNH}_2 + \text{NH}_4\text{Cl} \qquad [8.55]$$

Reaction [8.55] may continue until $B(NH_2)_3$ is formed, but it is still solvolysis (ammonolysis) because the acid character of the solvent is being increased. Ammonolysis reactions do have a practical purpose, i.e., a path to make anhydrous hydrazines is provided!

$$Cl_2 + 2NH_3\,(l) \longrightarrow ClNH_2 + NH_4Cl \qquad [8.56a]$$

$$ClNH_2 + 2NH_3\,(l) \longrightarrow N_2H_4 + NH_4Cl \qquad [8.56b]$$

LIQUID HYDROGEN FLUORIDE

Conductivity measurements indicate a relatively large degree of auto-ionization:

$$2HF \rightleftarrows H_2F^+ + F^-, \qquad K \approx 10^{-10} \qquad [8.57]$$

The availability of the proton in this solvent system may cause strong acids in other solvent systems to become weak acids in this system; and it may cause weak acids in other systems to behave as bases in this system. Generally, if an anion is added to liquid hydrogen fluoride, a reaction occurs that may be represented by the equation

$$X^- + HF\,(l) \longrightarrow HX + F^- \qquad [8.58]$$

Therefore, with the exception of fluoride ion, other anions are usually present in only small amounts.

The dissolution process in liquid hydrogen fluoride is believed to take place by one of four different mechanisms:

(1) Dissociation into ions, as in

$$KF + HF \longrightarrow K^+ + HF_2^- \qquad [8.59]$$

(2) Chemical addition of hydrogen fluoride to the solute, followed by dissociation to give fluoride ion, as in

$$CH_3COOH + HF \rightarrow CH_3COOH_2F \rightarrow CH_3COOH_2^+ + F^- \qquad [8.60]$$

In general, liquid hydrogen fluoride is an exceptionally good solvent toward organic compounds, and the organic molecule will generally act as a base toward hydrogen fluoride; even methyl benzenes may be dissolved in this manner:

$$CH_3C_6H_5 + HF \longrightarrow CH_3C_6H_6^+ + F^- \qquad [8.61]$$

As would be expected, removal of fluoride ion will enhance the solubility of the preceding species.

(3) Chemical replacement of the anion by fluoride ion, as in

$$KCN + HF \longrightarrow HCN + KF \qquad [8.62]$$

(4) Chemical reactions involving more than simple replacement, such as,

$$H_2SO_4 + 2HF \longrightarrow HOSO_2F + H_3O^+ + F^- \qquad [8.63]$$

The foregoing examples indicate that simple dissolution in liquid hydrogen fluoride is not common. It appears that of the common anions, only ClO_4^-, BF_4^-, and HF_2^- exist in both water and liquid hydrogen fluoride. Some generalities concerning the solubilities of certain anions can be made: (a) oxides and hydroxides usually react violently to give fluoride ion and water; (b) water yields hydronium ion and fluoride ion; (c) the majority of the halides (except fluorides) are either insoluble and unaffected or react to give KF, as was shown previously for KCN [equation (8.62)].

Liquid hydrogen fluoride is such an acidic solvent that few acids are capable of donating a proton to it. The few that can do so include HPF_6, $HOsF_6$, and $HSbF_6$ (strongest). Considering the autoionization of liquid hydrogen fluoride, we see that anything increasing the concentration of H_2F^+ will act as an acid. Therefore, introducing a molecule with empty low-energy orbitals (Lewis acid), such as BF_3, will cause the removal of fluoride ion from solution in the formation of BF_4^-. The equilibrium expressed in equation [8.57] will shift to the right and the concentration of H_2F^+ will increase. Fluoride acceptors such as these are called *coacids*; these coacids include molecules such as SbF_5, BiF_5, TiF_4, and TaF_5.

Substances acting as bases in liquid hydrogen fluoride include the ionic fluorides. These fluorides increase the concentration of the fluoride ion, shifting the equilibrium expressed in equation [8.57] to the left and decreasing the concentration of H_2F^+.

Precipitation reactions are more difficult to discover in liquid hydrogen fluoride than in other solvents, due to the absence of the common anions. Some do occur, however:

$$KClO_4 + TlF \longrightarrow TlClO_4 \downarrow + KF \qquad [8.64]$$

A few other anions, such as SO_4^{2-} and IO_4^-, are also stable enough in liquid hydrogen fluoride to allow this to occur:

$$KIO_4 + AgF \longrightarrow AgIO_4 + KF \qquad [8.65]$$

LIQUID SULFUR DIOXIDE

Liquid oxide solvent systems have been studied to some extent. Liquid sulfur dioxide is one of the few liquid systems that is a useful solvent, yet contains no hydrogen.

Liquid sulfur dioxide would be expected to be a covalent material; consequently, it should dissolve preferentially covalent (nonionic) material. Therefore, it is reasonable that ionic inorganic substances are insoluble in liquid sulfur dioxide. In considering a given alkali metal halide—say, the chloride—we would expect more ionic character as we descend the group; therefore, the solubility of the alkali metal chloride would decrease as we descend this periodic group. If we consider the sodium halides (except fluoride ion), the polarizability of the halide ion increases as we descend the group; hence, we would expect an increase in covalent character and an increase in solubility. In general, covalent inorganic compounds would be expected to be soluble ($SnCl_4$, Br_2, BCl_3, SO_2X_2, etc.). Organic molecules are generally covalent, so these would be expected to be soluble also. However, saturated hydrocarbons are not as soluble as are the aromatic. This fact is frequently used for separating aromatic from aliphatic hydrocarbons in the processing of crude oils.

Many solutes are colored in liquid sulfur dioxide solutions. The most common color is yellow, despite the color of the pure solute, thus indicating that perhaps there is some similarity in the solution process.

Some of the general types of chemical reactions in the liquid sulfur dioxide system follow.

(1) *Acid-base*: At first glance, one might expect equation [8.66] to represent the autoionization of liquid sulfur dioxide:

$$2SO_2 \rightleftarrows SO^{2+} \text{ (thionyl = acid)} + SO_3^{2-} \text{ (sulfite = base)} \qquad [8.66]$$

The actual ionization reactions in sulfur dioxide have been investigated using tagged oxygen and sulfur in a mixture of sulfur dioxide and sulfur trioxide. With ^{18}O, there is found to be rapid exchange between SO_2 and SO_3. In the same system, using labeled ^{35}S, no exchange is observed. Also, a solution of SO_3 in liquid sulfur dioxide will conduct a current, indicating that some ions must be present. These facts strongly suggest oxide ion transfer [equation (8.67)]. Thionyl ion is not the result of the autoionization of liquid sulfur dioxide, but is rather the result of the reaction

$$SO_2\,(l) + SO_3 \rightleftarrows SO^{2+} + SO_4^{2-} \qquad [8.67]$$

If appreciable autoionization of liquid sulfur dioxide were occurring, ^{35}S exchange should be observed,

$$2SO_2 \leftrightharpoons SO^{2+} + SO_3^{2-} \qquad [8.68a]$$

$$SO_3^{2-} + S^*O_3 \leftrightharpoons S^*{}_2O_6^{2-} \text{ (dithionite ion)} \qquad [8.68b]$$

because the breakdown of dithionite into its components should lead to ^{35}S existing in both products.

Experimental evidence points to the fact that free thionyl plays no appreciable role in the chemistry of liquid sulfur dioxide. Another important fact is that thionyl chloride has been shown to ionize in the following manner:

$$SOCl_2 \leftrightharpoons SOCl^+ + Cl^- \qquad [8.69]$$

Furthermore, it has been found that there is no ^{18}O exchange between SO_2 and $SOCl_2$. These data lead to the conclusion that pure, liquid sulfur dioxide ionizes very little if at all and that equation [8.66] is not reliable in its predictive ability. Despite all this work, thionyl chloride (SO_2Cl_2) is an acid and cesium sulfite is a base in the liquid sulfur dioxide system. They do undergo a neutralization reaction that can be followed conductimetrically.

(2) *Precipitation reactions*: They are rather extensive in liquid sulfur dioxide and do not necessarily involve thionyl or sulfite ions, i.e.,

$$AlCl_3 + 3NaI \leftrightharpoons 3NaCl + AlI_3 \qquad [8.70]$$

In general, most metal chlorides are insoluble in liquid sulfur dioxide. As in most nonaqueous solvents, precipitation reactions may produce some interesting compounds that would be difficult to prepare in any other way—e.g., the preparation of thionylthiocyanate,

$$2NH_4SCN + SOCl_2 \rightarrow SO(SCN)_2 + 2NH_4Cl\,(s) \qquad [8.71]$$

(3) *Redox reactions*: Liquid sulfur dioxide may act as an oxidizing or reducing agent depending on the circumstances. It has no pronounced character either way, however, and, in general, just acts as a medium in which the redox reactions may occur. Thus, the following type of redox reaction occurs in liquid sulfur dioxide:

$$6KI + 3SbCl_5 \longrightarrow 2K_3SbCl_6 + SbCl_3 + 3I_2 \qquad [8.72]$$

(4) *Acid-metal interactions*: Since thionyl chloride acts as an acid in liquid sulfur dioxide, we might expect that metals might be dissolved in such a solution, as in

$$Fe + SOCl_2 \longrightarrow FeCl_2 + SO \qquad [8.73a]$$

$$2SO \longrightarrow S\,(s) + SO_2 \qquad [8.73b]$$

(5) *Amphoteric behavior*: Aluminum is amphoteric in this solvent just as it is in aqueous medium [in the following, $X = (CH_3)_4N^+$].

$$AlCl_3 + 3KOH \longrightarrow 3KCl + Al(OH)_3\,(s) \qquad [8.74a]$$

$$2AlCl_3 + 3X_2SO_3 \xrightarrow{SO_2(l)} Al_2(SO_3)_3\,(s) + 6XCl \qquad [8.74b]$$

$$Al(OH)_3 + 3KOH \longrightarrow K_3[Al(OH)_6] \qquad [8.75a]$$

$$Al_2(SO_3)_3 + 3X_2SO_3 \longrightarrow 2X_3[Al(SO_3)_3] \qquad [8.75b]$$

$$K_3[Al(OH)_6] + 3HCl \longrightarrow Al(OH)_3\,(s) + 3KCl + 3H_2O \qquad [8.76a]$$

$$2X_3[Al(SO_3)_3] + 3SOCl_2 \longrightarrow Al_2(SO_3)_3 + 6XCl + 6SO_2 \qquad [8.76b]$$

(6) *Solvolysis and complex formation*: Solvolysis occurs in liquid sulfur dioxide:

$$PCl_5 + H_2O \longrightarrow POCl_3 + 2HCl \qquad [8.77a]$$

$$PCl_5 + SO_2\,(l) \longrightarrow POCl_3 + SOCl_2 \qquad [8.77b]$$

In general, solvolysis is more complex in liquid sulfur dioxide than in water. The solvolysis of potassium bromide occurs in three steps (the reaction took eight years to go to completion).

$$8KBr + 8SO_2 \longrightarrow 4K_2SO_3 + 4SOBr_2 \qquad [8.78a]$$

$$4SOBr_2 \longrightarrow 2SO_2 + S_2Br_2 + 3Br_2 \qquad [8.78b]$$

$$4K_2SO_3 + 2Br_2 \longrightarrow 2K_2SO_4 + 4KBr + 2SO_2 \qquad [8.78c]$$

The total reaction is

$$4KBr + 4SO_2 \longrightarrow S_2Br_2 + Br_2 + 2K_2SO_4 \qquad [8.78d]$$

LIQUID BROMINE TRIFLUORIDE

Liquid bromine trifluoride is a rather exotic solvent whose postulated autoionization is usually expressed as

$$2BrF_3 \rightleftarrows BrF_2^+ + BrF_4^- \qquad [8.79]$$

In this solvent, both KF and $AgBrF_4$ would be bases. Potassium fluoride is a base because it causes an increase in the concentration of BrF_4^-, as does $AgBrF_4$.

$$KF + BrF_3 \longrightarrow K^+ + BrF_4^- \qquad [8.80]$$

Antimony pentafluoride is an acid in bromine trifluoride, because it increases the BrF_2^+ concentration.

$$SbF_5 + BrF_3 \longrightarrow BrF_2^+ + SbF_6^- \qquad [8.81]$$

It should be noted that in reaction (8.81), antimony is achieving the maximum coordination number of six; other substances, such as AuF_3, SiF_4, and VF_5, tend to do the same—hence their classification as acids in liquid bromine trifluoride.

Equation (8.82) exemplifies neutralization in the liquid bromine trifluoride system

$$KF + SbF_5 \xrightarrow{BrF_3(l)} KSbF_6 \qquad [8.82]$$

Liquid bromine trifluoride has two useful properties as a solvent. It can fluorinate many materials and produce many compounds that would otherwise be very difficult to synthesize:

$$Au \xrightarrow{BrF_3(l)} BrF_2^+ + AuF_4^- \qquad [8.83a]$$

$$Ag \xrightarrow{BrF_3(l)} Ag^+ + BrF_4^- \qquad [8.83b]$$

$$BrF_2AuF_4 + AgBrF_4 \longrightarrow 2BrF_3 + AgAuF_4 \qquad [8.83c]$$

HARD AND SOFT ACIDS AND BASES (HSAB)

A somewhat different classification of acids and bases has been proposed by Pearson. The classification is qualitative; still, it is very useful because it correlates and explains many observations. In the equation

$$\underset{\text{base}}{N} + \underset{\text{acid-base complex}}{S - X} \leftrightarrows \underset{\text{acid-base complex}}{N - S} + \underset{\text{base}}{X} \qquad [8.84]$$

N is a nucleophilic reagent (Lewis base) and $S - X$ is a complex containing a replaceable group X (Lewis base) and an electrophilic atom S (Lewis acid). If $X - S$ is treated with a series of bases, the relative strengths of these bases toward the acid S can be compared. In a reaction such as (8.84), it is assumed that N acts, in part at least, as an electron donor, and that S acts as an electron acceptor, so that a coordinate covalent bond results. For N in combination with some acids (S), the equilibrium depends chiefly on the basicity of N toward the proton, whereas for other acids, the equilibrium depends on the polarizability of N or the basicity toward CH_3Hg^+. On this basis, acids and bases are divided into two classes. (1) A *hard acid* is one that binds strongly to bases which bind strongly to the proton. A hard acid will be small in size, high in positive charge, and, will possess no valence electrons that are easily distorted or removed. A *hard base* is one that links strongly to the proton. The donor atom is of low polarizability and high electronegativity, and the base is hard to oxidize and associated with full, low-energy orbitals. (2) A *soft acid* is one that binds strongly to highly polarizable or unsaturated bases. A soft acid is one in which the acceptor atom is of large size, small or zero positive charge, and has several valence electrons that are easily removed. A *soft base* (binds strongly to CH_3Hg^+) has a donor atom that is easily polarized, low electronegativity, and is easily oxidized and associated with empty, low-energy orbitals. The general principle for acid-base reactions in this system is that *hard acids prefer to coordinate with hard bases and soft acids prefer to coordinate with soft bases*. Thus, the metal ions that are hard acids—such as, Mg^{2+}, Al^{3+}—will tend to be found in nature with hard bases, such as oxides or carbonates, whereas soft acids—such as Hg^{2+} and Pb^{2+}—are commonly found as sulfides (sulfur is a soft base).

Which species are hard and which are soft? In general, for bases in which the donor atom is from Group V-A, VI-A, or VII-A, the donor atoms of nitrogen, oxygen and fluorine are the hardest in each group and are the most basic toward the proton. The atoms become progressively softer with increasing atomic weight and bind protons less effectively but increase their ability to coordinate with certain other Lewis acids. Therefore, hard acid metal ions will coordinate most strongly with NH_3, OH^-, F^-, etc. Chatt, Ahrland, and Davies formulated a rule that hard acids (or Class *a*, according to some terminology) coordinate best to the lightest atom of a family of elements in the periodic table, and soft acids (Class *b*) coordinate best to one of the heavier atoms. In other words, hard acids bind best to the least polarizable atoms of a family, whereas soft acids bind best to the more highly polarizable atoms in a family. However, soft acids do not always form stable compounds with the most polarizable species, because some of these, such as R_3Sb, are extremely weak bases toward *all* acids. The principle in this case is that R_3Sb will generally form more stable complexes with soft acids than with hard acids, if it forms acid-base complexes at all. Thus, the following sequences of stability of bases with Class *a* or *b* metal ions can be written:

(*Class a*)

$$N \gg P > As > Sb > Bi \qquad F \gg Cl > Br > I \qquad O \gg S > Se > Te$$

(*Class b*)

$$N \ll P > As > Sb > Bi \qquad F < Cl < Br \ll I \qquad O \ll S \approx Se \approx Te$$

Based on the preceding discussion, Ti(IV) (no *d* electrons) should be a hard acid or Class *a* metal ion (small size, high positive charge, no valence electrons that are easily removed) and should, therefore, prefer to coordinate to oxygen rather than sulfur. Giddings found, however, that he could not coordinate Ti(IV) to oxygen in alcohols, ROH, but that Ti(IV) did coordinate with thioalcohols.

$$(C_5H_5)_2TiCl_2 + RSH \xrightarrow{\text{base}} (C_5H_5)_2Ti(SR)_2 + 2HCl \cdot \text{base} \qquad [8.85]$$

This was felt to be possible only if the metal atom had an excess of *d* electrons to donate into the empty *d* orbitals on sulfur. Therefore, it was said that the π electrons from the cyclopentadienyl groups go into the empty *d* orbitals on titanium (making it Class *b* or *soft*), and these, in turn, are pushed into the empty *d* orbitals on sulfur. Thus, now, as later, we see that the hard and soft classification is not fixed but can vary in different situations.

Comparison of the relative stabilities of Group V-A against VI-A ligands toward a given metal ion becomes more difficult. For Class *b* or soft metal ions, the order is generally found to be $S > C \approx P > I > Br > Cl > N > O > F$. This order is the same as that of increasing electronegativity and hardness. For Class *a* or hard metals, an inversion occurs which is strong but not complete. The order is therefore not completely reversed. Some hard metal ions will then form complexes with only oxygen and fluorine. The proton is the typical hard ion, and other hard metal ions will bind strongly to ligands that are basic to the proton, whether the ligands are hard or soft. Tables 8.6 and 8.7 contain a listing of some acids and bases and their classification.

The classification of metal ions is generally made by comparing the stabilities of fluorine vs. iodine, oxygen vs. sulfur, and nitrogen vs. phosphorus complexes. In cases where such a comparison is not possible,

TABLE 8.6 ***Classification of Acids***

Hard (Class *a*) *Acids*	*Soft* (Class *b*) *Acids*
H, Li^+, Na^+, K^+, Be^{2+}, Al^{3+}, In^{3+}, I^{7+} Cl^{7+}, Cr^{3+}, Co^{3+}, Fe^{3+}, $(CH_3)_3Al_2$, BF_3, RCO^+	Cu^+, Ag^+, Au^+, Tl^{3+}, Cs^+, Au^{3+}, I^+, Br^+, HO^+ I_2, O, Cl, Br, $(CH_3)_3Tl$, RO^+ M^0 (metal atom, bulk metal)

Borderline

Fe^{2+}, Co^{2+}, Ni^{2+}, Bi^{3+}, $(CH_3)_3B$, Cu^{2+}

TABLE 8.7 ***Classification of Bases***

Hard (Class *a*) *Bases*	*Soft* (Class *b*) *Bases*
H_2O, OH^-, F^-	R_2S, RSH, RS^-
$CH_3CO_2^-$, PO_4^{3-}, SO_4^{2-}	I^-, SCN^-, $S_2O_3^{2-}$
Cl^-, CO_3^{2-}, ClO_4^-, NO_3^-	R_3P, R_3As, $(RO)_3P$
NH_3, RNH_2, N_2H_4	CN^-, RNC, CO, H^-, R^-

Borderline

$C_6H_5NH_2$, C_5H_5N, N_3^-, Br^-, NO_2^-, SO_3^{2-}, N_2

soft acids will generally complex readily with soft bases, such as carbon monoxide, aromatic hydrocarbons, etc., that are of negligible proton basicity.

Some general principles may be used to predict hard or soft character. Hard character would be expected to increase with increasing positive oxidation state, and vice versa for soft character. Increase of positive charge makes the acceptor less polarizable and thus weakens its soft character. However, the increased positive charge will polarize the ligand more and make the soft character of the acceptor more pronounced.

The following general rules are given:

(1) For acceptors of identical electronic configuration, an increase in charge means strengthening the soft properties due to ability to polarize the ligand, providing that the accompanying decrease of polarizability of the acceptor is not too large. The number of *d* electrons in the metal ion is important in judging whether it is hard or soft. The soft character increases on going from calcium to zinc, because the number of *d* electrons increases.

(2) Groups that transfer charge to the central atom will increase the soft character of that atom, since such transfer of charge is equivalent to a reduction in valence state. Thus, the more polarizable is the donor, the more pronounced is the soft character of the acceptor. Soft bases (H^- is particularly good because of its high polarizability) will most easily transfer negative charge. The fact that BH_3CO exists and BF_3CO does not exemplifies this concept. Carbon monoxide and hydride ion are soft bases, whereas fluoride ion is a hard base. The soft hydride ions in BH_3 make

boron a soft acid, enabling it to form a stable complex with carbon monoxide. Actually, many hydride and organic compounds of the transition metals are known, since hydride ion can enhance the soft properties of the metal.

(3) A soft ligand (base) combines high polarizability with empty *d* orbitals in its outer shell. The more available these *d* orbitals are, the softer is the ligand. Stronger bonds will form when the ligands have orbitals available to accommodate extra electrons from the acceptor *d* shell, such as is the case with $Ni(CO)_4$. Coordination of very soft ligands to a soft acceptor may turn it into a hard acceptor toward another ligand, since it has no *d* electrons remaining for coordination to other soft ligands. These general principles indicate that there are no hard and fast rules in judging hard and soft acids and bases!

The solvent can have a great effect on hard and soft properties also, but these effects will be ignored to maintain simplicity of presentation. It will suffice to mention that hard solutes will tend to dissolve in hard solvents, and vice versa. Water is a hard solvent, but solvents that do not contain acidic hydrogen for hydrogen bond formation are generally much softer.

Two of the simple theories used to explain the preference of hard acids for hard bases and soft acids for soft bases are the following (these are actually used to explain Class *a* or Class *b* behavior; however, as we have seen, Class *a* and Class *b* are identical to hard and soft, respectively):

(1) *Ionic and covalent bonding*: This theory is the oldest, and it says that hard complexes use primarily ionic bonding and soft complexes use primarily covalent forces. The high positive charge and small size of hard ions favor ionic bonding, and bases of high negative charge and small size should be held most strongly by these acids. Hard and soft combinations then suffer from a mismatch in their bonding tendencies and are generally unstable.

(2) *Pi bonding theory of Chatt*: An important feature of soft acids is the presence of loosely held outer *d* electrons, which can be donated to ligands. Thus, most soft bases are π bond acceptors and most soft acids are π bond donors. Soft bases then include such species as phosphorus, arsenic, iodine, and unsaturated ligands, such as carbon monoxide. Hard acids in contrast are π bond acceptors and hard bases (nitrogen, oxygen, and fluoride) are π bond donors. Again, if the two are mixed (hard acid, soft base, etc.), there would be a mismatch in bonding tendencies. Soft acids would have full or nearly full *d* orbitals, and therefore would have a repulsive effect on the filled orbitals of the hard bases oxygen and fluorine.

Some of the applications to which the concepts of hard and soft acids and bases have been applied now follow.

(1) *Formation of M—M bonds*: Zero valent or bulk metals have been shown to be soft. They can be considered to act as either acids or bases. Therefore, strong metal–metal bonds should be formed when both atoms are soft, and such a situation would occur when both are in a low valence state (Hg_2^{2+}, $Mn_2(CO)_{10}$, etc.). To stabilize the metal–metal bond, the ligands attached to M should be soft bases (CO, R_3P, etc.). This stabilizes each metal in a low valence state and increases softness by increasing the electron density on the metal atom. Thus, one generally observes metal–metal bonding when the metal is attached to these ligands.

The HSAB treatment has also been used to explain the poisoning of metal catalysts. Certain metals are catalysts, since species such as carbon monoxide and olefins are absorbed on the metal surface and then undergo reaction at the metal surface, subsequently being desorbed. Bases containing phosphorus, arsenic, antimony, tellurium, and selenium in low oxidation states act as poisoners for these metallic catalysts, since they are strongly absorbed on the surface and thus block the active sites. Soft metal ions are also expected to act as poisons. Hard bases, such as nitrogen, oxygen, and fluorine, would then not be expected to act as poisons.

(2) *Stabilization of valence states*: It is well known that ligands of large size, low electronegativity, and low charge are good for stabilizing metal ions in low oxidation states [i.e., $Ni(CO)_4$]. Organic ligands are generally soft; to obtain stable organometallic compounds, the transition metals should be in a zero or low valence state and the ligand should be soft to preserve the metal in the low oxidation state [i.e., $HCo(CO)_4$]. To stabilize an element in a high oxidation state, we should use ligands or bases containing oxygen or fluorine. To obtain Ni^{5+}, we use oxygen as a stabilizing agent in NiO_4^{3-}.

(3) *Relative stabilities of compounds and complexes*: Which is more stable, CH_3COF or CH_3COI? The CH_3CO^+ ion is a hard acid, and therefore its fluoride would be expected to be more stable. Using the same reasoning, compounds of the type RSI and RSSR would be predicted to be stable; on the other hand, compounds of the type RSF and RSOR would be predicted to be unstable. Similarly, AgI_2^- and I_3^- are predicted to be stable, whereas AgF_2^- and I_2F^- are predicted to be unstable.

We should emphasize that this last concept is not applicable in some situations and gives erroneous results in others. Thus, there is some question about the validity of these concepts; however, they do appear to explain a great deal of experimental data, and thus are entitled to close scrutiny.

These concepts of hard and soft acids and bases have been criticized as being merely qualitative and as working well on the extreme of hardness and softness. Hardness and softness are not mutually exclusive properties, however, and in the more ambiguous situations, one can be (and is sometimes) rather arbitrary in explaining results.

A more quantitative assessment of the essential factors that contribute to donor and acceptor strength has been made by Drago. The enthalpy of adduct formation

$$A + B \rightarrow AB \tag{8.86}$$

in the gas phase or in poor solvating solvents has been written as

$$-DH = E_aE_b + C_aC_b \tag{8.87}$$

where a and b indicate acceptor and donor, respectively, and E and C are interpreted as being related to the electrostatic and covalent bond-forming abilities of the acids and bases. The values of E and C are calculated empirically. (By setting E and C for $I_2 = 1$ and then by using enthalpy data for interactions of I_2 with bases such as NH_3, $(CH_3)_2NH$, and CH_3NH_2, in the appropriate equations E_b and C_b could be calculated.) Then using these E_b and C_b values, we determine the E_a and C_a values for new acids from enthalpy data. Finally, these E and C values are used to calculate enthalpies that are checked with experimentally determined values and then added together to give the enthalpy of formation for the reaction in the gas phase or in a poorly solvating solvent according to equation [8.8b]. Results of these studies are listed in Table 8.8.

TABLE 8.8 ***Some Enthalpy Values for Acid-Base Reactions****

Acid	Base	$-H$ (kcal/mole)	
		Calculated	*Observed*
ICl	$CH_3C{\equiv}N$	5.0	4.9
	C_6H_6	2.8	2.9
	$P(CH_3)_2C_6H_4$	4.0	3.8
$(CH_3)_3B$	NH_3	13.7	13.7
	$C_2H_5NH_2$	18.1	18.0
SO_2	C_6H_6	1.1	1.0
	C_5H_5N	6.0	6.0

*Table taken from Drago, R. S., and B. Wayland, *J. Am. Chem. Soc.*, 87, 3571 (1965).

Such concepts as these may prove to be very valuable in the future. It has already been used in one interesting manner. Using the calculated E and C values, we calculate the enthalpy of formation of the $(CH_3)_3BN(CH_3)_3$ adduct to be 25.8 kcal/mole, compared with the experimental value of 17.6 kcal/mole. The difference of 8.2 kcal/mole is attributed to steric strain. Using combustion data on $(CH_3)_3CC(CH_3)_3$, H. C. Brown has found the strain in $(CH_3)_3CC(CH_3)_3$ to be 7.8 kcal/mole.

Suggested Readings

1. Day, M. C., and J. Selbin, *Theoretical Inorganic Chemistry*, Reinhold Publishing Corp., New York, 1962.
2. Douglas, B. E., and D. H. McDaniel, *Concepts and Models of Inorganic Chemistry*, Blaisdell Publishing Co., New York, 1965.
3. Drago, R. S., *Chem. Brit.*, **3,** 516 (1967).
4. Giddings, S. A., *Inorg. Chem.*, **6,** 849 (1967).
5. Holliday, A. K., and A. G. Massey, *Inorganic Chemistry in Non-aqueous Solvents*, Pergamon Press, Oxford, 1965.
6. Pearson, R., authored the following articles on the HSAB theory:
   (a) *J. Am. Chem. Soc.*, **85,** 3533 (1963).
   (b) *Chem. Eng. News*, **90,** May 31 (1965).
   (c) *Chem. Brit.*, **103,** March (1967).
   (d) *J. Am. Chem. Soc.*, **89,** 1827 (1967).
   (e) *J. Chem. Educ.*, **45,** 581 (1968).
   (f) *J. Chem. Educ.*, **45,** 643 (1968).
7. Young D. E., G. E. McAckran, and G. S. Shore, *J. Am. Chem. Soc.*, **88,** 439 (1966).

Study Questions

8.1. Arrange in order of increasing acid strength:
$HClO_2$, H_2SO_4, H_3AsO_3, HBrO, H_2TeO_3, HClO, H_2SeO_4
Explain your reasoning and predict a pK_1 value for H_2SeO_4.

8.2. Complete and balance the following equations:

(a) $Zn + NOCl \xrightarrow{N_2O_4(l)}$

(b) $Li + N_2O_4 \xrightarrow{N_2O_4(l)}$

(c) $KCl + N_2O_4 \xrightarrow{N_2O_4(l)}$

(d) $(NH_4)_2S + Li \xrightarrow{NH_3(l)}$

(e) $SIF_4 + NOF \xrightarrow{BrF_3(l)}$

(f) $Zn(NH_2)_2 + NH_4Cl \xrightarrow{NH_3(l)}$

8.3. Give the structures of the following acids and tell why you chose the structure:
(a) H_3PO_2, $pK_1 = 2$
(b) H_6TeO_6, $pK_1 = 7$

8.4. Use the π bonding theory (with examples) to explain why soft acids prefer to bind to soft bases and will not bond strongly to hard bases.

8.5. In liquid H_2S, As_2S_3 was very insoluble and CS_2 was very soluble. With labeled S in the solution, neither showed exchange. When Et_3N was added to the solution, exchange was rapid in both. What is the function of the Et_3N, and postulate an equation to explain this exchange. [*Hint*: $As_2S_3 \rightarrow AsS_3{}^{3-}$; *Inorg. Chem.*, **5,** 911 (1966).]

8.6. Very briefly discuss the validity of the following statements.
(a) J. P. Slipshod says that hydrogen fluoride is the strongest binary hydrogen acid, because fluorine has such a great tendency to go to fluoride ion.
(b) Inductive effects are able to explain the following orders of acid strength:
$$(CH_3)_3B < BH_3 < BF_3 \quad \text{and} \quad BBr_3 < BCl_3 < BF_3.$$
(c) The dielectric constant, E, of a solvent permits us to infer whether polar or nonpolar solutes are more soluble.
(d) In liquid sulfur dioxide as a solvent, the sulfur dioxide undergoes a self-ionization as follows:
$$2SO_2 \longrightarrow SO_3^{2-} + SO^{2+}$$
Thus, sulfites are bases and thionyls are acids.

8.7. Give short answers to the following:
(a) Define and illustrate F strain.
(b) The solubility of the silver halides in liquid ammonia and water is not the same. What are the solubilities and why? (i.e., $x > y \gg z$, etc.)
(c) What would you predict concerning the stability of the following complexes, given that AlF_6^{3-} and TlI_6^{3-} are very stable compounds: $Al(H_2S)_6^{3+}$, $Tl(Me_3P)_3^{3+}$, TlF_6^{5-}, $Al(H_3N)_6^{3+}$. Explain your choice.
(d) Define and illustrate the leveling effect using liquid ammonia as the solvent.
(e) Which is hydrolyzed more and why—Na^+ or Mg^{2+}?
(f) Outline a method to calculate the proton affinity for a halogen X.
(g) Explain: SH^- has a lower proton affinity than F^-, but pK_a for HF is lower than that for H_2S.
(h) Which is the stronger base, $(CH_3)_3P$ or $(CH_3)_3As$?
(i) Which is the harder acid, Be^{2+} or Cs^+?

8.8. Explain the function of the aluminum sulfate that is generally added to baking powder (carbonates or bicarbonates).

8.9. Use the following information to calculate the proton affinity for ammonia:

	ΔH_f (kcal/mole)
NH_3	-11.04
NH_4Cl	-75.38
N	$+112.98$
Cl	$+29.01$
H	$+52.09$

Ionization potential of H = 315
Electron affinity of Cl = -85
Lattice energy of NH_4Cl = 163 kcal/mole

8.10. What factors are most important in determining the acid strength of binary hydrohalic acids?

8.11. Explain the following observations:
(a) With boron trifluoride, the donor capacity is

$$(CH_3)_2O > (CH_3)_2S > (CH_3)_2Se$$

(b) With either BH_3 or $B(CH_3)_3$, the order of stability is

$$(CH_3)_2S > (CH_3)_2Se \sim (CH_3)_2O$$

8.12. Explain the following experimental observations:
(a) In the conductimetric titration of tetramethylammonium sulfite by thionyl chloride in liquid sulfur dioxide, there are two breaks in the titration curve—one close to 0.5 mole of sulfite per mole of thionyl chloride and one at 1 mole of sulfite per mole of thionyl chloride.
(b) Solutions of sodium in liquid ammonia are paramagnetic. The magnetic susceptibility decreases with increasing metal concentration.

8.13. When gallium metal is dissolved in 11 *M* perchloric acid (72% wt.), hydrogen gas is produced together with a significant amount of chloride ion [Schug, K. and Sadowski, A. *J. Am. Chem. Soc.*, **83,** 3538 (1961)]. No other reduction products are formed in significant amounts. The results from a typical experiment are

Gallium metal used	3.10 mmoles (0.2164 g)
Hydrogen gas produced	1.55 mmoles (34.8 ml, STP)
Chloride produced	0.78 mmoles

Given a mechanism that is quantitatively consistent with the observed stoichiometry.

8.14. Write equations for the following reactions:
(a) Iron metal is added to a solution of $SOCl_2$ in liquid SO_2.
(b) Tin metal is added to a solution of $[(CH_3)_4N]_2SO_3$ in liquid SO_2.
(c) Gaseous HCl is bubbled into a solution of HNO_3 in anhydrous H_2SO_4.
(d) Phosphine, PH_3, is bubbled into a solution of sodium in liquid ammonia.

8.15. If 40.0 ml of a 0.150 *N* NaOH solution is required to neutralize the solution made by dissolving in water a 0.912-g sample of a solid acid, what is the equivalent weight of the acid?

8.16. The ether, $H_3C—O—SiH_3$, has been synthesized and it has been found to be a much weaker Lewis base than $(CH_3)_2O$.
(a) On the basis of the relative electronegativity values of carbon and silicon, is that what is to be expected?
(b) Discuss the nature of the bonding involved in the C—O—Si linkages and relate it to the observed base strength.
(c) Would one expect the C—O—Si bond angle to be greater or less than the C—O—C bond angle in $(CH_3)_2O$?
(d) Predict the relative dipole moments of the two ethers mentioned and explain your prediction.

8.17. Discuss, using balanced equations, substances that act as amphoteric compounds in liquid ammonia and in liquid sulfur dioxide. Illustrate your answers by known reactions.

8.18. Which solution would you expect to have the higher pH—0.1 M $FeCl_3$ or 0.1 *M* $ScCl_3$? Explain.

8.19. Predict the products of the reaction of CaO, NaOH, concentrated H_2SO_4, and H_2O, each reacting with liquid N_2O_4. Give a reaction mechanism for the reaction of H_2O with N_2O_4, explaining each step.

8.20. The following reactions are known to take place:

$$TiO_2 + Na_2S_2O_7 \longrightarrow Na_2SO_4 + (TiO)SO_4$$
$$3SiO_2 + Ca_3(PO_4)_2 \longrightarrow 3CaSiO_3 + P_2O_5$$
$$SiO_2 + CaSO_4 \longrightarrow CaSiO_3 + SO_3$$
$$ZnO + S_2O_7^{2-} \longrightarrow Zn^{2+} + 2SO_4^{2-}$$
$$ZnO + Na_2O \longrightarrow 2Na^+ + ZnO_2^{2-}$$

Explain these reactions in terms of Lux-Flood and Franklin systems. What special point do the last two equations illustrate?

8.21. Explain the following experimental observations:

(a) An NH_3 (l) solution of phenolphthalein is colorless, but becomes intensely red on addition of KNH_2.

(b) Radioactive thionyl chloride ($SOCl_2$) dissolved in liquid SO_2 is found to exchange its ^{35}S only very slowly with the solvent.

(c) There is a break in the conductometric titration curve of $SbCl_3$ with $SOCl_2$ in liquid SO_2 at a reactant ratio of 2:3.

(d) Addition of triethylammonium cyanide to a solution of $FeCl_3$ in liquid HCN produces a precipitate of $Fe(CN)_3$, which redissolves in an excess of the reagent.

(e) Liquid $COCl_2$ has a low electrical conductivity, which is very much increased on addition of $AlCl_3$. The resulting solution attacks metals with the evolution of CO.

CHAPTER 9

Coordination Chemistry

INTRODUCTION

Perhaps the area of chemistry most widely developed in the last two decades has been that of coordination chemistry, applied particularly to complexes of transition metals. Not only has an upsurge taken place in experimental studies, but there has also been a renewed interest in the structure of these complexes and in corresponding advances in the elucidation of structure and bonding. Even though much progress has been made, no work has equaled that of Alfred Werner in the 1890's. Complexes are playing increasingly important roles in industry, ranging from anticorrosion agents and soil-treatment agents to medicinal agents, which certainly testify for their importance in contemporary life. The biologically important complexes, such as chlorophyll, hemoglobin, and vitamin B_{12}—all with very complex structures—suggest or demand a need by all chemists to undertake at least a fundamental study of the principles of coordination chemistry. Such an exposition is the purpose of this chapter.

The phenomenon of complex formation is really a very general one, but it is especially noted among the transition metal ions. The formation of every complex is the result of a Lewis acid-base type of interaction, in which one atom with a vacant orbital (generally the metal) attracts the electron pair on another atom (generally a nonmetal). For bonding to occur as a result of such an interaction, the metal must possess vacant orbitals and these orbitals must be symmetrically correct, sterically available, and of reasonably low energy. Since transition metal ions generally meet these requirements best, it is not surprising that they form complexes so readily. Since the coordinate covalent bonds can form between substances capable of independent existence [e.g., Ag(I) and NH_3; NH_3 and HCl; BF_3 and CH_3—O—CH_3], the product of the interaction is necessarily more complex. If the product of the interaction is an ion in solution, e.g., $Co(NH_3)_6^{3+}$, it is called a complex ion. The neutral species, $[Co(NH_3)_6]Cl_3$, is generally called a coordination compound, a complex compound, or simply a complex.

We must now define a number of terms to make our discussion of complexes more meaningful. Many of these terms are peculiar to the jargon of coordination chemistry.

(1) *Ligands*: The groups attached to the central metal atom or ion in a complex are called ligands. Ligands may be ions or neutral molecules.

(2) *Donor atom*: Within the ligand, the atom attached directly to the metal is called the donor atom. In $[Ag(NH_3)_2]^+$, nitrogen is the donor atom on the ammonia ligand.

(3) *Primary coordination sphere*: Usually enclosed in brackets in the formula, it represents the central metal ion plus the ligands immediately attached to it.

(4) *Coordination number*: The coordination number is the number of coordinate bonds (or number of ligands in the primary coordination sphere) or the number of electron pairs the metal attracts in forming a complex ion. In $[Ag(NH_3)_2]^+$, the coordination number of Ag(I) is two; in $[Cu(NH_3)_4]^{2+}$, the coordination number of copper is four, whereas in $[Fe(CN)_6]^{3-}$, the coordination number of iron(III) is six. In each of these examples, the coordination number is identical to the number of ligands attached to the metal ion, but such is not necessarily always the case. Some ligands, such as ethylenediamine, furnish two pairs of electrons per molecule (this

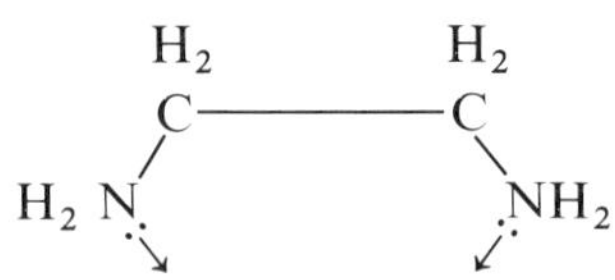

molecule is referred to as a bidentate ligand). Thus, the coordination number for cobalt in $[Co(H_2NCH_2CH_2NH_2)_3]^{3+}$ is six, although there are only three ligands attached to the central cobalt ion. The most common coordination numbers are 2, 4, and 6; in some complex ions, coordination numbers of 3, 5, 7, or 8 occur. We might also note that the coordination number is frequently twice the oxidation number of the central metal ion, but this is not necessarily the case. For example, iron(II) exhibits a coordination number of 6 in $[Fe(CN)_6]^{4-}$, but an oxidation number of (+2).

(5) *Polydentate ligand or chelating agent*: When a molecule attaches itself to more than one coordination position of a given central metal ion to form a closed ring, it is referred to as a polydentate ligand or chelating agent. The specific terms used for a polydentate liquid are bidentate, tridentate, quadridentate, etc., ligands, depending on the number of donor atoms in the ligand. Bidentate ligands, such as ethylenediamine ($H_2NCH_2CH_2NH_2$, abbreviated en) are very common. Polydentate ligands, such as ethylenediamine, that are chelating agents surround a metal ion; these complexes are sometimes referred to as *metal chelates*.

(6) *Labile and inert complexes*: Complexes that exchange ligands rapidly are generally referred to as *labile* complexes, whereas those that exchange ligands more slowly are called *nonlabile* or *inert* complexes. One must guard against confusing thermodynamic stability with kinetic stability when using these terms.

NOMENCLATURE

Due to the great number of complicated and structurally intricate complexes that might arise, we must have a systematic means of naming such compounds to discuss them intelligently. Such a system was originally devised by Werner to describe more precisely the unique structural features of complexes first noted by him. Werner's rules, with certain simplifying modifications, still serve as the basis for the present-day system of nomenclature, as recommended by the Inorganic Nomenclature Committee of the International Union of Pure and Applied Chemistry (IUPAC). The basic IUPAC rules are summarized as follows:

1. The cation is named first and then the anion, in accordance with usual nomenclature rules.
2. In naming the coordination sphere, whether it be the cation, anion, or neutral molecule, the ligands are named before the central metal.
3. The names of negative ligands end in -o; neutral ligands are named as the molecule (except water and ammonia), and positive ligands (rare) end in -ium. A list of ligands appears in Table 9.1.

TABLE 9.1 ***A Listing of Some of the Common Ligands***

Monodentate Ligands

Cl^-	Chloro	H^-	Hydrido	CO_3^{2-}	Carbonato*
Br^-	Bromo	O^{2-}	Oxo	CN^-	Cyano
I^-	Iodo	OH^-	Hydroxo	OCN^-	Cyanato
F^-	Fluoro	S^{2-}	Thio	$C_2O_4^{2-}$	Oxalata*
SCN^-	Thiocyanato	SO_4^{2-}	Sulfato*	H_2O	Aquo
NCS^-	Isothiocyanato	SO_3^{2-}	Sulfito	NH_3	Ammine
NH_2^-	Amido	HS^-	Thiolo	CO	Carbonyl
N_3^-	Azido	$S_2O_3^{2-}$	Thiosulfato	NO	Nitrosyl
NHOH	Hydroxylamido	ClO_3^-	Chlorato	N_2	Nitrogeno
NH^-	Imido	ClO_2^-	Chlorito	$(C_6H_5)_3P$	Triphenylphosphine
NO_3^-	Nitrato*	O_2^-	Peroxo	$NH_2NH_3^+$	Hydrazinium*
ONO^-	Nitrito	Ac^-	Acetato		
NO_2^-	Nitro				

* On occasion, these species have been known to behave as bidentate (or polydentate) ligands.

Polydentate Ligands

Abbreviation	*Formula*	*Name*
$AcAc^-$	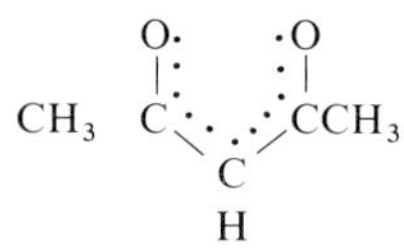	Acetylacetonato ion
Dien	$H_2\overset{\uparrow}{N}CH_2{-}CH_2\overset{\uparrow}{\underset{H}{N}}CH_2{-}\overset{\uparrow}{N}H_2$	Diethylenetriamine
Trien	$H_2\overset{\uparrow}{N}CH_2CH_2{-}\overset{\uparrow}{\underset{H}{N}}CH_2CH_2{-}\overset{\uparrow}{\underset{H}{N}}CH_2CH_2\overset{\uparrow}{N}H_2$	Triethylenetetramine
Phen	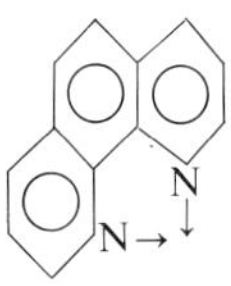	1-10-Phenanthroline
EDTA	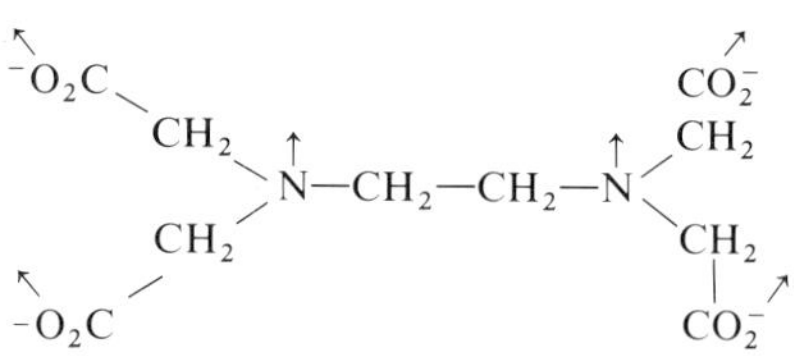	The anion of ethylenediamine-tetraacetic acid

TABLE 9.2 *Names and Formulas of Some Complexes*

Formula	*Name*
$[Ag(NH_3)_2]^+$	Diamminesilver(I) ion
$[Co(NH_3)_6]Br_3$	Hexaamminecobalt(III) bromide
$[Cr(H_2O)_4Cl_2]NO_3$	Tetraaquodichlorochromium(III) nitrate
$[Co(en)_3]_2(SO_4)_3$	Tris(ethylenediamine)cobalt(III) sulfate
$[Pt(NH_3)_4(NO_2)Cl]CO_3$	Tetraamminechloronitroplatinum(IV) carbonate
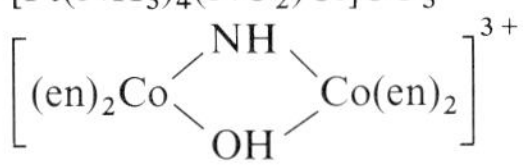	Bis(ethylenediamine)cobalt(III)-μ-imido-μ-hydroxobis (ethylenediamine)cobalt(III) ion
$[Pd(H_2O)_2(ONO)_2I_2]$	Diaquodiiododinitritopalladium(IV)
$[Cu(AcAc)_2]$	Bis(acetylacetonato)copper(II)
$K_3[Fe(CN)_5CO]$	Potassium carbonylpentacyanoferrate(II)
$NH_4[Cr(SCN)_4(NO)_2]$	Ammonium dinitrosyltetrathiocyanatochromate(III)
$[Pt(phen)_2][Pt(CO_3)_2]$	Bis(1-10-phenanthroline)platinum(II); bis(carbonato) platinate(II)
$Na_2[CrOF_4]$	Sodium tetrafluorooxochromate(IV)
$K_3[Ag(S_2O_3)_2]$	Potassium bis(thiosulfato)argentate(I)
$[Fe(C_5H_5)_2]$	Bis(cyclopentadienyl)iron(II); common name: ferrocene

4. The ligands in a complex are named in the order (a) negative, (b) neutral, and (c) positive, without separation by hyphens. Within each of these categories, the ligands are named in the order of their increasing complexity. However, the alphabetical ordering of ligands, irrespective of charge, is now recommended.
5. The prefixes di-, tri-, tetra-, etc., are used before the names of simple ligands when more than one ligand of a particular kind occur within the coordination sphere. Prefixes bis-, tris-, tetrakis-, etc., are used before more complex names such as ethylenediamine (en) and ethylenediaminetetraacetic acid (EDTA).
6. The oxidation number of the central metal is indicated by a roman numeral in parentheses after the name of the metal.
7. When the coordination sphere is an anion, the name of the central metal ends in -ate and latin stem names of the metal are often used. In neutral or cationic complexes, the name of the central metal remains intact.
8. A bridging group (one joining two or more metal atoms) is indicated by the Greek letter μ (mu) preceding its name. A bridging group is shown in the formula by writing an expanded structural formula. (See Table 9.2, sixth example from top.)

A list of complexes with names and formulas is presented in Table 9.2 to illustrate the preceding IUPAC rules.

ISOMERISM IN COMPLEXES

Isomerism is a phenomenon that arises when two or more structures have the same number and kinds of atoms and the same molecular weight (same empirical formula) but at the same time different physical and chemical properties. The roots of isomerism may be traced to the isolation by Pasteur in 1848 of two optically active isomeric forms of sodium ammonium tartrate. *Optical activity* was subsequently discovered in other organic compounds, and this activity gradually became associated with the presence of a particular type of carbon atom in a given compound. It was not until the elucidation of Werner's coordination theory in 1893 that it was recognized that certain inorganic complexes also could exhibit the property of optical isomerism. Since Werner's time, it has been verified experimentally that there exists a greater variety, if not number, of isomeric compounds, including optical isomers, in inorganic complexes due to the various coordination numbers and configurations encountered, than in the case of carbon compounds, which are largely limited to the tetrahedral configuration. The number and type of isomers of a given substance that can be synthesized can provide chemical evidence as to the structure of the complex, and they are an important adjunct to the more definitive diffraction techniques for elucidating structure.

Isomerism can be broken down into two principal types: structural isomerism and stereoisomerism. Each of these can then be further subdivided (Fig. 9.1).

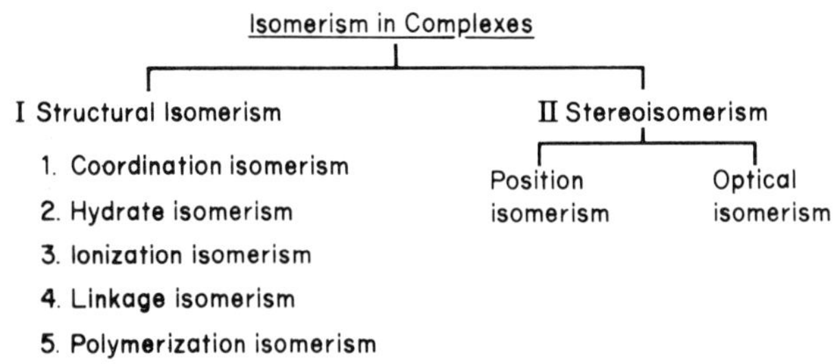

Figure 9.1. Tree chart of isomerism in complexes.

Of the two principal types, stereoisomerism is the more interesting and more important. Stereoisomers differ from each other only in the intramolecular orientation of the molecule in space, i.e., the spatial arrangement of the atoms around a central atom. The properties of stereoisomers are quite similar, and optical isomers, in particular, differ only in their optical activity. Structural isomers, on the other hand, differ from each other in various ways, which are often unique to complexes, resulting in properties that may be markedly disparate. A description of the more common types of structural isomerism will be offered first.

Structural Isomerism

Coordination isomerism. Isomers of this type contain a complex anion in which there is a reversal of the coordinating groups, e.g.,

$$[Co(NH_3)_6]\,[Cr(CN)_6] \quad \text{and} \quad [Cr(NH_3)_6]\,[Co(CN)_6]$$

These are obviously different compounds and only accidentally have the same empirical formula. A total of 12 isomers of this complex are possible.

Hydrate isomerism. In certain crystalline complexes, water may be bound in more than one chemically unique form. In the complex $CrCl_3{\cdot}6H_2O$, the water molecules may all be coordinated directly to the cation, or some may be found outside the coordination sphere without being bound to any specific ion. The three chemically different forms of this complex are

$[Cr(H_2O)_6]Cl_3$	violet
$[Cr(H_2O)_5Cl]Cl_2{\cdot}H_2O$	blue-green
$[Cr(H_2O)_4Cl_2]Cl{\cdot}2H_2O$	green

In addition to differing in color, they also differ in the amount of chloride ion readily ionizable per mole of complex. The first yields three equivalents of Cl^-, the second yields two, and the third yields one equivalent of Cl^- per mole. They differ in other respects as well.

Ionization isomerism. This type of isomerism is similar to hydrate isomerism in that the isomeric forms yield different ions in solution; e.g.,

$$\underset{\text{red-violet}}{[Co(NH_3)_5Br]^{2+}SO_4^{2-}} \quad \text{and} \quad \underset{\text{red}}{[Co(NH_3)_5SO_4]^{+}Br^{-}}$$

The sulfate ion in the latter example is acting as a monodentate ligand rather than as a bidentate ligand.

Linkage isomerism. There are certain ligands that can conceivably coordinate to the central atom in more than one way due to the existence of more than one donor atom. Examples of such ligands are nitro groups ($-NO_2$), which coordinate through nitrogen, and nitrito groups ($-ONO$), which coordinate through the oxygen atom. Other examples are the cyano ($-CN$) and isocyano ($-NC$) forms, and the thiocyanato ($-SCN$) and isothiocyanato ($-NCS$) forms. Although one complex isomeric form often predominates or exists to the exclusion of the other, some examples of *linkage* isomers, in which both forms have been isolated, are as follows:

(a)	$[H_3N)_5Co(NO_2)]Cl_2$	Nitropentaamminecobalt(III)chloride (yellow-brown, stable in acids)
	$[(H_3N)_5Co(ONO)]Cl_2$	Pentaamminenitritocobalt(III)chloride (red, unstable to acids)
(b)	$[Pd(dipy)\ (SCN)_2]$	Dipyrdyldithiocyanatopalladium(II) (stable at room temperature in solid)
	$[Pd(dipy)\ (NCS)_2]$	Dipyridyldiisothiocyanatopalladium(II) (stable at elevated temperatures and in solution)

Polymerization isomerism. Isomers of this type have the same empirical formula but differ in the ratio of the true formula to empirical formula. Examples are $[Pt(NH_3)_2Cl_2]$ and $[Pt(NH_3)_4]\,[PtCl_4]$. The term polymerization isomerism is actually a misnomer, since the second form of the preceding pair is not really a polymer of the first because it consists of both a complex cation and a complex anion.

Stereoisomerism

One of the fundamental postulates of Werner's coordination theory was that the ligands surrounding a central metal atom either in the solid state or in solution would be oriented in certain ways to lead to regular geometric patterns. Thus, a grouping of six similar ligands would surround a central metal so as to yield a regular octahedral arrangement. That is, the ligands would occupy the apices of the octahedron with directed valencies from the metal atom located at the center. In addition, Werner showed that a grouping of four ligands around a central metal would result in either a square planar or a tetrahedral configuration. Other geometrical arrangements have since been found. These involve the less important coordination numbers of 3, 5, 7, 8, 9, and 12.

For four-coordinated complexes, position (cis-trans) isomerism is observed in square planar complexes but not in tetrahedral ones, whereas optical isomerism is possible in tetrahedral complexes but not in square planar ones. The reason for this mutual exclusion of isomerism types in the two kinds of complexes lies in the symmetry elements inherent to a tetrahedral structure as opposed to a square planar arrangement of ligands.

In a tetrahedral complex, all groups are cis (adjacent) to all others, hence the absence of position isomerism; but if all four ligands are different, there is no symmetry in the complex and optical isomers are possible.

Unfortunately, because of the lability of tetrahedral complexes containing only monodentate ligands, no optical isomers have yet been isolated. The only tetrahedral complexes demonstrated to be optically active are those containing unsymmetrical chelating ligands. Thus, bis(benzoylpyruvato)beryllium(II) (Fig. 9.2) can be separated into *d* and *l* forms. This complex has no center of symmetry, no plane or axes of symmetry, and the two forms are nonsuperimposable.

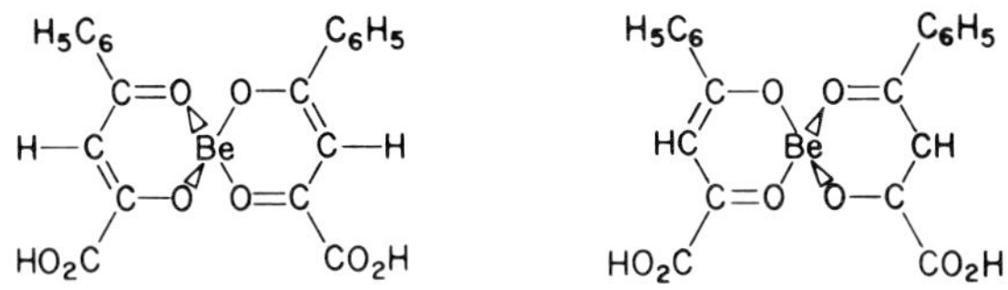

Dextro Levo

Figure 9.2. Optical isomers of bis(benzoylpyruvato)beryllium(II).

For a square planar complex, cis-trans isomers are possible, but because all the ligands and the metal atom lie in the same plane, the complex has a plane of symmetry and cannot exhibit optical isomerism even if all four groups are different.

Only those square planar complexes of formula type MA_2B_2, MA_2BC, and $MABCD$ exist in isomeric forms of the cis-trans type, where A, B, C, and D are monodentate ligands.

1. MA_2B_2 type complexes:

   Complexes of this type may occur in the cis and trans forms:

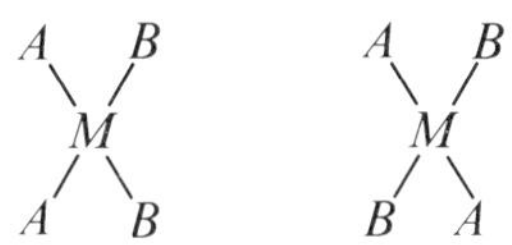

cis-isomer (similar ligands occupy adjacent positions) *trans*-isomer (similar ligands occupy opposite positions)

An example is the dichlorodiammineplatinum(II) complex, which, because it was found to exist in two forms, was considered by Werner to be of the square planar type.

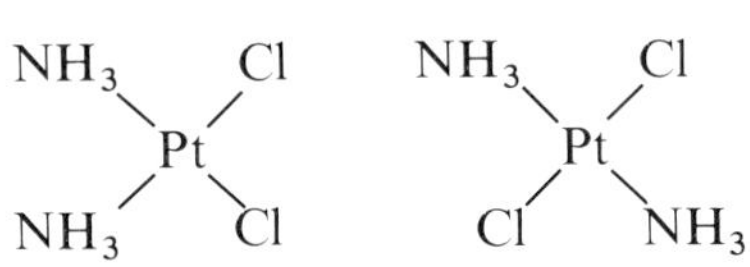

cis-dichlorodiammineplatinum(II) (pale yellow, insoluble in water) *trans*-dichlorodiammineplatinum(II) (dark yellow, more water soluble)

Other examples of complexes of this type are $[Pt(NO_2)_2(NH_3)_2]$ and $[Pt\,Cl_2(py)_2]$.

2. *MA_2BC* type complexes:

   Complexes of this type may occur in either of the following forms:

cis-isomer

trans-isomer

   An example of this type of complex is $[PtCl(NH_3)(py)_2]$.

3. *MABCD* type complexes:

   The possible isomeric forms of this type of complex are

trans-A-C

trans-A-B

trans-A-D

Three isomers of $[PtClBr(NH_3)(py)]$ have been isolated corresponding to the preceding three types of cis-trans isomers.

Because they possess a different molecular arrangement, position (cis-trans) isomers will usually differ to some extent in most of their physical properties and to a lesser extent in certain chemical properties, particularly in reactions with bidentate ligands. Certain typical methods are used for distinguishing between cis and trans isomers. (1) Measurement of the dipole moment in those cases when the complex is a neutral molecule will often distinguish between cis and trans isomers. As expected, the cis isomer will have a larger dipole moment. For example, the cis isomer of $[PtBr_2(Et_3P)_2]$ has a dipole moment of 11.2 D, whereas the trans isomer has a dipole moment of 0. (2) Various diffraction techniques, such as X-ray or electron diffraction, yield information about the relative positions of the atoms in the complex. (3) A chemical method of distinguishing between cis and trans isomers involves a reaction with a bidentate ligand. The basis of this method is the fact that a bidentate ligand can generally span cis positions but not trans positions. Using models, we see that the spanning of trans positions involves distances so large that it is very energetically unfavorable and is less likely than the spanning of cis positions. This is illustrated by the reaction of excess glycine with $[Pt(NH_3)_2Cl_2]$. The cis isomer reacts in a 1:1 mole ratio with glycine to produce

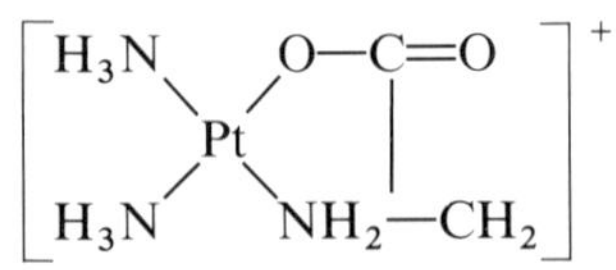

whereas the trans isomer reacts in a 1:2 mole ratio to produce

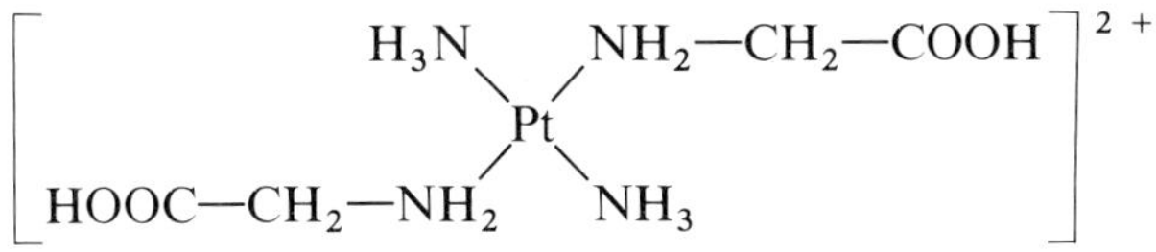

Thus, a stoichiometric evaluation of the reaction would lead to identification of an unknown substance as either the cis or trans isomer. Caution must be used in interpreting such data, however, because of the possibility of rapid rearrangement between the cis and trans forms.

Octahedral complexes may exhibit both position and optical isomerism in certain formula types.

1. Neither complexes of formula type MA_6 or MA_5B can exhibit stereoisomerism.
2. Type MA_4B_2 complexes can exist only in the cis-trans isomeric forms:

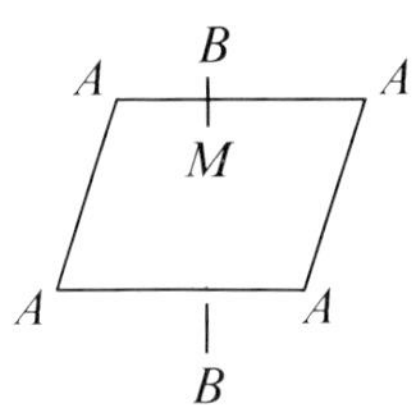

trans-isomer

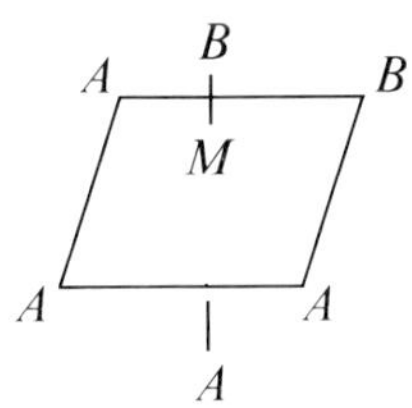

cis-isomer

An example is $[Cr(NH_3)_4Cl_2]^+$ which exists as

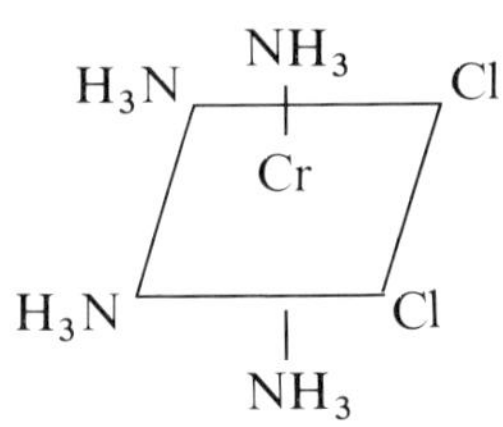

cis-tetramminedichlorochromium(III) ion (violet)

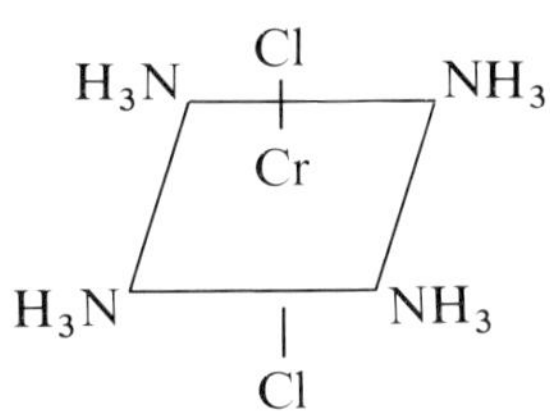

trans-tetramminedichlorochromium(III) ion (green)

3. A few isomers of type MA_3B_3 are known. The two forms of $[Rh(py)_3Cl_3]$ are

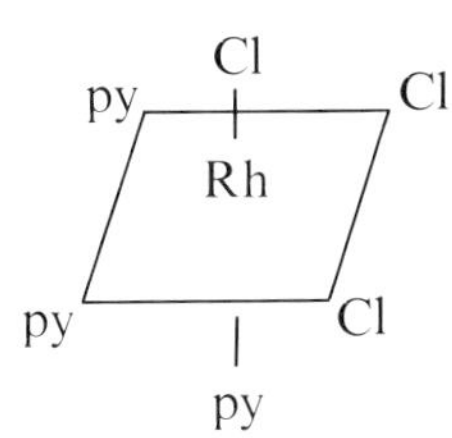

cis-trichlorotripyridinerhodium(III)

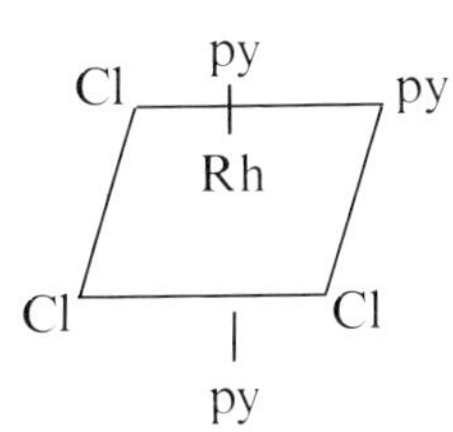

trans-trichlorotripyridinerhodium(III)

In the cis isomer, all the chloride ligands are cis to each other and are on the same octahedral face, as are also the pyridine ligands. In the trans isomer, however, two of the chloride ligands are trans to each other and two of the pyridine ligands are trans to each other. Another example of a slightly different type is the compound $[Co(NH_3)_3C_2O_4Cl]$, in which the oxalate ion acts as a bidentate ligand. The cis and trans forms are

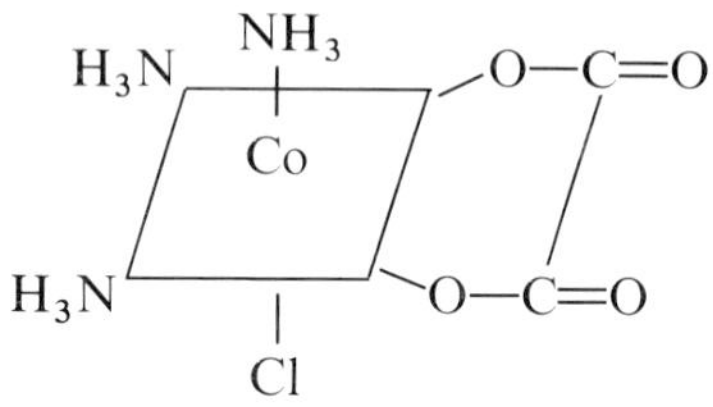

cis-chlorooxalatotriamminecobalt(III)

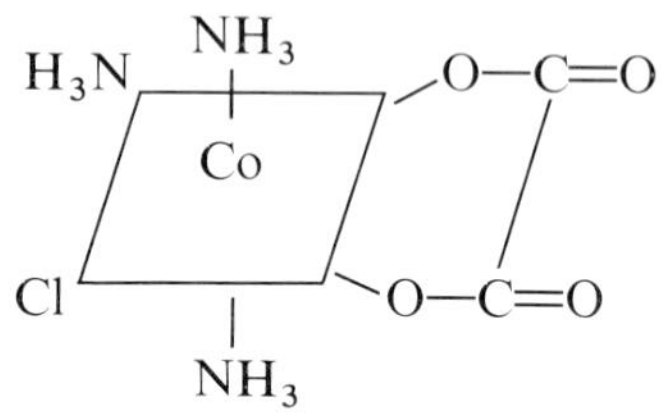

trans-chlorooxalatotriamminecobalt(III)

4. A complex of the type $MA_2B_2C_2$ is $[Cr(NH_3)_2(H_2O)_2Br_2]^+$, which exists in a variety of isomeric forms.
   (a) Each ligand can be trans to one like it:

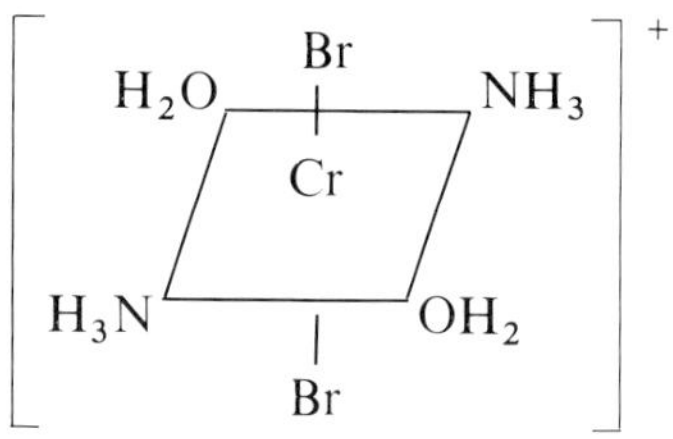

trans-dibromo-*trans*-diaquo-*trans*-diamminechromium(III) ion

   (b) The ammonia molecules can be trans to each other, whereas the water and bromine molecules are cis to ligands like themselves:

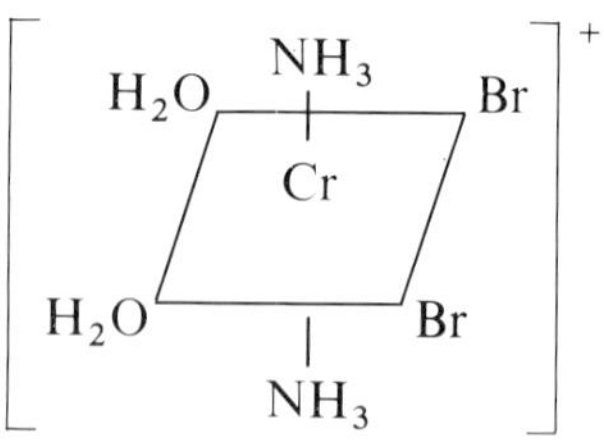

cis-dibromo-*cis*-diaquo-*trans*-diamminechromium(III) ion

   (c) Similarly, the water molecules or the bromine atoms can be trans to each other with the other ligands cis to ligands like themselves:

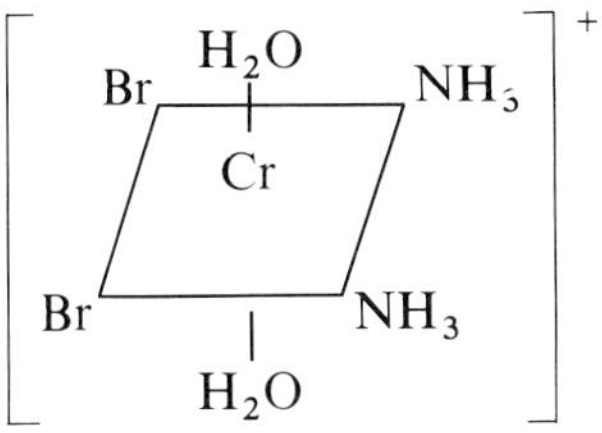

cis-dibromo-*trans*-diamminediaquochromium(III) ion

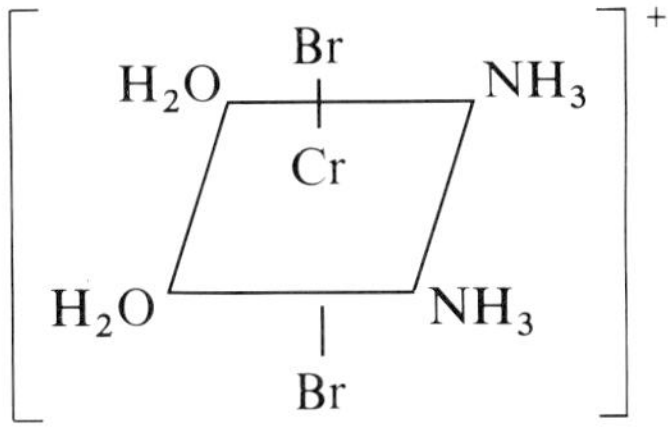

trans-dibromo-*cis*-diamminediaquochromium(III) ion

(d) Finally, each ligand can be cis to one like itself. In this latter case, there are actually two arrangements, differing only in that they are optical isomers:

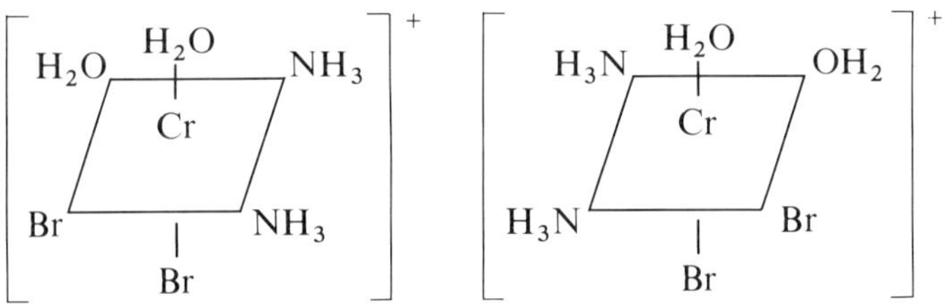

optical isomeric forms of the *cis*-dibromo-*cis*-diaquo-*cis*-diamminechromium(III) ion

In addition to meeting the requirement that optically active substances possess no S_n symmetry element, the preceding mirror image forms of the cis isomers meet the more common and perhaps more recognizable test that they are not superimposable on each other and therefore not identical.

The $[Cr(NH_3)_2(H_2O)_2(Br)_2]^+$ anion thus has a total of six stereoisomers, two of which are optical isomers of each other. The reader should realize that the rotation of any of the six isomers will not produce a new and unique form or any of the remaining five isomers. This can be done with sketches but is demonstrated best with the use of models.

5. The largest number of geometrical isomers would exist for a complex of the type *MABCDEF*, wherein each ligand is different. Such a species can exist in a total of 15 different cis-trans isomeric forms, with each form also having an optical isomer. The only compound of this type that has been prepared is $[Pt(py)(NH_3)NO_2(Cl)(Br)(I)]$. Three different forms were obtained but no attempt was made to isolate all 15 isomers.

6. The presence of a bidentate ligand in a complex can increase the

possibility of optical isomerism. The *tris*-(ethylenediamine)-cobalt(III) ion, $[Co(en)_3]^{3+}$, has two optical isomers:

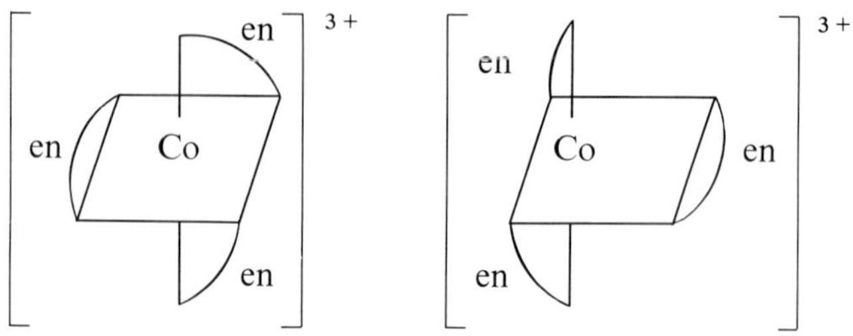

The $[Co(en)_2Cl_2]^+$ ion has two cis isomers, which are also optical isomers, and one trans isomer. The trans form has a plane of symmetry and cannot exhibit optical isomerism. Its mirror image is superimposable on it and is therefore identical with it. The three isomers are

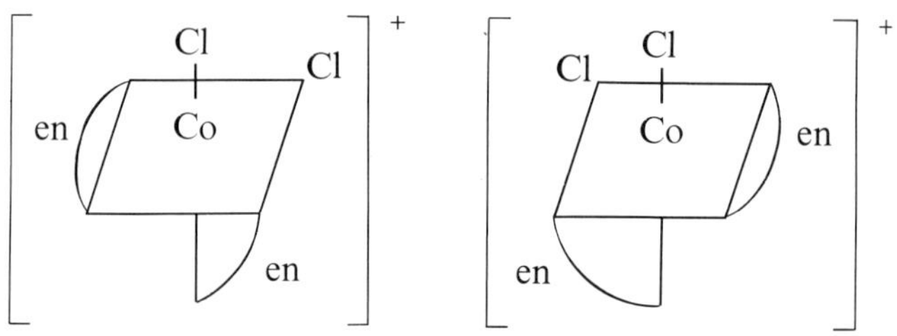

cis-forms (optical isomers) dichlorobis(ethylenediamine)cobalt(III) ion

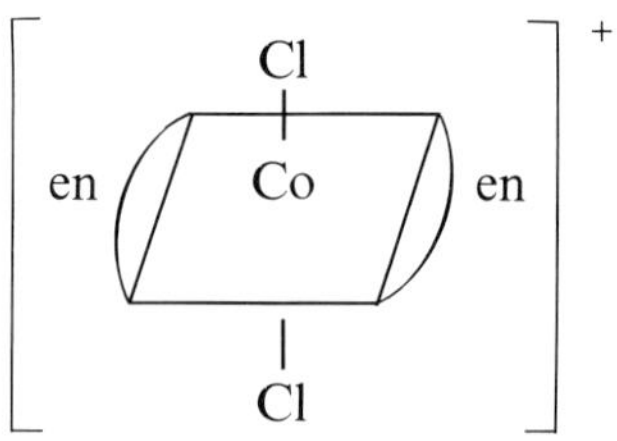

trans form

As is the case with four coordinate square planar complexes, the cis and trans isomers will differ to some extent in their physical and chemical properties. For example, $[cis\text{-}Co(en)_2Cl_2]^+$ will form an oxalate derivative, whereas trans-$[Co(en)_2Cl_2]^+$ will not. Also, only the cis form can be converted to the tetramminecarbonatecobalt (III) ion, $[Co(NH_3)_4CO_3]^+$:

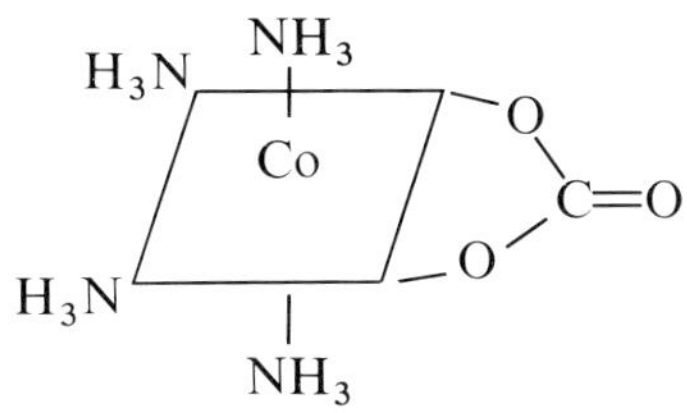

THEORIES OF THE COORDINATE BOND IN METAL COMPLEXES

Three important theories of bonding in complexes of the transition metals have evolved during the last four decades. They are the valence bond theory, crystal field theory (and its modification, ligand field theory), and the molecular orbital theory.

The listing of these theories is basically chronological in the order of their acceptance, although the crystal field theory was originally proposed around 1930, a few years prior to the elucidation of the valence bond theory by Pauling. The crystal field theory, developed by the physicists Bethe and van Vleck to explain certain anomalous magnetic effects in crystals, was not discovered by chemists until the early 1950's. The application of molecular orbital theory to bonding in metal complexes has only taken place in recent years.

The listing of the theories is also in the order of an intuitive, qualitative, and limited method in the case of the valence bond theory to the more general and quantitative molecular orbital theory. The delay in the acceptance of the molecular orbital theory undoubtedly lies in its inability to provide easily available numerical results of the sort provided by the crystal field theory. Nor does the molecular orbital theory present the simple and often more satisfying picture of bonding in complexes that the valence bond theory does. Clearly, each theory has its advantages and disadvantages, although the pendulum is clearly swinging away from the imprecise valence bond approach to the more sophisticated and explicit crystal (ligand) field and molecular orbital approaches.

VALENCE BOND THEORY

The application of valence bond theory to coordination compounds has been discussed briefly before (see Chapter 4) in conjunction with its other applications. Recall that in the application to complexes, we initially decide how many and what type of metal orbitals are needed to form the required number of σ bonds with the ligands. Since the ligands are considered to be electron pair donors, the metal must have low energy orbitals available to accept these electron pairs. The appropriate orbitals (s, p, d) on the metal are then hybridized to provide the new hybrid orbitals, which are directed to the ligand sites according to the symmetry requirements of the complex. In octahedral structures, d^2sp^3 hybridization usually occurs and the d orbitals that are used are the d_{z^2} and $d_{x^2-y^2}$. These hybrid orbitals then overlap the proper ligand orbitals to form coordinate covalent bonds, the electron pair being donated by the ligand. The concept of maximum effective orbital overlap as a criterion of bond strength is a basic tenet of valence bond theory.

Of the various types of hybrid orbital combinations that can arise, the most important, for the purpose of this discussion, are the ones associated with four and six coordination. The advantages and drawbacks of each can best be illustrated by the use of examples.

1. Six coordinate complexes are octahedral, and involve a hybridization of s, p, and d orbitals on the metal to produce d^2sp^3 hybrid orbitals. An example is found in the octahedral complexes of iron(II), with water and cyanide ligands. Iron(II) has a d^6 configuration, where d^6 indicates that there are six electrons remaining in the valence d level of iron after ion formation.

The cyanide complex, $[Fe(CN)_6]^{4-}$, has been shown by magnetic susceptibility measurements to have no unpaired electrons (low spin case), whereas the aquo-complex, $[Fe(H_2O)_6]^{2+}$, has four unpaired electrons (high spin case). To explain this difference by use of the valence bond theory, one is forced to use different orbitals in the hybridization process. The cyanide complex is represented as

	3*d*	4*s*	4*p*	4*d*
$[Fe(CN)_6]^{4-}$	×× ×× ×× ·· ··	··	·· ·· ··	_ _ _ _ _

d^2sp^3 hybridization
(octahedral)

where × represents an electron originating with Fe(II), and · represents an electron donated to Fe(II) by CN^-.

However, to represent the aquo-complex as being covalently bonded and still show four unpaired electrons, we must use the outer 4*d* orbitals:

	3*d*	4*s*	4*p*	4*d*
$[Fe(H_2O)_6]^{2+}$	×× × × × ×	··	·· ·· ··	·· ·· _ _ _

sp^3d^2 hybridization
(octahedral)

where × represents an Fe(II) electron, and · represents an H_2O electron donated to Fe(II).

An alternative representation is to consider the aquo-complex as being ionically bonded. That is, the negative end of the water dipole would be electrostatically bound to the positive iron(II) ion. In this case, the electronic structure of iron(II) in the complex would be the same as that of the free metal ion:

	3d					4s	4p			4d				
$[Fe(H_2O)_6]^{2+}$ or Fe(II)	××	×	×	×	×	—	—	—	—	—	—	—	—	—

Such a differentiation between the cyanide complex and the aquo-complex is not unreasonable, inasmuch as the aquo-complex, utilizing outer orbital (4*d*) covalent bonding, or being ionic, would be expected to be more reactive (more labile) than the cyanide complex. This is what is found experimentally. The valence bond theory, however, cannot satisfactorily explain why the differences in magnetic susceptibility exist in the two complexes. The crystal field theory, on the other hand, offers a satisfactory explanation of this not uncommon problem.

2. Some of the transition metals, particularly those with d^8 or d^9 electronic configurations, characteristically adopt the lower coordination number of four in their complexes and exhibit either tetrahedral or square planar geometry. The square planar complexes are usually of the low-spin type, whereas the tetrahedral complexes are high spin.

Most, if not all, of the complexes of Pd(II) and Pt(II), both with a d^8 configuration, are square planar, and at the same time are diamagnetic (no unpaired electrons). An example is the tetrachloroplatinate(II) complex, $[PtCl_4]^{2-}$:

	5d					6s	6p		
$[PtCl_4]^{2-}$	××	××	××	××	··	··	··	··	—

dsp^2 hybridization

where × is the Pt(II) electron, and · is the Cl^- electron.

A number of complexes of Ni(II), Cu(II), Ag(II), and Au(III) are also square planar. Examples are $K_2[Ni(CN)_4]$, $[Ag(py)_4]^{2+}$, and $[Cu(NH_3)_4]^{2+}$. The d^9 configuration, exhibited by Cu(II) and Ag(II), poses a problem that is difficult to explain by the valence bond theory. Consider the tetraamminecopper(II) complex as an example:

	3d					4s	4p		
$[Cu(NH_3)_4]^{2+}$	××	××	××	××	··	··	··	··	×

dsp^2 hybridization
(square planar)

where × is the Cu(II) electron and · is the N electron donated.

To utilize one of the *d* orbitals in hybridization, we must promote the odd electron to the vacant 4*p* orbital. Pauling suggests that the energy requirements involved in the promotion of one *d* electron are more than compensated for by the increased stability of a dsp^2 hybridized configuration, as opposed to the unhybridized situation. This seems reasonable, but it might then be expected that Cu(II) complexes would be strong reducing agents as a result (losing the promoted, *unstable* electron in the process). Such is, of course, not so experimentally. This question, as we shall see, is resolved more satisfactorily by both the crystal field theory and the molecular orbital theory.

A certain few of the complexes of Ni(II) exhibit tetrahedral geometry rather than square planar. An example is $[NiCl_4]^{2-}$, which is paramagnetic, having two unpaired electrons. It has the valence bond structure

	3*d*					4*s*	4*p*		
$[NiCl_4]^{2-}$	× ×	× ×	× ×	×	×	· ·	· ·	· ·	· ·

sp^3 hybridization
(tetrahedral)

where × is the Ni(II) electron and · is the Cl^- electron.

In this example, the magnetic susceptibility evidence correlates quite well with the expectation of tetrahedral geometry and sp^3 hybridization, but in any d^9 case, with a single unpaired electron being necessarily present, the magnetic moment of the complex cannot be used as a criterion for bonding.

The pros and cons of the valence bond theory as applied to bonding in complexes can now be summarized. The positive feature is that for many years it has provided a very convenient model of bonded atoms based upon a set of simple rules. The disadvantages are that the valence bond theory (1) is limited to qualitative explanations, (2) places undue reliance on magnetic susceptibility evidence but does not account for or predict magnetic behavior beyond specifying the number of unpaired electrons, (3) resorts to artificial or ad hoc explanations, such as *ionic* or *outer orbital covalent bonding*, or *promotion* of electrons when there are more electrons present than can be accommodated in the designated bonding orbitals, and (4) ignores the possibility of excited states or that the energy levels of the metal in a complex may be different from those of the free ion and, as a result, can offer no interpretation or prediction of the spectra of complexes.

CRYSTAL FIELD THEORY AND LIGAND FIELD THEORY

The most salient feature of the crystal field theory is its assumption that all bonds between the central metal atom and the ligands are totally ionic. This is in complete contrast to the valence bond theory, which assumes, at least initially, that the same bonds are covalent in character. Although the ionic bonding feature of crystal field theory is quite unrealistic, and is its principal drawback, its interpretation of the spectra of metal complexes and simple numerical treatment of many aspects of electronic structure are very positive features. When the crystal field theory is modified to allow for the effects of covalent character in the bonds, it becomes even more useful and is known as the ligand field theory. Note that the ligand field theory is often construed to mean the entire gradation of theories from crystal field theory, with its totally ionic concepts, to molecular orbital theory, with its totally covalent bond concepts.

When it is said that the bonding is ionic in the metal complex, it is meant that the complex ion is composed of a central metal ion surrounded by negative ions or the negative ends of polar molecules, all being treated as though they were point charges in space. It is the interaction of the *d* orbitals on the metal with the various electric fields of different ligands that produces the crystal field effects. Therefore, to understand the effects, we must first have a clear picture of the disposition of the five *d* orbitals in space (see Fig. 4.25, Chapter 4).

For any isolated metal ion there is no energy difference in the five *d* orbitals, i.e., they are degenerate. However, this degeneracy is destroyed as the ligands approach the central metal ion.

Consider the formation of the octahedral complex $[Fe(H_2O)_6]^{3+}$ from an isolated iron(III) ion (a d^5 case) and water molecules. As the six water molecules approach the iron(III) ion, the negative ends of the water dipoles will be repelled by the five *d* electrons. The energy of the entire system is now greater because of the repulsion. If one first imagines the hypothetical case in which the approaching water molecules are not disposed in a symmetrical arrangement about the Fe(III) ion, but rather have their electric charge dispersed uniformly in a spherically symmetrical fashion, then all the *d* levels will be raised an equal amount in energy and will still be degenerate (Fig. 4.25). If we now allow the water ligands to be disposed in a regular octahedral configuration about the iron(III) ion, a splitting of the *d* energy levels takes place and they are no longer degenerate. If one considers that the six water molecules, in arranging themselves octahedrally about the iron(III) ion, are located on the x, y, and z axes of a cartesian coordinate system, then the d_{z^2} and $d_{x^2-y^2}$ orbitals which are directed to the vertices of the octahedron will be repelled the most by the approaching ligands and will be raised in energy, whereas the d_{xy}, d_{xz}, and d_{yz} orbitals will be lowered in energy. The $d_{x^2-y^2}$ and d_{z^2} orbitals are commonly referred to by the symbols d_γ or e_g, whereas the d_{xy}, d_{xz}, and d_{yz} orbitals are known by the symbols d_ε or t_{2g}. In other words, the original d orbital degeneracy is split into two parts by the electric field influence of the ligands.

Now, according to a postulate of quantum mechanics, the average energy of the d levels must remain unchanged, so that the doubly degenerate e_g orbitals must be raised in energy exactly as much as the triply degenerate t_{2g} orbitals are lowered in energy. This crystal field splitting, or the energy difference between the e_g and t_{2g} orbitals, is frequently measured in terms of a parameter, Dq. The parameter Dq is a measure of the field strength of the ligands, the stronger the field, the larger is Dq. The magnitude of the splitting is arbitrarily set at $10Dq$. (Equivalent terms for Dq are Δ_0 and ΔE.) Thus, if we take as the zero of energy for the crystal field effect, the energy of the hypothetical case involving no splitting, the energy level of each of the two e_g orbitals would be $6Dq$ above the zero of energy, whereas the energy level of each of the three t_{2g} orbitals would be $4Dq$ below the zero of energy (Fig. 9.3). If each orbital is occupied equally by an electron, as is possible

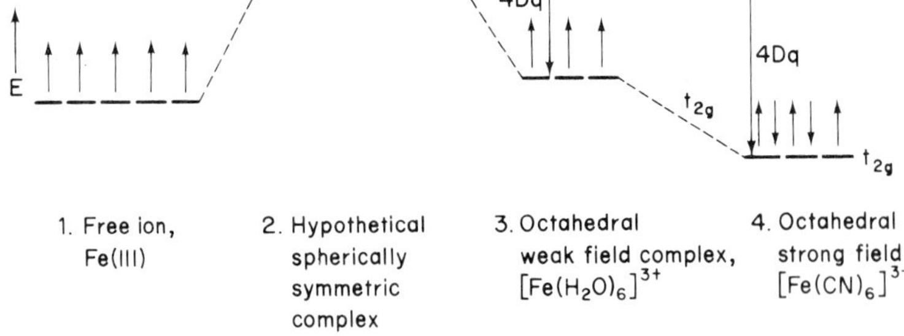

Figure 9.3. The energies of the d orbitals in (1) an isolated free metal ion Fe(III), (2) the hypothetical spherically symmetric complex, (3) an octahedral weak field, and (4) the octahedral strong field. The magnitude of $10Dq$ is shown to be greater in the cyano complex than in the aquo complex, causing the preferential filling of the t_{2g} orbitals.

for d^5 iron(III) ion, the resultant energy change is then zero and no gain in stability is achieved. Each electron in the t_{2g} orbitals stabilizes the system by $-4Dq$, and each electron in the e_g orbitals raises the energy of the system by $6Dq$. Thus, in iron(III), if the five electrons occupied the five d orbitals equally according to Hund's rule, the three t_{2g} electrons would stabilize the system by $-12Dq$, whereas the filling of the e_g orbitals by the remaining two electrons would destabilize the system by $+12Dq$. This is what happens in $[Fe(H_2O)_6]^{3+}$.

On the other hand, if the electrons in the d level preferentially fill the lower-lying t_{2g} orbitals, a gain in stability is achieved. This is what happens in $[Fe(CN)_6]^{3-}$. The difference in d orbital occupancy in the two complexes can be attributed to the much greater splitting in $[Fe(CN)_6]^{3-}$ caused by the much stronger electric field imposed by the cyanide ions. Since the splitting is greater in $[Fe(CN)_6]^{3-}$, the tendency for all five electrons to preferentially fill the more stable t_{2g} orbitals will also be greater. No such preferential filling of the t_{2g} orbitals occurs in $[Fe(H_2O)_6]^{3+}$, because the electric field strength of the water molecules is not large enough to cause a sufficiently large splitting of the d orbitals, so that electrons in the e_g orbitals preferentially fall to the t_{2g} orbitals despite the energy required to pair with other electrons present there. Thus, only when the crystal field splitting is large is the gain in stability achieved by preferentially filling the t_{2g} orbitals sufficient to compensate for the energy required in pairing of electrons (Hund's rule). This gain in stability is known as the crystal field stabilization energy (CFSE). For $[Fe(CN)_6]^{3-}$, it is calculated to be $5(-4Dq) = -20Dq$. For $[Fe(H_2O)_6]^{3+}$, the CFSE is zero. Thus, according to crystal field theory, with a knowledge of the relative field strength of various ligands, one can predict which complexes will be more stable than others, and why in some complexes maximum pairing occurs ($[Fe(CN)_6]^{3-}$), whereas others ($[Fe(H_2O)_6]^{3+}$) prefer to adhere to Hund's rule. This is one of the greatest achievements of crystal field theory.

The method for calculating the CFSE for any octahedral complex has been illustrated. A value of $-4Dq$ is assigned to each electron in a t_{2g} orbital and a value of $+6Dq$ is assigned to each electron in an e_g orbital. The algebraic sum of the total values is the CFSE. Table 9.3 contains the CFSE's for various ions with given numbers of d electrons in both a weak field and a strong field.

In Table 9.3, the CFSE values for octahedral cases in both strong and weak fields differ only for the d^4, d^5, d^6, and d^7 cases. Also, the CFSE is zero for all complexes involving d^0, d^{10}, and d^5 (weak field) configurations in which there is equal filling of all the d orbitals.

TABLE 9.3 *Crystal Field Stabilization Energies for Metal Ions in Octahedral Complexes*

Electronic Configuration	Examples	Octahedral Weak Field			Octahedral Strong Field		
		e_g Electrons	t_{2g} Electrons	*CFSE* $(-Dq)$	e_g Electrons	t_{2g} Electrons	*CFSE* $(-Dq)$
d^0	Ca(II), Sc(III)	0	0	0	0	0	0
d^1	Ti(III), V(IV)	0	1	4	0	1	4
d^2	Ti(II), V(III)	0	2	8	0	2	8
d^3	V(II), Cr(III)	0	3	12	0	3	12
d^4	Cr(II), Mn(III)	1	3	6	0	4	16
d^5	Mn(II), Fe(III), Os(III)	2	3	0	0	5	20
d^6	Fe(II), Co(III), Ir(III)	2	4	4	0	6	24
d^7	Co(II), Ni(III), Rh(II)	2	5	8	1	6	18
d^8	Ni(II), Pd(II), Au(III)	2	6	12	2	6	12
d^9	Cu(II), Ag(II)	3	6	6	3	6	6
d^{10}	Ag(I), Hg(II)	4	6	0	4	6	0

In summary, complexes of the type $[Fe(H_2O)_6]^{3+}$ are known as *weak field complexes*, because the electric field strength of the ligands is not strong enough to cause a large splitting of the d orbitals such that the electrons in the e_g orbitals will be caused to fall down to the more stable t_{2g} orbitals. Such weak field complexes are also known as *high spin* or *spin free complexes*, because the electron distribution of the complexed metal ion is similar to that of the free ion in which the spin quantum number is at a maximum. Conversely, complexes such as $[Fe(CN)_6]^{3-}$ are known as *strong field complexes*, because the electrons have been forced to pair up to a maximum degree. Since the spin quantum number in these latter complexes is at a minimum, they are also known as *low spin* or *spin paired complexes*.

Thus far, only six-coordinate octahedral complexes have been considered. For four-coordinate square planar or tetrahedral complexes, the different geometrical arrangements of ligands around the metal ion produce different splitting patterns, with resultant different CFSE values.

If we consider a square planar complex to be formed from an octahedral complex by the removal of the ligands along the z axis, which point directly at the d_{z^2} orbital, the splitting is quite different and greatly increased. Since the ligands along the z axis have been removed, the energy of the d_{z^2} orbital is greatly decreased relative to that in the octahedral field. At the same time, the metal-ligand bonds along the x and y axes are shortened so that the $d_{x^2-y^2}$ orbital is increased in energy, as is the d_{xy} orbital. The d_{xz} and d_{yz} orbitals are lowered in energy, because the repulsion effects along the z axis are no longer present. The resulting splitting pattern is shown in Fig. 9.4.

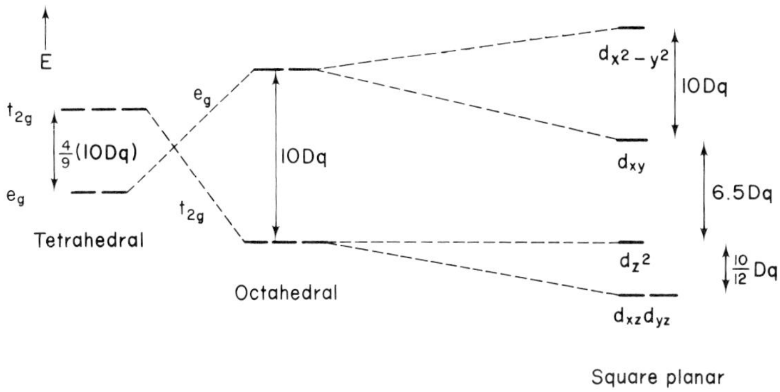

Figure 9.4. The relative energies of the *d* orbitals of a metal ion in complexes having different geometries.

The square planar arrangement is encountered only in those cases in which a greater splitting results in larger CFSE values (more stability), or where the planar configuration is imposed by the geometry of the ligand. The CFSE values for weak and strong field, square planar complexes are listed in Table 9.4. No weak field, square planar complexes have been reported to date. These CFSE values are not obtained as easily as those for octahedral complexes, because quantum mechanical treatments of each configuration are necessary.

TABLE 9.4 *Crystal Field Stabilization Energies for Square Planar Complexes, Weak and Strong Field in Terms of —Dq Units*

Electronic Configuration	*CFSE*, Weak Field	*CFSE*, Strong Field
d^0	0	0
d^1	5.14	5.14
d^2	10.28	10.28
d^3	14.56	14.56
d^4	12.28	19.70
d^5	0	24.84
d^6	5.14	29.12
d^7	10.28	26.84
d^8	14.56	24.56
d^9	12.28	12.28
d^{10}	0	0

The crystal field splitting for a tetrahedral complex is smaller than that for the octahedral complex and the splitting pattern is inverted (Fig. 9.4) relative to the octahedral case. To visualize the splitting pattern, first picture the tetrahedron inside a cube with the four corners of the tetrahedron located at four alternate corners of the cube (see Fig. 9.5). If the x, y, and z axes of our Cartesian coordinate system are then inserted through the center of each cube, the position of the four ligands with respect to the d orbitals can be visualized (Fig. 9.5). The d_{xy}, d_{xz}, and d_{yz} orbitals,

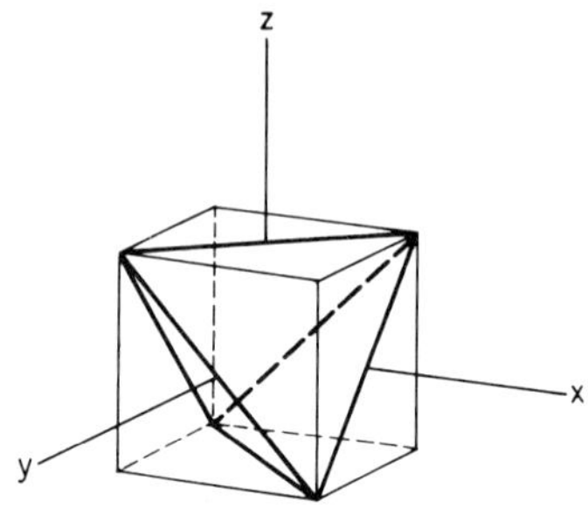

Figure 9.5. A tetrahedral arrangement of the ligands around a metal ion in a cube, with respect to the cartesian coordinate system.

which lie midway between the axes, although not pointing directly at the ligands, will be directed closest to the ligands and will be raised in energy relative to the hypothetical spherically symmetric case. At the same time, the $d_{x^2-y^2}$ and d_{z^2} orbitals, which point directly along the axes, will be lowered in energy. Since in a tetrahedral configuration none of the orbitals points directly at a ligand, the repulsion effects are less and the splitting is lower. In fact, the energy separation ($10Dq$) between the e_g and t_{2g} orbitals has been shown to be about $\frac{4}{9}$ that of octahedral complexes when the metal-ligand separations are the same. No low spin tetrahedral complex has been reported to date. Tetrahedral complexes are most likely for nontransition metals and those transition metals with no CFSE (d^0, d^{10}, and d^5 high spin), since in no case does the CFSE aid in the stability, and since ligand-ligand interactions are least in a tetrahedral complex. Because the CFSE for a square planar complex is invariably much greater, those transition metal ions for which there is CFSE possible will prefer the more stable square planar arrangement. Figure 9.4 summarizes the preceding thoughts concerning the splitting patterns and their relative CFSE's, with regard to complex species of different symmetries.

USES OF CRYSTAL FIELD THEORY

Spectroscopic Applications

Perhaps the greatest achievement of crystal field theory is its interpretation of the colors exhibited by complexes. Most complexes of the transition metals absorb electromagnetic radiation rather strongly in the ultraviolet region of the spectrum. In addition, much less intense absorption bands are found in the visible region for those complexes which are colored. It has been shown by the use of crystal field theory that the absorption bands in the visible region correspond to transitions between the split *d* levels in the complex. In other words, the comparatively small energy separations in the split *d* levels are usually such that absorption of visible light is sufficient to cause excitation of an electron from a lower to a higher *d* level. The result is a colored complex. For an octahedral complex, this absorption band often corresponds to a transition from a t_{2g} orbital to an e_g orbital. In many cases, a measurement of the frequencies of the absorption bands leads to the energy separation, $10Dq$, of the split *d* levels. The following is an example of how the energy separation, $10Dq$, is obtained from a spectrum of $[Ti(H_2O)_6]^{3+}$.

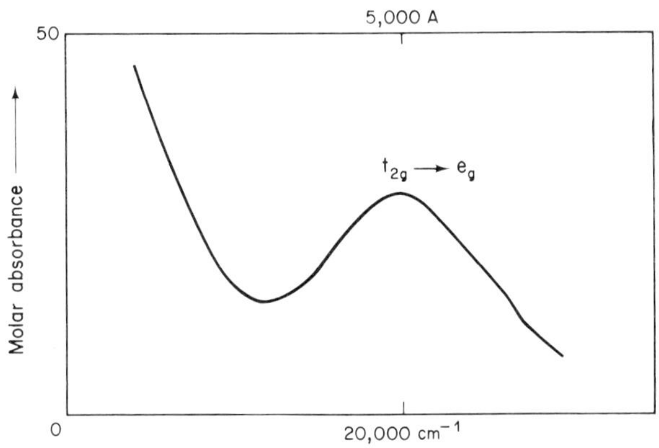

Figure 9.6. The absorption spectrum of $[Ti(H_2O)_6]^{3+}$.

From the spectrum presented in Fig. 9.6, we see that $[Ti(H_2O)_6]^{3+}$ has an absorption band at about 5000 A or 20,000 cm^{-1}. Now the titanium(III) ion has a d^1 configuration and the aqueous solution is purple. This implies that some of the white light hitting the solution is absorbed and only part of the initial white light is reflected. The mechanism suggested for this absorption of light is that the lone *d* electron jumps from the t_{2g} orbital up to an e_g orbital. From these data, the value of $10Dq$, Δ_o, or ΔE may be calculated from the relationship

$$\Delta E\text{ (photon)} = h\nu = \frac{hc}{\lambda} \qquad [9.1]$$

where h is Planck's constant, 6.62×10^{-27} erg-sec, ν is the frequency in

cm^{-1}, c is the speed of light, 2.99×10^{10} cm/sec, and λ is the wavelength of light in centimeters.

Since, in equation (9.1), h and c are constants, the frequency also has the units of energy. Thus ΔE can be equated to $10Dq$. For $[Ti(H_2O)_6]^{3+}$, $\nu = 10Dq = 20{,}000\ cm^{-1}$. Using the conversion factor, 1 kcal/mole = 350 cm^{-1}, $10Dq$ for $[Ti(H_2O)_6]^{3+}$ is equal to approximately 57 kcal/mole, which is in the range of values for energies of chemical bonds.

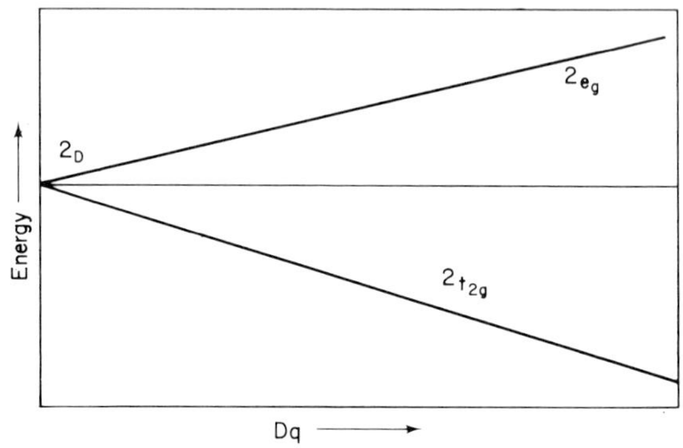

Figure 9.7. The theoretical effect of increasing field strength on the energy levels of a d^1 ion in an octahedral environment. Such a plot which defines the energy levels in terms of their spectroscopic terms is known as an Orgel diagram. The Russell-Saunders state of the free ion (^2D) is at the left.

A theoretical consideration of the splitting of the d orbitals is contained in what is called an Orgel diagram (see Fig. 9.7). This is a plot of the energy of the d orbitals against Dq. The Russell-Saunders term is at the left. The energy separation between the t_{2g} and e_g orbitals increases as the field strength increases. This has been pointed out before. The Orgel diagram in Fig. 9.7 corresponds to a d^1 octahedral case. The reader should remember the concept of *hole formalism* from the presentation of the Russell-Saunders spectroscopic states of the atom (see Chapter 1). Consequently, the d^9 case may be considered the same as the d^1 case, because in the d^9 case one may consider the vacancy created by the absence of an electron (hole) as a single positron moving around. Of course consistency of environment must be maintained—i.e., a d^1 and a d^9 case are comparable only if they are both in an octahedral environment.

Figure 9.8 shows an absorption spectrum, $[Ni(H_2O)_6]^{2+}$. The Ni(II) ion exemplifies a d^8 case. This spectrum is more complicated than the d^1 case. The spectrum shows three distinct bands in the 8000 to 25,000 cm^{-1} region. The assignments indicated on the spectrum may be arrived at, or

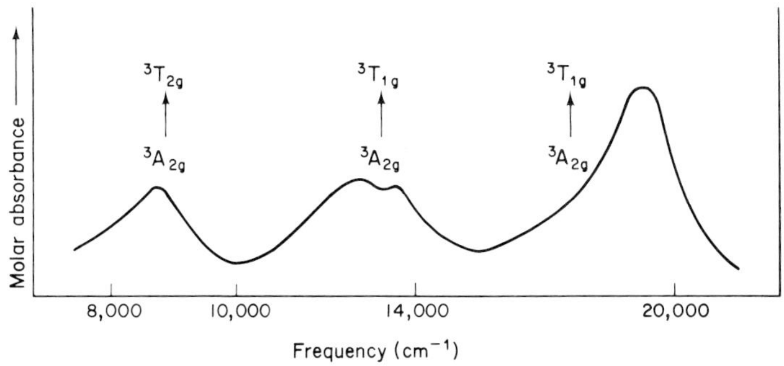

Figure 9.8. The absorption spectrum of $[Ni(H_2O)_6]^{2+}$, showing probable assignments.

at least accounted for, from the d^2-d^8 Orgel diagram presented in Fig. 9.9. The meaning of this Orgel diagram is identical to the one presented for the d^1-d^9 case. On the left of the diagram are the Russell-Saunders states for a d^2-d^8 case, the ^3F state being the ground state. As the field surrounding the free ion increases, the degeneracy of the d orbitals is removed and the orbitals are split. The electronic transitions are from one orbital to another. A quantum mechanical restriction forbids the transition between levels of different multiplicity; therefore, the only transitions that may occur from

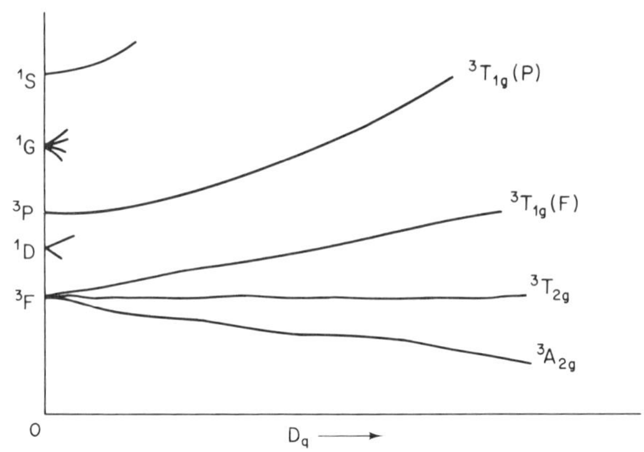

Figure 9.9. Orgel energy level diagram for a d^8 ion in an octahedral field, showing the triplet states and the origin of the singlet states from those of the free ion at the left.

the ground state, $^3A_{2g}$, are those to other triplet states indicated in Fig. 9.9. As the number of d electrons increases, the spectra and Orgel diagrams also increase in complexity.

If the same metal ion is used as a central metal ion and the ligand is changed, the magnitude of the splitting also changes and light of a different frequency will be required to cause the same $t_{2g} \rightarrow e_g$ transition. Measurements of such absorption bands for various complexes of the transition metals have led to the establishment of an order of field strength of various ligands. This *spectrochemical series* includes the following ligands arranged in order of decreasing field strength:

$$CN^- > NO_2^- > \text{o-phen} > \text{dipy} > \text{en} > NH_3 > \text{py} > SCN^- > H_2O > C_2O_4^{2-} > OH^- > Cl^- > Br^- > I^-$$

From the listing given, the CN^- ion and NO_2^- ion would be strong field ligands, whereas I^- and Br^- would be weak field ligands. The ligands in the middle of the series would be ligands of intermediate strength.

Variations in field splitting are caused by a number of factors. Among these are the following: (1) degree of polarizability of the ligand (a readily polarized electron pair on the ligand will be drawn more closely to the metal ion); (2) a small ligand will allow closer approach to the metal; (3) ability of ligand to form double bonds, such as CN^- and NO_2^-. Each of these factors will cause greater repulsion of the *d* electrons on the metal and thus increase the degree of splitting. Note, however, that these factors all involve increasing the covalent character of the bonding, and thus they cannot be considered a part of crystal field theory, which assumes totally ionic bonding. The more realistic ligand field theory would, of course, allow for such covalent bond formation.

Some illustrative examples of the *spectrochemical series* and the relative magnitude of $10Dq$ in different octahedral complexes follow. For various ligands attached to Cr(III), $10Dq$ in cm^{-1} is Cl^-, 13,800; H_2O, 17,400; NH_3, 21,600; en, 21,900; CN^-, 26,300. For Co(III) complexes, the values of Dq in cm^{-1} are H_2O, 18,600;NH_3, 23,000; en, 23,300; and CN^-, 34,000.

With care, this series may be used to predict the *d* orbital splittings and hence the relative frequencies of visible absorption bands for two complexes of the same metal with different ligands. Comparisons should be limited to metals in their most common oxidation states, but even in these cases inversions of the series order for certain members are sometimes found.

With the accumulation of spectrochemical data on a large number of complexes, certain periodic relationships have been observed. Two of these relationships follow.

1. Values of $10Dq$ for octahedral complexes of the first transition series range from 7500 to 12,500 cm^{-1} for divalent metal ions, and from 14,000 to 25,000 cm^{-1} for trivalent ions. For example, the

isoelectronic complexes $[Co(H_2O)_6]^{3+}$ and $[Fe(H_2O)_6]^{2+}$ have $10Dq$ values of 18,600 and 10,400 cm^{-1}, respectively. This can be explained by the fact that the smaller, more highly charged Co(III) ion allows closer approach of the ligands with resulting greater repulsions introduced, increasing the value of $10Dq$.

2. The degree of splitting for similar complexes of metals in the same oxidation state increases by 30 to 50% on descending a group from the first transition series to the second, and a like amount in descending from the second transition series to the third. For example, the values of $10Dq$ for the complexes $[Co(NH_3)_6]^{3+}$, $[Rh(NH_3)_6]^{3+}$, and $[Ir(NH_3)_6]^{3+}$ are 23,000, 34,000, and 41,000 cm^{-1}, respectively. This can be attributed to the fact that the $4d$ and $5d$ orbitals on Rh and Ir extend farther into space and can thus interact more strongly with the ligands and increase the degree of splitting.

There are certain transitions observed in the X-ray region of the spectrum for metal complexes. These may be best explained if a degree of covalent bonding is allowed to enter the complex along with the completely ionic concepts of the crystal field theory. The approach will now be that of the *ligand field theory* or *modified crystal field theory*. In the X-ray region, a number of complexes show definite transitions from the $1s$ to the $4s$ or $4p$ orbitals. This would indicate that the bonding in such complexes is primarily ionic in character and that the ligand electrons are not drawn very completely into the $4s$ or $4p$ orbitals of the metal. Such an observation is inconsistent with the valence bond theory, which assumes covalent bonding in such complexes. The ligand field theory, on the other hand, would explain the phenomenon by showing that orbital overlap is so poor that the electron density in the $3d$, $4s$, and $4p$ orbitals of the metal is sufficiently small to accommodate an electron from the lower $1s$ orbital.

Other complexes, however, indicate a considerable amount of electron density in the metal orbitals, inferring a considerable amount of covalent character in the bonds. The X-ray absorption spectrum of $[Ni(CN)_4]^{2-}$, for example, shows the following electron transitions:

$1s$ to $4s$	does not appear
$1s$ to $4p$	weak transition
$1s$ to $5p$	strong transition

This would indicate that the $4s$ orbital is completely filled with electrons, that the $4p$ orbitals are partly filled, and that the $5p$ orbitals are empty. This is consistent with the hypothetical 100% covalent form:

	3*d*				4*s*	4*p*			5*p*			
$[Ni(CN)_4]^{2-}$	××	××	××	××	··	··	··	··	—	—	—	—

*dsp*² hybridization

where × is the electron originating with nickel(II) and **.** is the electron donated to nickel(II) by cyanide ion.

The crystal field theory would account for the 100% ionic form, which would make a small contribution, in which the bonding forces are due only to electrostatic forces of attraction between nickel(II) and four cyanide ions.

The Jahn-Teller Effect

Occasionally an absorption spectrum of an octahedral complex reveals more transitions than one would expect from two degenerate groups of *d* orbitals (t_{2g} and e_g). This fact implies that the degeneracy of orbitals in the complex is not what we expect. In certain octahedral complexes, especially in d^9 cases, the removal of the *d* orbital degeneracy remaining after crystal field splitting occurs results in distortion of the octahedral configuration and additional stabilization as electrons are redistributed into lower energy orbitals. As one must imagine, removal of the degeneracy of an octahedral splitting pattern requires a new splitting pattern. This is shown in Fig. 9.10,

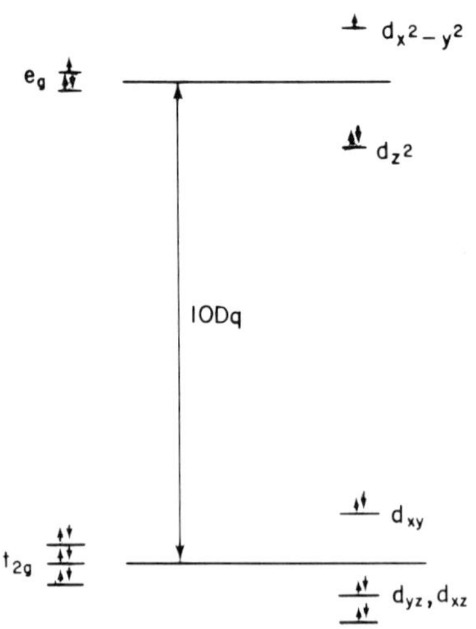

Figure 9.10. Schematic diagram of the orbital splittings caused by an elongation of an octahedral complex along one axis. The magnitude of the Jahn-Teller splitting relative to the $10Dq$ has been emphasized in the diagram.

which illustrates the Jahn-Teller effect and theorem. The theorem reads, *Distortions in configurations will occur whenever the resulting splitting of degenerate energy levels leads to lower energy orbitals and additional stabilization.* In octahedral symmetry, the electron vacancy in the d^9 case is either in the d_{x^2}-d_{y^2} or d_{z^2} orbital. The Jahn-Teller distortion occurs to remove the degeneracy of these orbitals. A possible distortion would be to move the two ligands along the z axis away from the central metal ion. Consequently, the d_{z^2} orbital becomes lower in energy and the degeneracy of the e_g orbitals is removed. The electron vacancy would then be more favorably placed if it were in the $d_{x^2-y^2}$ orbital. For $[Cu(H_2O)_6]^{2+}$, the extra stabilization achieved from the removal of this distortion has been estimated to be about 8 kcal/mole.

For the configurations d^4 high spin and d^7 low spin, distortion of the octahedral complex takes place by lengthening the z axis. For those complexes in which there is equal filling of the orbitals, Jahn-Teller distortions are not expected to take place. Configurations of this type include d^3, d^5, and d^8, all high spin, and d^3 and d^6, both low spin. Usually the distortions involving the movement of ligands in the xy plane are smaller and less important. There is one general restriction on such distortions—namely, that if the undistorted configuration has a center of symmetry, so also must the distorted configuration.

Relative Labilities of Complexes

The crystal field theory can be used to obtain information about the relative labilities of certain complexes. By lability of a complex is meant the rate with which it will act to exchange ligands. The greater is the lability the greater is the rate of exchange. For example, it can be shown by calculations that the two complexes $[Fe(H_2O)_6]^{3+}$ and $[Cr(H_2O)_6]^{3+}$ are nearly equally stable, thermodynamically speaking. The heats of hydration of iron(III) and chromium(III) ions to form the preceding aquo-complexes are 654 and 666 kcal/mole, respectively. If both species are assumed to be of the same type and to exchange with water by the same mechanism, then it might be expected that they would also exchange with water at the same rate. The crystal field theory can be used to show that one of these species would be expected to be more labile than the other and to verify the experimental facts about the two complexes.

In exchange reactions, one of the following two mechanisms is likely to occur: either an S_{N^2} mechanism or an S_{N^1} mechanism. An S_{N^2} mechanism is a substitution that is nucleophilic and bimolecular. That is, two species must unite to form the activated complex. For octahedral complexes, the exchange thus takes place by having the complex and the species that is substituting collide and form a pentagonal bipyramidal intermediate. This intermediate rapidly collapses to form the new complex, which includes the new substituent. On the other hand, the S_{N1} mechanism is a substitution that is nucleophilic and unimolecular, which means that the activated complex is made up of a less complex species, a pentacoordinated square pyramidal intermediate. Thus, as the activated complex ion loses a ligand by dissociation, the substituting ligand very rapidly replaces the dissociated ligand. This is illustrated in Fig. 9.11.

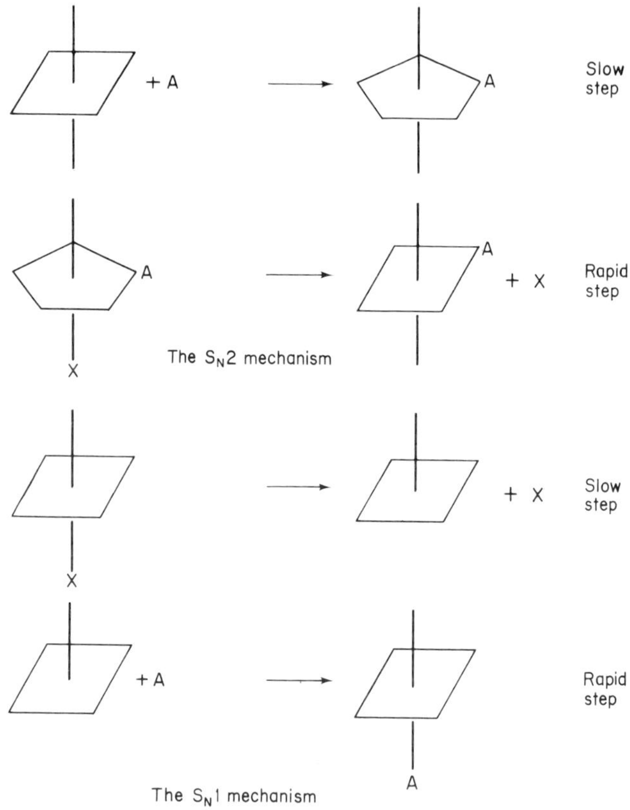

Figure 9.11. Pictorial representation of the S_N1 and S_N2 mechanisms during ligand exchange.

The CFSE values for the preceding types of complexes have been calculated and are shown in Table 9.5 for different d^n systems. Consequently, the *change* in CFSE on going from an octahedral complex to, say, a pentagonal bipyramidal complex can be readily determined. For example, a d^3 weak field octahedral system with a CFSE of $12Dq$ loses $4.26Dq$ in going to the pentagonal bipyramidal system, which has a CFSE of $7.74Dq$. This loss in stability of $4.26Dq$ units represents an activation energy for the conversion of the octahedral to the pentagonal bipyramidal configuration, and will be represented by ΔE_A. Referral to Table 9.5a will allow computation of the ΔE_A values for various configurations. For the d^5 weak field complex $[Fe(H_2O)_6]^{3+}$, the CFSE for *any* geometrical configuration is zero. Thus, there is no CFSE activation energy required for the process

$$[M(H_2O)_6]^{3+} + H_2O^{18} \rightarrow [M(H_2O)_6H_2O^{18}]^{3+} \rightarrow [M(H_2O)_5H_2O^{18}]^{3+} + H_2O \quad [9.2]$$

if M is iron(III); there would be if M were chromium(III). If the complexes are similar, then, to a first approximation, the smaller is the ΔE_A the more labile is the complex. Consequently, the crystal field theory would predict that $[Fe(H_2O)_6]^{3+}$ would exchange more rapidly with H_2O^{18} than $[Cr(H_2O)_6]^{3+}$, if the exchange processes both occurred by a displacement mechanism. This is what is observed experimentally. If a dissociation mechanism is assumed instead, involving the formation of a square pyramidal intermediate, referral to Table 9.5b leads one to the same conclusion—namely, that the $[Fe(H_2O)_6]^{3+}$ is more labile. Similar predictions may be made for other complexes, assuming either a displacement or dissociation mechanism, but it must be kept in mind that the predictions are crude, and only approximate information can be obtained from them.

Stabilization of Oxidation States

Another interesting application of the crystal field theory is in explaining why certain oxidation states are apparently stabilized by coordination to certain ligands. A good example lies in the relative stability of cobalt(II) vs. cobalt(III) in the presence of water and ammonia. Water stabilizes cobalt(II), whereas ammonia stabilizes cobalt(III). Realizing that ammonia has a greater tendency to form strong field complexes than does H_2O and using the following information obtained from Table 9.3,

	CFSE ($-Dq$) *Weak Field* (H_2O)	*CFSE* ($-Dq$) *Strong Field* (NH_3)
Co(II), d^7	8	18
Co(III), d^6	4	24

we find that cobalt(II) is favored with a weak field ligand, whereas cobalt(III) is favored with the strong field ligand, explaining the experimental data.

Another example is found with silver(II) complexes, which are greatly stabilized when forming a weak or strong field square planar structure, as in $[Ag(py)_4]^{2+}$. Silver(II), d^9 case, has $6Dq$ units CFSE in a weak or strong octahedral field, approximately $(\frac{4}{9} \times 6)Dq$ units CFSE in a weak or strong

TABLE 9.5a *Crystal Field Activation Energies for Displacement Mechanism*

	Strong Fields			Weak Fields		
System	Octahedral	Pentagonal Bipyramid	ΔE_A	Octahedral	Pentagonal Bipyramid	ΔE_A
d^0	$0Dq$	$0Dq$	$0Dq$	$0Dq$	$0Dq$	$0Dq$
d^1	4	5.28	−1.28	4	5.28	−1.28
d^2	8	10.56	−2.56	8	10.56	−2.56
d^3	12	7.74	4.26	12	7.74	4.26
d^4	16	13.02	2.98	6	4.93	1.07
d^5	20	18.30	1.70	0	0	0
d^6	24	15.48	8.52	4	5.28	−1.28
d^7	18	12.66	5.34	8	10.56	−2.56
d^8	12	7.74	4.26	12	7.74	4.26
d^9	6	4.93	1.07	6	4.93	1.07
d^{10}	0	0	0	0	0	0

TABLE 9.5b *Crystal Field Activation Energies for Dissociation Mechanism*

	Strong Fields			Weak Fields		
System	Octahedral	Square Pyramid	E_A	Octahedral	Square Pyramid	E_A
d^0	$0Dq$	$0Dq$	$0Dq$	$0Dq$	$0Dq$	$0Dq$
d^1	4	4.57	−0.57	4	4.57	−0.57
d^2	8	9.14	−1.14	8	9.14	−1.14
d^3	12	10.00	2.00	12	10.00	2.00
d^4	16	14.57	1.43	6	9.14	−3.14
d^5	20	19.14	0.86	0	0	0
d^6	24	20.00	4.00	4	4.57	−0.57
d^7	18	19.14	−1.14	8	9.14	−1.14
d^8	12	10.00	2.00	12	10.00	2.00
d^9	6	9.14	−3.14	6	9.14	−3.14
d^{10}	0	0	0	0	0	0

tetrahedral field, but a CFSE of $12.28Dq$ for a weak or strong square planar field. Thus, in this case, silver(II) is stabilized over silver(I), which has zero CFSE in any complex. Similarly, nickel(IV) is stabilized over nickel(II) by forming a strong field octahedral complex:

	CFSE Octahedral Strong Field $(-Dq)$	*Square Planar Strong Field* $(-Dq)$
Ni(IV), d^6	24	29.12
Ni(II), d^8	12	24.56

Therefore, there is a smaller tendency for nickel(IV) to revert to nickel(II) spontaneously when it is in an octahedral complex than when it is in a square planar complex.

In general, ions with smaller CFSE values will be more readily oxidized. For example, under certain conditions iron(II) in $[Fe(H_2O)_6]^{2+}$ is oxidized rapidly, whereas under the same conditions iron(II) in $[Fe(CN)_6]^{4-}$ is oxidized only very slowly. The less stable e_g electrons in the weak field, high spin, aquo-complex would be expected to be lost more readily than the strong field, low spin, cyanide complex d electrons, all of which occupy the more stable t_{2g} orbitals.

A corollary to the preceding discussion of cobalt(II) and cobalt(III) complexes involves a discussion of why certain metal ions favor coordination to nitrogen containing ligands such as ammonia and amines, whereas others favor coordination to oxygen containing ligands. Generally, a metal ion favors coordination with those ligands which produce the largest CFSE values. Cobalt(III) prefers to be six-coordinated to nitrogen rather than to oxygen to a much greater extent than, e.g., chromium(III). This is expected, since chromium(III) has a CFSE of $12Dq$, regardless of whether it is coordinated to a weak or strong field ligand. Cobalt(III), on the other hand, can obtain much increased stability if it coordinates with a strong field ligand; CFSE $= 24Dq$ for a strong field and $4Dq$ for a weak field octahedral complex. The most stable, most easily isolated complexes of cobalt(III) tend to involve nitrogen ligands because they generally produce stronger fields than do oxygen-containing ligands.

Ionic Radii

Certain changes in ionic radii that take place across the transition metal series can be explained as due to the effect of d orbital splittings. The octahedral radii of the divalent ions of the first transition series are presented in Fig. 9.12 to illustrate the problem. Such changes in radius may be expected also with trivalent ions and in other geometrical environments. A smooth curve (dashed line) is drawn through the points for Ca(II), Mn(II), and Zn(II), which have the electronic configurations $(t_{2g})^0(e_g)^0$, $(t_{2g})^3(e_g)^2$, and $(t_{2g})^6(e_g)^4$, respectively. A uniform distribution of charge exists around these ions because of the equal filling of orbitals. This curve thus represents the *ideal* contraction in radius with atomic number for this series. All the other ions have radii that fall below the *ideal* curve because the distribution of electrons is not uniform. For example, the titanium(II) ion has the configuration $(t_{2g})^2(e_g)^0$. Because the concentration of negative

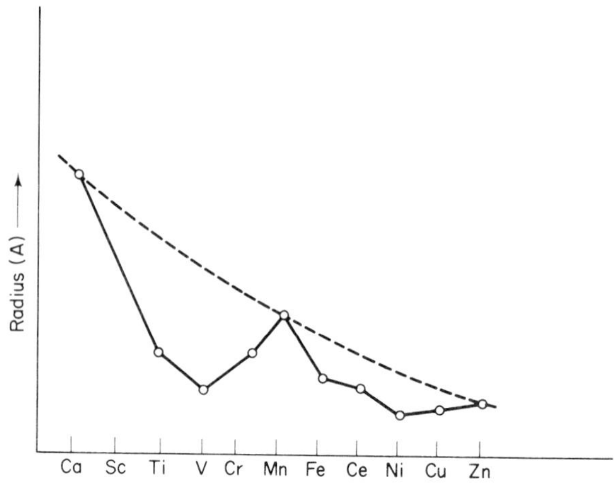

Figure 9.12. The relative ionic radii of the divalent ions of the first transition series. The dashed line is an "ideal" curve explained in the text. The radii of chromium(II) and copper(III) are estimated since Jahn-Teller effects disallow the possibility of octahedral ions in these cases.

charge from the two t_{2g} electrons is in the region of space between the metal-ligand axes, there is little repulsion of the ligands and they are drawn in more closely to the titanium(II) ion. The result is a reduction in size of the titanium(II) ion over the *ideal* or spherically symmetric charge distribution. For vanadium(II), the effect is even more pronounced, because there is one more t_{2g} electron. For chromium(II) and manganese(II), an increase in radius occurs because of the presence of electrons in the e_g orbitals, which increase the repulsion effect on the ligands so that the radii actually increase. A similar effect occurs in the second half of the series—first a decrease and then a slight increase to the last ion, zinc(II).

Structure

Crystal field theory suggests why copper(II) forms square planar complexes rather than tetrahedral or octahedral complexes. In the previous discussion of copper(II) complexes under valence bond theory, it was necessary to resort to artificial concepts, such as the promotion of an electron from the $3d$ to the $4p$ level. Referring to Table 9.3, one finds that copper(II) has a much higher CFSE ($12.28Dq$) in a square planar configuration than in an octahedral (CFSE $= 6Dq$) or tetrahedral configuration (CFSE $= \frac{4}{9} \times 6Dq$). There is no need to promote an electron to make room for ligand electrons because of the assumption of totally ionic bonding in crystal field theory.

In the ligand field theory modification, which allows covalent bonding, this problem is solved by using an antibonding e_g orbital that can accommodate the extra electron.

Another striking structural result of d orbital splittings is found among the class of compounds known as *spinels*. These are oxides whose most general formula is $ABCO_4$, where A, B, and C are cations with a sum of charges equal to $+8$. The mineral spinel, $MgAl_2O_4$, gives its name to the class where $B = C$. Other examples include compounds containing a single metallic element, such as Fe_3O_4, Mn_3O_4, and Co_3O_4. The spinel structure can be described as a cubic close-packed array of oxide ions with one-third of the cations occupying *tetrahedral holes* and the remaining two-thirds occupying *octahedral holes*. The *normal* spinel structure, when the charges on the cations are $+2$, $+3$, and $+3$, has the $+2$ ions in the tetrahedral holes and the $+3$ ions in the octahedral holes, an arrangement that has been shown to be slightly favored for spherical ions.

There are cases, however, in which the divalent ions occupy octahedral holes along with half the trivalent ions, whereas the other half of the trivalent ions go into the tetrahedral holes. In every case where such an inversion has been found, it can be explained in terms of CFSE's. For instance, Fe_3O_4 is inverted whereas Mn_3O_4 and Co_3O_4 are normal. To invert Mn_3O_4, we would have to transfer manganese(III) (d^4) ions, which are in octahedral holes, to tetrahedral holes, with a loss in CFSE (recall that the value of $10Dq$ for a tetrahedral complex is about $\frac{4}{9}$ that of the octahedral complex), whereas we would shift manganese(II) (d^5) ions, which have no CFSE in either environment, from tetrahedral to octahedral holes. Thus, the overall result would be a loss in CFSE, and inversion does not occur. For Co_3O_4, inversion would be highly unfavored, due to the large CFSE values of both the low spin cobalt(III) ions in the octahedral holes and the high spin cobalt(II) ions occupying the tetrahedral holes. Thus, inversion does not occur in this case either. For Fe_3O_4, however, the inversion is favorable insofar as CFSE goes, since transfer of high spin d^5 iron(III) from octahedral to tetrahedral holes carries no penalty, whereas transfer of high spin d^6 iron(II) from tetrahedral to octahedral holes results in a net gain in CFSE.

Another example is provided by $NiAl_2O_4$, which is an inverse spinel. Here, transfer of aluminum(III) does not change the CFSE, but transfer of nickel(II) from the tetrahedral to the octahedral holes is strongly favored.

Such predictions about the structures of spinels could not otherwise be made without a more detailed understanding of all the energy terms than is available. However, given the empirical fact that the normal and inverse structures are similar in stability in the absence of CFSE effects, one obtains very good indications of which compounds of this type will exhibit the unusual inverse structure.

MOLECULAR ORBITAL THEORY

The increasing popularity of the molecular orbital theory is due not only to its more general application but also to its flexibility in the use of all degrees of ionic character in the bonds. In the previous discussion of molecular orbital theory (Chapter 4), the basic rules for formulating molecular orbitals of the LCAO type were described. These rules will now be extended to the bonding in complexes.

Initially, it must be determined which orbital overlaps are and are not possible because of the inherent symmetry requirements present in the complex. Since the application to octahedral complexes is the simplest because of the symmetry involved, these complexes will be used in the following illustrations.

For the metal, there are nine valence orbitals to consider—one s, three p, and five d orbitals. The t_{2g} orbitals that do not point directly at the ligands are not suitable for sigma type bonding. These t_{2g} orbitals usually become nonbonding orbitals. There will be six bonding molecular orbitals derived from the two e_g orbitals, an s orbital, and three p orbitals on the metal (all of which have lobes lying along the x, y, and z axes) and appropriate ligand sigma orbitals. Remember that there is an antibonding orbital associated with each bonding orbital. There also may be π orbitals on the ligands that may effectively overlap with one or more of the t_{2g} orbitals on the metal to produce π bonding and π antibonding molecular orbitals.

The complexes of cobalt(III)—$[CoF_6]^{3-}$ and $[Co(NH_3)_6]^{3+}$—will be considered first, as examples of complexes exhibiting no π bonding. The energy level diagram of these complexes is shown in Fig. 9.13. On the left are the $3d$, $4s$, and $4p$ atomic orbitals of the metal. The ligand σ orbitals are located on the right, showing one energy level for the six ligands. The ligand orbitals are generally of lower energy than the metal orbitals, resulting in some ionic character in the metal-ligand bonds. The σ bonding molecular orbitals lie lower in energy than either the metal or ligand atomic orbitals indicating stable bonds, whereas the antibonding molecular orbitals lie higher in energy. The bonding molecular orbitals are filled with electrons, thus corresponding to six metal-ligand bonds. The remaining electrons are distributed among the t_{2g} nonbonding orbitals and the $\sigma d^*(e_g^*)$ antibonding molecular orbitals, which arise from the interaction of the e_g

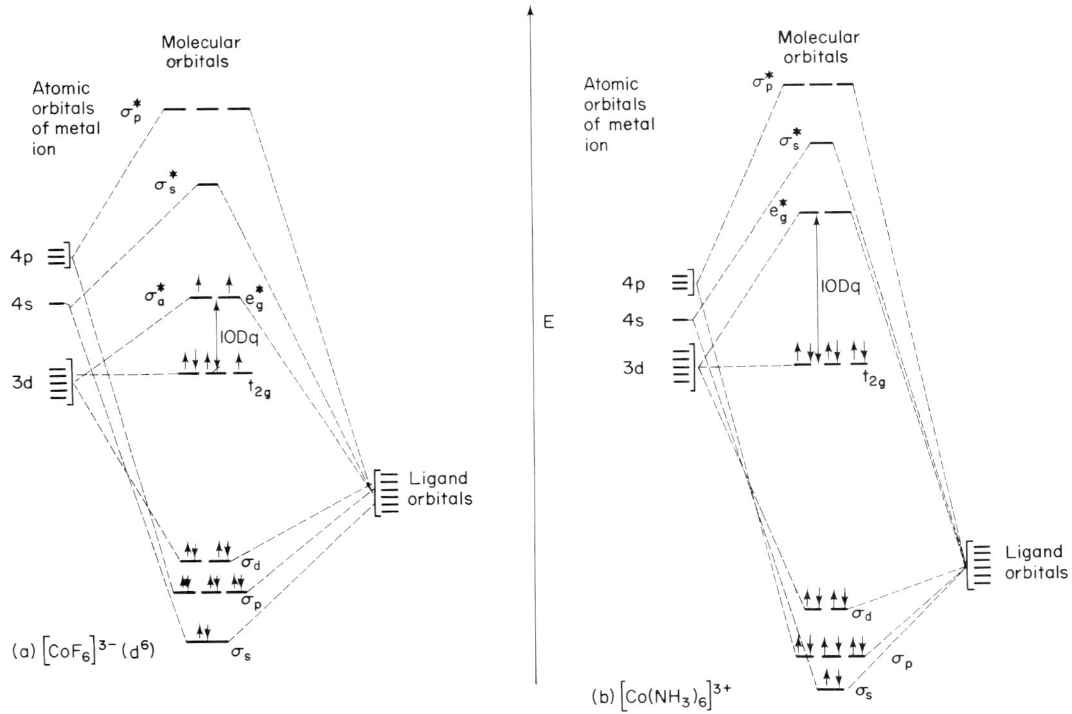

Figure 9.13. Molecular orbital energy diagrams for (a) high spin $[CoF_6]^{3-}$ and (b) low spin $[Co(NH_3)_6]^{3+}$.

metal orbitals and ligand σ orbitals. The placement of these remaining electrons in the nonbonding and antibonding molecular orbitals is quite analogous to the arrangement predicted by crystal field theory, where the same number of electrons are distributed in the same way between the t_{2g} and e_g orbitals. The larger separation between these orbitals in $[Co(NH_3)_6]^{3+}$ causes all the electrons to enter the nonbonding orbitals, whereas Hund's rule is obeyed in $[CoF_6]^{3-}$. The presence of the two antibonding electrons in $[CoF_6]^{3-}$ effectively cancels out the contribution of two electrons in the bonding σ_d orbitals and results in a weakening of the cobalt-fluorine bonds. Whereas the differences in splitting in the two complexes are attributed to electrostatic repulsion of the d electrons by the ligands in accord with the concepts of crystal field theory, the molecular orbital theory explains the same differences as due to differences in energy of the ligand atomic orbitals (fluoride ion and ammonia) contributing to the molecular orbitals.

The combination of a metal ion with ligands, such as cyanide or carbon monoxide, which are capable of forming π bonds in addition to the σ bonds, can strengthen the metal-ligand bonds and contribute to the unusual stability of ions such as $[Fe(CN)_6]^{4-}$. In such cases, the crystal field theory and valence bond theory are grossly inadequate, whereas the molecular orbital theory gives a quite satisfactory explanation of those complexes that do contain π bonds. The ligand orbitals used for π bonding are generally derived from empty atomic p orbitals which are mutually perpendicular and which also are perpendicular to the metal-ligand σ bond. The nonbonding t_{2g} metal orbitals, which have the proper symmetry for π bonding may combine with the empty π orbitals of the ligand to produce π molecular orbitals, the electrons in the π molecular orbitals originating with the metal. The procedure followed in setting up π molecular orbitals

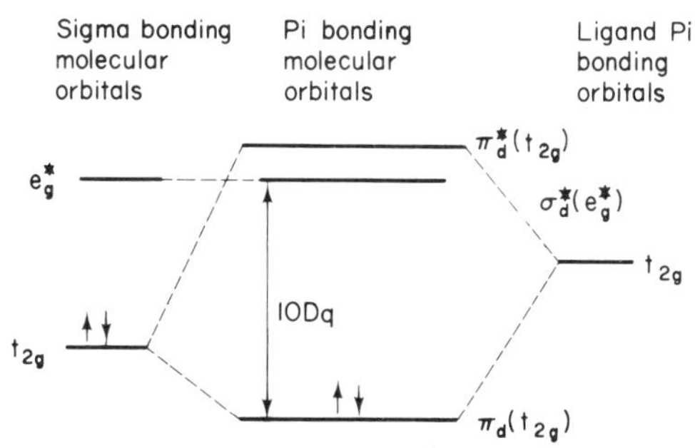

Figure 9.14. Energy level diagram illustrating the formation of a π molecular orbital from π-bonding orbitals of t_{2g} symmetry located on the metal ion and on the ligands. The ligand orbitals are empty and of higher energy than the metal orbitals available for π-bonding.

from π bonding orbitals is quite similar to that employed in forming σ type molecular orbitals. The metal and ligand orbitals are split into bonding and antibonding molecular orbitals. If the ligands have unfilled π orbitals of higher energy than the metal t_{2g} orbitals, the depression in energy of the resulting π_d molecular orbital will increase the magnitude of $10Dq$, and thus increase the stability of the complex. This situation is depicted in Fig. 9.14.

The increased energy difference between π_d and σ_d^*, according to the molecular orbital theory, causes the increased tendency to form low spin complexes. This situation is found in complexes involving ligands such as phosphines or arsines. If, on the other hand, the ligands have filled π orbitals of lower energy than the t_{2g} orbitals of the metal, the result of the π interaction is a destabilization of the π_d orbitals relative to the σ_d^* orbitals and a decrease in the value of $10Dq$. This situation, which is considered to occur with fluorine and oxygen complexes of metals in their normal oxidation states, especially the lower ones, is depicted in Fig. 9.15. There are also important cases involving ligands which have both empty and

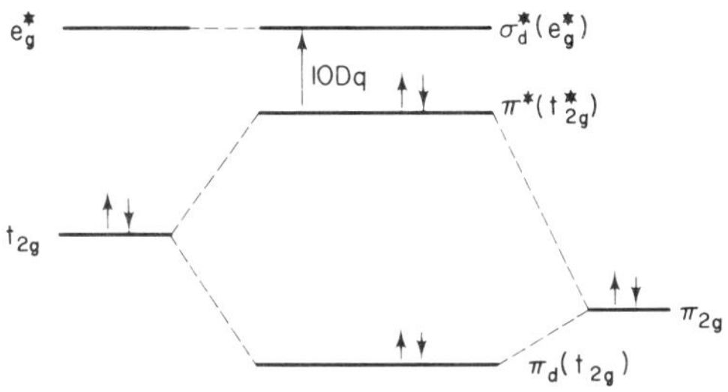

Figure 9.15. Energy level diagram illustrating the formation of a π molecular orbital from appropriate metal and ligand π-bonding orbitals. The ligand π orbitals are filled and of lower energy than the metal π orbitals.

filled π orbitals. For ligands such as carbon monoxide and cyanide ion, the filled π orbitals are the bonding π orbitals and the empty π orbitals are the π^* orbitals. The resulting interactions with the metal t_{2g} orbitals are not simple, and predictions about the relative energies of the orbitals in the complex are not as easily made.

Of the three theories of bonding in complexes, there is no doubt that the molecular orbital theory has the most potential for advancing our understanding of coordination. It can be modified to include any degree of covalent or ionic bond character, and it can accommodate π bonding. However, it is also the most complicated and does not lend itself to a pictorial representation of chemically bonded atoms.

Polarization of Cation or Anion. A cation, or the acceptor atom in the complex, may become deformed or polarized under the influence of an external electrical field, such as is imposed by the ligands. Such deformations may involve a general distortion of the valence orbitals or a specific effect on certain valence energy levels. In either case, the result may be a lower energy content of the complex and a resultant strengthening of the bonds.

In addition, a cation can exert a more significant polarizing effect on the ligands, causing an increase in bond stability. For those ligands which are anions, such deformation occurs more readily than for a smaller, neutral, or cationic species, since the interelectronic repulsions in anions force the electrons into average positions that are farther from the nucleus. Such a system can be adjusted more readily to a more stable arrangement in the presence of an external polarizing influence, thereby increasing the bond strength. The external polarizing ability of a cation on a ligand, in general, is greater the higher is the charge density on the cation. Transition metal ions tend to be of this type, explaining in part the tendency of transition metal ions to form stable complexes.

The Chelate Effect. For multidentate ligands, the factors influencing stability are more complicated due to ring size, ring strain, conjugation, π bonding, and the like. But in general, if the donor atom remains the same and its properties nearly so, the formation of chelate rings enhances the stability of the complex (the chelate effect). The greater the number of chelate rings are in the complex, the more stable it is. The greatest stability appears to exist with five- and six-membered rings. Ligands such as acetylacetone form particularly stable six-membered chelate structures.

Steric Strain. Because of steric strain, large bulky ligands form less stable complexes than do analogous smaller ligands. For example, $H_2NCH_2CH_2NH_2$ forms more stable complexes than its substituted derivative, $(CH_3)_2NCH_2CH_2N(CH_3)_2$. The strain is sometimes due to the geometry of the ligand coupled with the stereochemistry of the complex. For example, $H_2NCH_2CH_2NHCH_2CH_2NHCH_2CH_2NH_2$ can coordinate its four nitrogens at the corners of a square, but $N(CH_2CH_2NH_2)_3$ cannot; thus, the straight-chain tetramine forms more stable complexes with copper(II) than does the latter branched-chain amine, which is unable to assume the preferred square planar arrangement.

Lability or Kinetic Stability of Complexes

The lability of a complex refers to the rate at which it reacts, substitutes, or exchanges with other chemical species. As suggested by Taube, a labile complex would have a rapid rate of reaction (less than one minute), whereas an inert or nonlabile complex would have a slower rate (more than one minute) under similar conditions of 25°C and 0.1 *M* reactants. There is no real correlation between the thermodynamic stability of a complex and its lability, although it is often found that a stable complex is inert and a labile complex is unstable. For example, the stability constants for $[Ag(CN)_2]^-$ and $[Fe(CN)_6]^{4-}$ are 10^{21} and 10^{37}, respectively, indicating that although both complexes are quite stable, $[Fe(CN)_6]^{4-}$ is much the more stable complex. In exchange reactions with $^{14}CN^-$, the silver complex exchanges CN^- for $^{14}CN^-$ instantaneously and is labile, whereas the iron complex does so only very slowly and is inert. As another example, cobalt(III) ammine complexes, such as $[Co(NH_3)_6]^{3+}$, are unstable in acid solution. At equilibrium, which occurs very slowly, complete conversion to $[Co(H_2O)_6]^{2+}$, NH_4^+, and O_2 is observed, according to the equation:

$$4[Co(NH_3)_6]^{3+} + 2OH^+ + 26H_2O \rightarrow 4[Co(H_2O)_6]^{2+} + 24NH_4^+ + O_2 \qquad [9.6]$$

However, $[Co(NH_3)_6]^{3+}$ can be kept in acid solution for days at room temperature without noticeable decomposition. Thus, although it is unstable, it is also inert; its rate of decomposition is very low.

The first recognition that the lability of a complex depends primarily on the electronic structure of the complex was made by Taube in 1950. Until this time, it was believed that the lability of a complex depended upon the extent of ionic character in the bonds. A labile complex was *ionic*, whereas an inert complex was thought to be covalently bonded.

We may now make a fairly reliable prediction of the relative lability of octahedral complexes based on electronic structure of the complex. If the complexes are classified as inner orbital and outer orbital, depending on whether the metal *d* orbitals involved in bonding are of the e_g type outer orbital (an outer orbital of e_g type would have the same principal quantum number, *n*, as the *s* and *p* orbitals) or e_g type inner orbital (an inner orbital of e_g type has a principal quantum number of $n - 1$, with reference to the *ns* and *np* valence orbitals), then certain distinctions can be recognized. Inner orbital complexes are found to be either labile or inert. There are very few complexes with an intermediate lability. However, in the case of outer orbital complexes, there is a gradual change from those of the metals, which are labile, to those of the nonmetals, which are inert. These differences are illustrated in Table 9.6. For the latter five complexes in Table 9.6, the *d* orbitals involved in bonding must be of the outer orbital type.

TABLE 9.6 ***Labilities of Octahedral Complexes Versus Electronic Configuration of Metal***

Electronic Configuration of Metal		Examples	Lability	Classification
nt_{2g} Electrons	ne_g Electrons			Inner or Outer Orbital
0	0	Sc(III), Ti(IV)	Labile	Inner
1	0	Ti(III), V(IV)	Labile	Inner
2	0	Ti(II), V(III)	Labile	Inner
3	0	V(II), Cr(III)	Inert	Inner
4	0	Cr(II), Mn(III)	Inert	Inner
5	0	Mn(II), Fe(III)	Inert	Inner
6	0	Fe(II), Co(III)	Inert	Inner
3	2	Mn(II)	Labile	Outer
4	2	Fe(II)	Labile	Outer
5	2	Co(II)	Labile	Outer
6	2	Ni(II)	Labile	Outer
6	4	Zn(II), Cd(II)	Labile	Outer

With labile inner orbital complexes, valence bond theory suggests that the exchange reaction proceeds via an S_N2 mechanism involving the formation of a heptacoordinated intermediate (see section on crystal field theory). For example, the reaction

$$[V(NH_3)_6]^{3+} + 6HCl \xrightarrow{H_2O} [V(H_2O)_6]^{3+} + 6NH_4Cl \qquad [9.7]$$

is postulated as proceeding via six steps, each similar to

$$[V(NH_3)_6]^{3+} + H_2O \longrightarrow \underset{\text{unstable}}{[V(NH_3)_6]^{3+}} \longrightarrow [V(NH_3)_5(H_2O)]^{3+} + NH_3 \qquad [9.8]$$

Furthermore, the unstable intermediate in equation [9.8] can only be formed easily if there is a vacant d orbital present. This can be seen more readily if we picture the valence bond configuration of $[V(NH_3)_6]^{3+}$:

	3*d*					4*s*	4*p*		
$[V(NH_3)_6]^{3+}$	·	·	__	××	××	××	××	××	××
				d^2sp^3 octahedral					

where · denotes V(III) electron and × denotes electron donated to V(III) by NH_3. The heptacoordinated species could then be pictured as

	3d					4s	4p		
$[V(NH_3)_6(H_2O)]^{3+}$	·	·	∘∘	××	××	××	××	××	××
d^3sp^3	(pentagonal bipyramid)								

where ∘ denotes the electron donated to V(III) by H_2O. In the latter case, the vacant t_{2g} orbital is filled readily in the intermediate.

With chromium(III) complexes, magnetic susceptibility measurements do not show whether a complex is inner or outer orbital. However, the fact that silver nitrate solution precipitates AgCl only very slowly from solutions of $[Cr(H_2O)_5Cl](NO_3)_2$ indicates that the complex is inert and is therefore inner orbital:

	3d					4s	4p		
$[Cr(H_2O)_5Cl]^{2+}$	·	·	·	××	××	××	××	××	∘∘

where · denotes Cr(III) electron, × denotes H_2O electron donated to Cr(III), and ∘ denotes Cl electron donated to Cr(III). There are no vacant *d* orbitals in this configuration, thus prohibiting the easy formation of a hepta-coordinated intermediate.

Where magnetic susceptibility data are useful, they invariably support the chemical evidence in classifying a complex as inner or outer orbital. For example, both magnetic susceptibility and chemical measurements indicate that $[Fe(CN)_6]^{4-}$ is inert and should be classified as an inner orbital complex, whereas $[Fe(H_2O)_6]^{2+}$ is labile and should be classified as outer orbital:

	3d					4s	4p			4d				
$[Fe(H_2O)_6]^{2+}$	··	·	·	·	·	××	××	××	××	××	××			

where · is the Fe(II) electron and × is the H_2O electron donated to Fe(II).

Outer orbital complexes with metals are generally found to be labile, and there is a gradual change toward inertness as the central atom becomes less metallic. For example, outer orbital complexes of the nonmetals phosphorous(V) and sulfur(VI) are inert, but complexes of aluminum(III) and silicon(IV) have intermediate labilities. The lability decreases as the oxidation number of the central atom increases, which indicates increased polarization of the ligand by the smaller, more highly charged central atom resulting in more covalent character in the bonds. If, as has been suggested, outer orbital complexes substitute via an S_{N^1} dissociation mechanism with a pentacoordinated intermediate being formed, then the more covalent is the bond, the less readily dissociation occurs and the more inert is the complex.

Although the preceding picture of lability in complexes from the valence bond viewpoint is satisfying to a considerable degree, it has some serious drawbacks. Some of these follow.

1. The evidence on which lability or inertness is assigned in Table 9.6 is often only qualitative.
2. There is experimental evidence that bond strengths are very much the same for inner and outer orbital complexes of a given type. The variations that exist can be explained in terms of crystal field theory but not in terms of the valence bond theory.
3. Differences between a d^1 and a d^2 system, or between an inner orbital d^3 or d^4 system, cannot be predicted.
4. All outer orbital systems are grouped together except as charge and size of the central atom is varied. No distinction is made based on the number of d electrons.
5. The valence bond theory favors an S_{N^2} mechanism for inner orbital complexes, yet there is little experimental evidence supporting the postulate of such a mechanism being operative in complexes. On the other hand, considerable evidence exists for an S_{N^1} mechanism being operative for a great many complexes.
6. Where the valence bond and crystal field theories agree, the correspondence to experiment is perfect. Where they disagree, the evidence is clearly on the side of crystal field theory.

The arguments pro and con the two theories have been presented previously. As described in the preceding section, the crystal field theory explanation is quite satisfying.

In general, four-coordinated complexes (both tetrahedral and square planar) react more rapidly than analogous six-coordinated systems. For example, the rate of exchange of $^{14}CN^-$ with octahedral cyanide complexes such as $[Mn(CN)_6]^{4-}$ and $[Co(CN)_6]^{3-}$ is invariably slow, but the very stable square planar complex $[Ni(CN)_4]^{2-}$ undergoes rapid exchange with $^{14}CN^-$. The greater lability of four-coordinated complexes may be due to the fact that there is enough room around the central ion, and there are orbitals available of sufficiently low energy for a fifth group to enter and form an unstable intermediate. The presence of such an additional group would aid in the release of one of the original ligands.

Preference for Coordination Numbers and Stereochemistry of Transition Metal Ions

As stated previously, the most common coordination numbers of transition metal ions are four and six. Those four-coordinated complexes for which there is no CFSE will prefer the tetrahedral arrangement, since the ligand-ligand repulsions are minimized in those cases (d^0, d^5 weak field, and d^{10}). Those complexes of coordination number four in which there is CFSE possible will prefer the square planar arrangement, since the CFSE is always larger for a square planar than for a tetrahedral complex. However, the question of why a metal ion is found to have a coordination number of four rather than a coordination number of six has not been discussed. The following factors can be seen to be important in explaining the stability of a square planar complex in certain cases.

Increased σ Bond Strength in Certain Square Planar Complexes. In both the octahedral and square planar complexes, the $d_{x^2-y^2}$ orbital is raised in energy because of repulsive forces with reference to the ligands. Since this orbital is empty in square planar complexes, except for a d^9 case, there is an increase in σ bond strength that contributes to the stability of the four-coordinated complex because of this vacancy. The only loss of energy to the system lies in the fact that when the $(d_{z^2})^1$ $(d_{x^2-y^2})^1$ configuration exists, energy must be expended to pair the electrons in the d_{z^2} orbital. The configuration will then be $(d_z)^2$ $(d_{x^2-y^2})^0$. For example, in the case of several nickel(II) complexes, it has been found that the metal-ligand bonds are 0.15 to 0.20 A shorter in square planar than in octahedral complexes, indicating an increase in bond energy in the square planar complex.

Increased π Bond Strength in the Square Complex. By utilizing the d_{xz}, d_{yz}, and p_z orbitals, a higher degree of metal-ligand out-of-plane π bonding can be established in square planar complexes than in octahedral ones. Presumably, this factor is particularly important for the heavier elements, due to the ability of their more extended orbitals to overlap more effectively with ligand π orbitals. The great stability of platinum(II) square planar complexes supports this factor.

Since covalent bonding theories predict that in general the greater is the number of bonds formed to a central element the greater is the stability of the resulting compound, large coordination numbers are favored by the higher electrostatic attraction of many negatively charged ligands for a positive metal ion. This tendency for large coordination numbers is opposed by steric and electrostatic repulsion factors among the ligands. Thus, it is difficult to make predictions from these criteria. However, in the first series of transition elements, the metal ions are frequently six-coordinated with four-coordination predominant in certain cases as already described when other factors become important. Ions of the second and third series of transition metals have a greater tendency to exhibit the higher coordination numbers of seven, eight, and even nine, with coordination number eight being the most frequently observed.

Suggested Readings

1. AC_3 Resource Paper: Ligand Field Theory. F. A. Cotton, *J. Chem. Educ.*, **41,** 9 (1964).
2. Basolo, F., and R. G. Johnson, *Coordination Chemistry*, W. A. Benjamin, Inc., New York, 1964.
3. Cotton, F. A., and G. Wilkinson, *Advanced Inorganic Chemistry*, 2nd ed., Interscience Publishers, New York, 1966.
4. Day, M. C., and J. Selbin, *Theoretical Inorganic Chemistry*, 2nd ed., Reinhold Publishing Corp., New York, 1962.
5. Douglas, B. E., and D. H. McDaniel, *Concepts and Models of Inorganic Chemistry*, Blaisdell Publishing Co., New York, 1965.
6. Drago, R. S., *Physical Methods in Inorganic Chemistry*, Reinhold Publishing Corp., New York, 1965.
7. Harvey, K. B., and G. B. Porter, *Introduction to Physical Inorganic Chemistry*, Addison-Wesley Publishing Co., Inc., Reading, Mass., 1963.
8. Jones, M., *Elementary Coordination Chemistry*, Prentice-Hall, Inc., Englewood Cliffs, N. J., 1964.
9. *Nomenclature of Inorganic Chemistry*, *J. Am. Chem. Soc.*, **82,** 5223 (1960).
10. Sanderson, R. T., *Inorganic Chemistry*, Reinhold Publishing Corp., New York, 1964.

Study Questions

9.1. For each of the following compounds, give the name of the compound and coordination number of the central element, the number and type of stereoisomers possible, and a sketch of each isomer and its name.

(a) $[Zn(NH_3)_2Cl_2]$
(b) $[V(H_2O)(NH_3)_3Br_2]_2SO_4$
(c) $Na_2[Pt(CN)_2NO_2I]$

9.2. Give the systematic names (IUPAC) for the following:
(a) $[Cr(en)_3](SO_4)_3$
(b) $K_3[Fe(CN)_5CO]$
(c) $[Cr(NH_3)_6](NO_3)_3$
(d) $[Pt(NH_3)_4(NO_2)Cl]SO_4$
(e) $K_2[PtCl_6]$

9.3. Give the formula (structure where necessary) of the following compounds:
(a) diamminesilver(I) hexacyanoferrate(II)
(b) μ-amido-μ-nitrooctaamminedicobalt(III) sulfate
(c) *cis*-aquochlorobis(ethylenediamine)rhodium(III) chloride
(d) μ-octaaquodihydroxodiiron(III) sulfate

9.4. Sketch and name the isomeric species possible for $M[a_2b_2cd]$.

9.5. The compound $Co(en)_2(NO_2)_2Cl$ has been prepared in a number of isomeric forms. One form undergoes no reaction with either $AgNO_3$ or en and is optically inactive. A second form reacts with $AgNO_3$, but not with en, and is optically inactive. A third form is optically active and reacts with both $AgNO_3$ and en. Identify each of the three isomeric forms by name and sketch each of the structures.

9.6. You are given the following data for a complex compound:
(a) Empirical formula—$M(CrO_4)Cl_2(NH_3)_2$.
(b) Two distinct crystals: *A*, red; *B*, blue.
(c) One mole of *A* and/or *B* + one mole of $AgNO_3$ yields immediately $\frac{1}{2}$ mole of *red* precipitate (*only*).
(d) One mole of *A* in solution from (c) + one mole of $Ag_2C_2O_4$ yields very slowly 2 moles of white precipitate *only*.
(e) One mole of *B* in solution from (c) + one mole of $Ag_2C_2O_4$ yields no reaction.

Use this information to determine:

(a) The hybrid orbitals on *M* (show diagram).
(b) The coordination number of *M*.
(c) Which groups are bonded ionically and which are bonded covalently.
(d) The stereochemistry of the red and blue forms.

9.7. Would the $[Cu(CN)_4]^{3-}$ ion be expected to be (a) square planar or tetrahedral? (b) colored or colorless? Explain.

9.8. What is the crystal field stabilization energy (in *Dq* units) of the Mn(III) ion in an (a) octahedral weak field? (b) octahedral strong field? What differences in magnetic properties would you expect the Mn(III) ion to exhibit in the two cases?

9.9. Would Au(III) be expected to form a six-coordinate octahedral, a four-coordinate square planar, or a four-coordinate tetrahedral complex when coordinated to nitrite ions? Explain.

9.10. From reasoning based on crystal field concepts alone, would you expect $K_3[Co(CN)_6]$ to be paramagnetic or diamagnetic? Explain.

9.11. From reasoning based solely on crystal field concepts, predict with justification which ion would be more labile, $[Mn(CN)_6]^{4-}$ or $[Fe(CN)_6]^{4-}$. Assume a mechanism based on a seven-coordinate intermediate species.

9.12. What type of structure would you predict for the compounds Cr_2CoO_4 and Fe_2CoO_4? Explain any differences.

9.13. What magnetic moment would you predict for (a) $[NiCl_6]^{4-}$ and (b) $[NiCl_4]^{2-}$? Explain how the square planar case arises from the octahedral, and explain the effects, etc., of this change.

9.14. Without using crystal field concepts, predict with justification whether:

(a) Mn(II) or Ni(II) would be expected to form a more stable octahedral complex with a given ligand.

(b) Pt(II) would be expected to form a more stable complex with F^- or I^-.

(c) Ti(II) or Cr(II) cyanide complexes would be expected to be more labile.

9.15. Why are compounds of Mn(II) only very faintly colored?

9.16. What specific energy level transition resulting from absorption of visible radiation produces the violet color of the hexaaquotitanium(III) ion?

9.17. Which of the ions, Mn(III), Fe(III), Cr(III), or Ti(IV), would be considered to be a Jahn-Teller ion when located on an octahedral site?

9.18. Give the molecular orbital energy diagram for the octahedral complex $[Fe(H_2O)_6]^{2+}$. Label all orbitals and indicate the electrons present in each. Would you expect this complex to exhibit π bonding? Explain.

9.19. (a) Explain and predict the shape of the absorption spectrum for $[Fe(H_2O)_6]^{2+}$.

(b) Would you expect the same spectrum for $[Fe(CN)_6]^{4-}$? Why or why not? Draw your proposed spectrum.

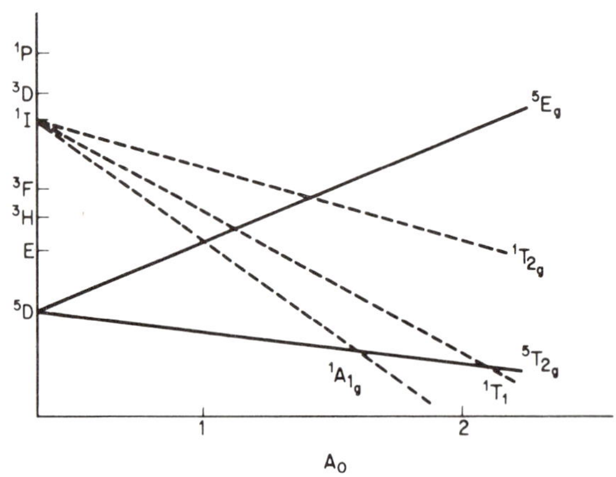

In the case of Fe^{+2}, H_2O as a ligand = point 1

CN^- as a ligand = point 2

(c) Cobalt complexes in a certain oxidation state will also obey this diagram. For all of these CoL_6^{X+} complexes known (oxidation state of Co = +X), the $^1A_{1g}$ crosses the $^5T_{2g}$ almost instantaneously with any Dq. Explain the significance of this. What type of spectrum would you predict for $Co(H_2O)_6^{X+}$ and $Co(CN)_6]^{-6+x}$?

9.20. Give a clear but brief explanation of each of the following:

(a) The CN^- ion produces a larger degree of splitting of the d orbitals in transition metal complexes than the F^- ion does.

(b) $[Co(NH_3)_6]^{3+}$ is diamagnetic, whereas $[CoF_6]^{3-}$ is paramagnetic.

(c) $[FeF_6]^{3-}$ has a normal octahedral structure, whereas $[CuF_6]^{3-}$ has a distorted octahedral structure.

9.21. The periodate (IO_6^{5-}) bidentate ligand forms a stable complex with Ag(III) ion.

(a) Using crystal field concepts, discuss the structure that the complex would most likely prefer (i.e., a six-coordinate octahedral, a four-coordinate planar, or a four-coordinate tetrahedral structure). (b) Would magnetic susceptibility measurements clarify whether the complex is octahedral, square planar, or tetrahedral? (c) Would magnetic susceptibility data indicate whether the periodate group is producing a strong or weak crystal field?

9.22. Give complete but brief explanations of the following. (a) Practically all Cr^{3+} complexes undergo very slow substitution reactions, whereas practically all V^{3+} complexes undergo very rapid substitution reactions. (b) Two complexes are known that have the formula $Pt(NH_3)_2Cl_2$, whereas only one complex is known with the formula $Zn(NH_3)_2Cl_2$.

9.23. Using the following subject headings, state why, and show how, the ligand field theory is important in inorganic chemistry.

(a) Describe qualitatively the ligand field theory and any closely related theory or theories on which it is based.

(b) Draw and explain the five d atomic orbitals.

(c) Show diagrammatically the splitting of the d orbitals, energywise, in octahedral, tetrahedral, and square planar fields.

(d) What are strong field, weak field, high spin, and low spin complexes?

(e) What is crystal field stabilization energy?

(f) Calculate the crystal field stabilization energy in the following complexes: $[Fe(CN)_6]^{3-}$, $[V(NH_3)_6]^{3+}$ and $[HgI_4]^{2-}$.

(g) Show why it is to be expected that the MnF_6^{3-} ion should have a distorted, rather than a regular, octahedral structure. In what manner would the ion be distorted?

9.24. From reasoning based solely on crystal field concepts, predict which would be the more labile ion: $[Mn(CN)_6]^{3-}$ or $[Fe(CN)_6]^{3-}$. Assume a mechanism based on a five-coordinate intermediate species.

9.25. Izzat, R. M., Watt, G. D., Bartholomew, C. H., and Christensen, J. J., [*Inorg. Chem.*, **7**, 2236 (1968)] report that

$$Co(CN)_5^{3-} + H_3O^+ \rightarrow \tfrac{1}{2}H_2\ (g) + Co(CN)_5H_2O^{2-} \qquad (1)$$
$$[\Delta H^\circ = -58.4 \text{ kcal/mole of } Co(CN)_5H_2O^{2-}]$$

$$Fe(CN)_6^{4-} + H_3O \rightarrow \tfrac{1}{2}H_2 + H_2O + Fe(CN)_6^{3-} \quad [\Delta H^\circ = +26.7 \text{ kcal/mole of } Fe(CN)_6^{4-}] \qquad (2)$$

$10Dq$ (kcal/mole)	96	99.5	93.8	91.2
Complex	$Fe(CN)_6^{4-}$	$Fe(CN)_6^{3-}$,	$Co(CN)_5^{3-}$,	$Co(CN)_5H_2O^{2-}$

CFSE of $Co(CN)_5^{3-} = 19.14Dq$

(a) Calculate the change in CFSE for these reactions and compare with ΔH° (short discussion).
(b) Using molecular orbital or ligand field concepts, briefly discuss why (1) appears to be more favorable than (2).
(c) Using molecular orbital concepts, briefly discuss the lability of the two iron complexes in regard to ligand exchange.

9.26. In *Inorg. Chem.,* **7,** 1273 (1968), Caulton, K. G., and Fenske, R. F., report the following data and state that *normally* frequency increases with increasing oxidation state of the metal:

(1)	$V(CO)_6^-$	$Cr(CO)_6$	$Mn(CO)_6^+$
CO *stretch* (cm^{-1})	1859	1981	2101
MC stretch (cm^{-1})	460	441	416
(2)		$MnCl^{2-}$	$FeCl^-$
M—Cl *stretch* (cm^{-1})		3784	4824

(a) *Briefly* explain the change in stretching frequency in (b) for the two isoelectronic metal ions.
(b) *Briefly* discuss the change in CO stretch and also its relation to the change in *MC* stretch.
(c) *Briefly* explain how CO can stabilize low oxidation states of metals.

9.27. There is little information available concerning the electronic spectrum of the high spin M^{x+} complex. In a ligand field of octahedral symmetry, the $^5D_4$ free ion ground state (same ground state term symbol for the free atom) splits into a t_{2g} ground state and an e_g excited state. Accordingly, the spectra show one spin-allowed band. It has been observed, however, when the six ligands bound to M^{x+} are identical, $M(H_2O)_6^{x+}$ or $M(NCSe)_6^{x-6}$, that this band often suffers appreciable broadening or is split into two maxima [Goodgame, D., Goodgame, M., Hitchman, M., and Weeks, M., *Inorg. Chem.,* **5,** 635 (1966)]. Explain all this information (why splitting, which orbitals split, where, why broadening, etc.), and tell with what element and which oxidation state these authors were concerned.

CHAPTER 10

Applications of Electrode Potentials in Inorganic Chemistry

ELECTRODE POTENTIALS

The vast variety of inorganic reactions demands the systematizing of the huge quantity of data that is available as a consequence of these reactions. This classification of reactions is best accomplished by the use of thermodynamic data and, in the case of oxidation and reduction reactions, by standard electrode potentials. One disadvantage in the use of thermodynamic data is that information is obtained only on the relative stability of the products compared with reactants in chemical systems. Such data thus give no indication of the path of the reaction, or of the reaction rate.

Under agreement of the International Union of Pure and Applied Chemistry (IUPAC), the *standard electrode potential*, E°, has been made synonomous with the electromotive force value for the *reduction half-cell reaction*. We shall adhere to this convention in this chapter. It is worth noting, however, that wide disparity still exists in the various areas of chemistry regarding the use of reduction potential, oxidation potential, and the sign convention for the E° values. Particularly in inorganic chemistry has oxidation potential been used with some regularity in measuring the ionization tendency of the elements in solution and correlating this tendency with other properties of the atom, such as ionization potential. Under the concept of oxidation potential, the tendency for reaction

$$Li^\circ\,(s) \rightarrow Li^+\,(aq) + e^-, \qquad E^\circ = +3.04\,V \qquad [10.1]$$

to take place is measured directly by the magnitude of the oxidation potential, $E^\circ = 3.04$ V. Under the IUPAC agreement, the equation written for the reduction process is

$$Li^+ (aq) + e^- \rightarrow Li^\circ (s) \qquad E^\circ = -3.04 \text{ V} \qquad [10.2]$$

Under this agreement correlations are still readily made with adjustments. Thus lithium, with a high ionization potential, has the lowest reduction potential, and fluorine, with a low ionization potential, has the largest reduction potential. Some references still use the oxidation potential convention, and the reader should always note the convention when using tabulated values.

Initially, some basic concepts from physical chemistry will be reviewed. The change in standard free energy (ΔG°), which is negative when products are more stable than reactants, can be simply related to the standard electrode potential (E°) and the equilibrium constant (K) of a reaction. In equation [10.3],

$$\Delta G^\circ = -n\mathbf{F}E^\circ \qquad [10.3]$$

n refers to the number of electrons gained or lost by a species and **F** is the Faraday constant (23.1 kcal/V equiv); E° refers to the electrode potential under standard conditions, which for liquids and solids represents the standard state at 298°K and 1 atm. Ions and gases have activities[1] and fugacities of unity, respectively, as their standard states. In equation [10.4],

$$\Delta G^\circ = -2.30RT \log K \qquad [10.4]$$

R and T refer to the gas constant (1.99×10^{-3} kcal/deg mole) and the temperature in °K, respectively. With the given constants, ΔG° will have kilocalorie units in equations [10.3] and [10.4].

Under nonstandard conditions at 298°K, the electrode potential is given by the Nernst equation [equation (10.5)].

$$E = E^\circ - \frac{0.0592}{n} \log Q \qquad [10.5]$$

In equation [10.5], E is the electrode potential under nonstandard conditions and Q (the reaction quotient) has the same form as the equilibrium constant, but the activities do not represent equilibrium values. Table 10.1 gives relationships among free energy, electrode potential, and spontaneity of a reaction.

[1] In all future discussion, the term *concentration* or *molarity* will be used as an approximation of activity.

TABLE 10.1 ***Relationships Among ΔG, $E°$, and Condition of the Reaction***

ΔG	$E°$	*Reaction*
−	+	Spontaneous
+	−	Nonspontaneous
0	0	At equilibrium

To understand periodic trends involving standard electrode potentials, it is well to break down the electrochemical process into several steps, using Born-Haber cycles. Figure 10.1 shows a cycle for the conversion of a monovalent aqueous cation to an atom. In Fig. 10.1, ΔG_A°, ΔG_I°, and ΔG_H° refer to the free energies of atomization, ionization, and hydration,

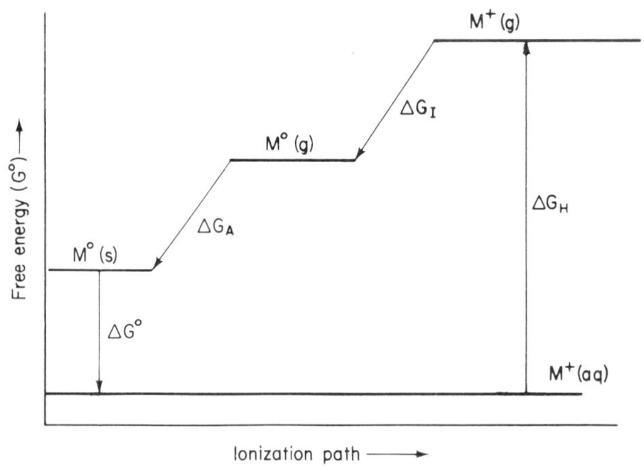

Figure 10.1. Born-Haber cycle showing energy contributions for the reduction of an aqueous cation to the metal.

respectively. The ΔG_A° term includes processes required to change an element in its standard state to gaseous atoms, i.e., processes such as sublimation of solids, vaporization of liquids, and bond dissociation of gaseous molecules. The magnitude of the hydration energy term for metals, ΔG_H°, is such that it usually far overshadows the contribution of the ΔG_A° or ΔG_I° terms for the ionization process in aqueous solution. In this case, the magnitude of ΔG° is negative corresponding to an overall spontaneous process. The energy terms in Fig. 10.1 are related by

$$\Delta G^\circ = -\Delta G_H^\circ - \Delta G_I^\circ - \Delta G_A^\circ \qquad [10.6]$$

Figure 10.2 shows a similar cycle for the conversion of a halogen to a halide ion in aqueous solution. In Fig. 10.2, the ΔG_{EA} term refers to the electron

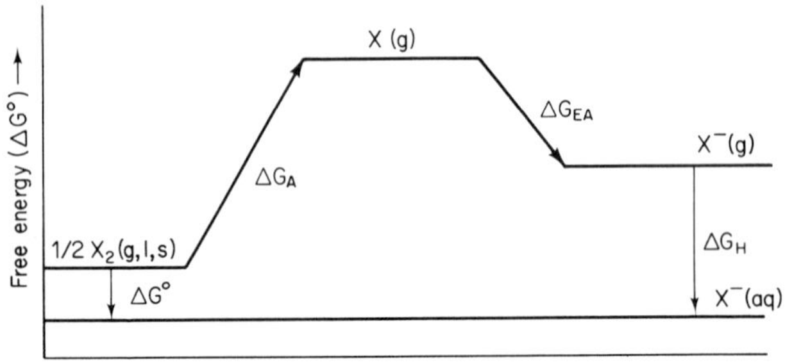

Figure 10.2. Born-Haber cycle showing energy contributions for conversion of a halogen atom to its monovalent aqueous anion.

affinity free energy, whereas the other terms are as defined previously. The equation relating these terms is

$$\Delta G^\circ = \Delta G_A^\circ + \Delta G_{EA}^\circ + \Delta G_H^\circ \qquad [10.7]$$

The ΔG_H° term plays a less significant role here than in the ionization of a metal. This is not unexpected, considering the lower hydration energies of the larger anions compared with those of smaller cations.

The energy of the process,

$$e^- \text{ (electrode)} \rightarrow e^- \text{ (g)} \qquad [10.8]$$

has been ignored on the assumption that it is constant for all electrodes and that it will cancel when standard potentials are put on the relative hydrogen scale.

Standard potentials calculated from the Born-Haber cycle can be related to the usual hydrogen scale using equation (10.9),

$$E^\circ = \frac{\Delta G^\circ}{-23.06n} - 4.44 \qquad [10.9]$$

where E° is the standard electrode potential; ΔG_A°, ΔG_I°, and ΔG_H° have already been defined. However, for a metal, ΔG_I° is the ionization potential, and for a nonmetal, ΔG_{EA}° is the electron affinity; $23.06n$ converts the kilocalorie units to volt equivalents, n being the number of electrons transferred and -4.44 the ΔG° value for $H^+/\frac{1}{2}H_2$ (g). Calculated values of E° using equation [10.9] are compared with the experimentally observed values of E° in Table 10.2. The comparison is quite favorable. Data in this table will be used in an attempt to present periodic trends among families of elements.

TABLE 10.2 *Contributions (kcal) to Standard Electrode Potentials**

	ΔG_A°	ΔG_I°	ΔG_H°	E° (V, calc.)	E° (V, exptl.)
Li^+, Li	30.6	125	−122	−3.0	−3.04
Na^+, Na	18.6	119	−98	−2.7	−2.71
K^+, K	14.6	100	−80.3	−3.0	−2.92
Rb^+, Rb	12.9	96.7	−75.2	−3.0	−2.93
Cs^+, Cs	12.2	90.2	−67.5	−2.9	−2.92
Ca^{2+}, Ca	34.1	415	−380	−2.9	−2.87
Sr^{2+}, Sr	26.3	386	−340	−2.9	−2.89
Ba^{2+}, Ba	34.2	352	−315	−2.9	−2.90
Ag^+, Ag	59.1	175	−114	+0.75	+0.799
$\frac{1}{2}F_2$, F^-	14.5	81.0†	−103.1	+2.9	+2.87
$\frac{1}{2}Cl_2$, Cl^-	25.2	85.7†	−74.8	+1.4	+1.36
$\frac{1}{2}Br_2$, Br^-	19.7	80.6†	−67.9	+1.2	+1.09
$\frac{1}{2}I_2$, I^-	16.8	73.4†	−59	+0.58	+0.54

* R. T. Sanderson, *Inorganic Chemistry*, Reinhold Publishing Corp., New York, 1967.
† The negative value of these terms is the electron affinity of X (g).

First, we shall consider the reducing ability of the members of Group I–A of the periodic table. According to experimental standard potentials, lithium and sodium are the best and poorest reducing agents, respectively, in this family of elements. Based on ionization potential values alone, it is predicted that the reducing ability of the members will increase as one descends Group I–A. The free energy of atomization values, ΔG_A° (sublimitation in this example), are in agreement with the expected trend; However, the free energy of hydration values, ΔG_H°, becomes *less* favorable regarding reducing ability as we descend the group. Because the standard potential is the sum of three terms, it is not surprising that the ionization potential and standard electrode potential differ in their prediction of reducing abilities. Lithium ion has the smallest ionic radius and has a ΔG_H° value that is 54 kcal more negative than the larger cesium ion value. It is, then, the favorable ΔG_H°, value which accounts for the stronger reducing ability of lithium relative to the other alkali metals.

Among the alkaline earth metals (Group II–A), the strong reducing ability of the latter members of this family—calcium, strontium, and barium—can be attributed to the fact that large, negative, free energy of hydration values effectively cancel large, unfavorable, ionization potentials.

Because of the opposing trends of ionization potentials and energies of hydration, the standard potentials for these three elements are nearly identical.

From a comparison of the reducing ability of rubidium and silver, the effect of the addition of 10 inner 4*d* electrons can be observed. Rubidium is a far better reducing agent than silver, despite the relative free energy of hydration values, because the free energies of atomization and ionization favor rubidium by about 124 kcal. Unlike the situation when we compare Groups I–A and II–A, the higher ionization potential is not balanced by a sufficiently negative free energy of hydration of silver.

Lastly, consider the superior oxidizing ability of fluorine compared with that of the other halogens. This difference is sometimes erroneously attributed to the high electron affinity of fluorine. The values of electron affinity and ionization potential are opposite in sign [equation (10.10)].

$$X\,(g) + e^- \rightleftarrows X^-\,(g) \qquad [10.10]$$

For the halogens, free energy of atomization and free energy of electron affinity are relatively invariant. The powerful oxidizing ability of fluorine can be attributed mainly to the highly negative free energy of hydration of the small fluoride ion.

In general, we conclude that trends of standard potentials of a given family of elements can be rationalized using thermodynamic data and a Born-Haber cycle. The remainder of this chapter will emphasize predictions of whether or not a particular inorganic reaction will occur. Because water is the common solvent for redox reactions, it is first necessary to determine which species are stable with respect to oxidation or reduction of the solvent.

OXIDATION-REDUCTION REACTIONS IN WATER AS A FUNCTION OF pH

Figure 10.3 shows plots of the potential of the reduction and oxidation of water as a function of pH. The potentials are plotted for the following half-reactions:

$$H^+ + e^- \longrightarrow \tfrac{1}{2}H_2\,(g) \qquad [10.11]$$

$$\tfrac{1}{2}O_2\,(g) + 2H^+ + 2e^- \longrightarrow 2H_2O \qquad [10.12]$$

The symbols E_A°, E_N, and E_B° refer to hydrogen ion concentrations, $[H^+]$, in moles per liter, of 1, 1×10^{-7}, and 1×10^{-14} (or $[OH^-] = 1\ M$), respectively. The solid lines are based on calculations using the Nernst equation.

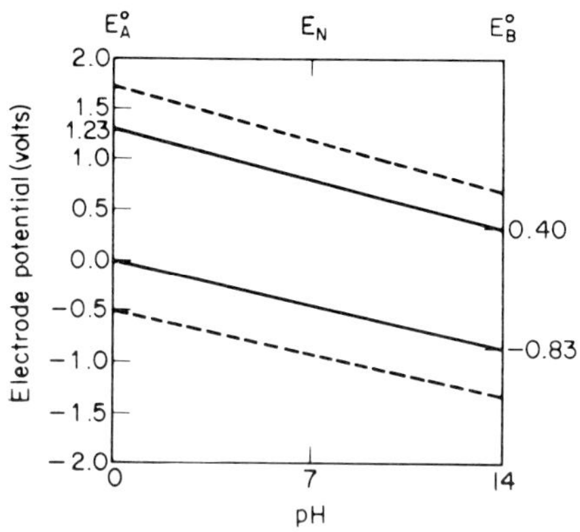

Figure 10.3. The electrode potential of the H^+/H_2 and the O_2/H_2O couples as a function of pH.

Any oxidizing or reducing agent whose potential does not fall between the solid lines at the pH of the reaction is predicted to oxidize or reduce the solvent. Consider the oxidizing ability of the permanganate ion at pH = O [equation (10.13)].

$$8H^+ + MnO_4^- + 5e^- \longrightarrow Mn^{2+} + 4H_2O, \qquad E_A^\circ = 1.51 \text{ V} \qquad [10.13]$$

Because its E_A° value is higher than that of the O_2/H_2O couple, it is predicted to be a poor oxidizing agent to use in water. Actually, potassium permanganate is commonly used as an oxidizing agent in analytical determinations. Based on experimental evidence, the region of stability of water should be extended about 0.5 V above the O_2/H_2O couple and about 0.5 V below the H^+/H_2 couple. Therefore, all but a few powerful oxidizing agents, such as fluorine, can be used as reactants in water. In contrast, many reducing agents, i.e., the Group I–A metals, reduce water at any pH. Liquid ammonia or ethers, such as tetrahydrofuran, are often used as solvents when powerful reducing agents are required.

As suggested, whenever hydrogen ion or hydroxyl ion is involved in a reaction or half-reaction, the potential will be a function of pH and can be calculated using the Nernst equation. For example, for the reaction

$$2H^+ + ClO_4^- + 2e^- \rightarrow ClO_3^- + H_2O, \qquad E_A^\circ = 1.19 \text{ V} \qquad [10.14]$$

the appropriate expression of the Nernst equation is

$$E = E_A^\circ - \frac{0.0592}{2} \log \frac{[ClO_3^-]}{[ClO_4^-][H^+]^2} \qquad [10.15]$$

Under standard acidic conditions, all concentrations are unity, the log term is zero, and $E = E_A^\circ$. However, in determining the potential in basic solution, E_B°, from E_A°, the concentrations of ClO_3^-, ClO_4^-, and OH^- are unity but the concentration of the hydrogen ion is 1×10^{-14} *M*. Therefore, the Nernst equation is

$$E_B^\circ = 1.19 - \frac{0.0592}{2} \log \frac{1}{(1)(1 \times 10^{-14})^2} = 0.36 \text{ V} \qquad [10.16]$$

As expected from the application of LeChatelier's principle to equation [10.14], the oxidizing ability of the couple decreases as the hydrogen ion concentration decreases.

The calculation of electrode potentials as a function of pH becomes slightly more complicated in most cases involving metal ions because of metal hydroxide precipitation. The oxidizing ability of copper(II) will be considerably different under acidic and basic conditions [see equations (10.17) and (10.18)].

$$Cu^{2+} + 2e^- \rightarrow Cu\ (s), \qquad E_A^\circ = 0.34 \text{ V} \qquad [10.17]$$

$$Cu(OH)_2\ (s) + 2e^- \rightarrow Cu\ (s) + 2OH^-, \qquad E_B^\circ = -0.22 \text{ V} \qquad [10.18]$$

In the latter case, $[Cu^{2+}]$ will be considerably less than 1 *M*. The calculation must take into account the solubility product constant (K_{sp}) of the copper(II) hydroxide. Under conditions of E_B°, $[Cu^{2+}] = 1.6 \times 10^{-19}$ and E_B° may be calculated from E_A° using the Nernst equation.

$$Cu(OH)_2\ (s) \rightleftarrows Cu^{2+} + 2OH^- \qquad [10.19]$$

$$K_{sp} = [Cu^{2+}][OH^-]^2 = 1.6 \times 10^{-19} \qquad [10.20]$$

$$E_B^\circ = 0.34 - \frac{0.0592}{2} \log \frac{1}{1.6 \times 10^{-19}} = -0.22 \text{ V} \qquad [10.21]$$

ELECTROMOTIVE FORCE DIAGRAMS

Several methods of presenting standard electrode potentials exist. The most common presentation is a table based on the relative oxidizing (or reducing) abilities of the couples arranged—e.g., from the best to the poorest oxidizing couple. This usual tabulation of data is possibly the least useful to inorganic chemists, who are very likely to be interested in all oxidation–reduction behavior of an element or family of elements.

The electromotive force (emf) diagram is often more useful than a table. Consider as an example the oxidation-reduction relationships involving bromine in basic solution; E_B° represents standard conditions at which hydroxyl ion concentration is 1 *M*.

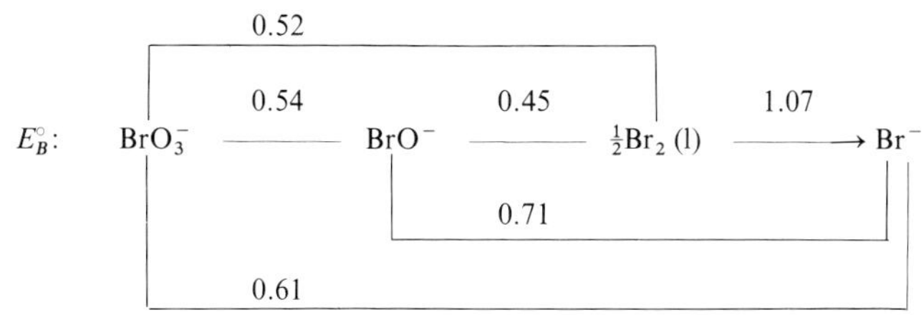

The diagram abbreviates the half-reactions, and the signs of the potential designate the direction of the reaction corresponding to standard half-cell reduction potentials; for example, $BrO_3^- \xrightarrow{0.54} BrO^-$ is an abbreviation of the following half reaction:

$$BrO_3^- + 4H_2O + 4e^- \rightarrow BrO^- + 4OH, \qquad E_B^\circ = 0.54 \text{ V} \tag{10.22}$$

Often in diagrams of this type, only the standard half-cell reduction potentials of adjacent species are given. It is then necessary to calculate new standard half-cell potentials from given half-cell values. Since ΔG° values of half-cells are additive and E° values are not, all calculations must be made using ΔG° values. The following relationships may be used:

$$\Delta G^\circ = \Sigma\, \Delta G_i^\circ = \tag{10.23}$$

$$(\Sigma n_i)E^\circ \mathbf{F} = \Sigma\, (n_i E_i^\circ)\, \mathbf{F} \tag{10.24}$$

$$E^\circ = \frac{\Sigma\, (n_i E_i^\circ)}{\Sigma\, n_i} \tag{10.25}$$

In the preceding equations, ΔG° and E° represent the free energy change and standard potential values of the new half-cell; ΔG_i° and E_i° represent the respective free energy changes and individual half-cell potentials, that are summed to calculate ΔG° and E° for the new half-cell; n_i is the electron change for E_i°; and **F** is the Faraday which is cancelled during the calculation. As an example, suppose the following information were given:

$$E_B^\circ\text{:} \quad BrO_3^- \xrightarrow{0.54} BrO^- \xrightarrow{0.45} \tfrac{1}{2}Br_2\text{ (l)} \xrightarrow{1.07} Br^-$$

Then one was asked to calculate E_B° for the half-cell $BrO_3^- \rightarrow Br^-$. The reaction is

$$BrO_3^- + 3H_2O + 6e^- \rightarrow Br^- + 6OH^-.$$

By equation [10.25],

$$E_B^\circ = \frac{4(0.54) + 1(0.45) + 1(1.07)}{6} = 0.61 \text{ V}$$

In an analogous fashion, E_B° for the half-cell $BrO^- \rightarrow Br^-$ can be ascertained to be 0.76 V.

The susceptibility of any species to disproportionation or self-oxidation reduction can easily be obtained from an emf diagram. From the E_B° diagram for the oxidation states of bromine, we may ascertain that bromine disproportionates into bromide ion and hypobromite ion:

$$\tfrac{1}{2}Br_2 + e^- \rightarrow Br^-, \qquad E_{B(1)}^\circ = 1.07 \text{ V}$$
$$H_2O + BrO^- + e \rightarrow \tfrac{1}{2}Br_2 + 2OH^-, \qquad E_{B(2)}^\circ = 0.45 \text{ V}$$

$$2OH^- + Br_2 \rightarrow Br^- + BrO^- + H_2O, \qquad E_B^\circ = E_{B(1)}^\circ - E_{B(2)}^\circ = 0.62 \text{ V} \qquad \text{(A)}$$

The positive reduction potential indicates that the reaction will proceed spontaneously to the right, as written. Also, BrO^- disproportionates into Br^- and BrO_3^-:

$$2BrO^- + 2H_2O + 4e^- \rightarrow 2Br^- + 4OH^-, \quad E_{B(1)}^\circ = 0.71 \text{ V}$$
$$BrO_3^- + 2H_2O + 4e \rightarrow BrO^- + 4OH^-, \quad E_{B(2)}^\circ = 0.54 \text{ V}$$

$$3BrO^- \rightarrow 2Br^- + BrO_3^-, \qquad E_B^\circ = 0.17 \text{ V} \qquad \text{(B)}$$

Therefore, a combination of equations (A) and (B) represents the reaction predicted when bromine is in basic solution:

$$6OH^- + 3Br_2\,(l) \rightarrow 5Br^- + BrO_3^- + 3H_2O, \qquad E_B^\circ = 0.55 \text{ V} \qquad [10.26]$$

The student should confirm the result in equation [10.26]. It has been observed that basic solutions of bromine form quantitative amounts of bromide ion and bromate ion at 50°C.

It is instructive to compare the emf diagram for bromine in acidic solution ($[H^+] = 1\ M$), E_A°, with that of the E_B° diagram.

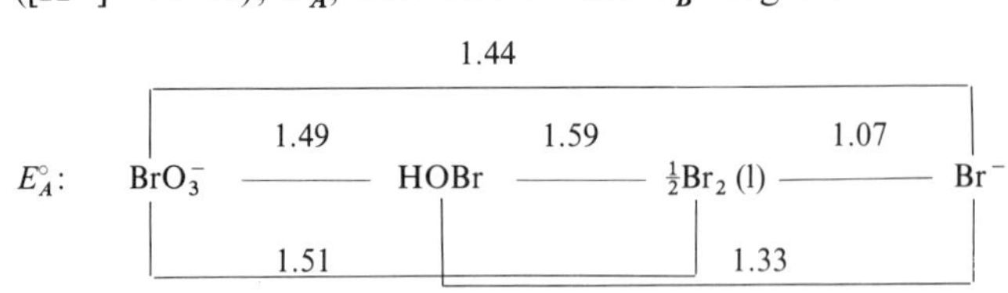

In the comparison, three classes of behavior are observed. (1) For $\tfrac{1}{2}Br_2\,(l) + e^- \rightarrow Br^-$, $E^\circ = 1.07$ V. This potential is independent of pH because the couple involves no hydrogen or hydroxyl ions. (2) The BrO_3^- is a better oxidizing agent (more easily reduced) in an acidic solution than in a basic solution. (3) The potential changes as a function of pH when the nature of one or more of the species in the couple is pH dependent. For example, HOBr, a weak acid, exists in its molecular and ionized forms in acidic and basic solutions, respectively.

From the E_A° diagram for bromine, we see that it will not disproportionate in acidic solution to Br^- and HOBr ($E_A^\circ = 1.07 - 1.59 = -0.52$ V) or to Br^- and BrO_3^- ($E_A^\circ = 1.07 - 1.51 = -0.44$ V). This contrast in behavior in the two solutions, acidic and basic, occurs because the oxidizing ability of Br_2 is independent of pH. However, from the acidic medium diagram for bromine, it can be predicted that HOBr disproportionates into BrO_3^- and Br_2 [equation (10.27)].

$$5HOBr \rightarrow 2Br_2\,(l) + BrO_3^- + H^+ + H_2O, \qquad E_A^\circ = 0.10\text{ V} \qquad [10.27]$$

However, E_A° for reaction [10.27] is very near zero, and it is not surprising that some HOBr exists in solution.

From the emf diagrams presented and from those of the other halogens, we predict correctly that chlorine, bromine, and iodine disproportionate in basic solution:

$$3X_2 + 6OH^- \rightarrow XO_3^- + 5X^- + 3H_2O \qquad [10.28]$$

But the conditions are difficult to anticipate. Bromine and iodine rapidly disproportionate to XO_3^- and X^- at room temperature, but the rapid reaction of chlorine to ClO_3^- and Cl^- requires a temperature of 75°C. From emf diagrams, we predict that chlorine, bromine, and iodine are stable in acidic medium. The prediction is correct, although measurable amounts of HOX are observed in saturated solutions of the halogens. In strongly acidic solutions, the HOX concentration would be considerably decreased relative to concentrations near pH = 7, as can be deduced by the application of LeChatelier's principle to equation [10.29].

$$H_2O + X_2\,(aq) \rightarrow H^+ + X^- + HOX \qquad [10.29]$$

Thermodynamic predictions are not always kinetically reliable. The equilibrium constant for the disproportionation of chlorate ion is calculated to be 10^{20} from standard potential data; however, the reaction is so slow

$$4ClO_3^- \rightarrow Cl^- + 3ClO_4^-, \qquad K = 10^{29} \qquad [10.30]$$

that even at 100°C it is not useful for preparative purposes.

Table 10.3 contains the standard half-cell information in diagrammatic form.

TABLE 10.3 ***Standard Half-Cell emf Data****

E°_A, ACID SOLUTION

$$H^+ \xrightarrow{0.00} H_2 \xrightarrow{-2.23} H^-$$

$$O_2 \xrightarrow{-0.13} HO_2 \xrightarrow{1.5} H_2O_2 \xrightarrow{0.72} H_2O + OH \xrightarrow{2.82} 2H_2O$$

$O_2 \to 2H_2O$: 1.229; $O_2 \to H_2O_2$: 0.67; $H_2O_2 \to 2H_2O$: 1.77

$$2O_2 \xrightarrow{-0.13} O_2 + HO_2 \xrightarrow{0.89} H_2O + O_3$$

$2O_2 \to H_2O + O_3$: 0.38

GROUP I-A

M^+	$\rightarrow$	M°
Li		−3.04
Na		−2.71
K		−2.92
Rb		−2.92
Cs		−2.92

GROUP II-A

M^{2+}	$\rightarrow$	M°
Be		−1.85
Mg		−2.37
Ca		−2.87
Sr		−2.89
Ba		−2.90
Ra		−2.92

GROUP III-A

$$H_3BO_3 \xrightarrow{-0.90} B \xrightarrow{-0.14} B_2H_6 \xrightarrow{-0.36} BH_4^-$$

$$Al^{3+} \xrightarrow{-1.66} Al^\circ$$

$$Ga^{3+} \xrightarrow{-0.65} Ga_2^{4+} \xrightarrow{-0.45} Ga^\circ$$

$Ga^{3+} \to Ga^\circ$: −0.53

$$In^{3+} \xrightarrow{\text{ca. } -0.45} In_2^{4+} \xrightarrow{\text{ca. } -0.35} In^+ \xrightarrow{\text{ca. } -0.25} In^\circ$$

$In^{3+} \to In^\circ$: −0.342

$$Tl^{3+} \xrightarrow{1.25} Tl^{+1} \xrightarrow{-0.3363} Tl^\circ$$

$$TlCl_3 \xrightarrow{0.89} TCl\ (c) \xrightarrow{-0.557} Tl^\circ$$

*The abbreviation "ca." means that the potential has been calculated. Values in parentheses are uncertain. These emf data were gathered from W. M. Latimer, *Oxidation Potentials*, 2nd ed., Prentice-Hall, Inc., Englewood Cliffs, N.J., 1952; B. E. Douglas and D. H. McDaniel, *Concepts and Models of Inorganic Chemistry*, Blaisdell Publishing Co., Waltham, Mass., 1965; J. R. Brand and S. A. Buick, *J. Am. Chem. Soc.*, **91**, 6500 (1969).

TABLE 10.3 *Standard Half-Cell emf Data (cont.)*

GROUP IV-A

CO_2	−0.116	CO	0.51	C	0.13	CH_4
SiO_2	−0.86	Si	0.102	SiH_4		
SiF_6^{2-}	−1.2 (→ Si)					
GeO_2	−0.3	Ge^{2+}	0.0	Ge	< −0.3	GeH_4
Sn^{4+}	0.15	Sn^{2+}	−0.136	Sn		
PbO_2	1.455	Pb^{2+}	−0.126	Pb		

GROUP V-A

NO_3^-	0.94	HNO_2	1.45	N_2	0.27	NH_4^+
H_3PO_4	−0.276	H_3PO_3	−0.50	P	−0.065	PH_3
H_3AsO_4	0.56	$HAsO_2$	0.247	As	−0.60	AsH_3
Sb_2O_5	0.581	SbO^+	0.212	Sb	−0.51	SbH_3
Bi_2O_5	1.6	BiO^+	0.32	Bi	< −0.8	BiH_3

GROUP VI-A

O_2	1.23	H_2O				
SO_4^{2-}	0.17	H_2SO_3	0.45	S	0.14	H_2S
SeO_4^{2-}	1.15	H_2SeO_3	0.74	Se	−0.40	H_2Se
H_6TeO_6 (c)	1.02	TeO_2 (c)	0.529	Te	0.72	H_2Te
PoO_3 (c) (?)	1.5	PoO_2 (c)	(0.74)	Po	< −1.0	H_2Po

GROUP VII-A

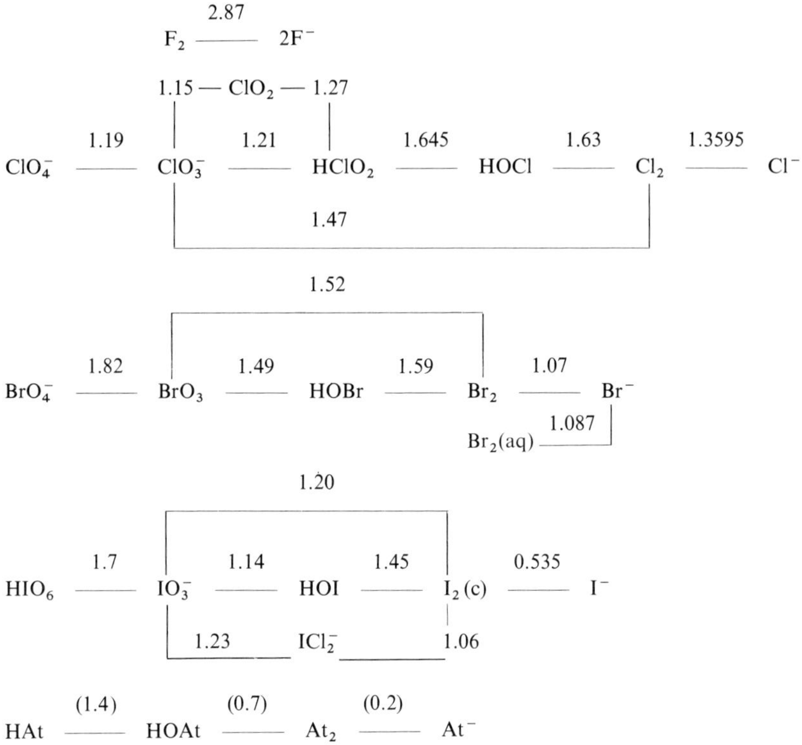

GROUP VIII-A

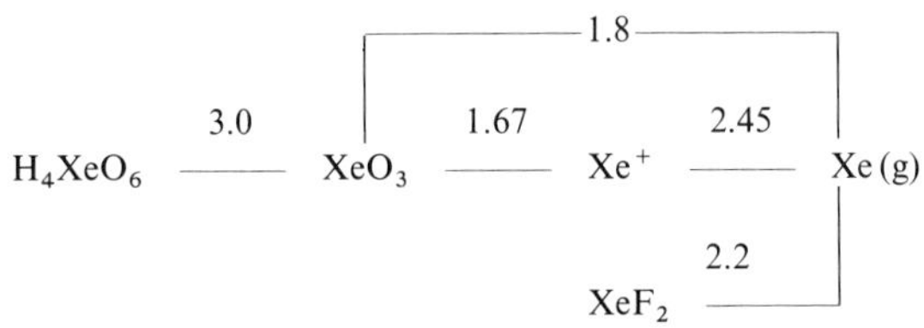

TRANSITION METALS

GROUP III-B

Sc^{3+} —(−208)— Sc°

Y^{3+} —(−2.37)— Y°

GROUP IV-B

TiF_6^{2-} —(−1.19)— Ti

TiO^{2+} —(0.1)— Ti^{3+} —(−0.37)— Ti^{2+} —(−1.63)— Ti

TiO^{2+} —(−0.86)— Ti

Zr^{4+} —(−1.56)— Zr°

Hf^{4+} —(−1.70)— Hf°

GROUP V-B

$V(OH)_4^+$ —(1.0)— VO^{2+} —(0.36)— V^{3+} —(−0.25)— V^{2+} —(−1.2)— V°

$V(OH)_4^+$ —(−0.253)— V°

Nb_2O_5 —(ca. −0.1)— Nb^{3+} —(ca. −1.1)— Nb°

Nb_2O_5 —(−0.65)— Nb°

Ta_2O_5 —(−0.81)— Ta°

GROUP VI-B

$Cr_2O_7^{2-}$ —(1.33)— Cr^{3+} —(−0.41)— Cr^{2+} —(−0.91)— Cr°

Cr^{3+} —(−0.74)— Cr°

$CrCl_2^+$ —(−0.76)— Cr°

$Cr_2O_7^{2-}$ —(0.295)— Cr°

H_2MoO_4 —(+0.4)— MoO_2^+ —(0.0)— Mo^{3+} —(−0.2)— Mo°

H_2MoO_4 —(0.0)— Mo°

WO_3 —(−0.03)— W_2O_5 —(−0.04)— WO_2 —(−0.15)— W^{3+} —(−0.11)— W°

WO_2 —(−0.12)— W°

GROUP VII-B

$MnO_4^- \xrightarrow{0.564} MnO_4^{2-} \xrightarrow{2.26} MnO_2 \xrightarrow{0.95} Mn^{3+} \xrightarrow{1.51} Mn^{2+} \xrightarrow{-1.18} Mn^\circ$

$MnO_4^{2-} \xrightarrow{(2.0)} MnO_3^- \xrightarrow{(2.5)} MnO_2$; MnO_3^- (0.6); $MnO_2 \xrightarrow{1.23} Mn^{2+}$

$TcO_4^- \xrightarrow{(0.7)} TcO_2 \xrightarrow{(0.6)} Tc^{2+} \xrightarrow{(0.4)} Tc^\circ$

$TcO_4^- \rightarrow Tc^\circ$ (bracket, no value shown); $TcO_2 \xrightarrow{(0.5)} Tc^\circ$

$ReO_4^- \xrightarrow{(0.4)} ReO_3 \xrightarrow{(0.6)} ReO_2 \xrightarrow{(0.2)} Re^{3+} \xrightarrow{(0.3)} Re \xrightarrow{(-0.4)} Re^-$

$ReO_4^- \xrightarrow{0.51} ReO_2$; $ReO_2 \xrightarrow{0.252} Re$

$ReO_4^- \xrightarrow{(0.5)} ReCl_6^- \xrightarrow{(0.3)} ReCl_4^- \xrightarrow{(0.15)} Re$

GROUP VIII-B

$FeO_4^{2-} \xrightarrow{>1.9} Fe^{3+} \xrightarrow{0.771} Fe^{2+} \xrightarrow{-0.440} Fe^\circ$

$Fe(CN)_6^{3-} \xrightarrow{0.36} Fe(CN)_6^{4-} \xrightarrow{-1.5} Fe^\circ$

$RuO_4 \xrightarrow{1.00} RuO_4^- \xrightarrow{0.59} RuO_4^{2-} \xrightarrow{0.3} Ru^{2+} \xrightarrow{0.45} Ru^\circ$

$RuO_4 \xrightarrow{1.5} RuCl_5OH^{2-} \xrightarrow{1.3} RuCl_5^{2-} \xrightarrow{0.4} Ru^\circ$

$OsO_4\,(s) \xrightarrow{1.0} OsCl_6^{2-} \xrightarrow{0.85} OsCl_6^{3-} \xrightarrow{(0.4)} Os^{2+} \xrightarrow{(0.85)} Os^\circ$

$OsO_4\,(s) \xrightarrow{(0.97)} OsCl_6^{3-}$; $OsCl_6^{3-} \xrightarrow{(0.71)} Os^\circ$; $OsO_4\,(s) \xrightarrow{0.85} Os^\circ$

$CoO_2 \xrightarrow{>1.8} Co^{3+} \xrightarrow{1.82} Co^{2+} \xrightarrow{-0.277} Co^\circ$

$RhO_4^{2-} \xrightarrow{(1.5)} RhO^{2+} \xrightarrow{(1.4)} Rh^{3+} \xrightarrow{(1.2)} Rh^{2+} \xrightarrow{0.6} Rh^+ \xrightarrow{(0.6)} Rh^\circ$

$Rh^{3+} \xrightarrow{(0.8)} Rh^\circ$

TABLE 10.3 *Standard Half-Cell emf Data (cont.)*

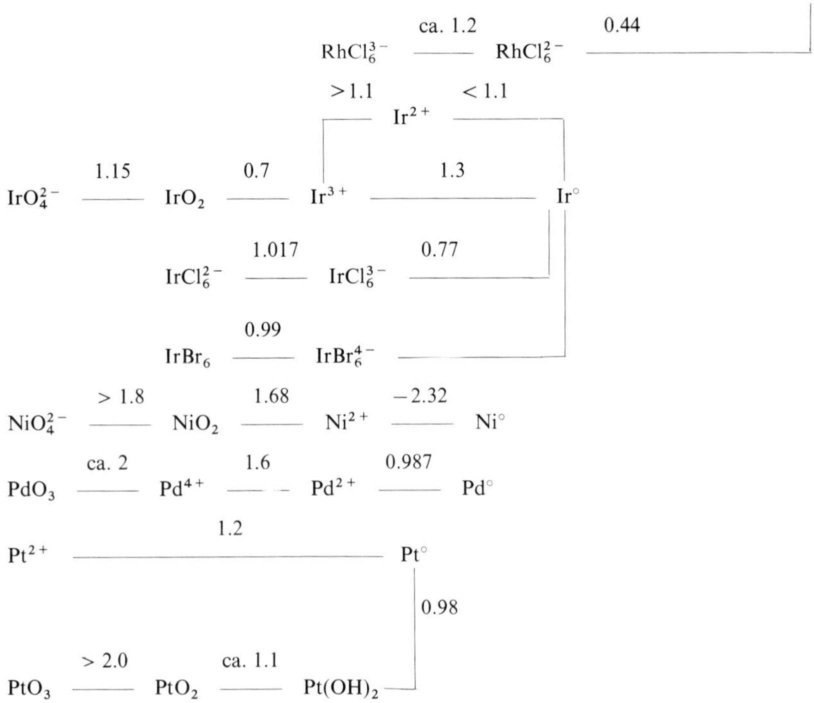

GROUP I-B

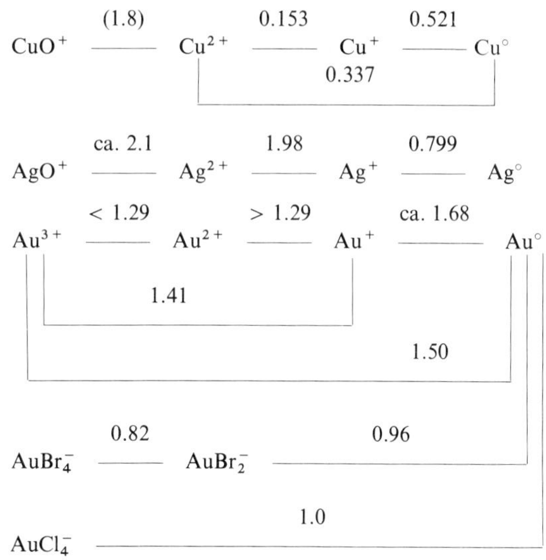

GROUP II-B

$$Zn^{2+} \xrightarrow{-0.763} Zn^{\circ}$$

$$Cd^{2+} \xrightarrow{< -0.6} Cd_2^{2+} \xrightarrow{> -0.2} Cd^{\circ}$$

$$Cd^{2+} \xrightarrow{-0.403} Cd^{\circ}$$

$$Hg^{2+} \xrightarrow{0.920} Hg_2^{2+} \xrightarrow{0.789} Hg^{\circ}$$

$$Hg^{2+} \xrightarrow{0.854} Hg^{\circ}$$

$$HgCl_2\,(sat) \xrightarrow{0.53} Hg_2Cl_2 \xrightarrow{0.2676} Hg^{\circ}$$

E_B°, BASIC SOLUTION

$$H_2O \xrightarrow{-0.828} OH^- + H_2 \xrightarrow{-2.23} H^-$$

$$O_2 \xrightarrow{-0.56} O_2^- \xrightarrow{0.41} HO_2^- \xrightarrow{-0.24} OH^- + OH \xrightarrow{2.0} 2OH^-$$

$$O_2 \xrightarrow{-0.8} HO_2^-$$

$$HO_2^- \xrightarrow{0.87} 2OH^-$$

GROUP I-A Potentials are the same as in acidic solutions.

GROUP II-A

$$Be_2O_3^{2-} \xrightarrow{-2.62} 2Be$$

$$Mg(OH)_2 \xrightarrow{-2.69} Mg$$

$$Ca(OH)_2 \xrightarrow{-3.03} Ca$$

$$Sr(OH)_2 \xrightarrow{-2.99} Sr$$

$$Ba(OH)_2 \xrightarrow{-2.97} Ba$$

GROUP III-A

$$H_2AlO_3^- \xrightarrow{-2.35} Al$$

$$H_2GaO_3^- \xrightarrow{-1.22} Ga$$

$$In(OH)_3 \xrightarrow{1.0} In$$

$$Tl(OH)_3 \xrightarrow{0.05} Tl(OH) \xrightarrow{-0.344} Tl$$

TABLE 10.3 *Standard Half-Cell emf Data (cont.)*

GROUP IV-A

$$CO_3^{2-} \xrightarrow{-1.01} HCO_2^- \xrightarrow{-0.52} C^\circ \xrightarrow{-0.70} CH_4$$

$$SiO_3^{2-} \xrightarrow{-1.73} Si^\circ \xrightarrow{-0.73} SiH_4$$

$$HGeO_3^- \xrightarrow{-1.0} Ge^\circ \xrightarrow{1.1} GeH_4$$

$$Sn(OH)_6^{2-} \xrightarrow{-0.90} HSnO_3^{3-} \xrightarrow{-0.91} Sn^\circ$$

$$PbO_2 \xrightarrow{0.28} PbO \xrightarrow{-0.54} Pb^\circ$$

GROUP V-A

$$NO_3^- \xrightarrow{-0.86} N_2O_4 \xrightarrow{0.88} NO_2^- \xrightarrow{-0.46} NO \xrightarrow{0.76} N_2O \xrightarrow{0.94}$$

$$N_2 \xrightarrow{-3.04} NH_2OH \xrightarrow{0.73} N_2H_4 \xrightarrow{0.1} NH_4OH$$

$$PO_4^{3-} \longrightarrow H_2PO_6^{2-} \longrightarrow HPO_3^{2-} \xrightarrow{-1.57} H_2PO_2^- \xrightarrow{-2.05} P^\circ \xrightarrow{(-0.9)} P_2H_4 \xrightarrow{(-0.8)} PH_3$$

$H_2PO_2^- \to PH_3$: -1.18; $PO_4^{3-} \to HPO_3^{2-}$: -1.12; $HPO_3^{2-} \to P^\circ$: -1.73; $P^\circ \to PH_3$: -0.89

$$AsO_4^- \xrightarrow{-0.67} AsO_2 \xrightarrow{-0.68} As^\circ \xrightarrow{-1.43} AsH_3$$

$$H_3SbO_6^{4-} \xrightarrow{(-0.40)} SbO_2^- \xrightarrow{-0.66} Sb^\circ \xrightarrow{(-1.34)} SbH_3$$

$$Bi_2O_4 \xrightarrow{0.56} Bi_2O_3 \xrightarrow{0.46} Bi^\circ \xrightarrow{< -1.6} BiH_3$$

GROUP VI-A

$$O_2 \xrightarrow{0.401} OH^-$$

$$SO_4^{2-} \xrightarrow{-0.91} SO_3^{2-} \xrightarrow{-0.61} S^\circ \xrightarrow{-0.48} S^{2-}$$

$$SeO_4^{2-} \xrightarrow{0.05} SeO_3^{2-} \xrightarrow{-0.366} Se^\circ \xrightarrow{-0.92} Se^{2-}$$

$$TeO_4^{2-} \xrightarrow{< 0.4} TeO_3^{2-} \xrightarrow{-0.57} Te^\circ \xrightarrow{-1.44} Te^{2-}$$

$$Po^{2-} \xrightarrow{< 1.4} Po^\circ$$

GROUP VII-A

ClO_4^- —0.36— ClO_3^- —0.33— ClO_2^- —0.66— ClO^- —0.40— Cl_2^- —1.3595— Cl^-

ClO_3^- —(−0.50)— ClO_2 —1.16— ClO_2^-

ClO_3^- —0.50— ClO^-

ClO^- —0.88— Cl^-

BrO_3^- —0.54— BrO^- —0.45— Br_2° —1.07— Br^-

BrO_3^- —0.61— Br^-

BrO^- —0.76— Br^-

$H_3IO_6^{2-}$ —(0.7)— IO_3^- —0.14— IO^- —0.45— $I_2^\circ(c)$ —0.535— I^-

IO^- —0.49— I^-

AtO_3^- —(0.5)— AtO^- —(0.0)— At_2° —(0.2)— At^-

GROUP VIII-A

$HXeO_6^{3-}$ —0.9— $HXeO_4$ —0.9— Xe° (g)

TRANSITION METALS

GROUP III-B

$Sc(OH)_3$ —ca. −2.6— Sc°

$Y(OH)_3$ —(−2.8)— Y°

GROUP IV-B

TiO_2 —(−1.69)— Ti

H_2ZrO_3 —(−2.36)— Zr

H_2HfO_3 —(−2.50)— Hf

GROUP V-B

$HV_6O_{17}^{3-}$ —(−1.15)— 6 V

GROUP VI-B

CrO_4^{2-} —(−0.13)— $Cr(OH)_3$ —(−1.1)— $Cr(OH)_2$ —(−1.4)— Cr°

$Cr(OH)_3$ —(−1.3)— Cr°

CrO_2^- —(−1.2)— Cr°

MoO_4^{2-} —(−1.4)— MoO_2 —(−0.87)— Mo

WO_4^{2-} —(−1.25)— W

GROUP VII-B

$$MnO_4^- \xrightarrow{0.564} MnO_4^{2-} \xrightarrow{0.60} MnO_2 \xrightarrow{-0.2} Mn(OH)_3 \xrightarrow{0.1} Mn(OH)_2 \xrightarrow{-1.55} Mn^\circ$$

$$MnO_4^{2-} \xrightarrow{0.34} MnO_3^- \xrightarrow{0.84} MnO_2$$

$$MnO_4^- \xrightarrow{0.588} MnO_2 \qquad MnO_2 \xrightarrow{0.05} Mn(OH)_2$$

$$ReO_4^- \xrightarrow{(-0.7)} ReO_4^{2-} \xrightarrow{(-0.5)} ReO_2 \xrightarrow{(-0.53)} Re(OH)_3 \xrightarrow{(-0.6)} Re^\circ \xrightarrow{(-0.4)} Re^-$$

$$ReO_4^- \xrightarrow{-0.595} ReO_2 \qquad ReO_2 \xrightarrow{-0.576} Re^\circ \qquad ReO_4^- \xrightarrow{-0.584} Re^\circ$$

GROUP VIII-B

$$FeO_4^{2-} \xrightarrow{>0.9} Fe(OH)_3 \xrightarrow{-0.56} Fe(OH)_2 \xrightarrow{0.887} Fe^\circ$$

0.58

$$RuO_4 \xrightarrow{(0.9)} RuO_4^- \xrightarrow{0.6} RuO_4^{2-} \xrightarrow{0.4} RuO_2 \xrightarrow{0.1} Ru_2O_3 \xrightarrow{0.1} Ru^\circ$$

$$RuO_4 \xrightarrow{(0.75)} RuO_4^{2-} \qquad RuO_2 \xrightarrow{-0.04} Ru^\circ \qquad RuO_4 \xrightarrow{0.3} Ru^\circ$$

$$HOsO_5^- \xrightarrow{0.3} OsO_4^{2-} \xrightarrow{0.1} OsO_2 \longrightarrow Os_2O_3 \longrightarrow Os^\circ$$

$$HOsO_5^- \xrightarrow{0.2} OsO_2 \qquad OsO_2 \xrightarrow{-0.15} Os^\circ$$

$$CoO_2 \xrightarrow{0.7} Co(OH)_3 \xrightarrow{0.14} Co(OH)_2 \xrightarrow{-0.72} Co^\circ$$

$$RhO_4^{2-} \xrightarrow{0.9} RhO_2 \xrightarrow{0.9} Rh_2O_3 \xrightarrow{0.04} Rh^\circ$$

$$IrO_4 \xrightarrow{0.4} IrO_2 \xrightarrow{0.1} Ir_2O_3 \xrightarrow{0.1} Ir^\circ$$

$$NiO_4^{2-} \xrightarrow{0.4} NiO_2 \xrightarrow{0.49} Ni(OH)_2 \xrightarrow{-0.72} Ni^\circ$$

$$Pd(OH)_4 \xrightarrow{0.73} Pd(OH)_2 \xrightarrow{0.07} Pd^\circ$$

$$PtO_4^{2-} \xrightarrow{0.4} Pt(OH)_6^{2-} \xrightarrow{0.2} Pt(OH)_2 \xrightarrow{0.15} Pt^\circ$$

GROUP I-B

$$Cu(OH)_2 \xrightarrow{-0.08} Cu_2O \xrightarrow{-0.358} Cu^\circ$$

$$Ag_2O_3 \xrightarrow{0.74} AgO \xrightarrow{0.57} Ag_2O \xrightarrow{0.344} Ag^\circ$$

$$H_2AuO_3^- \xrightarrow{0.7} Au^\circ$$

GROUP II-B

ZnO_2^{2-} $\xrightarrow{-1.216}$ Zn°

$Cd(OH)_2$ $\xrightarrow{-0.809}$ Cd°

HgO $\xrightarrow{0.098}$ Hg°

THE VOLT EQUIVALENT-OXIDATION NUMBER DIAGRAM

The volt equivalent (VE) of a substance is the product of the oxidation state and the reduction potential relative to the element in its standard state. This diagram relates the volt equivalent to the oxidation state and is called a Frost diagram. Figure 10.4 is the Frost diagram for nitrogen under basic (E_B°) conditions.

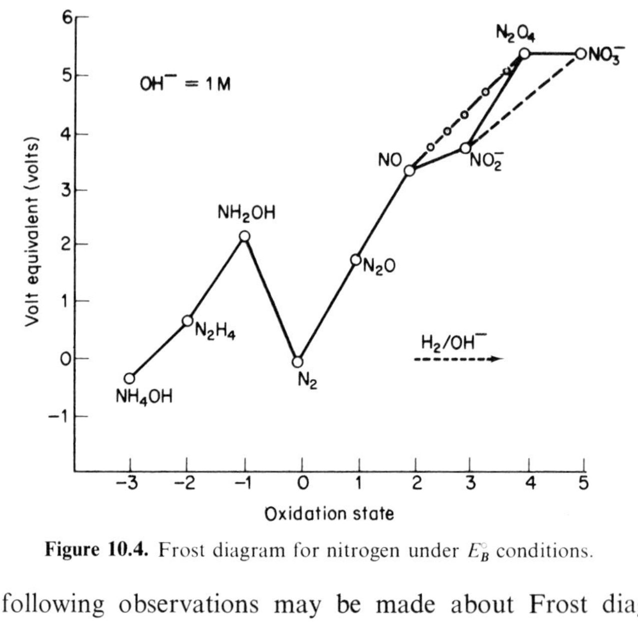

Figure 10.4. Frost diagram for nitrogen under E_B° conditions.

The following observations may be made about Frost diagrams in general. The stability of the species increases as the volt equivalent becomes less positive or more negative. The volt equivalent scale is actually a relative free energy scale. The slope of the line connecting any two species is equal to the oxidation-reduction potential of the half-cell involving the two species. Of course, the more positive the slope, the greater is the oxidizing power of the couple, and the more negative the slope, the greater is the reducing power of the couple. The diagrams are commonly plotted so that the H^+/H_2 or OH^-/H_2 couple has a slope of zero. Under acidic conditions ($E_A^\circ = 0.0$ V), no correction to standard potentials is necessary. However, when $[OH^-] = 1\ M$, the standard electrode potential, E_B°, is -0.83 V for the OH^-/H_2 couple. A correction of $+0.83$ V must be made so that the couple has a slope of zero (see Fig. 10.4).

The Frost diagram has limitations. It is useful only for qualitative discussion as the slopes of couples, i.e., electrode potentials are not accurately measured from the diagram. Actually, this is not disadvantageous to our purposes, since our predictions are based on a series of different approximations, e.g., the use of concentrations in place of activities. To make quantitative predictions, we must use activities in all considerations.

Frost diagrams can be calculated easily from available standard electrode potentials. For example, the volt equivalent for N_2H_4 may be calculated with the aid of equation [10.31].

$$\tfrac{1}{2}N_2(g) + 2H_2O + 2e^- \rightarrow \tfrac{1}{2}N_2H_4(l) + 2OH^-, \qquad E_B^\circ = -1.16\,V \qquad [10.31]$$

The volt equivalent is the standard potential multiplied by the oxidation state of nitrogen in N_2H_4, which is -2. The E_B° value for the N_2/N_2H_4 couple is corrected by $+0.83$ V, so that the OH^-/H_2 couple will have a slope of zero. The volt equivalent of N_2H_4 is $(-1.16 + 0.83)(-2) = 0.66$ V. In like fashion, the volt equivalent of NO_3^- is $(0.25 + 0.83)(5) = 5.40$ V. For species such as nitrogen, nitrogen(I) oxide, nitrogen(II) oxide, and dinitrogen tetroxide, which are neither acids nor bases, the volt equivalent will have the same value in both acidic and basic media.

Figure 10.4 will serve as an example of the qualitative information that may be gleaned from Frost diagrams. Relative thermodynamic stability can easily be predicted from the diagram. Any species above a line connecting two species is predicted to disproportionate. In basic solution, dinitrogen tetroxide is above the line (– – –) connecting the nitrate and nitrite ions and, as predicted, disproportionates into nitrate and nitrite.

$$N_2O_4 + 2OH^- \rightarrow NO_2^- + NO_3^- + H_2O \qquad [10.32]$$

In a similar manner, the reverse of disproportionation may be predicted. The reaction between nitrogen(II) oxide and dinitrogen tetroxide to form nitrite ion is predicted to be thermodynamically feasible because nitrite ion is below the line connecting nitrogen(II) oxide and dinitrogen tetroxide.

$$2OH^- + NO + \tfrac{1}{2}N_2O_4 \rightarrow 2NO_2^- + H_2O \qquad [10.33]$$

The oxidizing and reducing ability of any couple on the diagram depends on the sign and magnitude of its slope. Because they have large positive slopes the N_2O_4/NO_2^-, NO/N_2O, and NO_3^-/NO_2^- couples, among others, are powerful oxidizing agents. In fact, each species containing nitrogen in a positive oxidation state is a better oxidizing agent than the OH^-/H_2 couple, which has zero slope. Both the N_2H_4/N_2 and the NH_2OH/N_2 couples are strong reducing couples, and their tie lines have large negative slopes.

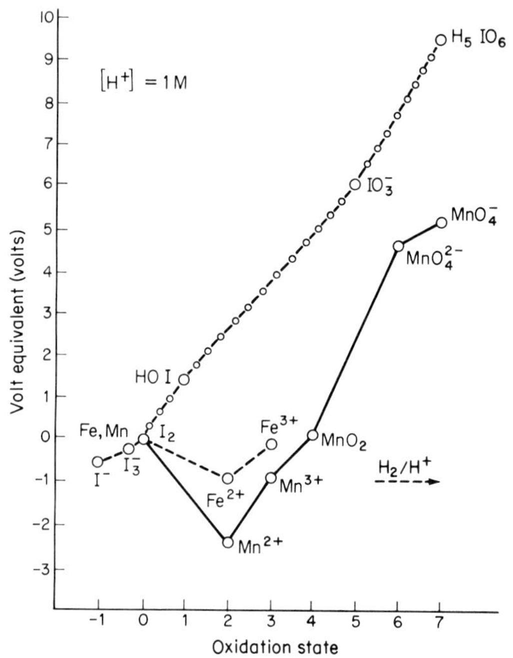

Figure 10.5. Frost diagram for iron (---), manganese (——), and iodine (.–.–.) under E_A° conditions.

Figure 10.5, which contains Frost diagrams of iron, manganese, and iodine in acidic solution (E_A°), can be used to describe the redox reactions of these systems. In the consideration of the reaction between ferrous ions and permanganate ions, the oxidized product of the ferrous ion must be ferric ion but several reduced products of permanganate ion are possible. Initially, consider the manganese system by itself to discover which reduced manganese species are stable at pH = O. Both manganate ion and manganese(III) ion are unstable with respect to disproportionation [see equations (10.34) and (10.35)].

$$4H^+ + 3MnO_4^{2-} \rightarrow MnO_2\,(s) + 2MnO_4^- + 2H_2O \qquad [10.34]$$

$$2H_2O + 2Mn^{3+} \rightarrow Mn^{2+} + MnO_2\,(s) + 4H^+ \qquad [10.35]$$

Manganese metal is oxidized by H^+ at pH = O. This reaction can be predicted by translating the H^+/H_2 couple so that the potential reactants H^+ and Mn coincide (see Fig. 10.6). Consider the possibilities that H^+

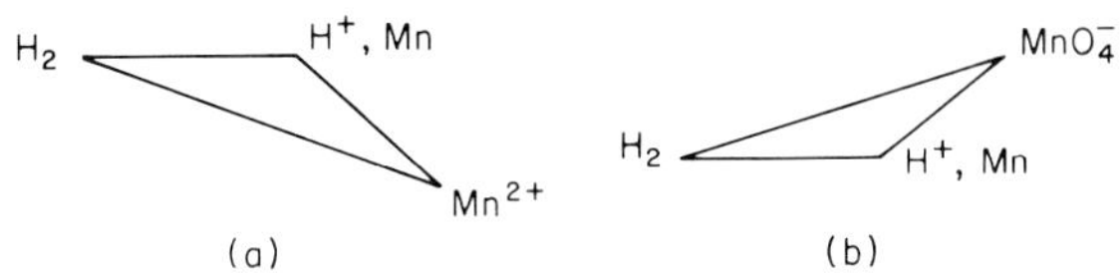

Figure 10.6. The reaction of the H^+/H_2 couple with the Mn/Mn^{2+} and $Mn/Mn0_4^-$ couples in acidic solution.

will oxidize manganese to manganese(II) or permanganate ion. Since the reactants are above the line connecting the potential products, hydrogen and manganese(II), in Fig. 10.6(a), that reaction will occur. Hydrogen ion will not oxidize manganese to permanganate ion because the line connecting the potential products, hydrogen and permanganate ion, is above the reactant point of hydrogen ion and manganese in Fig. 10.6(b). At pH = O, therefore, the only stable, reduced manganese products are manganese dioxide and manganese(II).

Consider, now, the reaction in which a potassium permanganate solution is slowly added to a solution containing ferrous ion, Fe^{2+}. The potential reduction products of permanganate ion are manganese(II) ion or manganese dioxide, but manganese dioxide must be resistant to reduction by ferrous ion if it is to be a product. The determine if the ferrous ion-manganese dioxide reaction is thermodynamically favorable, translate the ferrous ion of the Fe^{2+}/Fe^{3+} couple to the manganese dioxide position of the MnO_2/Mn^{2+} couple (see Fig. 10.7). Since the MnO_2, Fe^{2+} point is above the Mn^{2+}-Fe^{3+} tie line, the reaction will occur. The reaction observed when

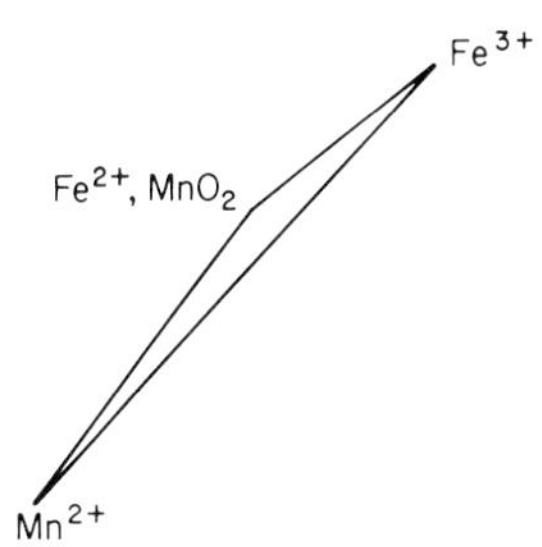

Figure 10.7. The reaction of the Fe^{2+}/Fe^{3+} and the MnO_2/Mn^{2+} couples in acidic solution.

ferrous ion is titrated with permanganate ion in acidic solution is as predicted:

$$8H^+ + MnO_4^- + 5Fe^{2+} \rightarrow Mn^{2+} + 5Fe^{3+} + 4H_2O \qquad [10.36]$$

The Frost diagrams for iodine and manganese are also given in Fig. 10.5. The products formed when a potassium permanganate solution is slowly added to a solution of HI at pH = O can be predicted from the diagram. As previously discussed, the reduced product must be manganese(II) or manganese dioxide. On scrutinizing the Frost diagram for iodine, we see that hypoiodous acid will disproportionate into iodate ion and triiodide ion or elemental iodine. In the presence of excess iodide ion, iodine reacts with the iodide ion and forms the triiodide ion, I_3^-.

$$I_2 + I^- \rightarrow I_3^-, \qquad K = 700 \qquad [10.37]$$

The reaction is predicted because triiodide ion is below the line connecting iodide ion and iodine in Fig. 10.5. Neither IO_3^- nor H_5IO_6 are produced because the I^--IO_3^- and I^--H_5IO_6 tie lines are above the triiodide ion's position (see Fig. 10.8). The potential reduction product, manganese

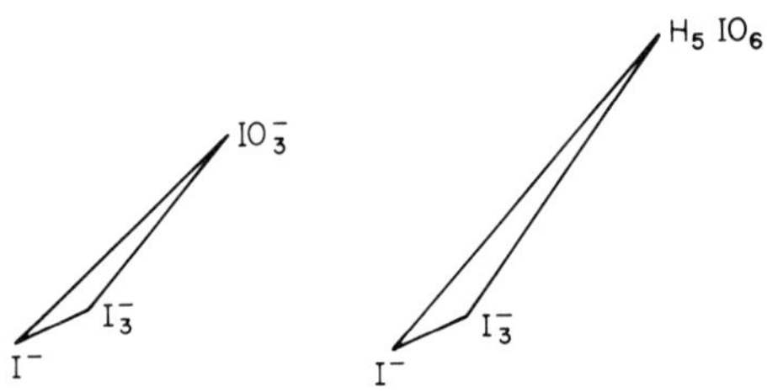

Figure 10.8. The reaction of I^- with IO_3^- or H_5IO_6 to form I_3^- in acidic solution.

dioxide, must be stable to iodide ion if it is to occur as a product in this reaction. From Fig. 10.9, we see, to the contrary, that iodide ion reduces manganese dioxide to manganese(II). The predicted and observed reaction is

$$15I^- + 2MnO_4^- + 16H^+ \rightarrow 5I_3^- + 2Mn^{2+} + 8H_2O \qquad [10.38]$$

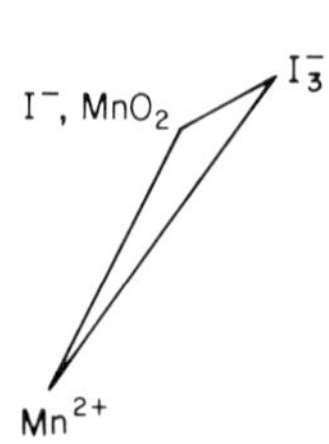

Figure 10.9. The reaction between the I^-/I_3^- and MnO_2/Mn^{2+} couples in acidic solution.

We conclude from this section that Frost diagrams allow prediction of certain redox reaction products. However, the relative thermodynamic stability of products cannot be predicted from the diagram if the slopes are similar. Even if thermodynamic stability can be predicted, the expected products may not be observed because the reactions are too slow. In other words, electrode potentials will give you the direction of the reaction, but will not tell you how long it will take to occur!

Suggested Readings

1. Frost, A. A., *J. Am. Chem. Soc.*, **73,** 2680 (1951).
2. Jolly, W. L., *J. Chem. Educ.*, **43,** 198 (1966).
3. Lattimer, W. M., *Oxidation Potentials*, 2nd ed., Prentice-Hall, Inc., Englewood Cliffs, N. J., 1952.
4. Phillips, C. S. G., and J. P. Williams, *Inorganic Chemistry*, vol. 1, Oxford University Press, New York, 1966, pp. 314–21.
5. Sanderson, R. T., *J. Chem. Educ.*, **43,** 584 (1966).
6. Sanderson, R. T., *Inorganic Chemistry*, Reinhold Publishing Corp., New York, 1967.

Study Questions

10.1. Under standard conditions, Li is a better reducing agent than Cs by 0.12 V. The first ionization potential of Li is higher than that of Cs by 35 kcal. Reconcile this discrepancy in reducing ability using a thermochemical cycle.

10.2. Which thermodynamic term is most responsible for the superior reducing ability of K relative to that of Ag?

10.3. Manganese(III) acetate dihydrate is synthesized by the reaction between manganese(II) acetate tetrahydrate and potassium permanganate in acetic acid. Write a balanced equation for the reaction. Why is this an elegant synthetic method?

10.4. Manganese(III) acetate dihydrate can be used to prepare manganese(III) complexes only in acidic media. How do you account for this?

10.5. Why is calcium about equivalent to potassium as a reducing agent, despite calcium's unfavorable ionization potential?

10.6. What thermodynamic term is most responsible for the high oxidizing ability of fluorine relative to bromine?

10.7. (a) Calculate the potentials for the $H^+ + e \rightarrow \frac{1}{2}H_2$ (g) and $\frac{1}{2}O_2$ (g) $+ 2H^+ + 2e \rightarrow H_2O$ half-reactions at pH 6.
(b) Is water easier to oxidize as the pH increases?
(c) Is water easier to reduce as the pH increases?

10.8. Calculate $E^\circ{}_B$ and compare it with $E^\circ{}_A$ for the following couples:
(a) Co^{2+}/Co
(b) Na^+/Na
(c) $Fe^{3+}Fe$

10.9. Latimer gives the following E° values:

$$E_A^\circ:\quad SeO_4^{2-} \xrightarrow{+1.15} H_2SeO_3 \xrightarrow{+0.74} Se \xrightarrow{-0.40} H_2Se$$

$$E_B^\circ:\quad SeO_4^{2-} \xrightarrow{+0.05} SeO_3^{2-} \xrightarrow{-0.37} Se \xrightarrow{-0.92} Se^{2-}$$

(a) Calculate E_A° for SeO_4^{2-} ——— Se.
(b) Calculate E_B° for Se^{2-} ——— SeO_3^{2+}.
(c) Is SeO_4^{2-} more stable to reduction in an acidic or basic medium?
(d) Calculate E_A° or E_B° for any species that disproportionates.

10.10. Calculate the Frost diagram as in Fig. 10.5 for the manganese and iodine systems under E_A° conditions from Latimer's data.

$$E_A^\circ:\quad H_5IO_6 \xrightarrow{ca.\ +1.7} IO_3^- \xrightarrow{+1.14} HOI \xrightarrow{+1.45} I_2 \xrightarrow{+0.54} I_3^- \xrightarrow{+0.54} I^-$$

(IO_3^- to I_2: +1.20; below IO_3^-: +1.70)

$$E_A^\circ:\quad MnO_4^- \xrightarrow{+0.56} MnO_4^{2-} \xrightarrow{+2.26} MnO_2 \xrightarrow{+0.95} Mn^{3+}$$

$$MnO_2 \xrightarrow{+1.51} Mn^{2+} \xrightarrow{-1.18} Mn$$

(MnO_4^- to MnO_2: +1.70)

10.11. Using the following data from Latimer, predict the products of the reaction that occurs when potassium iodide is added slowly to basic potassium permanganate:

$$E_B^\circ:\quad H_3IO_6^{2-} \xrightarrow{ca.\ +0.7} IO_3^- \xrightarrow{+0.14} OI^- \xrightarrow{+0.45} I_2 \xrightarrow{+0.54} I^-$$

(IO_3^- to I^-: +0.26; OI^- to I^-: +0.49)

$$E_B^\circ:\quad MnO_4^- \xrightarrow{+0.56} MnO_4^{2-} \xrightarrow{+0.60} MnO_2 \xrightarrow{-0.2} Mn(OH)_3 \xrightarrow{+0.1} Mn(OH)_2 \xrightarrow{-1.55} Mn$$

(MnO_2 to $Mn(OH)_2$: −0.05; MnO_4^- to MnO_2: +0.59)

10.12. Give an explanation for each of the following statements, using the data found in problems 10.10 and 10.11:
(a) The potential of the I_2/I^- couple is the same under E_A° and E_B° conditions.

(b) I_2 disproportionates in basic, but not in acidic, media.
(c) The IO_3^-/I^- couple oxidizes much better in acidic media than in basic media.
(d) Mn is more easily oxidized to the divalent state in basic than in acidic media.
(e) MnO_4^{2-} is less stable in acidic than in basic media.
(f) Mn^{2+} is stable to air oxidation in acidic media, but not in basic media. (Calculate the O_2-H_2O potentials, using the data of Fig. 10.4.)

10.13. Frost, A. A., [*J. Am. Chem. Soc.*, **73,** 2680 (1951)] discusses a volt equivalent-oxidation state diagram for the $H^+ + e \rightarrow \frac{1}{2}H_2$ and $\frac{1}{2}O_2 + 2e \rightarrow H_2O$ couples at pH values of 0, 7, and 14. Calculate this diagram including 0.5 V for overvoltage. The diagram is useful for estimating when oxygen and hydrogen will be evolved during aqueous redox reactions.

10.14. Calculate the equilibrium constants from electrode potentials for the following reactions, using data from Latimer:

$$E_A^\circ:\quad Cu^{2+} \xrightarrow{+0.15} Cu^{+} \xrightarrow{+0.52} Cu \qquad (Cu^{2+} \xrightarrow{+0.34} Cu)$$

$$E_B^\circ:\quad Cu(OH)_2 \xrightarrow{-0.08} Cu_2O \xrightarrow{-0.36} Cu$$

(a) $2Cu^+ \leftrightarrows Cu^{2+} + Cu$
(b) $Cu_2O \rightleftarrows Cu(OH)_2 + Cu$

10.15. Calculate the solubility product from the given E° values:
(a) $Cd(OH)_2 + 2e \rightarrow Cd$; $E_B^\circ = -0.81$ V, $E_A^\circ = -0.40$ V
(b) $In(OH)_3 + 3e \rightarrow In$; $E_B^\circ = -1.0$ V, $E_A^\circ = -0.34$ V
(c) $Pd(OH)_2 + 2e \rightarrow Pd$; $E_B^\circ = +0.07$ V, $E_A^\circ = +0.99$ V

10.16. Using the Frost diagrams of iron, manganese, and iodine (Fig. 10.5), predict either the products of the following reactants, or that no reaction will occur:
(a) 1 *N* H^+ is slowly added to Fe.
(b) IO_3^- is slowly added to Fe^{3+}.
(c) 1 *N* HCl is slowly added to K_2MnO_4 (Cl^- is oxidized to Cl_2).
(d) Fe^{2+} is slowly added to MnO_2.
(e) Mn is slowly added to 1 *N* HCl.

10.17. Latimer lists the following E_A° data:

$$Cr_2O_7^{2-} \xrightarrow{+1.33} Cr^{3+} \xrightarrow{-0.41} Cr^{2+} \xrightarrow{-0.91} Cr \qquad (Cr_2O_7^{2-} \xrightarrow{+0.30} Cr;\ Cr^{3+} \xrightarrow{-0.74} Cr)$$

$$H^+ \xrightarrow{0.00} H_2 \qquad O_2 \xrightarrow{+1.23} H_2O \qquad Cl_2 \xrightarrow{+1.36} Cl^-$$

Predict the chromium products of the following assuming that oxygen will not oxidize chromium beyond Cr^{3+}:
(a) Cr is placed in 1 *N* HCl.
(b) Cr^{3+} and Cr come into contact in aerated 1 *N* HCl.

(c) Cr^{3+} and Cr are mixed under anaerobic conditions in 1 *N* HCl.
(d) Cr and $Cr_2O_7^{2-}$ are mixed in aerated 1 *N* H^+.
(e) Cr and $Cr_2O_7^{2-}$ are mixed anaerobically in 1 *N* H^+

10.18. Latimer lists the following data for vanadium:

$$\overbrace{E_A^\circ:\quad VO_2^+ \xrightarrow{+1.0} VO^{2+} \xrightarrow{+0.36} V^{3+} \xrightarrow{-0.25} V^{2+} \xrightarrow{\text{ca. } -1.2} V}^{-0.25}$$

Additional data are given in problem 10.17. Assume anaerobic conditions for the following questions:
(a) What species are stable in 1 *N* HCl?
(b) What are the products when Cr^{2+} and VO_2^+ are mixed in 1 *N* HCl?
(c) What products occur when V and Cl_2 are mixed in 1 *N* HCl?
(d) What species disproportionate in 1 *N* HCl?

10.19. Calculate the potential of the Fe(III)/Fe° couple at pH 7.
$E_A^\circ = +0.036$ V and $K_{sp} = 6 \times 10^{-38}$

10.20. Consider the following data from Latimer:

$$N_2\,(g) + 4H_2O + 2e \rightarrow 2NH_2OH + 2OH^-, \qquad E_B^\circ = -3.04 \text{ V}$$
$$NH_2OH + 2H_2O + 2e \rightarrow NH_4OH + 2OH^-, \qquad E_B^\circ = +0.42 \text{ V}$$
$$ClO_2^- + H_2O + 2e \rightarrow ClO^- + 2OH^- \qquad E_B^\circ = +0.66 \text{ V}$$

(a) Is hydroxylamine generally an oxidizing or reducing agent in basic solution?
(b) Would it be more or less effective as an oxidizing agent when OH^- is 2 *M* rather than under E_B° conditions?
(c) Write a balanced equation for the expected reaction when NH_2OH is added to a solution containing ClO^- and ClO_2^- ions.
(d) Calculate E_B° for the reaction in part (c).

10.21. Write balanced equations for the following systems, or write "no reaction" when applicable. Use the data of Table 10.3, assuming standard conditions.
(a) A solution of V^{3+} is added to an H_2SeO_3 solution in acidic medium.
(b) Solid NiO_2 is added to an acidic solution of Pb^{2+}
(c) Selenium metal is added to an acidic solution of ReO_4^-.
(d) Rhenium metal is added to nitric acid.

10.22. Consider the following standard state data from Latimer:

$$ClO_4^- \xrightarrow{+1.19} \underline{ClO_3^-} \xrightarrow{+1.15} ClO_2 \xrightarrow{+1.27} \underline{HClO_2} \xrightarrow{+1.64} HOCl \xrightarrow{+1.63} \underline{Cl_2} \xrightarrow{+1.36} Cl^-$$

$$NO_3^- \xrightarrow{+0.79} \underline{N_2O_4} \xrightarrow{+1.07} HNO_2 \xrightarrow{+1.00} NO \xrightarrow{+1.59} \underline{N_2O} \xrightarrow{+1.77} N_2 \xrightarrow{+0.27} NH_4^+$$

(a) Predict the stability of each underlined species relative to disproportionation to its nearest neighbors.
(b) Write a balanced equation for the reaction expected when Cl_2 is bubbled into a HNO_2 solution.
(c) Write a balanced equation for the reaction expected when $(NH_4)_2SO_4$ is added to a $HClO_4$ solution.

CHAPTER 11

Inorganic Reaction Mechanisms The Isoelectronic Principle

INORGANIC REACTION MECHANISMS

The study of reaction mechanisms began before 1900. Early investigations were generally performed by physical chemists who were testing the then current theories. Before 1915, these chemists studied a large number of inorganic systems, such as the decomposition of ozone and the reaction of hydrogen gas with chlorine gas. In recent years, many new inorganic reactions have been studied and a wide variety of instrumental methods have become available. These new methods have given a firmer foundation to the study of kinetics.

A *reaction mechanism* is the detailed, stepwise movement of atoms and electrons that takes place while reactants become products. Probably the most useful technique in determining a reaction mechanism is the study of chemical kinetics, which is the determination of the rate of a reaction and its rate law. Other clues to mechanisms are obtained through isotope tracer studies, stereochemical investigations, and solvent effects.

For the present purpose, chemical reactions are best viewed from the standpoint of transition state theory. Consider the following simple, bimolecular reaction:

$$AB + C \rightleftharpoons AC + B \qquad [11.1]$$

The approach of C to AB causes an increase in the energy of the system until a critical energy state is reached. In this critical energy state the $A—B$ bond is weakened and the $B—C$ bond is partially formed; at this point, the maximum energy of the system is obtained. This increase in energy is called the *free energy of activation* (ΔG^*) and is the energy that colliding reactants must possess to form the species $A \cdots B \cdots C$. This species is known as the *transition state* or *activated complex* and is to be regarded as a moiety with a special type of bonding. In Fig. 11.1 the free energy of the activated complex is at the maximum of the curve, and consequently it has a short lifetime. The activated complex seeks out a lower energy state; it may dissociate into either new products or reactants.

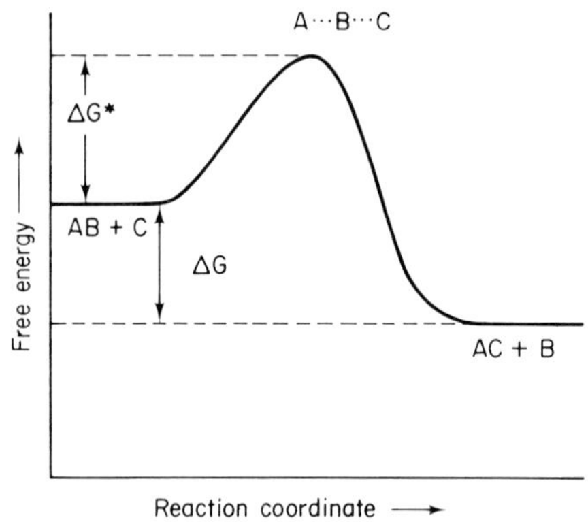

Figure 11.1. A reaction coordinate diagram.

Intermediates, which sometimes occur in chemical reactions, differ from the activated complex because they occur at free energy minima and can sometimes be isolated. In Fig. 11.2, *D* is an intermediate.

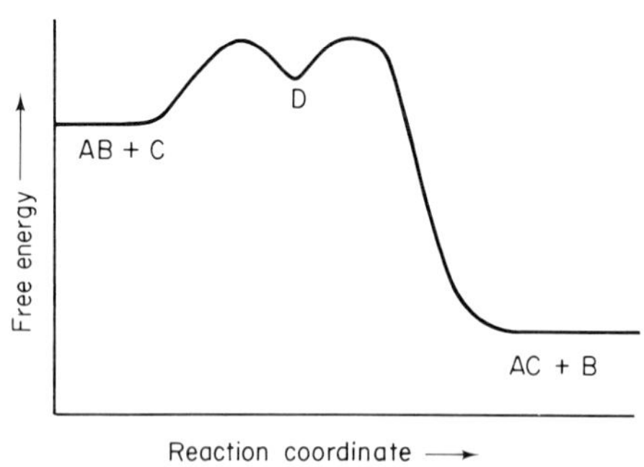

Figure 11.2. A reaction coordinate diagram that includes an intermediate of the reaction.

The free energy increase required to convert the reactants to the transition state is the free energy of activation, ΔG^*. The relationship between the free energy of activation and the rate constant is

$$k = \frac{BT}{h} e^{-\Delta G^*/RT} \qquad [11.2]$$

The rate constant is k, B is the Boltzmann constant, R is the gas constant, h is Planck's constant, and T is the absolute temperature.

In theory, one should be able to calculate rate constants, but these calculations prove to be very difficult. The energy of activation, E_a is usually determined experimentally using the Arrhenius equation and plotting log k against $1/T$.

$$k = Ae^{-E_a/RT} \qquad [11.3a]$$

or

$$\log k = \frac{E_a}{2.303} \cdot \frac{1}{T} + \log A \qquad [11.3b]$$

where A is a constant. Now E_a is related to the enthalpy of activation, ΔH^*, and its contribution to ΔG^*.

$$\Delta H^* = E_a - nRT \qquad [11.4]$$

$$\Delta G^* = \Delta H^* - T\Delta S^* \qquad [11.5]$$

In equation [11.4], n represents the molecularity of the reaction. Care must be taken to avoid confusing E_a with ΔG^*.

The free energy of the reaction, ΔG°, which is the difference in free energy between the reactants and the products, is related to the equilibrium constant K [equation (11.6)]

$$\Delta G^\circ = -RT \ln K \qquad [11.6]$$

The study of kinetics demands a distinction between *order* and *molecularity*. Molecularity refers to the number of reacting species (molecules) involved in forming the transition state complex. Figure 11.1 is illustrative of a bimolecular mechanism—i.e., two species unite to form a transition state complex. The term *order*, in contrast to molecularity, has an experimental connotation. Order refers to the sum of powers to which each reactant is raised in the rate law expression. For example, a rate law expression may be

$$\text{Rate} = \frac{d[A]}{dt} = -k[A][B] \qquad [11.7]$$

where a second-order reaction is designated. No indication of the molecularity is given in this expression. In the basic hydrolysis of $[Co(NH_3)_5Cl]^{2+}$ to form $[Co(NH_3)_5OH]^{2+}$, the rate expression is second order but the reaction appears to be unimolecular in the transition state.

The slowest reaction in a series of reactions is called the *rate determining step*. The net reaction rate is equal to the rate of this slow step, which may be preceded and/or followed by rapid steps.

It is very difficult to classify the mechanisms of inorganic reactions. An arbitrary classification of reaction types follows. The classification chosen has five reaction types. After they are defined, all five will be discussed in more detail.

1. *Dissociation and addition.* In this type of reaction, two species combine to form one species, or a species decomposes into two species [equation (11.8)].

$$(CH_3)_3P + BF_3 \rightarrow (CH_3)_3PBF_3 \qquad \textit{addition} \qquad [11.8a]$$

$$CaCO_3\,(s) \rightarrow CO_2\,(g) + CaO\,(s) \qquad \textit{dissociation} \qquad [11.8b]$$

2. *Free radical.* A free radical may be defined as a particle with one or more unpaired electrons. In a free radical reaction, one or more of the reactants has an unpaired electron.

$$\cdot NO_2 + \cdot NO_3 \rightarrow N_2O_5 \qquad [11.9]$$

3. *Substitution.* Substitution is the replacement of a group or ion in a molecule or ion by another ion or molecule.

$$[Fe(CN)_5OH_2]^{2-} + CN^- \rightarrow [Fe(CN)_6]^{3+} + H_2O \qquad [11.10]$$

4. *Exchange.* Exchange refers to redistribution of groups between two molecules [intermolecular, equation (11.11)]

$$BCl_3 + BF_3 \rightarrow BClF_2 + BFCl_2 \qquad [11.11]$$

or between nonequivalent positions in the same molecule [intramolecular, equation (11.12)].

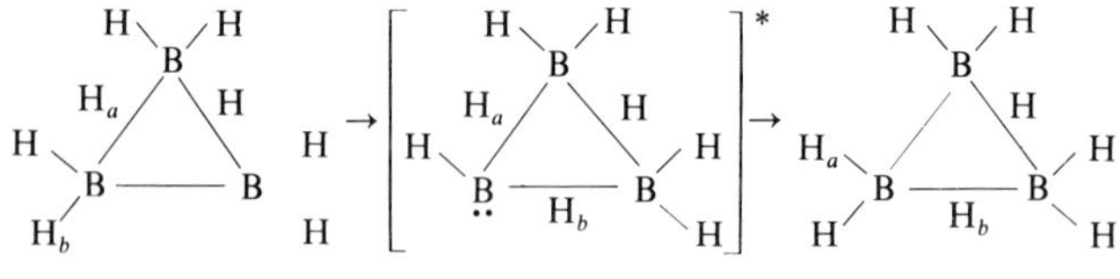

11.12

5. *Oxidation-reduction.* An oxidation-reduction reaction may be thought of as the transfer of an electron or atom from one molecule or ion to another.

$$Sn(II) + Hg(II) \rightarrow Hg^\circ\,(l) + Sn(IV) \qquad [11.13]$$

Dissociation and Addition Reactions

The formation of a donor-acceptor complex (Lewis acid-base reaction) may be regarded as an addition reaction and its reverse as a dissociation reaction. These ideas will be exemplified by the use of the reaction between boron trifluoride and trimethylamine, which has a rate constant of 3×10^9 liter/mole sec. No groups are lost, but there is significant rearrangement within the donor and acceptor when the adduct is formed (Fig. 11.3). The boron-fluorine and nitrogen-carbon bond distances increase, the F—B—F

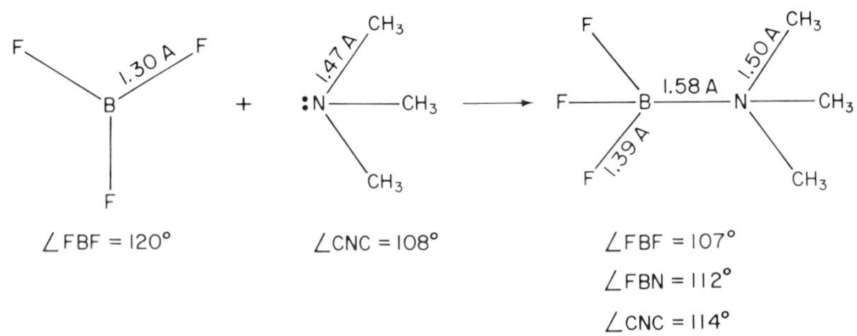

Figure 11.3. Molecular parameters of BF_3, $N(CH_3)_3$, and $F_3B{:}N(CH_3)_3$.

bond angles decrease, and the C—N—C bond angles increase upon adduct formation. This reaction and similar reactions are thought to occur by the following mechanism:

$$F_3B + \text{donor} \xrightarrow{k_1} (F_3B \cdot \text{donor})^* \xrightarrow[\text{fast}]{Mk_2} F_3B \cdot \text{donor} \qquad [11.14]$$

M is a third body with which the activated complex collides and which accepts its excess energy.

Reactions of this type have a low activation energy of about 3 kcal/mole and an exothermic enthalpy of about 25 ± 5 kcal/mole. The calculated energy for reorganization of bond lengths and angles of boron trifluoride and trimethylamine prior to electron transfer is much higher than the activation energy. It is also energetically unfeasible that electron transfer from nitrogen to boron should occur before intramolecular rearrangement. Neither extreme case occurs; electron transfer and intramolecular reorganization are concerted to keep the activation energy low.

Free Radical Reactions

In many cases, rate-determining steps involve ions. Such reactions are called polar or ionic. In other cases, reactants, postulated intermediates, and products are electrically neutral. The intermediates often contain unpaired electrons and are called free radicals. Some free radicals, such as nitrogen(II) oxide and chlorine(IV) oxide, are stable enough to be isolated easily, but many, such as hydroxyl radicals and halogen atoms, are not. Most free radical reactions occur in the gas phase but solution analogs are known. This type of reaction is often initiated by thermal or photolytic processes.

Consider, for example, the decomposition of ozone.

$$2O_3\ (g) \rightarrow 3O_2\ (g) \qquad [11.15]$$

In the presence of excess oxygen, the rate of the decomposition of ozone is second order in ozone and inversely proportional to the concentration of oxygen. This system is complicated also by a dependence of rate on the vessel wall, (M).

The *steady state approximation* is useful in translating experimental data into a mechanism. In the use of the steady state approximation, it is assumed that I is some intermediate or activated complex in a reaction. The steady state approximation is that $d[I]/dt$ is zero. This approximation is based on the assumption that the concentration of I is small and constant. This occurs when I is formed and consumed immediately in a fast step.

The suggested free radical mechanism involves two steps:

$$O_3 + M \underset{k_b}{\overset{k_a}{\rightleftarrows}} O_2 + O^{\cdot} + M \qquad [11.16a]$$

$$O^{\cdot} + O_3 \xrightarrow{k_c} 2O_2 \qquad [11.16b]$$

If the steady state approximation is applied to the oxygen atom, the following rate expression is determined:

$$-\frac{d[O_3]}{dt} = \frac{2k_a k_c [O_3]^2 [M]}{k_b[O_2] + k_c[O_3]} \qquad [11.17]$$

This rate law is in agreement with (a) the second-order dependence on ozone, (b) the inverse dependence on the concentration of oxygen, and (c) the heterogeneous reaction involving the wall.

Substitution Reactions

Consider the following general reaction:

$$AB + C \rightarrow AC + B \qquad [11.18]$$

In a substitution reaction, an ion or group is replaced by another. The substituting agent, C, in equation [11.18] may be a *nucleophile* or an *electrophile*. A nucleophile is a species that reacts by donating electrons to a substrate (AB), whereas an electrophile is a species having vacant orbitals of low energy which can accept electrons from an electron-rich substrate. In equation [11.18], B is called the *leaving group*. The leaving group must be the same type (nucleophile or electrophile) as the substituting or entering group. It is important here to differentiate between the terms *nucleophile* and *base*. The effectiveness of a nucleophile depends upon how rapidly it attacks the substrate, i.e., its effectiveness is based on kinetic factors. In contrast, the effectiveness of a base depends upon equilibrium considerations.

In a very simplified sense, we can consider a broad classification of substitution reactions in two categories: *electrophilic attack* and *nucleophilic attack*. An electrophilic mechanism is denoted by S_E. An electrophilic mechanism occurs, for example, when a metal ion displaces another metal ion in a substrate. Nucleophilic reactions are denoted by S_N. In nucleophilic reactions, an electron-rich group or nucleophile displaces another nucleophile in a substrate.

There are two extremes of nucleophilic substitution reactions: the unimolecular nucleophilic substitution reactions (S_N1), and the bimolecular nucleophilic substitution reactions (S_N2). The S_N1 reaction is exemplified by equations [11.19] and its rate expression is given in equation [11.20].

$$AB \xrightarrow[\text{slow}]{k} A^* + B \qquad [11.19a]$$

$$A^* + C \xrightarrow[\text{fast}]{} AC \qquad [11.19b]$$

$$\text{Rate} = k[AB] \qquad [11.20]$$

In this unimolecular reaction, the activated complex (designated by *) is a species of reduced coordination number. Equations [11.21] and [11.22] denote the S_N2 reaction and its rate expression, respectively.

$$AB + C \xrightarrow[\text{slow}]{k} \begin{bmatrix} & & C \\ & \cdot & \\ \cdot & & \\ A\ .\ . & & .\ B \end{bmatrix}^* \qquad [11.21a]$$

$$\begin{bmatrix} & & C \\ & \cdot & \\ \cdot & & \\ A\ .\ . & & .\ B \end{bmatrix}^* \xrightarrow{\text{fast}} AC + B \qquad [11.21b]$$

$$\text{Rate} = k[AB][C] \qquad [11.22]$$

The bimolecular S_N2 reaction has a first-order dependence upon the concentrations of AB and C, and is second order overall. In this mechanism, an increased coordination number occurs in the activated complex.

We cannot overstress that reactions seldom show clearcut S_N1 or S_N2 behavior, since these are limiting cases. In reality, at least to a certain extent, both mechanisms occur simultaneously. That is, as the $A—B$ bond lengthens, the nucleophile C aids in its weakening. When considering experimental data, it is important in differentiating between the mechanism extremes to notice whether the rate is independent of the nucleophile C. If the rate is independent of the nucleophile C, an S_N1 dissociation mechanism is probably operative and a bond-breaking process is predominant. Strong dependence upon the concentration and nature of C suggests an S_N2 mechanism.

Substitution reactions have been observed in substrates of a variety of geometries. Consider, first of all, an example of nucleophilic substitution at silicon(IV), specifically the hydrolysis of triethylsilane, $(C_2H_5)_3SiH$, which has a measurable rate in aqueous solution. The stoichiometry of the reaction is

$$(C_2H_5)_3SiH + OH^- + HOH \rightarrow (C_2H_5)_3SiOH + H_2 + OH^- \quad [11.23]$$

The observed rate law for this reaction is

$$\text{Rate} = k[C_2H_5)_3SiH][OH^-] \quad [11.24]$$

and, as can be observed, it is first order in hydroxide ion and the silane.

In establishing a mechanism, the consideration of the rate law leads to the conclusion that the rate determining step is the substitution of hydride ion by hydroxide ion in the substrate. The second step is the rapid reaction of hydride ion with the solvent, H_2O [see equation (11.25)].

$$(C_2H_5)_3SiH + OH^- \xrightarrow[k]{\text{slow}} (C_2H_5)_3SiOH + H^- \quad [11.25a)$$

$$H^- + HSol \xrightarrow{\text{fast}} H_2\,(g) + OH^- \quad [11.25b]$$

The intermediate containing the silane and the hydroxide ion has a coordination number of 5. It will be assumed that a trigonal bipyramidal geometry is generated by the intermediate, with bonding involving silicon d orbitals. Two possible intermediates are shown (Fig. 11.4). One isomer occurs when the nucleophile hydroxide ion attacks the tetrahedral face opposite the leaving group hydride ion [Fig. 11.4(a)]. A second intermediate occurs [Fig. 11.4(b)] if the nucleophile attacks a face of the tetrahedron containing the leaving group, the hydride ion. If the three R groups are

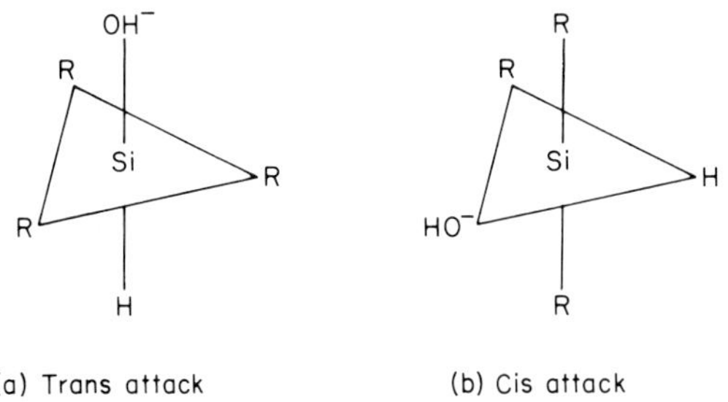

Figure 11.4. Two possible intermediates for nucleophilic attack on trialkyl silicon hydrides.

different, optically active silicon substrates are possible. In analogy to the Walden inversion in carbon chemistry, *trans* attack will invert the optical configuration, whereas *cis* attack will result in retention of the optical configuration. In silicon chemistry, retention usually occurs for leaving groups that are poor nucleophiles. Leaving groups that are good nucleophiles often result in inversion. Because partial inversion is common, simultaneous *cis* and *trans* attack must sometimes occur.

A very important reaction in the chemistry of octahedral complexes is nucleophilic substitution by water. This reaction is called aquation or acid hydrolysis. Consider, for example, the aquation of the following complex [equation (11.27)]:

$$[Co(NH_3)_5Cl]^{2+} + H_2O \xrightarrow{H^+} [Co(NH_3)_5(OH_2)]^{3+} + Cl^- \qquad [11.26]$$

In this reaction, and in general for aquation reactions, the experimental rate law is first order [equation (11.27)].

$$\text{Rate} = k[\text{complex}] \qquad [11.27]$$

The rate law suggests that an S_N1 reaction predominates, and that the following mechanism occurs [equation (11.28)]:

$$[Co(NH_3)_5Cl]^{2+} \xrightarrow[k]{\text{slow}} [Co(NH_3)_5]^{3+} + Cl^- \qquad [11.28a]$$

$$[Co(NH_3)_5]^{3+} + H_2O \xrightarrow{\text{fast}} [Co(NH_3)_5OH_2]^{3+} \qquad [11.28b]$$

There is a wide variety of experimental data which shows that the mechanism is indeed S_N1 for most aquation reactions. Consider first of all the effect of the charge on the complex. The rate is slower for *cis*-$[Coen_2NH_3Cl]^{2+}$ than for *cis*-$[Coen_2Cl_2]^+$ by a factor of 100. This is evidence for an S_N1 mechanism because the increased charge on the complex would result in a stronger attraction for the negative leaving group chloride ion, making the bond-breaking less favorable energetically. In contrast, an S_N2 mechanism would be expected to be faster for the complex with the higher positive charge, because the nucleophile would be more strongly attracted to the substrate. A second item of evidence that aquation reactions are generally S_N1 is based upon the effect of the size of nonleaving groups. The larger nonleaving group results in a faster rate. This evidence agrees with an S_N1 mechanism, because the lower coordination of five in the activated complex results in decreased steric strain relative to the octahedral complex. The coordination number of seven for an S_N2 mechanism would become energetically less feasible as the bulk of the nonleaving groups increases.

Among the few substitution reactions of cobalt(III) complexes showing second-order behavior is the basic hydrolysis of chloropentammine cobalt(III) ion,

$$[Co(NH_3)_5Cl]^{2+} + OH^- \longrightarrow [Co(NH_3)_5OH]^{2+} + Cl^- \qquad [11.29]$$

The experimental rate law,

$$\text{Rate} = k'[\text{complex}][OH^-] \qquad [11.30]$$

is second order and in agreement with an S_N1 mechanism:

(1) The reaction is nearly insensitive to the bulk of the leaving group and the nonleaving groups. An S_N2 reaction should become slower as the bulk of the ligands in the original octahedral complex increases.

(2) The rate is sensitive to the nature of the leaving group. If labeled water is used, the amount of ^{18}O in the product is virtually independent of the leaving group, and is the quantity expected for an S_N1 mechanism.

(3) The basic hydrolysis rate is fast only if a protonic nonleaving group is present.

The following mechanism, called an S_N1CB (CB mechanism signifies conjugate base), has been postulated for the basic hydrolysis of ammine complexes:

$$[Co(NH_3)_5Cl]^{2+} + OH^- \underset{K}{\overset{\text{fast}}{\rightleftharpoons}} [Co(NH_3)_4NH_2Cl]^+ + H_2O \qquad [11.31a]$$

$$[Co(NH_3)_4NH_2Cl]^+ \xrightarrow[\text{slow}]{k} [Co(NH_3)_4NH_2]^{2+} + Cl^- \qquad [11.31b]$$

$$[Co(NH_3)_4NH_2]^{2+} + H_2O \xrightarrow[\text{fast}]{} [Co(NH_3)_5OH]^{2+} \qquad [11.31c]$$

The rate law for the preceding mechanism is

$$\text{Rate} = \frac{kK_a}{K_w}[\text{complex}][\text{OH}^-] \qquad [11.32]$$

where K_a is the acid dissociation constant for the original complex and K_w is the ion product for water. The ratio K_a/K_w is the equilibrium constant K [equation (11.31a)]. The expression could be rewritten so that k, K_a and K_w are combined into one constant, k'; then the rate expression agrees with the experimental expression, equation [11.30]. The validity of this mechanism is substantiated by the fact that the basic hydrolysis of $[Co(CN)_5Cl]^{3-}$ is slow and first order. Thus, apparently the fast, basic hydrolysis mechanism depends upon the existence of a protonic nonleaving group in agreement with the S_N1CB mechanism.

Exchange Reactions

Exchange reactions are of two types: intermolecular and intramolecular. Intermolecular reactions involve ligand exchange between central atoms of the same element or different elements. Consider the general case for intermolecular exchange:

$$MX_2 + MR_2 \longrightarrow 2XMR \qquad [11.33]$$

At least four different mechanisms are possible for this intermolecular exchange reaction:

1. Electrophilic attack on $R(S_E2)$:

$$MX_2 \xrightleftharpoons{\text{fast}} MX^+ + X^- \qquad [11.34a]$$

$$MX^+ + MR_2 \xrightleftharpoons{\text{slow}} [XM \cdots R \cdots MR^+]^* \xrightarrow{\text{fast}} XMR + MR^+ \qquad [11.34b]$$

$$MR^+ + X^- \xrightarrow{\text{fast}} XMR \qquad [11.34c]$$

2. Nucleophilic attack on $M(S_N2)$:

$$MX_2 \xrightleftharpoons{\text{fast}} MX^+ + X^- \qquad [11.35a]$$

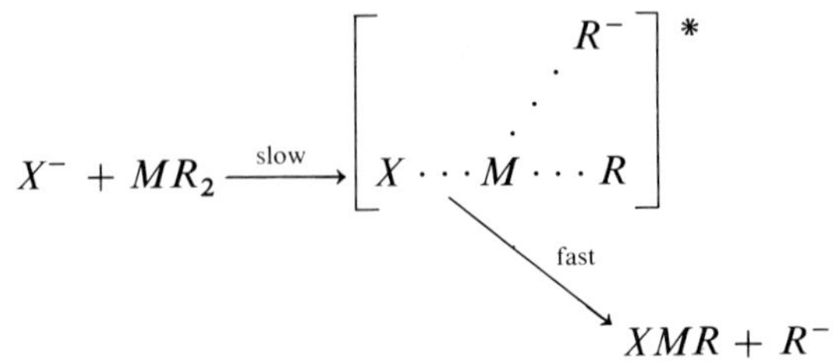

[11.35b]

$$MX^+ + R^- \xrightarrow{\text{fast}} XMR \qquad [11.35c]$$

3. Free radical attack on M:

$$MX_2 \xrightarrow{\text{fast}} XM^{\cdot} + X^{\cdot} \qquad [11.36a]$$

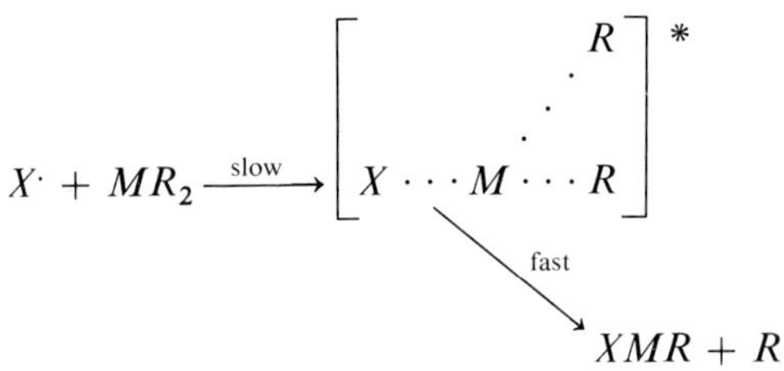

[11.36b]

$$R^{\cdot} + XM^{\cdot} \xrightarrow{\text{fast}} RMX \qquad [11.36c]$$

4. Bridge mechanism:

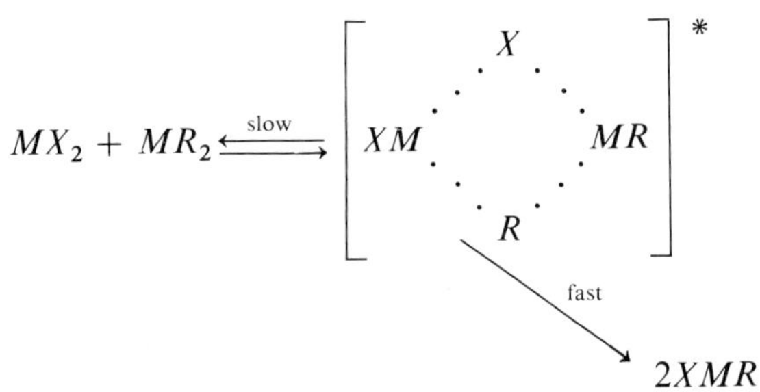

[11.37]

Equation [11.37], the bridge mechanism, combines elements of the nucleophilic and electrophilic mechanisms. The major difference between the previous mechanisms [equations (11.34) and (11.35)] and the bridge mechanism is that electrophilic attack of MX^+ on R *and* nucleophilic attack of X^- on M are slow steps in the bridge mechanism.

Consider the following reaction as an intermolecular exchange example:

$$HgBr_2 + Hg(s-C_4H_9)_2 \longrightarrow 2HgBr(s-C_4H_9) \qquad [11.38]$$

This reaction has a measurable rate at room temperature in ethanol and is first order in $Hg(s-C_4H_9)_2$ and in $HgBr_2$. The rate is greatly reduced by excess lithium bromide, which assists in the formation of the complex $HgBr_3^-$ while retaining an abundance of free bromide ion. The rate of reaction decreases in the order $Hg(NO_3)_2 > Hg(C_2H_3O_2)_2 > HgBr_2$. It is also experimentally known that expected products of a free radical mechanism, e.g., bromine, are not observed.

We shall now compare the experimental information relating to the four possible mechanisms for intermolecular exchange. The observed second-order rate law agrees with all four mechanisms. The free radical mechanism [equation (11.36)] can be discarded, because products such as bromine (formed by $Br\cdot + Br\cdot \rightarrow Br_2$) are not observed. The nucleophilic attack of Br^- on $Hg(s-C_4H_9)_2$ can also easily be ruled out. The addition of excess Br^- with resultant formation of $HgBr_3^-$ would result in an increased rate in a nucleophilic mechanism [equation (11.35)]. The experimental data show that excess Br^- decreases the reaction rate rather than increases it.

All the experimental results agree with the electrophilic mechanism [equation (11.34)]. Since the solubility of the mercury(II) salts decreases in the order $NO_3^- > C_2H_3O_2^- > Br^-$, the equilibrium concentration of the electrophilic HgX^+ [see equation (11.39)] decreases in the same order.

$$HgX_2\,(s) \rightleftarrows HgX^+ + X^- \qquad [11.39a]$$

$$HgX^+ \rightleftarrows Hg^{2+} + X^- \qquad [11.39b]$$

The solubility order of HgX_2 is the same as the order of the reaction rates. That is, increased concentration of the electrophile HgX^+ increases the rate in agreement with the electrophilic mechanism of equation [11.34]. Also in agreement with the electrophilic mechanism is the observation that excess Br^-, which decreases the concentration of $HgBr^+$, slows the rate. In contrast, the presence of either $HgBr_3^-$ or large concentrations of HgX^+ would be expected to increase the rate in the bridge mechanism of equation [11.37], because both electrophilic and nucleophilic slow steps are involved. From the experimental observations, the electrophilic mechanism seems preferable. It is, however, difficult to rule out the bridge mechanism completely.

The ^{19}F NMR study of SF_4 is also in agreement with an intermolecular exchange mechanism. At room temperature the ^{19}F resonance consists of a single sharp resonance, but below $-98°C$ two triplets are observed. The sharp singlet is due to rapid fluorine exchange between axial (a) and equatorial (e) positions. The temperature at which the triplets coalesce is concentration dependent. In addition, the exchange kinetics are greater than first order. Both facts support the argument that the fluorine exchange is intermolecular.

Intramolecular exchange is a unimolecular process involving ligand exchange between non-equivalent sites on the same molecule. Several intramolecular exchange reactions have been elucidated by NMR studies. The ^{19}F NMR study of $(C_2H_5)_2NPF_4$ is an excellent example. At room temperature the ^{19}F spectrum consists of two singlets in agreement with equivalent fluorines split by the phosphorous (^{31}P nuclear spin = 1/2), or by rapid exchange. Below $-100°C$ the ^{19}F spectrum consists of four triplets which represent two triplets split by the ^{31}P. The latter spectrum suggests one pair each of axial and equatorial fluorine, and a trigonal bipyramidal structure in which $N(C_2H_5)_2$ is in an equatorial position. Since the temperature of coalescence of the triplets into singlets is independent of the concentration of $(C_2H_5)_2NPF_4$, the room temperature exchange is intramolecular. That is the axial and equatorial fluorine on a molecule exchange without interacting with another molecule.

The preceding discussion has been offered to illustrate that molecules are not static species, that molecular rearrangements do take place, and that rearrangements are both intramolecular and intermolecular.

Oxidation-Reduction Reactions

Oxidation-reduction mechanisms can be divided into two limiting categories: outer-sphere and inner-sphere mechanisms. An outer-sphere mechanism is said to occur when no substitution occurs in the first coordination sphere of the reactants before or during the oxidation-reduction process. This mechanism must be seriously considered when the reacting compounds are substitutionally inert, i.e., nonlabile. The second limiting mechanism is called the inner-sphere mechanism. In this case, substitution preceeds electron transfer.

An example of the outer-sphere mechanism is the reaction between $[Fe(CN)_6]^{3-}$ and $[Fe^*(CN)_6]^{4-}$, in which hexacyanoferrate(II) is isotopically labeled. Both the complexes are nonlabile, and electron transfer is much more rapid than their substitution reactions. The electron transfer between $[Fe(CN)_6]^{3-}$ and $[Fe^*(CN)_6]^{4-}$ shows second-order kinetics [equation (11.40)].

$$\text{Rate} = k[\text{Fe}^*(\text{CN})_6^{4-}][\text{Fe}(\text{CN})_6^{3-}] \qquad [11.40]$$

To understand the mechanism proposed for outer-sphere electron transfer, we must consider the Franck-Condon principle. The principle states that the motion of nuclei is slow compared with the motion of electrons and that electron transfer occurs without any appreciable movement of the nuclei. In applying the Franck-Condon principle to kinetic studies, we see that there are several requirements before an electron can be transferred. First, all metal-ligand bond distances must be identical. Second, all metal-ligand bonds must have the same mode of bonding. Third, the oxidation states of the metals in the reacting complexes must have the same number of unpaired electrons when the exchanging electron is excluded. It is the rearrangement suggested by these three points that comprises the slow step in the outer-sphere electron transfer mechanism [see equations (11.41)].

$$[Fe(CN)_6]^{3-} + [Fe^*(CN)_6]^{4-} \xrightarrow[k]{\text{slow}} [(NC)_6Fe \cdots Fe^*(CN)_6]^{7-} \quad [11.41a]$$

$$[(NC_6)Fe \cdots Fe^*(CN)_6]^{7-} \xrightarrow{\text{fast}} [Fe(CN)_6]^{4-} + [Fe^*(CN)_6]^{3-} \quad [11.41b]$$

The rearranged molecule is symbolized by $\cdots$ between the iron ions. This mechanism is in agreement with the observed second-order kinetics.

The other limiting mechanism for oxidation-reduction reactions is the inner-sphere mechanism. In this mechanism, it is proposed that the oxidation and reduction occur by the transfer of an atom or group from one coordination sphere to the other. Consider, as an example, the reaction between sulfur dioxide and chlorate ion:

$$SO_2 + ClO_3^- + H_2O \rightarrow HSO_4^- + H^+ + ClO_2^- \quad [11.42]$$

The direct transfer of an oxygen atom from the chlorate ion to sulfur dioxide is demonstrated by the observation that one oxygen in the bisulfate ion originated from the chlorate ion. It seems that most two-electron redox reactions involving nonmetal centers occur by an inner-sphere activated complex.

Consider the exchange reaction between $[Cr(OH_2)_5Cl]^{2+}$ and $[Cr^*(OH_2)_6]^{2+}$ as a second inner-sphere example. (For clarity, the water molecules will be omitted in the discussion.) The experimental rate law is second order:

$$\text{Rate} = k[CrCl^{2+}][Cr^{2+}] \quad [11.43]$$

The suggested mechanism [equations, (11.44)] is in agreement with the second-order kinetics.

$$[CrCl^{2+}]+[Cr^{*2+}] \xrightarrow{\text{slow}} [Cr \cdots Cl \cdots Cr^{*4+}] \quad [11.44a]$$

$$[Cr \cdots Cl \cdots Cr^{*4+}] \xrightarrow{\text{fast}} [Cr^{2+}] + [Cr^{*}Cl^{2+}] \quad [11.44b]$$

The first system involving metal complexes in which this kind of mechanism was demonstrated involves $[Co(NH_3)_5Cl]^{2+}$ and $[Cr(OH_2)_6]^{2+}$ The reaction is shown in equation [11.45]:

$$[Co(NH_3)_5Cl]^{2+} + Cr^{2+} \xrightarrow[H_2O]{H^+} CrCl^{2+} + Co^{2+} + 5NH_4^+ \quad [11.45]$$

When this reaction occurs in the presence of tagged chloride ion, ^{37}Cl, no tagged chloride ion is observed in the inert chromium(III) product. This conclusively demonstrates that the chlorine is transferred directly from one complex to the other. This experiment is beautifully designed, because the nonradioactive chlorine is transferred from a nonlabile cobalt(III) complex to a nonlabile chromium(III) complex.

THE ISOELECTRONIC PRINCIPLE

In Chapter 1, we showed that the elements are arranged in families or groups in the periodic table and that the elements in the ground state and in the same family have the same electronic configuration and, in general, the same properties differing in degree. A general statement of the isoelectronic principle is that *molecules having the same number of electrons and the same number of heavy atoms (any atom except lithium, helium, or hydrogen) have similar electronic structures and similar heavy atom geometries*. Now, hopefully, we shall show that molecules composing an isoelectronic family have relationships that go beyond simple geometric relationships, so that the application of the isoelectronic principle has predictive power in terms of both properties and behavior.

A simple example will serve as a good introduction to the isoelectronic principle. Elemental carbon exists in two allotropic forms: graphite and diamond, the physical properties of which are well known. Boron nitride, $(BN)_x$ (boron has one less proton and one less electron than carbon, and nitrogen has one more proton and one more electron than carbon) has the same number of electrons (*isoelectronic*) as $(C_2)_x$ and also exists in the diamond and graphite structures. The physical properties of both boron nitride structures are very similar to those of the analogous $(C_2)_x$ structures. Hexagonal boron nitride is a lubricant, as is graphite, and cubic boron nitride approaches the hardness of a diamond.

Some rather dissimilar molecules, such as allene, H_2CCCH_2, ketene, H_2CCO, carbon dioxide, OCO, and hydrazoic acid, HNNN, all contain three heavy atoms in a linear structure. These molecules can be considered as derivatives of allene by the transfer of the protons of the hydrogen nuclei into the carbon nucleus, thereby successively converting the CH_2 group to NH and then to oxygen. Similar transformations lead to other members of this "isoelectronic family" and to the concept of atom substitution that has no effect on the heavy atom geometry, viz.,

$$\begin{array}{l} -CH_3 \text{ for } -NH_2 \text{ for } -OH \text{ for } -F \\ =CH_2 \text{ for } =NH \text{ for } =O \\ \equiv CH \text{ for } \equiv N \end{array} \qquad [11.46]$$

Such families may also include molecules in which the atoms have the same number of valence electrons rather than the same total number of electrons, just as in families of elements where each member possesses the same number of valence electrons but not the same total number of electrons.

Further examples of structural and physical similarities in isoelectronic molecules are provided by the following families, which are formed by the hypothetical movement of a nuclear proton from the nonmetal to the metal:

$$\overset{p}{\curvearrowleft}\mathrm{CaO} \longrightarrow \overset{p}{\curvearrowleft}\mathrm{ScN} \longrightarrow \mathrm{TiC} \quad \text{all have rock salt structure} \qquad [11.47]$$

$$\overset{4p}{\curvearrowleft}\mathrm{F_2Ca} \longrightarrow \overset{2p}{\curvearrowleft}\mathrm{Na_2S} \longrightarrow \mathrm{Mg_2Si} \quad \text{all have fluorite structure} \qquad [11.48]$$

The *Grim-Sommerfeld compounds* (four valence electrons per heavy atom) can be considered in the same manner, because they all adopt the diamond, zinc blende, or wurtzite structure (see Table 11.1). In addition, all these species have melting points that decrease with an increase of pressure, and

TABLE 11.1 *Some Solid Phase Grim-Sommerfeld Compounds*

$\overset{p}{\curvearrowright}$ IV–IV	$\overset{p}{\curvearrowright}$ III–V	$\overset{p}{\curvearrowright}$ II–VI	$\overset{p}{\curvearrowright}$ I–VII
CC	BN	BeO	
SiSi	AlP		
GeGe	GaAs	ZnSe	CuBr
SnSn	InSb	CdTe	AgI

are similar to silicon or germanium in that many of them are found to be semiconductors. The compound indium antimonide, InSb (Table 11.1; isoelectronic with Sn—Sn), has received a great deal of attention, because it adopts the white tin metallic structure under high pressures. White tin becomes a superconductor at 3.7°K, whereas InSb in the white tin structure becomes superconducting at temperatures below 2°K.

In a rather similar manner, the anion lattice formed by isoelectronic species when a metal combines with a nonmetal is similar, in many cases, to the crystal structure of the next higher neighboring element of the anion. Each electron the anion has gained results in a change in the family of elements to which it is isoelectronic. Table 11.2 shows changes of isoelectronic families with the resultant change in nonmetal lattice.

TABLE 11.2 ***Some Examples of Isoelectronic Family Changes in Anions***

Anion Family Change	*Example*	
III → IV	MgB_2	B atoms form planar sheets similar to graphite
	LiGa, $CaGa_2$	Ga forms three-dimensional network as does elemental germanium
IV → V	$CaSi_2$	Si atoms form puckered layers similar to elemental As
V → VI	LiAs	As forms spiral chains similar to elemental selenium
VI → VII	Na_2O_2	O atoms dimerize as does elemental fluorine
VII → VIII	NaF	F atoms form a lattice similar to crystalline neon

In Chapter 7 the presentation dealing with the metallic bond, the Hume-Rothery rules for the three different structures of brass were discussed and were related to the electron/atom ratio of the species involved. Table 11.3 relates the isoelectronic principle with its now familiar alchemical shifting of nuclear protons to the different brass structures.

Some of the most interesting applications of the isoelectronic principle lie in the area of synthesis and chemical reactions. Extensive application of this principle has been in two areas: noble gas compounds and boron hydrides.

Consider the noble gas compounds first. We can see that iodide ion, I^-, is isoelectronic with xenon, Xe, and therefore molecules in which the iodide ion is replaced by a xenon atom might have somewhat the same properties, modified, of course, to account for the difference in size,

TABLE 11.3 ***Some Compounds with β, γ, and ε Brass Structures****

$\frac{3}{2}$ *Compounds* (*β Brass*):

$$\text{CuBe (I-II)} \xrightarrow{\text{double}} (\text{I}_2\overset{p}{\text{II-II}}) \longrightarrow (\text{I}_3\text{III}) \sim \text{CuAl}$$

$$\xrightarrow{\text{triple}} \text{I}_3\overset{2p}{\text{II}_2\text{II}} \longrightarrow (\text{I}_5\text{IV}) \sim \text{Cu}_5\text{Sn}$$

$\frac{21}{13}$*Compounds* (*γ Brass*):

$$\text{Cu}_5\text{Zn}_8 \ (\text{I}_5\text{II}_8) \xrightarrow{\text{double}} (\text{I}_{10}\overset{8p}{\text{II}_8\text{II}_8}) \longrightarrow (\text{I}_{18}\text{III}_8) = (\text{I}_9\text{III}_4) \sim \text{Cu}_9\text{Al}_4$$

$$(\text{I}_5\text{II}_8) \xrightarrow{\text{double}} (\overset{5p}{\text{I}_5\text{I}_5}\text{II}_{16}) \longrightarrow (\text{O}_5\text{II}_{21}) \sim \text{Fe}_5\text{Zn}_{21}$$

$\frac{7}{4}$*Compounds* (*ε Brass*):

$$\text{CuZn}_3 \ (\text{I-II}_3) \xrightarrow{\text{double}} (\text{I}_2\overset{\longrightarrow}{\text{II}_3\text{II}_3}) \longrightarrow \text{I}_5\text{III}_3 \sim \text{Ag}_5\text{Al}_3$$

$$(\text{I-}\overset{2p}{\text{II}_2\text{II}}) \longrightarrow (\text{I}_3\text{IV}) \sim \text{Cu}_3\text{Si}$$

*The Roman numerals refer to the group numbers of the periodic table and the subscript refers to the number of atoms taken from that group. The number of atoms of each group can be doubled or tripled, and then the nuclear protons shifted ($\overset{p}{\curvearrowright}$) to obtain the new compound.

electronegativity, etc. The structures of such species as the following are similar:

$$\begin{array}{ll} \text{XeF}_2, & \text{IF}_2^- \\ \text{XeF}_4, & \text{IF}_4^- \\ \text{XeO}_3, & \text{IO}_3^- \\ \text{XeO}_6^{4-}, & \text{IO}_6^{5-} \end{array}$$

One particularly interesting example of the utility of such ideas involves the preparation of some xenon chlorides. The known ions ICl_2^- and ICl_4^- are known to exist, and possibly $XeCl_2$ and $XeCl_4$ might also exist. It is known that ^{129}I is an unstable nuclide and that it undergoes beta decay, transforming into ^{129}Xe. Thus, iodate ion, $^{129}IO_3^-$, or periodate ion, $^{129}IO_4^-$, might be synthesized; the radioactive iodine would undergo beta decay to produce $^{129}XeO_3$ and $^{129}XeO_4$. This synthesis has been carried out, and the xenon oxides resulting from radioactive decay have proved by Mossbauer spectroscopy to be identical to the xenon oxides produced in direct synthesis using xenon. Consequently, $^{129}ICl_2^-$ and $^{129}ICl_4^-$ were used to synthesize the previously unknown xenon dichloride and xenon tetrachloride.

Oxygen and borane, BH_3, are isoelectronic. Therefore, pairs that might have similar properties include B_2H_6 and O_2; H_3BCO and CO_2; $H_3BCO_2^{2-}$ and CO_3^{2-}; F_3PO and F_3PBH_3. Some interesting relationships between H_3BCO and CO_2 include the following:

$$CaO + CO_2 \xrightarrow{\text{dry}} \text{no reaction} \quad [11.49a]$$

$$CaO + H_3BCO \xrightarrow{\text{dry}} \text{no reaction} \quad [11.49b]$$

$$CaO + CO_2 \xrightarrow{\text{trace } H_2O} CaCO_3 \quad [11.50a]$$

$$CaO + H_3BCO \xrightarrow{\text{trace } H_2O} CaH_3BCO_2 \quad [11.50b]$$

In fact, the two salts, K_2CO_3 and $K_2H_3BCO_2$, are prepared in an analogous manner:

$$CO_2 + 2KOH \xrightarrow{\text{alcohol}} K_2CO_3 + H_2O \quad [11.51a]$$

$$H_3BCO + 2KOH \xrightarrow{\text{alcohol}} K_2H_3BCO_2 + H_2O \quad [11.51b]$$

Other salts may be prepared by the use of suitable cation exchange columns. As is the case with the carbonates, the stability of these other metallic boranocarbonate salts is dependent on the size and charge of the cation. Thus, the alkali metal boranocarbonates are more stable than alkaline earth boranocarbonates.

The reactions of the boranocarbonates with acids have produced some rather interesting comparisons. Sodium boranocarbonate, $Na_2H_3BCO_2$, has been titrated with hydrochloric acid, with the first half of the titration curve showing great similarity to the first half of the titration curve of sodium carbonate. From the titration curve, the second ionization constant for the hypothetical acid, $H_2(H_3BCO_2)$, was estimated to be $6 \pm 3 \times 10^{-11}$. The second ionization constant for carbonic acid measured in the same fashion as for $H_2(H_3BCO_2)$ is observed to be 7.5×10^{-11}, but the accepted second ionization constant for carbonic acid is 4.5×10^{-11}. The estimated first ionization constant for $H_2(H_3BCO_2)$ is 10^{-8}, comparing favorably with 4.5×10^{-7}, the first ionization constant for carbonic acid. Since only a fraction of carbon dioxide dissolved in water is present as carbonic acid, the corrected first ionization constant for carbonic acid is 1.32×10^{-4}. The measurement of the amount of free $H_2(H_3BCO_2)$ in water is most difficult, because this acid reacts with water.

Hopefully, this presentation has shown that the isoelectronic principle does have some utility in structural and synthetic chemistry. We strongly stress that the results acquired with this reasoning by analogy must be used with caution. As with any qualitative concept, its use must be modified and directed by the chemical knowledge and intuition of the employer. Used in a proper manner, however, the isoelectronic principle can be a useful tool to classify and predict structures and to suggest possibly intriguing chemical reactions.

Suggested Readings

A. Reaction Mechanisms

1. Basolo, F., and R. G. Pearson, *Mechanisms of Inorganic Reactions,* 2nd ed., John Wiley & Sons, Inc., New York, 1967.
2. Edwards, J. O., *Inorganic Reaction Mechanisms,* W. A. Benjamin, Inc., New York, 1965.
3. Laidler, K. J., *Chemical Kinetics,* McGraw-Hill Book Company, New York, 1965.
4. Lockhart, J. C., *Introduction to Inorganic Reaction Mechanisms*, D. Van Nostrand Co., Inc., Princeton, N.J., 1966.
5. Sommer, L. H., *Stereochemistry, Mechanism, and Silicon*, McGraw-Hill Book Company, New York, 1965.
6. Sykes, A. G., *Kinetics of Inorganic Reactions,* Pergamon Press, Oxford, 1966.
7. Taube, H., *J. Chem. Educ.,* **45,** 452 (1968).

B. The Isoelectronic Principle

1. Bent, H. A., *J. Chem. Educ.,* **43,** 170 (1966).
2. Greenwood, N. N., *Chem. Brit.,* **3,** 66 (1967).
3. Malone, J., and R. W. Parry, *Inorg. Chem.,* **6,** 817 (1967).
4. Parry, R. W., *et al., Advan. Chem.,* **42,** American Chemical Society, Washington, D.C., 302 (1964).

Study Questions

11.1. A solution containing $[Coen_2Cl_2]^+$ is separated into three portions. The portions are treated with KNO_2, KI, and KSCN, yielding $[Coen_2ClNO_2]^+$, $[Coen_2ClI]^+$, and $[Coen_2Cl(SCN)]^+$, respectively. The rate is independent of the concentration and nature of the incoming anion. How can these observations be explained most simply?

11.2. For the reaction $[Co(NH_3)_6]^{3+} + Cl^- \rightleftarrows [Co(NH_3)_5Cl]^{2+} + NH_3$, $K_{eq} = 2 \times 10^4$. Why is it possible to reflux the hexamine complex for an hour in 6 *M* HCl without any NH_3 being displaced?

11.3. Burmeister, J. L., and Basolo, F., [*Inorg. Chem.*, **3,** 1587 (1964)] studied the linkage isomers of $Pd(Ph_3As)_2(SCN)_2$. They isolated the S-bonded form, which changed to the more stable N-bonded form in the solid state. Why is it possible to isolate the unstable form?

11.4. Posey, F. A., and Taube, H., [*J. Am. Chem. Soc.*, **75,** 4099 (1953)] studied:

$$[Co(NH_3)_4CO_3]^+ \xrightarrow[H_2O]{H^+} [Co(NH_3)_4(H_2O)_2]^{3+} + CO_2\ (g)$$

When this reaction is carried out in $H_2^{18}O$, about half the water in the complex contains ^{18}O. Suggest a mechanism for this result.

11.5. Retention of configuration (97%) occurs in the reaction

$$RR'R''Si^{*}H + Cl_2\ (g) \longrightarrow RR'R''Si^{*}Cl$$

For the observed rate law, rate $= k[RR'R''Si^{*}H][Cl_2]$. It has been suggested that a three-center bridge mechanism is responsible for these results. Suggest a transition state, explaining why retention occurs. The data are from the Sommer reference.

11.6. The reactions of $RR'R''Si^{*}F$ with $LiAlH_4$ in ether forms $RR'R''Si^{*}H$ with 95% inversion. The rate expression is first order in each reactant. Discuss the mechanism of this reaction, showing in detail the transition state. The data are from the Sommer reference.

11.7. Consider the following reaction (C_6H_{11} is cyclohexyl);

$$RR'R''Si^{*}Cl + C_6H_{11}NH_3^{36}Cl \xrightarrow{CHCl_3} RR'R''Si^{*36}Cl + C_6H_{11}NH_3Cl$$

The rate of racemization equals the rate of chloride exchange. Suggest a simple mechanism for the reaction. Why does the transition state lead to racemization? Data are from the Sommer reference.

11.8. Weingarten, H., and Van Wazer, J. C., [*J. Am. Chem. Soc.*; **88,** 2700 (1966)] studied the following reaction in dimethylformamide:

$$Sb(CH_3)_3 + SbCl_3 \rightarrow (CH_3)_2SbCl + CH_3SbCl_2$$

The reaction is first order in each antimony reactant, and the authors suggest that a four-center mechanism is in agreement with the data. Draw the four-center transition state, and explain why it implies an electrophilic *and* a nucleophilic attack.

11.9. Pearson, P. T., Boston, C. R., and Basalo, F., [*J. Am. Chem. Soc.*, **75,** 3089 (1953)] studied the acid hydrolysis of *trans*-$Co(AA)_2Cl_2^+$ to form $Co(AA)_2(OH_2)Cl^{2+}$ at 25°C, and observed first-order kinetics.

Diamine	$k \times 10^5$ (sec^{-1})
$H_2N\ CH_2CH_2NH_2$	3.2
$H_2N\ CH_2CH(CH_3)NH_2$	6.2
dl-$H_2N\ CH(CH_3)CH(CH_3)NH_2$	15
meso-$H_2N\ CH(CH_3)CH(CH_3)NH_2$	42

Why, in addition to the order of the reaction, do these data suggest an S_N reaction? Why is the *meso*-diamine faster than the *dl*-diamine? Use of molecular models is helpful. (Data are from Basolo, F., and Pearson, R. G., *Mechanisms of Inorganic Reactions*, 2nd ed., John Wiley & Sons, Inc., New York, 1967.)

11.10. Johnson, S. A., Basolo, F., and Pearson, R. G. [*J. Am. Chem. Soc.*, **85,** 1741 (1963)] studied at 80°C the following aqueous reaction:

$$trans\text{-}[Rh(en)_2Cl_2^+] + 2X \rightarrow trans\text{-}[Rh(en)_2X_2{}^{n+}] + 2Cl^-$$

X	$k \times 10^5$ (sec^{-1})	X	$k \times 10^5$ (sec^{-1})
OH^-(0.1 *M*)	5.1	I^-(0.05 *M*)	5.1
NO_2^-(0.1 *M*)	4.2	$^{36}Cl^-$(0.01 *M*)	4.0
NO_2^-(0.05 *M*)	4.2	NH_3(5.0 *M*)	4.0
I^-(0.1 *M*)	5.2		

For every nucleophile, first-order kinetics is followed. Write a mechanism in agreement with the given data. The authors suggest an S_N2 mechanism. What is the attacking agent in an S_N2 mechanism, and why are first-order kinetics observed?

11.11. Consider the cis attack of Y^- on *cis*- and *trans*-$Coen_2H_2OCl$. Assuming attack on a cis octahedral face, predict the geometrical configuration of $Coen_2H_2OY$ from (a) the cis reactant, and (b) the trans reactant.

11.12. Assuming (a) a trigonal bipyramidal and (b) a square pyramidal intermediate, predict the geometrical distribution of the product in

$$trans\text{-}Co(NH_3)_4(H_2O)Br^{2+} + Y^- \rightarrow Co(NH_3)_4(H_2O)Y^{2+} + Br^-$$

Assume that Y^- attacks an edge in the plane of the trigonal bipyramid because of the wider angle in the plane.

11.3. Burmeister, J. L., and Basolo, F., [*Inorg. Chem.*, **3,** 1587 (1964)] studied the linkage isomers of $Pd(Ph_3As)_2(SCN)_2$. They isolated the S-bonded form, which

$$\text{Rate} = k_1[Pt(NH_3)_3Cl^+] + k_2[Pt(NH_3)_3Cl^+][I^-]$$

It is thought that the reaction goes by two bimolecular paths. Explain the observation.

11.14. Consider the reaction $NO_2^- + HOCl \xrightarrow{H_2O} NO_3^- + HCl$. When $H^{18}OCl$ is used, the products are $N^{18}OO_2^+$ and HCl. Write a mechanism for this reaction.

11.15. Halperin, J., and Taube, H., [*J. Am. Chem. Soc.*, **74,** 375 (1952)] studied the reaction

$$HSO_3^- + Cl^{18}O_3^- \xrightarrow{H^+} HS^{18}OO_3^- + Cl^{18}O_2^-$$

At 0°C, the rate law is first order in both reactants. Devise a mechanism for the reaction.

11.16. Nordmeyer, F., and Taube, H., [*J. Am. Chem. Soc.*, **90,** 1162 (1968)] studied at high acidity the following reaction:

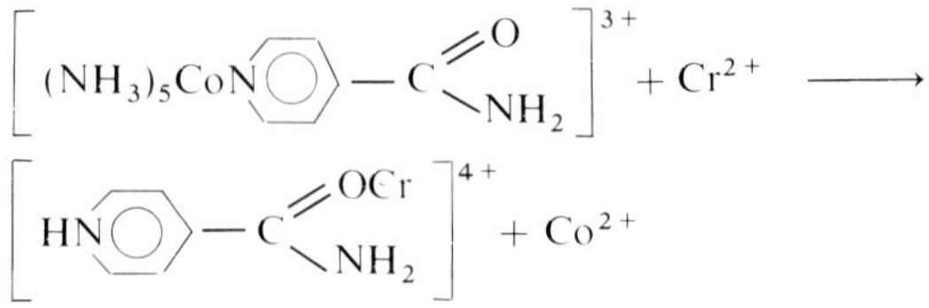

(Water has been omitted for simplicity.) $Cr(H_2O)_6^{3+}$ does not form a complex with 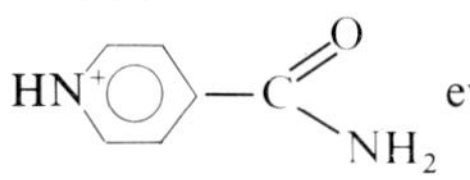 even in the presence of Cr^{2+}, but the preceding intermediate aquates slowly enough to be characterized. Discuss the mechanism of the reaction.

11.17. There are many redox reactions—e.g., between $Mo(CN)_8^{3-}$ and $Fe(CN)_6^{4-}$, or between $Fe(CN)_6^{3-}$ and $W(CN)_8^{4-}$—in which the redox reaction occurs faster than the substitution reaction. Explain this observation. [Data are from Taube, H., *J. Chem. Educ.*, **45,** 452 (1968).]

11.18. Espenson, J. H., [*Inorg. Chem.*, **4,** 1025 (1965)] studied the reduction of V^{3+} (aq) by Cr^{2+} (aq). The rate equation has the form, rate $= k[V^{3+}][Cr^{2+}]$; H^+ is released during the slow step. Further interpretation of the data leads to the conclusion than an unstable intermediate containing vanadium and chromium is formed. Suggest a simple mechanism for this reaction.

11.19. Snellgrove, R., and King, E. L., [*J. Am. Chem. Soc.*, **84,** 4609 (1962)] demonstrated that azide ions are transferred in pairs from $(H_2O)_4Cr(N_3)_2^+$ during reduction by Cr^{2+}. How would the experiments be designed to demonstrate the point?

11.20. Pearson, R. G., and Edgington, D. N., [*J. Am. Chem. Soc.*, **84,** 4607 (1962)] studied the rate of hydrolysis of 0.0029 M $Co(NH_3)_5Cl^{2+}$ by (a) 0.0029 M OH^-, and (b) 0.0029 M OH^- plus 0.018 M H_2O_2. A good nucleophile but a much weaker base than OH^-, HO_2^- decreases the rate of the reaction. How do the experiments with HO^{2-} aid in determining the mechanism of the basic hydrolysis of $Co(NH_3)_5Cl^{2+}$?

11.21. It is thought that the isomerism of *trans*-$[Co(H_2NCH_2CH_2NH_2)_2(H_2O)_2^{3+}]$ to the cis isomer occurs through two intermediates. Draw a reaction coordinate diagram that agrees with this assumption.

11.22. Explain the observation that, as the bulk of the leaving group is increased in a series of reactions, the rate of an S_N2 reaction decreases. In contrast, the rate of an S_N1 reaction often increases.

11.23. Figure 11.4 depicts two potential trigonal bipyramidal intermediates.

(a) Show that the cis intermediate [Fig. 11.4(b)] occurs by attack of three faces of the tetrahedron.

(b) Show that tetrahedral edge attack would result in two additional trigonal bipyramidal intermediates. Draw these intermediates.

11.24. Interpret the following NMR spectra of dimethylformamide in terms of the structure of dimethylformamide at the various temperatures.

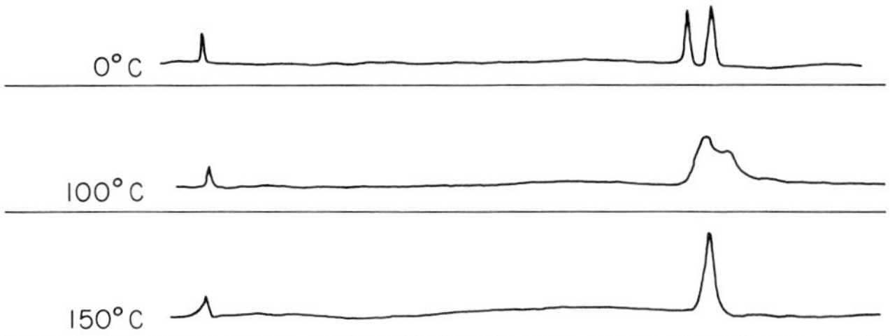

11.25. Harris and coworkers [*Inorg. Chem.*, **6**, 1052 (1967)] studied the kinetics of the following reaction in mixtures of $H_2{}^{18}O$ and H_2SO_4.

$$(NH_3)_5CoNO_2^{2+} + 2H^+ \xrightarrow[H_2SO_4]{H_2^{18}O} (NH_3)_5CoH_2O^{3+} + NO^+$$

The reaction is first order in H^+, and the product contains no appreciable $H_2{}^{18}0$. Postulate a mechanism for the reaction, explaining the significance of the experimental observations.

11.26. Meyer, T. J., and Taube, H., [*Inorg. Chem.*, **7**, 2369 (1968)] studied the kinetics of the following redox reaction:

$$Ru(NH_3)_6^{2+} + Ru(ND_3)_6^{3+} \leftrightarrows Ru(NH_3)_6^{3+} + Ru(ND_3)_6^{2+}$$

Although the Ru^{2+} and Ru^{3+} complexes are inert, the self-exchange rate constant for the preceding reaction has a value near 800/mole sec at 25°C. The presence of low concentrations of sulfate ion enhances the reaction. Does the reaction occur by an inner-sphere or outer-sphere mechanism? Why? Explain the effect of added sulfate ion.

11.27. Moore, P., Basolo, F., and Pearson, R. G., [*Inorg. Chem.*, **5**, 223 (1966)] studied the kinetics of the following reaction:

$$Cr(H_20)_5I^{2+} + 2H_2{}^{18}0 \xrightarrow{H^+} Cr(H_20)_4(H_2{}^{18}0)_2{}^{3+} + I^-$$

They observed that exchange of one water molecule occurs prior to the acid hydrolysis. In contrast, $Cr(H_20)_5Cl^{2+}$ undergoes water exchange very slowly. Explain the two observations, giving the structure of the complex $Cr(H_20)_4(H_2{}^{18}0)I^{2+}$.

11.28. N_2O_4 is isoelectronic with oxalate ion. When oxalate ion reacts with protons, the following is postulated to take place:

$$CO + CO_3^{2-} + 2H^+ \rightarrow CO + CO_2 + H_2O$$

Predict the products where N_2O_4 reacts with H^+ ions.

11.29. Parry, R. W., [*Inorg. Chem.*, **6**, 819, (1967)] said that the following reaction would not be expected to occur, and then proved it experimentally.

$$BF_3 + K_2H_3BCO_2 \rightarrow H_3B + K_2F_3BCO_2$$

Explain in some detail why this would not be expected. He titrated the $K_2H_3BCO_2$ with HCl and two equivalence points were noted. What happened and why?

PART III

The Behavior of Atoms and Molecules

B. The Studied Behavior of Selected Elements

CHAPTER 12

The Hydrides

The hydrogen atom is an atom consisting in a single proton and an electron. In the ground state, the lone electron is in the 1*s* orbital of the hydrogen atom. The 1*s* orbital has a maximum capacity of two electrons; this orbital may also contain no electrons. These possibilities define the chemical activity of hydrogen. When the hydrogen atom has two electrons, it becomes the hydride ion, H^-. The chemistry of the hydride ion may be summarized by saying that it is a Lewis base. The hydrogen ion, H^+, on the other hand, possesses no electrons at all, and its behavior may be summarized by saying that it is a Lewis acid. The hydrogen atom may simply share its valence electron with another atom that contributes an electron, thus forming a simple covalent bond.

Hydrogen has been placed in several positions in the periodic table, usually in Group I-A because of its lone valence electron. Occasionally, it has been placed in Group VII-A because of its ability to form a uninegative ion. Some even have positioned it near Group IV, because its valence shell is half-filled and it has an electronegativity of 2.1., similar to that of carbon.

There are three isotopes of hydrogen. The first is protium, ${}^1_1H$, or normal hydrogen. The second is deuterium, ${}^2_1H$, and the third is tritium, ${}^3_1H$. They are distinguishable by the number of neutrons in their nuclei, and hence by their masses. In a sample of naturally occurring hydrogen, 0.015% of the mass is due to deuterium and tritium occurs in 1 part per 10^{17} parts of hydrogen.

Deuterium is prepared by the electrolysis of water. Since deuterium is twice as heavy as protium, it is much slower to react (isotope effect).

Consequently, near the completion of the electrolysis, the solution is richer in deuterium than it was at the beginning. Needless to say, this type of fractionation is necessarily carried out many times to obtain a good concentration of deuterium.

The chemist's interest in deuterium lies in the synthesis of deuterated compounds. These compounds may be used for tracer studies of exchange reactions because the deuterium can be followed by physical methods. This following of the deuterium atoms during the course of a reaction gives insights into the structure of molecules and the mechanisms of chemical reactions.

Radioactive tritium has a half-life of 12.4 years. It is generally prepared in nuclear reactors and it decomposes in the following manner:

$$^{3}_{1}H \longrightarrow {}^{3}_{2}He + \beta \qquad [12.1]$$

Tritium is used in tracer studies, even though its soft beta rays are very hard to detect. Being three times heavier than hydrogen, it will have a pronounced isotope effect.

In its native state, under normal conditions, elementary hydrogen exists as a diatomic molecule, H_2, in the gaseous state. Here, two atoms are bound together by sharing a pair of electrons, exemplifying the "ideal" covalent bond.

Two types of molecular hydrogen have been distinguished in terms of nuclear spins. Ortho-hydrogen is hydrogen in which the two nuclei in the hydrogen molecule have parallel spins; para-hydrogen is hydrogen in which the two nuclei have opposed spins. The para type of hydrogen molecule seems to predominate at lower temperatures, whereas the ortho form predominates at higher temperatures. The ortho/para ratio has a maximum value of $\frac{3}{4}$ in any given sample of hydrogen gas. Homonuclear diatomic molecules possess this ortho-para property, but the ortho/para maximum ratios differ for each species.

Hydrogen is prepared by several methods that should already be familiar to the student: electrolysis of water, the reaction of metals above hydrogen in the electromotive series with dilute mineral acids, and the reaction of aluminum and zinc with strong aqueous bases.

In general, hydrogen forms binary compounds with almost every element of the periodic table. Admittedly, some of these compounds exhibit a pronounced instability, decomposing even at low temperatures. For the sake of discussion, all binary hydrogen compounds will be called *hydrides*. Technically, the term hydride connotes that the hydrogen atom exists in the compound as the more negative of the two species—namely,

H^-. In this discussion, HF will be considered the hydride of fluorine as well as the fluoride of hydrogen. This extension of the meaning of hydride allows one to distinguish among several classes of hydrides relative to their position in the periodic table (see Fig. 12.1). The classes of hydrides are as

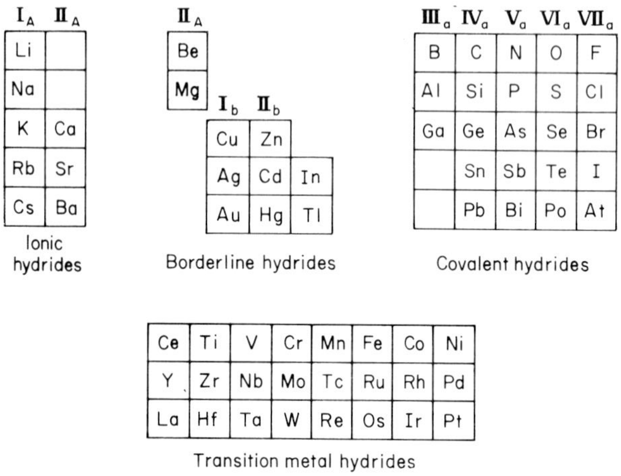

Figure 12.1. Classification of the binary hydrides as to type and location in the periodic table.

follows: (1) saltlike (saline) or ionic hydrides; (2) covalent hydrides; (3) borderline hydrides; (4) transition metal hydrides, interstitial hydrides, or simply metal hydrides. No hydrides of the noble gases have been reported.

THE IONIC HYDRIDES

Ionic hydrides are formed by the alkali metals and the alkaline earth metals (Groups I-A and II-A of the periodic table). They may be prepared by direct union of the elements with the help of heat in the 150 to 700° range. These species may be represented as M^+H^- or M^{2+} $2H^-$, where M^+ may be Li^+, Na^+, K^+, Rb^+ or Cs^+, and M^{2+} may be Ca^{2+}, Sr^{2+}, or Ba^{2+}.

The ionic hydrides have properties that are characteristic of an ionic-type compound. The hydrogen atom has a negative charge because it takes an electron from an element more electropositive than itself. These compounds are crystalline, have high melting points, conduct electricity in the molten state, yield H_2 at the anode with equivalent amounts of the metal at the cathode when the molten salts are electrolyzed, and are insoluble in liquid ammonia and organic solvents. Lithium hydride has relatively great solubility in organic solvents, indicating a great degree of covalency in the Li—H bond.

Stability of the hydrides depends on several factors. The Born-Haber treatment is usually presented to illustrate this fact (see Fig. 12.2). The variable terms in this consideration are S, the heat of sublimation of the metal; IP, the ionization potential of the metal; D, the bond dissociation

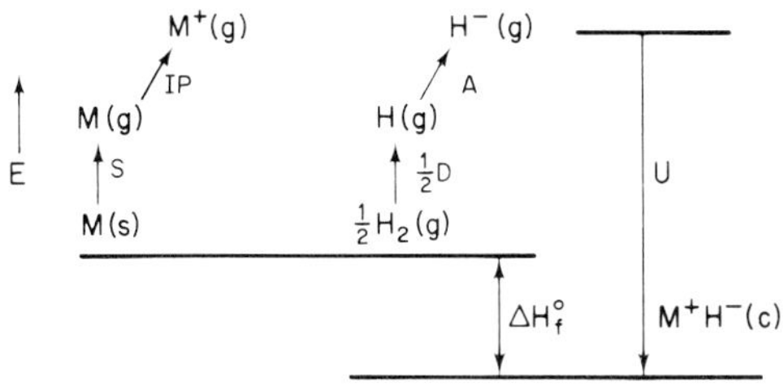

Figure 12.2. Born-Haber treatment of the energetics of saline hydride formation.

energy of hydrogen; A, the electron affinity of hydrogen; U, the crystal lattice energy; ΔH_f°, the standard heat of formation. From Fig. 12.2, we can see that $\Delta H_f^\circ = S + \frac{1}{2}D + A + U$. The preceding diagram shows the reaction to the exothermic only if U is large. The student should convince himself that A and U will be negative terms relative to the other terms in this presentation. The value of the crystal lattice energy is much larger numerically than the electron affinity, and one can say broadly that this term determines whether the reaction is exothermic.

The student should realize that the enthalpy term ΔH_f°, some of which are given in Table 12.1, is not the ultimate criterion used in predicting whether a system is thermodynamically favorable or not. The free energy, ΔG_f°, is the important factor. From the definition of entropy and the usual considerations of elementary thermodynamics,

$$\Delta G^\circ = \Delta H^\circ - T\Delta S^\circ \qquad [12.2]$$

In the preceding discussion, ΔS° is assumed to be negligible, and hence $\Delta G^\circ = \Delta H^\circ$. Also note that as a periodic group is descended, the thermal stability of the hydrides decreases.

TABLE 12.1 ***Some Standard Enthalpies of Formation Values (ΔH_F°) for Selected Binary Hydrides (kcal/mole)***

LiH	−21.6			B_2H_6	7.5	CH_4	−17.9	NH_3	−11.0	H_2O	−68.32	HF	−64.8
NaH	−13.7	MgH_2	−18.2			SiH_4	−11.2	PH_3	5.5	H_2S	−4.93	HCl	−22.06
KH	−13.6	CaH_2	−45.2	Ga_2H_6	−21.4	GeH_4	65	AsH_3	15.9	SeH_2	7.1	HBr	−8.66
RbH	−12	SrH_2	−42.4			SnH_4	20	SbH_3	34.7	TeH_2	23.8	HI	6.33
CsH	−19.9	BaH_2	−41.0										

Lithium hydride, calcium hydride, and strontium hydride are relatively stable in dry air, but some of the other hydrides may ignite spontaneously with dry air. Users of these compounds should treat them with great respect.

The chemistry of the alkali metal hydrides can be summarized as the chemistry of the hydride ion, H^-. The H^- has no existence in aqueous solution because it is a stronger base than OH^- and can take a proton away from the water molecule, leaving an OH^- ion.

$$H:^- + \overset{\delta^-}{H}-\overset{\delta^-}{O}H \longrightarrow H_2 + OH^- \quad [12.3]$$

Generally speaking, the H^- is a very strong Lewis base, and it will displace weaker bases by nucleophilic attack on a given acid-base adduct. The hydride ion will also attack acidic sites in molecules:

$$H:^- + O{=}C{=}O \rightarrow O{=}\underset{}{\overset{\displaystyle H \atop |}{C}}{-}O \quad [12.4]$$

$$2H:^- + 2BH_3 \text{ or } B_2H_6 \rightarrow 2BH_4^- \quad [12.5]$$

Finally, the alkali metal hydrides may act as reducing agents:

$$2NaH + \tfrac{1}{8}S_8 \rightarrow Na_2S + H_2 \quad [12.6]$$

THE COVALENT HYDRIDES (MOLECULAR HYDRIDES)

The covalent hydrides are also called molecular hydrides, because they generally exist as discrete molecules. Physically, they are distinct from other hydrides because they are usually gases or volatile liquids under normal conditions. The bonds are not strikingly polar; hence, electron sharing must play a predominant role in all these compounds. In general, as the molecular weight of these compounds increases, their boiling points and melting points increase. The notable exceptions to this general observation will be discussed with the concept of hydrogen bonding.

The molecular hydrides may be prepared by several methods. Direct combination of hydrogen with highly electronegative elements, such as fluorine, chlorine, oxygen, nitrogen, bromine, and iodine, is a relatively efficient method. Hydrides of carbon, silicon, phosphorus, arsenic, and selenium may be obtained by direct combination, but the reaction is either so slow or the yield so small that the reaction is considered to have little importance. Another method of preparation of the molecular hydrides consists in the hydrolysis of a *nonmetalide*. A *nonmetalide* is a binary compound of a metal and the element that one desires to convert to a hydride. Occasionally, acid is necessary to assist hydrolysis, e.g.,

$$Mg_2Si + H_2O \longrightarrow SiH_4 + \text{higher silanes} + Mg(OH)_2 \quad [12.7]$$

$$Mg_3N_2 + 6H_2O \longrightarrow 2NH_3 + 3Mg(OH)_2 \quad [12.8]$$

$$CaC_2 + 2H_2O \longrightarrow C_2H_2 + Ca(OH)_2 \quad [12.9]$$

$$FeS + 2HCl(aq) \longrightarrow H_2S + FeCl_2 \quad [12.10]$$

These methods are very useful in preparing deuterium and tritium derivatives of many elements.

Small yields of some hydrides can be prepared by the electrolysis of an acidic solution, where the metal of the desired hydride to be formed is set up as the cathode. Stibine (SbH_3) and bismuthine (BiH_3) are prepared in this manner.

Covalent hydrides may also be prepared by metathetical or reduction reactions, using ionic or complex hydrides.

$$Al_2Cl_6 + LiAlH_4 \xrightarrow{\text{ether}} (AlH_3)_x + H_2 + LiCl \quad [12.11]$$

$$SiCl_4 + LiAlH_4 \xrightarrow{\text{ether}} SiH_4 + SiH_3Cl + SiH_2Cl_2 + SiHCl_3 + AlCl_3 + LiCl \quad [12.12]$$

$$2Al + 3H_2 + 2BCl_3 \longrightarrow B_2H_6 + 2AlCl_3 \quad [12.13]$$

A few of the covalent hydrides exhibit the property of *catenation*, that is, the nonmetal atoms (other than hydrogen) are joined together by covalent bonds, as in $H_3C—CH_3$ and $H_3C—CH_2—CH_3$. Among the hydrides, this phenomenon is supremely exemplified by carbon in its various homologous series. More will be said about this property as the various periodic groups are discussed.

The covalent hydrides are susceptible to attack by oxygen, halogens, and water. This property decreases with an increase in electronegativity of the parent element—e.g., arsine (AsH_3) is more easily oxidized than ammonia. Caution must be used when employing this generalization, because although silane is more easily halogenated than methane, methane is less easily hydrolyzed than alane.

The strength of a covalent hydride as a reducing agent increases as the group is descended:

$$Ag^+ + NH_3 \longrightarrow Ag(NH_3)_2^+ \quad [12.14]$$

$$8Ag^+ + AsH_3 + 3H_2O \longrightarrow AsO_3^- + 8Ag^\circ + 6H^+ \quad [12.15]$$

From these equations, we see that ammonia will not reduce silver, but arsine will.

The strength of a covalent hydride as a reducing agent tends to decrease on traversing a period from left to right. Thus, hydrogen sulfide will reduce ferric ion to ferrous ion, but hydrogen chloride will not.

$$2Fe^{3+} + H_2S \rightarrow Fe^{2+} + 2H^+ + S \qquad [12.16]$$

The rationalization of many trends in this general introductory section may be based on several facts, e.g., atomic size, orbital overlap, and electronegativity.

These thoughts finish the introductory section on ionic and covalent hydrides. Note that these two types of hydrides comprise most of the binary compounds of the representative elements with hydrogen. At this point, we shall interrupt the general discussion to look at the hydrides of the groups of representative elements in more detail.

Group III hydrides

These include

$(BH_3)_2$ or B_2H_6	diborane
$(AlH_3)_x$	alane
$(GaH_3)_2$ or Ga_2H_6	digallane

Diborane (6), B_2H_6, is the simplest boron hydride. Many other boron hydrides (boranes) exist, and they follow two general compositions:

(1) The *normal boranes*, B_nH_{n+4}, which are relatively stable thermally and not always spontaneously inflammable in air; and (2) The *dihydroboranes*, B_nH_{n+6}, which are much less stable thermally than are the normal boranes (all spontaneously inflammable in air).

Diborane can be prepared by the reduction of a boron halide.

$$LiAlH_4 + BF_3 \xrightarrow{\text{ether}} LiBH_4 + AlF \qquad [12.17]$$

$$3LiBH_4 + BF_3 \longrightarrow 2B_2H_6 + 3LiF \qquad [12.18]$$

Although the formula of diborane is B_2H_6, diborane acts as a Lewis acid, and for all practical purposes we may write it as $2BH_3$. Actually, at elevated temperatures, a tiny amount of BH_3 is in equilibrium with its dimer, B_2H_6.

$$B_2H_6 \rightleftarrows 2BH_3 \qquad [12.19]$$
$$(K = 1.63 \times 10^{-5} \text{ at } 155^\circ\text{C, 1 atm})$$

Therefore, the principle behind reactions (12.17) and (12.18) is that BH_3 is a Lewis acid reacting with a Lewis base, H^-, to form the complex BH_4^-. Since BF_3 is a stronger acid than BH_3, it replaces BH_3 and $LiBH_4$ to produce B_2H_6:

$$3H^- + BF_3 \longrightarrow BH_3 + 3F^- \qquad [12.20a]$$

$$H^- + BH_3 \longrightarrow BH_4^- \qquad [12.20b]$$

$$BF_3 + BH_4^- \longrightarrow BH_3 + BF_3H^- \qquad [12.20c]$$

$$2BH_3 \longrightarrow B_2H_6 \qquad [12.20d]$$

Diborane is generally the source of the other boron hydrides. Figure 12.3 shows the relative interconversions. It also shows why the higher boron hydrides are referred to as condensation products of diborane.

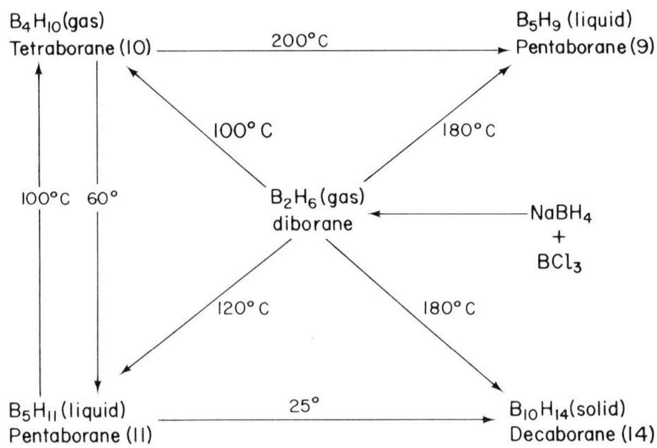

Figure 12.3. Boron hydride interconversions. Adapted from Massey, A. G., Scientific American, **210,** 88 (1964).

The structures of these species are best pictured in relation to the structure of elemental boron. The β-rhombohedral form of boron is a system of icosahedral units. Using the iscosahedron as the fundamental unit, we may picture the structures of the boron hydrides as fragments of icosahedra. The student should convince himself of this fact (see Fig. 12.4).

The structures of the boron hydrides pictured in Fig. 12.4 would certainly be difficult for the ordinary chemist to draw. Lipscomb has developed an empirical method of representing the boron hydride structures.

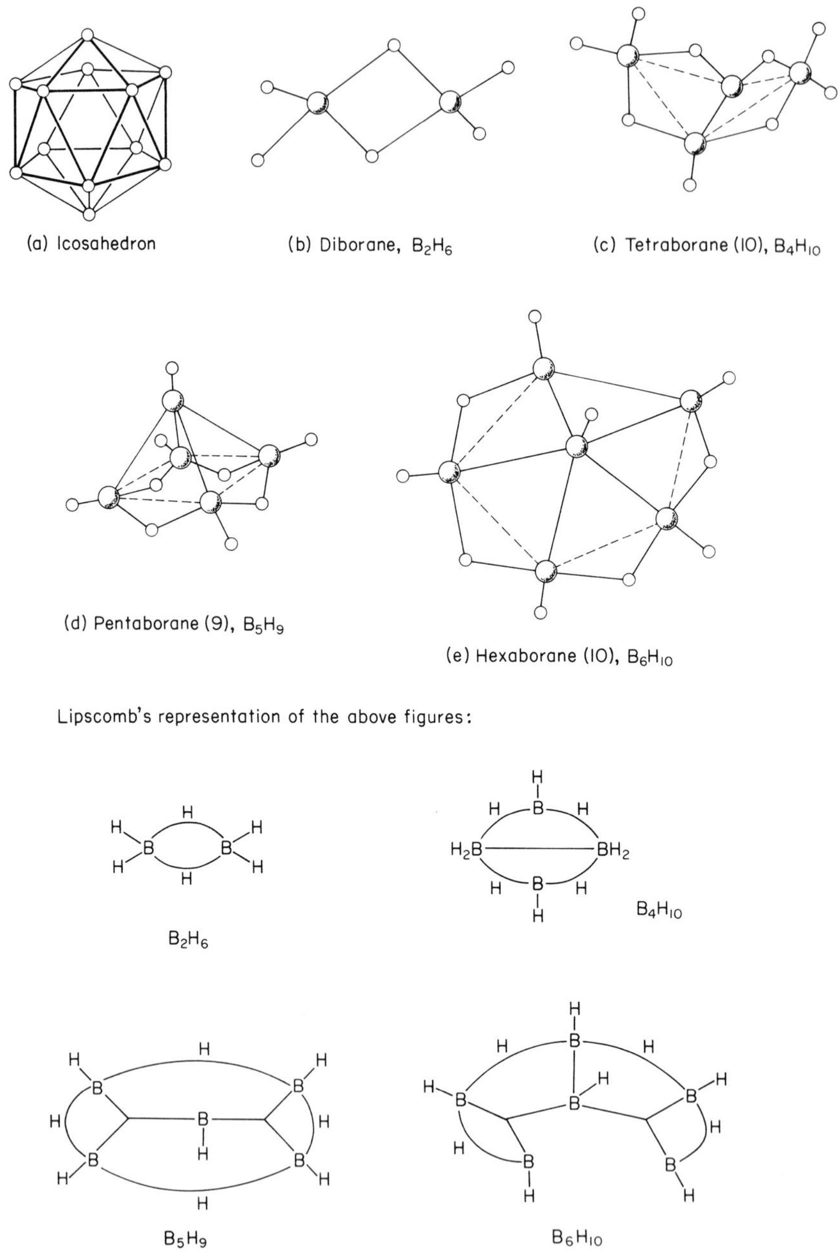

Figure 12.4. Various representations of some of the boron hydrides.

The symbols used in his approach are

B—H	terminal, 2-center, 2-electron, boron-hydrogen bond
B—B	2-center, 2-electron, boron-boron bond
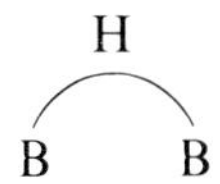	3-center, 2-electron, hydrogen-bridging bond
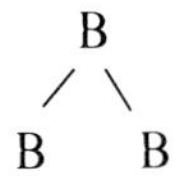	3-center, 2-electron, boron-bridging bond (open type)
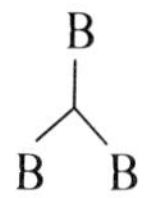	3-center, 2-electron, boron bond (closed type)

The definitions following the preceding symbols have not been explained; it will be done presently.

The structure of diborane is rather simple in the geometrical sense. Six of the eight atoms lie in a plane; the two boron atoms are at a given distance from each other, each having two hydrogen atoms attached to them with the H—B—H angle approximately 120°. There are two *bridging* hydrogen atoms, one above and one below the plane mentioned, holding the two boron atoms together. There are 8 *atoms* in this compound held together by 8 *bonds*, but there are only 12 electrons available for bonding. A system such as this is called *electron deficient*, because there is an insufficient number of electrons to account for all the bonds in terms of the two-electron covalent bond. The very fact that B_2H_6 has considerable stability indicates that there must be enough electrons present to hold the species together. The preceding definitions clarify the concept of the three-center bond and distinguish it from the normal two-electron two-center bond. This type of bonding has been discussed in the presentation of the molecular orbital theory.

Table 12.2 lists some properties of selected boron hydrides, and Table 12.3 organizes some of the various relationships concerning bonding in the boron hydrides, which have always been a challenge to theories of bonding. Three types of three-centered bonds have been postulated to explain the structure of the boron hydrides: (1) B—H—B bridge type; (2) B—B—B central three-center bond; and (3) B—B—B open three-center bond (see Fig. 12.5).

All boron hydrides react spontaneously with the free halogens, the vigor of the reaction decreasing, as might be expected, from fluorine to iodine. The higher boranes react less readily with halogens than the lower boranes.

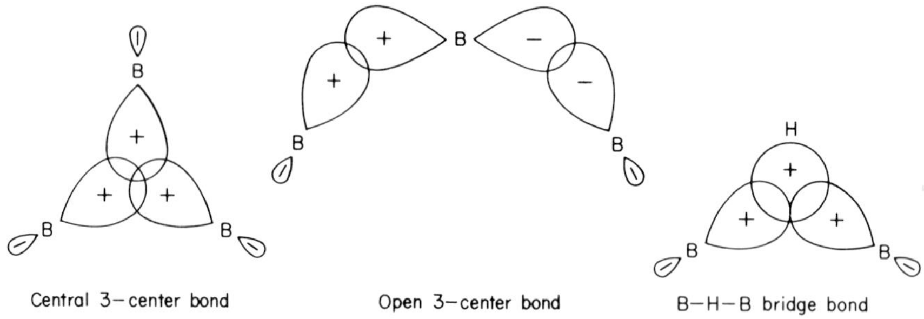

Figure 12.5. Diagrams representing orbital overlap in various three-center bonds.

With fluorine and chlorine, the only halogenated products formed are BF_3 and BCl_3. However, with bromine and iodine, partly halogenated products such as B_2H_5Br can be obtained.

Partly halogenated products can be synthesized much more conveniently

TABLE 12.2 ***Some Properties of the Boron Hydrides***

Formula	Name	B.P. (°C)	Reaction with Air	Reaction with Water
B_2H_6	Diborane (6)	−92.5	Spontaneously flammable	Instant hydrolysis
B_4H_{10}	Tetraborane (10)	18.0	Spontaneously flammable if impure	Instant hydrolysis
B_5H_9	Pentaborane (9)	62.0	Spontaneously flammable	Hydrolyzes on heating
B_5H_{11}	Pentaborane (11)	63.0	Spontaneously flammable	Hydrolyzes rapidly
B_6H_{10}	Hexaborane (10)		Stable	Hydrolyzes on heating
B_6H_{12}	Hexaborane (12)			
B_9H_{15}	Nonaborane (15)		Stable	
$B_{10}H_{14}$	Decaborane (14)	219	Very stable	Slow hydrolysis

TABLE 12.3 ***Number of Three-Centered Bonds in Some Boron Hydrides***

Hydride	No. of Atomic Orbitals	No. of Valence Electrons	No. of Orbitals in Excess	No. of Three-Centered Bonds
B_2H_6	14	12	2	2
B_4H_{10}	26	22	4	4
B_5H_9	29	24	5	5
$B_{10}H_{14}$	54	44	10	10

from the hydrogen halide in the presence of the appropriate anhydrous aluminum halide catalyst.

$$B_2H_6 + HCl \xrightarrow[\text{heat}]{Al_2Cl_6} B_2H_5Cl + H_2 \qquad [12.21]$$

The partly halogenated derivatives disproportionate readily, viz.,

$$6B_2H_5Cl \longrightarrow 5B_2H_6 + 2BCl_3 \qquad [12.22]$$

The rate of hydrolysis decreases on going from the lower to the higher boranes. Thus, B_2H_6 hydrolyzes in a few seconds, whereas $B_{10}H_{14}$ hydrolyzes only to the extent of 10% in 10 days:

$$B_2H_6 + 6H_2O \longrightarrow 2B(OH)_3 + 6H_2 \qquad [12.23]$$

Since most substances containing labile hydrogen tend to act as reducing agents, it is not surprising to find that the boron hydrides and their complex salts, such as $NaBH_4$, act as strong reducing agents. Diborane will reduce many metal ions to the metal when passed through solutions of their salts. Boron hydrides and their derivatives can act as Lewis acids and form coordination compounds with many Lewis bases:

		1s	2s		2p	
BL_3	boron	··	·×	·×	·×	∘∘
			L	L	L	

where the dots represent the boron electrons, the ×'s the electrons from L and ∘∘ the lone pair electrons from the Lewis base.
Ammonia forms white, nonvolatile solids with boron hydrides:

$$B_2H_6 + NH_3 \xrightarrow{-78^\circ} B_2H_6(NH_3)_2 \qquad [\text{possibly, } BH_2(NH_3)_2^+ BH_4^-] \qquad [12.24]$$

A compound $(HBNH)_3$ is formed on heating this diborane adduct, viz.,

$$3B_2H_6(NH_3)_2 \xrightarrow{200^\circ} 2(HBNH)_3 + 12H_2 \qquad [12.25]$$

The compound $(HBNH)_3$ is called borazole or borazine, and is frequently referred to as "inorganic benzene" because its structure is similar to that of benzene (six-membered B—N ring). Some of its physical properties also resemble those of benzene. The relationship ends there, because there is little delocalization in the ring. This statement is supported by the fact that borazole hydrolyzes readily and also reacts with 3 moles of alcohol. This type of reaction is characteristic of material containing bond polarity.

The higher boranes, specifically $B_{10}H_{14}$, also act as Lewis acids:

$$B_{10}H_{14} + 2L(\text{ligands}) \longrightarrow B_{10}H_{12} \cdot 2L + H_2 \qquad [12.26]$$

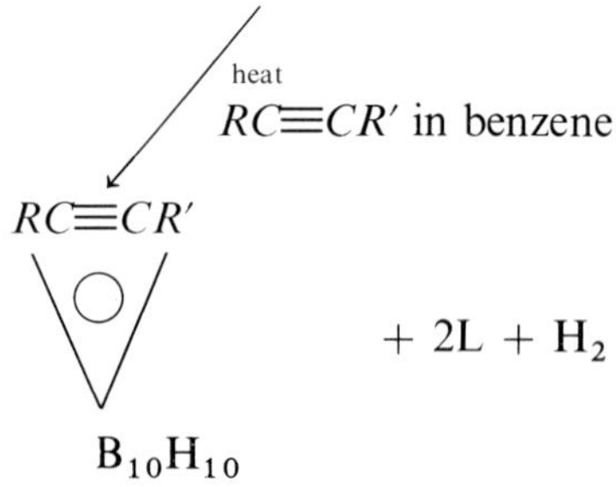

This type of compound is called a *carborane*. If one considers the 2 carbon atoms and the 10 boron atoms placed at the vertices of a polyhedron, the polyhedron defined would be the icosahedron. If the unit $RC{\equiv}CR'$ were acetylene, the reaction product would be $C_2B_{10}H_{12}$. The carboranes possess remarkable thermal stability (see Fig. 12.6).

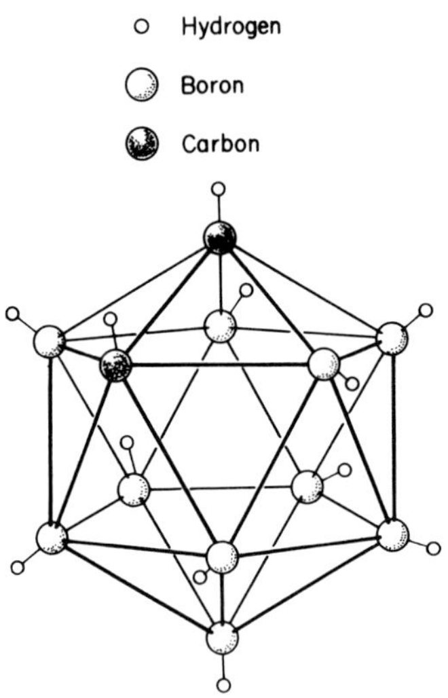

$B_{10}H_{14}C_2$

Figure 12.6. The carborane "nucleus."

An interesting derivative of the carborane family is the $B_9C_2H_{11}^{2-}$ anion. Compounds of this anion react with metal ions, and the boron, carbon, and metal ions define an icosahedron. The anion itself would have an open pentagonal face described by two carbon atoms and three boron atoms, with six electrons in delocalized orbitals directed to the open vertex of an icosahedron. The anion could be considered analogous to the cyclopentadienyl anion, $C_5H_5^-$. The compound $C_5H_5Fe(B_9C_2H_{11})$ has been reported; its structure, shown in Fig. 12.7, has been postulated.

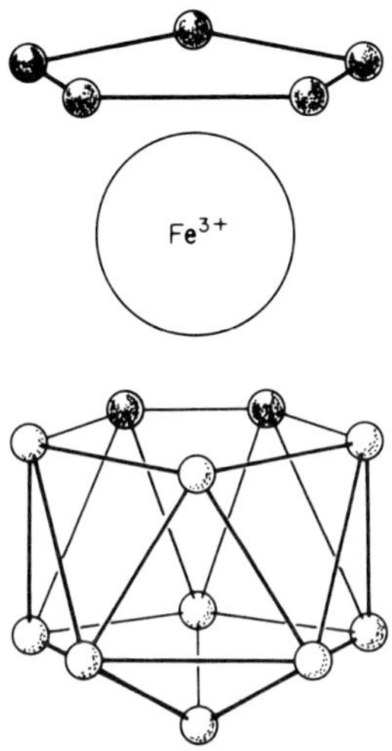

Figure 12.7. The proposed structure of $C_5H_5Fe(B_9C_2H_{11})$. The hydrogen atoms have been omitted for the sake of clarity.

Aluminum hydride, or *alane*, is a stoichiometric white solid, insoluble in all solvents. A convenient method of preparation of alane involves mixing ethereal solutions of LiH and $AlCl_3$, whereupon AlH_3 is formed as its ether complex:

$$3LiH + AlCl_3 \longrightarrow (AlH_3)_x + 3LiCl \qquad [12.27]$$

If an excess of LiH is used, then $LiAlH_4$ results:

$$LiH + (AlH_3)_x \longrightarrow LiAlH_4 \qquad [12.28]$$

The compound $LiAlH_4$ is a very convenient reducing agent because of its solubility in many organic ethers. Alane decomposes into its elements at approximately 100°C. It hydrolyzes explosively to yield aluminum hydroxide and hydrogen gas. It is a potent reducing agent. Alane also acts as a Lewis acid and its ammonia adducts are known.

Gallium hydride or *digallane*, Ga_2H_6, is a liquid at room temperature and has a boiling point of 139°C. If hydrogen and the vapor of $(CH_3)_3Ga$ are subjected to glow discharge, one of the products formed is $(CH_3)_4Ga_2H_2$. This acid, $(CH_3)_4Ga_2H_2$, reacts with triethylamine to give digallane and the adduct $(CH_3)_3Ga{:}N(C_2H_5)_3$. Lithium gallium hydride, $LiGaH_4$, has also been reported.

Hydrides of Group IV

Paraffins, olefins, acetylenes, aromatic hydrocarbons, etc., are not directly related to this text.

Si_nH_{n+2} from SiH_4 to Si_8H_{18}	silanes
GeH_4, Ge_2H_6 . . . Ge_9H_{20}	germanes
SnH_4	stannane
PbH_4	plumbane

The silicon hydrides and their derivatives may be regarded as comprising a field of *inorganic* organic chemistry, since a large number of derivatives of the silicon hydrides are analogous to simple organic compounds that have been prepared and studied, e.g., $(SiH_3)_3N$, silyl amine, and $(Si_2H_5)_2O$, bis-disilyl ether.

The preparation of the silicon hydrides or silanes may be accomplished by methods already mentioned. The silanes are all colorless gases or liquids with the exception of the solid, nonvolatile hydrides of compositions approaching $(SiH)_x$ or $(SiH_2)_y$. These species may be prepared by heating silane, SiH_4, or disilane, Si_2H_6. The polymeric hydrides are believed to consist in networks of Si—Si bonds.

Unlike the boron hydrides, the volatile silanes decrease in thermal stability on going toward the higher members. Thus, silane and disilane remain unchanged for several years at room temperature, whereas n-Si_4H_{10} decomposes to the extent of 20% after standing for several months at room temperature.

The silanes are spontaneously inflammable in air, but they are not affected by pure water or by slightly acid solutions. However, traces of alkali in water cause rapid hydrolysis:

$$SiH_4 + 2H_2O \xrightarrow{OH^-} SiO_2 + 4H_2 \qquad [12.29]$$

Free halogens readily attack the Si—H bond to give partially or completely halogentated materials. As with the boranes, the vigor of the reaction decreases on passing from fluorine to iodine.

Analogously with the boranes, halogens may be introduced into the silanes in a more controllable fashion by using an anhydrous hydrogen halide together with anhydrous aluminum halide catalyst at room temperature, or slightly above it. Usually, a mixture of all substituents is obtained, i.e.,

$$SiH_4 + HCl \xrightarrow{Al_2Cl_6} SiH_3Cl, SiH_2Cl_2, SiHCl_3, SiCl_4 + H_2 \qquad [12.30]$$

$$Si_2H_6 + HCl \xrightarrow{Al_2Cl_6} Si_2H_5Cl, Si_2H_4Cl_2, Si_2H_3Cl_3 \ldots + H_2 \qquad [12.31]$$

Pseudohalogenated silanes and polysilanes may be prepared in the following manner:

$$SiH_3I + AgCN \longrightarrow SiH_3CN + AgI \quad [12.32]$$

Both the halides and pseudohalides of silicon hydrides are susceptible to attack by water to form ethers.

$$2SiH_3CN + H_2O \longrightarrow (SiH_3)_2O + 2HCN \quad [12.33]$$

Sulfides and selenides of the silicon hydrides may be formed by the action of silver selenide or mercury(II) sulfide on the iodide of a given silicon hydride—i.e.,

$$2SiH_3I + HgS \longrightarrow (SiH_3)_2S + HgI_2 \quad [12.34]$$

From the relative electronegativity values of silicon (1.8) and carbon (2.5), it might be expected that $(SiH_3)_2O$ would be a stronger Lewis base than $(CH_3)_2O$. Experimentally, the reverse has been found to be true. Rationalization of this fact is based on the concept of back bonding or $(p \rightarrow d)$ π bonding. That is, the lone pairs of electrons on the oxygen atom are *back donated* to the low-lying, empty *d* orbitals of the silicon atom, adding double bond character to the Si—O bond with a decrease in the bond length. Experimental evidence cited for this explanation is that the Si—O—Si bond angle is greater than 155° in the preceding silyl ether, whereas the C—O—C bond angle is only 111°. Another piece of evidence is that the dipole moment of $(SiH_3)_2O$ is only 0.24 D. Therefore, $(p \rightarrow d)$ π bonding possesses a great deal of merit in this argument, but there remains no direct evidence for this type of rationalization.

The amines, phosphines, and arsines of the silicon hydrides are known. For example, $(SiH_3)_3N$, $(SiH_3)_3P$, and $(SiH_3)_3As$ have been reported. The silicon-halogen bond in simple silicon hydride derivatives reacts instantaneously (at or below room temperature) with N—H bonds in NH_3 or amines to give the corresponding silyl derivative:

$$3SiH_3Cl + 4NH_3 \longrightarrow (SiH_3)_3N + 3NH_4Cl(CH_3)_2NH_2^+ \quad [12.35]$$

$$SiH_3I + 2(CH_3)_2NH \longrightarrow (CH_3)_2NSiH_3 \quad [12.36]$$

However, if a tertiary amine is used, an ionic quaternary ammonium salt is formed:

$$SiH_3I + (CH_3)_3N \longrightarrow [SiH_3N(CH_3)_3^+]I^- \quad [12.37]$$

As might be expected from the preceding discussion, the base strength of the mixed silyl methylamines is just the opposite of what would be expected from electronegativity considerations:

$$\underrightarrow{(SiH_3)_3N < (SiH_3)_2NCH_3 < SiH_3N(CH_3)_2 < :N(CH_3)_3}$$
base strength increase

The silicon analogs of organometallic compounds are also known, such as $KSiH_3$ and KSi_2H_5. Potassium silyl may be prepared from disilane:

$$Si_2H_6 + KH \longrightarrow KSiH_3 + SiH_4 \qquad [12.38]$$

Potassium silyl reacts quantitatively with methyl iodide to form methyl silane and potassium iodide.

The properties of the silicon hydrides and their derivatives may be summarized as follows:

(1) They have the same types of formula as their carbon analogs, e.g., SiH_3Cl and CH_3Cl, etc.
(2) They may have the same or different molecular structure as the corresponding hydrocarbon; SiH_3NCS is linear, whereas CH_3NCS is angular.
(3) In comparable environments, they have somewhat less thermodynamic stability than their carbon analogs.
(4) They react more rapidly than their carbon analogs because (a) silicon is larger than carbon; (b) silicon can increase its coordination number to more than four because of vacant *d* orbitals, which carbon cannot do. Consequently, intermediate reaction complexes can be formed more readily with silicon intermediates.

The Germanes or Germanium Hydrides. The properties of the germanes and their derivatives are usually similar but by no means identical with those of the corresponding silanes and their derivatives. However, not as many divalent silicon compounds are known as divalent germanium compounds. This fact supports the trend that as a nontransitional group of the periodic table is descended, the lower valence state becomes more and more stable with respect to conversion to the higher valence state.

A mixture of the germanes is obtained by the hydrolysis of magnesium germanide or by the action of electrical discharge on germane. Low yields of germane are obtained from the reaction of germanium tetrachloride with lithium aluminum hydride, presumably because germanium tetrachloride is itself reduced to the divalent hydride, GeH_2. Sodium borohydride, a milder reducing agent, gives better yields of germane when it acts on germanium tetrachloride. The best preparation of germane is based on germane's stability in water. Germanium dioxide is sprinkled into a water solution of sodium borohydride, and this gives a fine yield of germane.

Germane is not spontaneously inflammable in air, and it is not attacked even by moderately strong aqueous alkali. This is reasonable if one considers the reaction mechanism to involve a rate-controlling step of coordination:

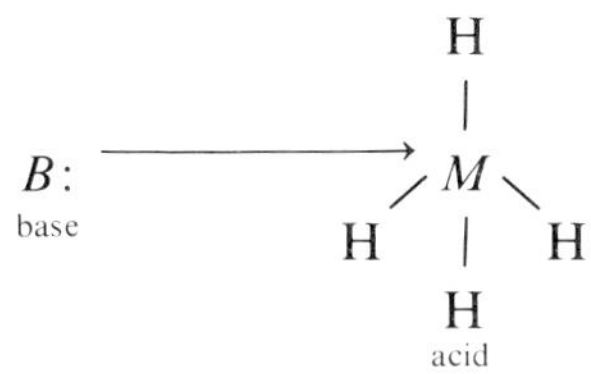

Also, since the electronegativity of germanium is reportedly larger than silicon, one cannot rely on the electronegativity concept for the explanation. At least qualitatively, we can state that the slower reaction rates are due to a lesser tendency to form a reaction intermediate on the part of germanium. In an analogous consideration, we might note that phosphine is spontaneously flammable in air but arsine is not.

If we wish to generalize the preceding concepts, we should stress that steric factors must be considered. The preceding argument applies only when the ligands attached to the central metal atom are small and the base, *B*, is small. If one examines the relative strengths of Lewis acids, the MCl_4 species ($R_{Cl} >> R_H$), then $GeCl_4$ is a stronger Lewis acid than $SiCl_4$. In this case, it is much more difficult for base *B* to attach itself to silicon, since the four atoms shield the silicon considerably. This effect is also observable in other Lewis acids. Trimethyl aluminum dimer, $[Al(CH_3)_3]_2$, for example, is usually a stronger Lewis acid than trimethyl boron. Using electronegativity concepts, we might expect the reverse.

Germyl compounds, compounds containing the $—GeH_3$ group, analogous to methyl compounds, have been synthesized by similar methods as those used for preparing silyl compounds. Among the species that have been isolated are GeH_3I, GeH_3CN, and $(GeH_3)_2S$. Mixed germyl and silyl compounds, such as H_3GeSiH_3, have also been reported. The halide derivatives of the hydrides of germanium do not form amines with NH_3, but complicated decomposition occurs to give elemental germanium.

In the basic solvent, liquid ammonia, the acidic properties of the Ge—H bond are enhanced, and the following reaction occurs:

$$GeH_4 + Na \longrightarrow NaGeH_3 + \tfrac{1}{2}H_2 \qquad [12.39]$$

With alkyl halides, $NaGeH_3$ undergoes the expected reaction

$$NaGeH_3 + RX \longrightarrow RGeH_3 + NaX \qquad [12.40]$$

but note that with aryl halides, the phenyl group is reduced to benzene:

$$NaGeH_3 + C_6H_5Br \longrightarrow C_6H_5GeH_3 + NaBr \qquad [12.41]$$

$$C_6H_5GeH_3 \longrightarrow C_6H_6 + GeH_2 \qquad [12.42]$$

Germanium(II) hydride is a white solid soluble in liquid ammonia.

Stannane or Tin Hydride, SnH_4. Stannane may be formed from $SnCl_4$ and $LiAlH_4$, or by reducing $SnCl_4$ in acid solution with zinc, i.e.,

$$SnCl_4 + 4H^+ + 4Zn \longrightarrow SnH_4 + 4Zn^{2+} + 4Cl^- \qquad [12.43]$$

It is a colorless gas that decomposes slowly at room temperature, and it is not decomposed by dilute acids or alkalis. It reacts with sodium in liquid ammonia in a manner similar to that of germane to give $NaSnH_3$, sodium stannyl. Sodium stannyl may react with methyl halides to form methyl stannane, CH_3SnH_3. As expected, CH_3SnH_3 is more stable thermally than SnH_4.

Reportedly, Sn_2H_6 is prepared from the reaction of SnO_2^{2-} and BH_4^- in basic solution.

Plumbane or Lead Hydride, PbH_4. Plumbane is a colorless gas that may be formed in small amounts by the electrolysis of sulfuric acid solutions using lead electrodes. It is also produced by the action of acid on magnesium-lead alloys:

$$4H^+ + 4e^- + Pb \rightarrow PbH_4 \qquad [12.44]$$

Plumbane is very unstable and decomposes rapidly at room temperature.

Hydrides of Group V

They include the following:

NH_3	ammonia	N_2H_4	hydrazine	HN_3	hydrazoic acid
PH_3	phosphine	P_2H_4	biphosphine		
AsH_3	arsine				
SbH_3	stibine				
BiH_3	bismuthine				

In general, this series of hydrides exhibits all the previously mentioned trends: their thermal stability decreases, and they become better reducing agents, as the group is descended.

In the simple hydrides listed, all the Group V elements are in oxidation state −3. They have a lone pair of electrons attached to the central element and hence the potentiality of behaving as Lewis bases. The strength of each species as a Lewis base decreases as the group is descended. The student should be able to rationalize this fact on his own.

Related to this trend is the fact that the solubility of the Group V hydrides decreases as we descend the group. Ammonia is the strongest base and is thus the most soluble in water, whereas, arsine is prepared in aqueous medium and is evolved. The solubility of ammonia in water involves a Lewis acid-base reaction:

$$H_3N: + H{-}OH \longrightarrow H_3NH^+{-}OH^- \qquad [12.45]$$

In the case of the other hydrides in the group, the acid-base interaction is reduced considerably and the solubility in water is smaller.

Intimately related to base strength is the ability to form *−onium* type compounds. The formula for amm*onium* chloride is NH_4Cl. This tendency also decreases as the group is descended:

PH_4I	phosphonium iodide	
AsH_4X	arsonium	only known when H is replaced by an organic group
SbH_4X	stibonium	

Ammonia, NH_3. Ammonia may be prepared by the general methods given previously. The direct combination of hydrogen and nitrogen under pressure in the presence of catalysts is the most important industrial process (the Haber process).

Ammonia burns in air. It can be oxidized by air or pure oxygen in a strongly exothermic reaction on the surface of platinum metal:

$$4NH_3 + 5O_2 \xrightarrow{\text{Pt}} 4NO + 6H_2O \qquad [12.46]$$

The hydrogen atoms of ammonia may be replaced one at a time to give amides, MNH_2, imides, M_2NH, or nitrides, M_3N (where M is a monovalent metal). When imides or amides are heated, nitrides finally result.

$$LiNH_2 \xrightarrow{\Delta} Li_2NH \xrightarrow{\Delta} Li_3N + NH_3 \qquad [12.47]$$

Hydrazine, N_2H_4. Hydrazine may be prepared by the partial oxidation of ammonia:

$$2NH_3 + NaOCl \longrightarrow N_2H_4 + NaCl + H_2O \qquad [12.48]$$

Hydrazine boils at 113.5°C, and usually explodes when heated to about 140°C. It is a fairly strong reducing agent:

$$N_2H_4 + 2S \longrightarrow N_2 + 2H_2S \qquad [12.49]$$

It is a weaker base than ammonia, but forms hydrazonium hydroxide in water:

$$N_2H_4 + H_2O \longrightarrow N_2H_5^+ + OH^- \qquad [12.50]$$

Hydrazine is capable of forming complexes—i.e., protonated cationic species. The complex species formed with sulfuric acid are the following:

$(N_2H_5^+)(HSO_4^-)$	monohydrogen sulfate
$(N_2H_5)_2SO_4$	dihydrazine sulfate
$(N_2H_6)(HSO_4)_2$	hydrazine bisulfate

The basic character of hydrazine is also shown by the existence of a few metal coordination compounds, such as $[Pt(NH_3)_2(N_2H_4)_2]Cl_2$. Here, the hydrazine molecule is bonded to the metal by only one of the lone pairs of electrons on the nitrogen atoms. However, in the compound $[Pt(NO_2)_2N_2H_4]$, it is believed that the hydrazine is a bidentate ligand. Many of the hydrazine-complexed species are polymeric and insoluble in water. The polymeric character can be rationalized by the fact that hydrazine may act as a bridging group because of its two active basic sites.

Hydrazoic Acid, HN_3. Hydrazoic acid, a weak acid with a boiling point at 37°C, may be prepared by the reaction of N_2O with $NaNH_2$ at 200°C:

$$N_2O + 2NaNH_2 \longrightarrow NaN_3 + NaOH + NH_3 \qquad [12.51]$$

$$H_2SO_4 + NaN_3 \longrightarrow HN_3 + NaHSO_4 \qquad [12.52]$$

The HN_3 is obtained by the drop-by-drop addition of H_2SO_4 to the alkali metal azide. Care should be taken because the pure acid is explosive. In fact, all azides should be respected. Many azides of the heavy metals have been used as detonators because they explode on impact. Two of the resonance structure of HN_3 are

The azide group is considered to be a pseudohalogen, and the metal azides resemble the metal halides in solubility. Hydrazoic acid may act as either an oxidizing agent or a reducing agent:

Oxidizing agent: $NH_4^+ + N_2 = HN_3 + 3H^+ + 2e, \quad E^\circ = -1.82\ V$ [12.53]

Reducing agent: $HN_3 = \frac{3}{2}N_2 + H^+ + e, \quad E^\circ = 3.1\ V$ [12.54]

Phosphorus Hydrides. The hydrolysis of Ca_3P_2 yields a mixture of PH_3 and P_2H_4. Pure phosphine may be prepared by boiling yellow phosphorus with strong aqueous alkali hydroxide:

$$4P + 6H_2O \xrightarrow{OH^-} PH_3 + 3H_3PO_2 \quad [12.55]$$

In water, phosphine differs from ammonia in showing only a slight tendency to form an *-onium* hydroxide, PH_4OH. However, with HBr and HI, the phosphonium halides are formed:

$$PH_3 + HX \longrightarrow PH_4X \quad [12.56]$$

Phosphonium bromide and iodide are white crystalline solids that appear to have a large amount of covalent character. They are hydrolyzed by water to phosphine:

$$PH_4X + H_2O \longrightarrow PH_3 + H_3O^+ + X^- \quad [12.57]$$

Biphosphine boils at 51.7°C and decomposes on heating to phosphine and phosphones. It is a strong reducing agent and is analogous to hydrozine in most properties. Organic derivatives of phosphine, analogous to amines, are known, i.e., CH_3PH_2, methyl phosphine. The hydrogen of a P—H bond can be replaced by an alkali metal in liquid ammonia:

$$C_6H_5PH_2 + Na \longrightarrow C_6H_5PHNa + C_6H_5PNa_2 \quad [12.58]$$

Arsenic Hydride or Arsine, AsH_3. The action of dilute acid on calcium arsenide gives 14% yield of arsine. As with stannane, arsine can also be prepared by the action of zinc and acid on almost any simple, soluble arsenic compound.

Arsine is relatively unstable thermally and decomposes rapidly at 300°C. This property is used as a test for traces of arsenic. The arsenic-containing compound is mixed with zinc and acid, and the mixture of hydrogen and arsine is passed through a heated glass tube. The heat decomposes the arsine into hydrogen and a finely divided arsenic "mirror," which is deposited on the side of the tube. The mirror may then be compared with mirrors found from known quantities of arsenic.

As with ammonia and phosphine, arsine reacts with sodium dissolved in liquid ammonia to give $NaAsH_2$. From this system, many compounds analogous to amines may be prepared:

$$KAsH_2 + CH_3Cl \longrightarrow KCl + CH_3AsH_2 \qquad [12.59]$$

Antimony Hydride or Stibine, SbH_3. Stibine is a colorless gas that decomposes slowly at room temperature. It is a strong reducing agent, reducing some heavy metal salts to the metals.

Stibine may be prepared by the action of dilute acid on a zinc-antimony alloy. A very recent method involves the reduction of antimony trichloride with a borohydride. From its method of preparation, we see that stibine is fairly stable in the presence of dilute acid or alkaline solutions.

Bismuth Hydride or Bismuthine, BH_3. Bismuthine, BH_3, may be prepared by treating an alloy or finely divided mixture of Bi and Mg with aqueous acid. It boils at 22°C, and at room temperature 20% of the original mixture will decompose in one hour to bismuth and hydrogen. It is only slightly soluble in water. No species such as BiH_4^+ has ever been noted. The chemistry of this species is very limited because of the great instability.

Hydrides of Group VI

They include the following:

H_2O	hydrogen oxide (water)	H_2O_2	hydrogen peroxide
H_2S	hydrogen sulfide		
H_2Se	hydrogen selenide		
H_2Te	hydrogen telluride		
H_2Po	hydrogen polonide		

Hydrides of Oxygen. The hydrides of this group may be prepared by the general methods previously described. As expected, the thermal stabilities of these hydrides decreases as the group is descended, hydrogen telluride decomposing slowly at room temperature to tellurium and hydrogen. The acid strengths of the hydrides increase as the group is descended. Their strength as reducing agents increases in the same fashion. Hydrogen sulfide *slowly* reduces moist oxygen, whereas hydrogen telluride undergoes the

$$2H_2S + O_2 \longrightarrow 2H_2O + 2S \qquad [12.60]$$

same type of reaction in a matter of minutes. As expected, polynuclear hydrides are formed most readily by elements near the top of the group—particularly with sulfur.

Hydrogen Peroxide, H_2O_2. The dimeric hydride of oxygen, H_2O_2, has the —O—O— linkage characteristic of peroxides. Hydrogen peroxide may be formed by the hydrolysis of metallic peroxide:

$$Na_2O_2 + 2H_2O \xrightarrow{0^\circ C} H_2O_2 + 2NaOH \qquad [12.61]$$

A commercial method for the production of hydrogen peroxide is the hydrolysis of peroxydisulfuric acid, $H_2S_2O_8$:

$$\begin{array}{c} \mathrm{O} \qquad\quad \mathrm{O} \\ \uparrow \qquad\quad \uparrow \\ \mathrm{HO{-}S{-}O{-}O{-}S{-}OH} \\ \downarrow \qquad\quad \downarrow \\ \mathrm{O} \qquad\quad \mathrm{O} \end{array} + 2H_2O \longrightarrow 2H_2SO_4 + 2H_2O_2 \qquad [12.62]$$

The peroxydisulfuric acid itself is prepared by electrolysis, using high current density in strong sulfuric acid solutions:

$$\text{Anode} \qquad 2HSO_4^- \longrightarrow H_2S_2O_8 + 2e^- \qquad [12.63]$$

Pure hydrogen peroxide is separated from the sulfuric acid by distillation. Remember that the pure hydrogen peroxide is an endothermic compound and a very strong oxidizing agent, and that the 100% material may explode violently in the presence of traces of impurities, especially transition metal salts and oxides. Organic materials are also oxidized very rapidly. Even dilute solutions are readily decomposed by either heat or catalysts:

$$2H_2O_2 \xrightarrow{MnO_2} 2H_2O + O_2 \qquad [12.64]$$

Dilute solutions of hydrogen peroxide in water may act either as oxidizing agents or as reducing agents:

$$2I^- + H_2O_2 + 2H^+ \longrightarrow I_2 + 2H_2O, \qquad E^\circ = 1.77\ V \qquad [12.65]$$

$$Ag_2O + H_2O_2 \longrightarrow 2Ag + H_2O + O_2, \qquad E^\circ = 0.87\ V \qquad [12.66]$$

If hydrogen peroxide is used in place of water in certain reactions, then peroxy compounds result:

$$\underset{\text{chlorosulfonic acid}}{\begin{array}{c} \mathrm{OH} \\ | \\ \mathrm{O \leftarrow S{-}Cl} \\ \downarrow \\ \mathrm{O} \end{array}} + H_2O \rightarrow \begin{array}{c} \mathrm{OH} \\ | \\ \mathrm{O \leftarrow S{-}OH} \\ \downarrow \\ \mathrm{O} \end{array} + HCl \qquad [12.67]$$

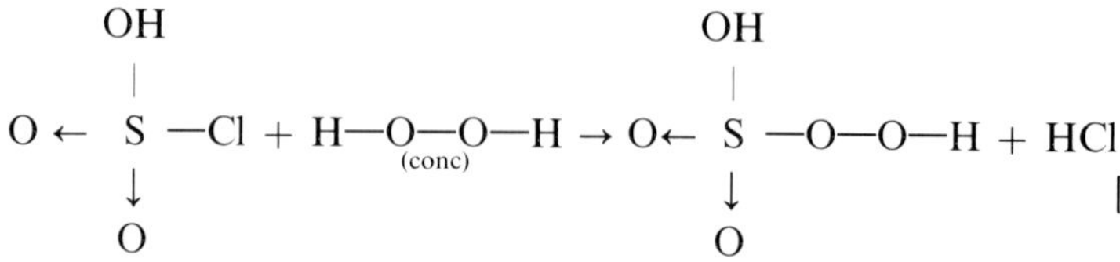

[12.68]

peroxymonosulfuric acid

Some $H_2S_2O_8$ may be formed in this reaction:

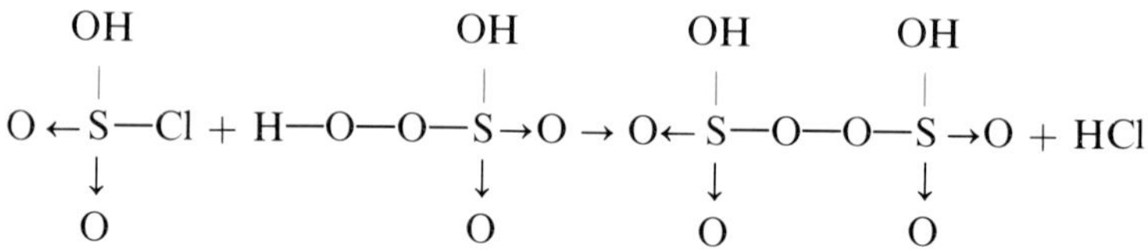

[12.69]

Hydrogen Sulfide and Polysulfides. Hydrogen sulfide may be prepared by the general methods previously mentioned. As expected, it is a stronger reducing agent than water and it reduces sulfur dioxide to sulfur:

$$2H_2S + SO_2 \longrightarrow 2H_2O + 3S \qquad [12.70]$$

Sulfide ion in solution readily dissolves elemental sulfur to form polysulfides,

$$:\ddot{\underset{..}{S}}:^{2-} + \ddot{\underset{..}{S}}: \longrightarrow :\ddot{\underset{..}{S}}:\ddot{\underset{..}{S}}:^{2-}, \quad \text{etc.} \qquad [12.71]$$

Polyanions containing at least six sulfur atoms have been observed. By carefully acidifying solutions of polysulfides with concentrated hydrochloric acid at approximately $-15°C$, we may obtain an oily mixture consisting in hydrogen polysulfides (H_2S_x up to H_2S_9). The structure of these species is simply a hydrogen atom attached to both ends of a sulfur atom chain, H—S—S_y—SH. When y is zero, the sulfur analog of H_2O_2 results; it boils at 70.7°C and is an oily liquid soluble in organic solvents. As expected, H_2S_2 is a less polar molecule than hydrogen peroxide and is therefore an excellent solvent for a large number of covalent compounds, including sulfur.

The question may now arise as to why the polysulfides exist in greater numbers than do the *polyoxides* (polywater has been reported but not characterized). There is no concrete answer, but sulfur is in the third row of the periodic table and has low-lying empty $3d$ orbitals available. There is no reason why lone pairs of electrons on neighboring atoms cannot back-bond into these empty $3d$ orbitals. This is another argument for $(p \rightarrow d)$ π bonding. Oxygen does not have the d orbitals available, and the atoms below sulfur (selenium and tellurium) possess orbitals too diffuse for effective overlap. This may be the rationalization for the preference in the formation of polysulfides.

Hydrogen Selenide. Hydrogen selenide is prepared according to the general methods described previously. It is a colorless gas with an even more unpleasant smell than that of hydrogen sulfide. Its properties are exactly those to be expected from the trends mentioned previously: it begins to decompose at about 160°C, it precipitates many metal selenides from aqueous solutions of their salts, it is a fairly strong reducing agent, and in moist air it rapidly reduces oxygen to water, itself being oxidized to selenium.

Hydrogen Telluride. The preparation and properties of hydrogen telluride are those expected from the previous discussion. It is a colorless gas with a disgusting smell, and it decomposes into its elements even at 0°C. It is rapidly oxidized in moist air to tellurium. It precipitates metallic tellurides from aqueous solution, but they are usually mixed with some elemental tellurium, which is formed at the same time.

Hydrogen Polonide. Very little is known about hydrogen polonide, H_2Po. It may be prepared by the action of acid on a finely divided mixture of polonium and magnesium. It decomposes to an extent of 50% during 4 min at room temperature, and it is oxidized extremely rapidly in moist air to polonium. Recall that polonium is a radioactive element, and all of its compounds will be likewise.

Hydrogen Halides. These compounds will be treated in more detail in Chapter 13. Hydrogen fluoride exists as a dimer due to hydrogen bonding, which is not an example of real catenation. The increasing strength of the hydrides as reducing agents as a group is descended is illustrated by the fact that hydrogen iodide slowly reduces moist oxygen to water:

$$2HI + \tfrac{1}{2}O_2 \rightarrow H_2O + I_2 \qquad [12.72]$$

THE BORDERLINE HYDRIDES

This class of hydrides seems to bridge the gap between the ionic-type hydrides and the covalent hydrides. The elements forming this set of hydrides are

II-A			*III-A*
Be			
Mg			
	I-B	*II-B*	
	Cu	Zn	
	Ag	Cd	In
	Au	Hg	Tl

These hydrides are all white, stoichiometric solids (with the exception of CuH, which is reddish-brown), and they are all more covalent than lithium hydride. No doubt their covalent character is due to a great deal of polarization in the metal-hydrogen bond. Their structure is not ionic, but polymeric in nature—i.e., the units of the hydride system seem to be held together by a great deal of hydrogen bridging.

The members of this class of compound may be prepared by several methods, most by the reaction of the dimethyl derivative of the element with lithium aluminum hydride:

$$(CH_3)_2Be + LiAlH_4 \longrightarrow BeH_2 + LiAlH_3CH_3 \quad [12.73]$$

$$(CH_3)_2Zn + LiAlH_4 \longrightarrow ZnH_2 + LiAlH_3CH_3 \quad [12.74]$$

$$(CH_3)_2Cd + LiAlH_4 \longrightarrow CdH_2 + LiAlH_3CH_3 \quad [12.75]$$

Thermal decomposition of an organometallic compound may also be employed to prepare this type of hydride. At 200°C, in vacuuo, the following reaction may take place:

$$(C_2H_5)_2Mg \longrightarrow MgH_2 + 2C_2H_4 \quad [12.76]$$

$$2C_2H_5MgI \longrightarrow MgH_2 + MgI_2 + 2C_2H_4 \quad [12.77]$$

Copper hydride is prepared in aqueous solution:

$$2Cu^{2+} + 3H_3PO_2 + 3H_2O \rightarrow 2CuH + 3H_3PO_3 + 4H^+ \quad [12.78]$$

The hydrides of silver, gold, and mercury have been prepared by the attack of atomic hydrogen on foils of these metals. As with the hydrides of indium and thallium, they are far from being well characterized.

Generally, borderline hydrides are strong reducing agents and react violently with water.

TRANSITION METAL HYDRIDES

By transition metal hydrides we mean the simple binary hydrides formed by transition metals, Groups I-B to VIII-B in the middle part of the long form of the periodic table. For convenience, we also include the hydrides of the lanthanide and actinide metals. Synonyms for this group of hydrides are *interstitial hydrides*, *nonstoichiometric hydrides*, or *simple metal hydrides*.

Transition metal hydrides usually have crystal structures similar to those of the parent metals, and most of their properties are consistent with their being formed by *alloying* of the metal with hydrogen to give a solid-solution structure. The hydrogen atoms may be considered as occupants of the spaces between metal atoms in the metallic crystal lattice. The compounds are usually nonstoichiometric in nature—i.e., the hydrogen/metal ratio is not generally a small integar; it usually varies for a given hydride from one preparation to another. The ability of the transition metals to react chemically with or to absorb hydrogen decreases on traversing a period from left to right; at the same time, the amount of heat liberated on reaction with hydrogen decreases.

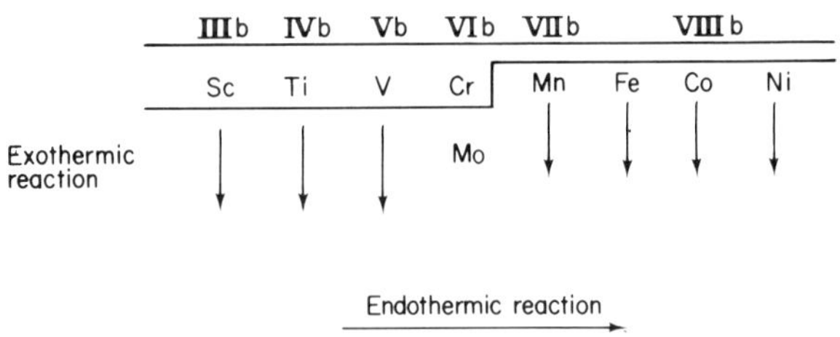

Figure 12.8. Arrows indicate the decreasing ability to react with molecular hydrogen. The exception is palladium.

Many of the transition metal hydrides may be prepared by the direct combination of the elements. The exceptions are nickel and chromium. The mechanism for hydride formation by this direct combination has been explained in four steps: (1) hydrogen gas is absorbed and then dissociates into hydrogen atoms; (2) the hydrogen atoms diffuse into the body of the metal to form a solution of hydrogen atoms in the metal; (3) the hydrogen in the saturated metal reacts to form a hydride phase; (4) hydrogen atoms continue to diffuse through the hydride phase. On heating titanium metal in an atmosphere of hydrogen gas, we obtain a gray powder, $TiH_{1.26}$. Scandium hydride, ScH_2, and uranium hydride, UH_3, have both been reported; the scandium is in the $+3$ oxidation state and uranium is in the $+6$ oxidation state.

Other hydrides have been prepared by electrolytic methods. For example, palladium hydrides can be prepared by electrolyzing a $2NH_2SO_4$ solution using a palladium cathode. As the hydrogen atoms are formed, they are adsorbed by the metal and then react.

Iron, cobalt, and nickel have small powers of occluding molecular hydrogen compared with the other transition metals. The occlusion is an endothermic process, and apparently no chemical bonds are formed. The *platinum metals* (ruthenium, rhodium, palladium, osmium, irridium, and platinum), with the exception of palladium, occlude only small amounts of hydrogen and no chemical bonding is believed to be involved. Palladium, however, absorbs up to 1000 times its own volume of hydrogen at STP when used as a cathode in electrolysis. During adsorption, the volume of palladium increases. Compositions ranging from $PdH_{0.7}$ to $PdH_{0.8}$ have been reported.

The transition metal hydrides, having many properties of the finely divided metal itself, also react with air or moisture to some degree. Note that the rare earth and actinide hydrides are pyrophoric. These species may also be used as reducing agents.

Three models are used to explain the bonding in the metal hydrides, based on the manner of existence of the hydrogen in the hydride: (1) the protonic model; (2) the covalent model; and (3) the hydride anion model. The protonic model has been used to explain the decrease in magnetic susceptibility. The covalent model has been employed to explain the internuclear distances and some structural problems of the metallic hydrides. The hydride anion model has helped to explain the large heats of formation (30 kcal/mole) and brittleness of the species that suggests ionic character. Also, metallic bonding has been invoked to explain the electrical conductivity of these species. It would seem that the range of these invocations indicate that the problem is far from being solved.

THE COMPLEX HYDRIDES OF NONTRANSITION METALS

By *complex hydride* is meant a hydride that is composed of species capable of independent existence, implying, of course, that the complex species are also capable of independent existence. The complex hydrides are usually prepared by the attack of a hydride ion, $H\colon^-$, on a binary, covalent hydride. This point may be illustrated by considering the reaction of lithium hydride on diborane:

$$2LiH + B_2H_6 \longrightarrow 2LiBH_4$$

or

$$H\colon^- + BH_3 \longrightarrow BH_4^- \qquad [12.79]$$

The most common of the complex hydride anions are BH_4^-, AlH_4^-, and

GaH_4^-. The BH_4^- anion is stable in aqueous solution, whereas its congeners, AlH_4^- and GaH_4^-, react dangerously and violently in water. The stability of the BH_4^- ion is generally attributed to the fact that the boron atom has reached its maximum coordination number of 4. Both Al^{3+} and Ga^{3+} are larger than B^{3+}, and it is easier to pack more hydride ions around them; therefore, their coordination number is higher. These ions have a coordination number of 6. Keep in mind that the smaller B^{3+} is also more difficult to attack and thus forms an unstable pentacoordinated intermediate. Being larger and having available *d* orbitals, it is easier for both AlH_4^- and GaH_4^- to form unstable intermediates and thus decompose rapidly. However, we should point out that Na_3AlH_6 is readily attacked by water and alcohol; therefore, the extent of coordination is not the only factor in explaining the stability of BH_4^- or instability of GaH_4^- and AlH_4^- in aqueous solution.

It is believed that all MH_4^- ions are tetrahedral in structure. These compounds are all solids and have at least some solubility in polar organic solvents. The more popular reagents are sodium borohydride, potassium borohydride, and lithium aluminum hydride, all of which are used as reducing agents in both organic and inorganic chemistry.

THE COMPLEX HYDRIDES OF TRANSITION METALS

By definition, we shall consider the hydrogen atoms in the complex transition metal hydrides to be bound to the transition metal atom by two-electron covalent bonds. Conveniently, we can distinguish between two types of complex transition metal hydrides: (1) the pure hydrides—i.e., those hydrides where hydrogen is the sole ligand, e.g., ReH_9^{2-}; and (2) mixed hydrides, the type where hydrogen is not the sole ligand. The ligands that are usually joined to a transition metal along with hydrogen are substituted phosphines, carbon monoxide (both mononuclear and polynuclear carbonyls), π-cyclopentadienyl groups, cyanide ions, and nitrogen-type ligands.

The complex hydrides may be prepared by direct hydrogenation of a carbonyl,

$$CO_2(CO)_8 + H_2\ (30\ \text{atm}) \xrightarrow{25^\circ} 2HCo(CO)_4 \qquad [12.80]$$

or by reducing the halide complexes with suitable borohydrides or aluminohydrides (or even hydrazine):

$$(C_5H_5)_2WCl_2^+Cl^- \xrightarrow[\text{THF}]{\text{LiAlH}_4} (C_5H_5)_2WH_2 \qquad [12.81]$$

These hydrides can also be prepared by the hydrolysis of alkali-metal salts of the complex carbonyls:

$$Fe(CO)_5 \xrightarrow[THF]{Na/Hg} Na_2[Fe(CO)_4] \qquad [12.82]$$

$$Na_2[Fe(CO)_4] \xrightarrow{H^+(ag)} Na^+[HFe(CO_4)] \xrightarrow{H^+} H_2Fe(CO)_4 \qquad [12.83]$$

The thermal stabilities of the complex hydrides vary widely. The complexes containing the phosphine derivatives as ligands are the most stable, e.g., $H_2Fe(C_2H_4PH_2)_2$, whereas the mononuclear carbonyls—$H_2Fe(CO)_4$—are the least stable.

It is also possible for the hydrogen ligand to exchange rapidly, as in

$$ReH_3[P(C_6H_5)_3]_4 \xrightarrow[ether]{HCl} ReCl_4[P(C_6H_5)_3]_2 \qquad [12.84]$$

or add across a double bond, as in

$$Co(CO)_4H + C_2H_4 \rightleftharpoons Co(CO)_4C_2H_5 \qquad [12.85]$$

The preceding reaction is reversible.

The hydrogen in the complex hydrides may be protonic or hydridic, depending on the conditions. No general rules have been established. The proton NMR data indicate that the hydrogen atoms in these complex species are well shielded because they resonate well above the frequency for hydrogen in tetramethylsilane, about 20 ppm higher.

The greatest uses that the complex hydrides have in the field of chemistry depend upon their ability to act as catalysts and to serve as intermediates in the formation of organometallic compounds.

THE HYDROGEN BOND

The *hydrogen bond* is a special case of chemical bonding that is important enough to warrant its being placed in a class by itself. The striking effects of this type of bonding are exemplified in Fig. 12.9. The boiling points and melting points of the Group IV hydrides vary regularly; therefore, it might be expected that the others should behave similarly. Careful observation of Fig. 12.9 indicates that the latter groups (Group V, VI, and VII) of hydrides do not follow this trend.

When hydrogen is bonded to an element that is distinctly more electronegative than itself, the electron pair of the bond must of necessity be pulled toward the more electronegative element. Because of the simplicity of the structure of the hydrogen atom—it contains only a 1*s* orbital—the bare proton of its nucleus is exposed. The small size of the proton would then give this species a high charge density and a great attraction for negative charge. Hence, molecules of this type would be expected to have great polar or electrostatic intermolecular attraction. This attraction is more commonly referred to as *molecular association*.

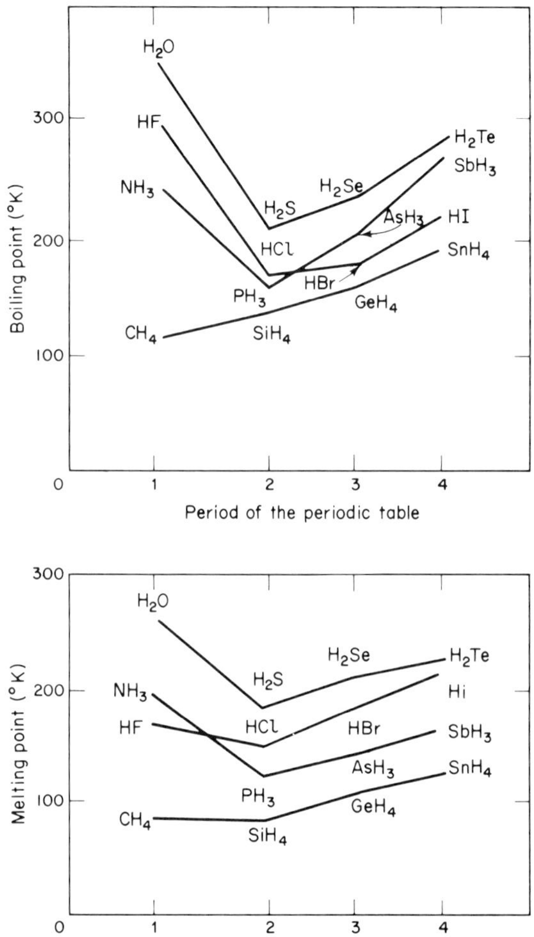

Figure 12.9. Comparison of melting points and boiling points of the covalent hydrides.

The concepts already developed should help us establish a set of conditions whereby the formation of a hydrogen bond might be expected to exist. To have a hydrogen bond, we must first have a positively charged hydrogen atom, which would exist in any polar H-*X* group. It is also necessary to have an electronegative atom or highly polarizable atom or group. The other electronegative atom—usually fluorine, oxygen, nitrogen, or chlorine—will have a strong attraction for its own electron pairs, and thus they will be very compact in space when compared with the diffusiveness one might expect from a less electronegative element. It would then be expected that the electrostatic attraction between the proton of the H-*X* group and the electronegative atom would be the greatest for the case of the most concentrated electronic charge. In view of the preceding discussion, apparently there is no need to state that the presence of an unused electron pair on the other atom is necessary for the formation of a hydrogen bond.

In the system $A—H \cdots B$, the electronegatives of A and B must be near each other in magnitude to achieve hydrogen bonding. If A is a much stronger base than B, it will be difficult to get proton transfer. If B is a much stronger base than A, then proton transfer will be essentially complete, even if the electronegativities of A and B are about equal, an asymmetric bond will be formed and the hydrogen bond formed will be longer than the normal covalent bond of hydrogen and the element in question. However, in $(F—H—F)^-$, the hydrogen is positioned symmetrically between the two fluorine atoms.

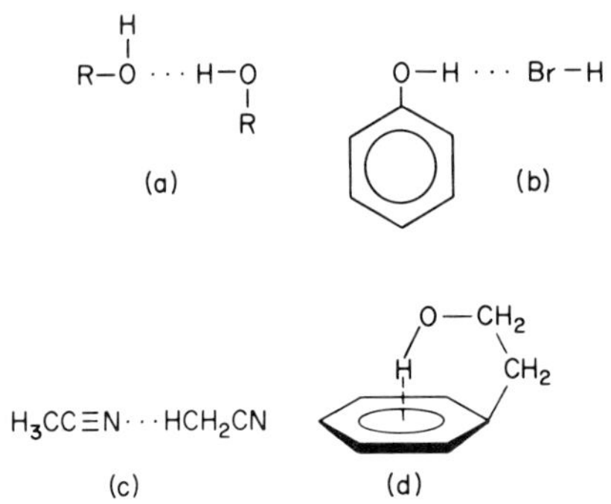

Figure 12.10. Molecules that exhibit hydrogen bonding.

Figure 12.10(a) illustrates the type of hydrogen bond that is probably the most familiar. Such bonds are common in species containing polar hydroxyl groups. Their strength is usually of the order of 5 kcal/mole. Hydrogen bonds in other compounds range from 1 to 10 kcal/mole in strength. If a normal covalent bond is generally rated from 50 to 100 kcal/mole, the strength of the hydrogen bond is about 10% of this value. In Fig. 12.10(b), we can note that HBr is not very polar. Hence, we would not expect HBr to form hydrogen bonds with itself. The hydroxyl hydrogen on phenol *is* polar, and since bromine is quite polarizable, a hydrogen bond is formed between the phenol and the hydrogen bromide. We turn to Fig. 12.10(c).

One normally thinks of the methyl group as being unable to form a hydrogen bond, since the C—H bond is nonpolar. In the case of CH_3CN, the C—H bond is polar due to the very high electron-withdrawing power of the CN group. In the molecule shown in Fig. 12.10(d), there is no nearby atom that is electronegative or highly polarizable. However, the π electron density in the ring is polarizable, and is able to form a hydrogen bond with the hydrogen from the same molecule.

The molecular orbital treatment of the hydrogen bond is rather interesting. Using $(FHF)^-$ as the model, and selecting the internuclear axis as the z axis, we see that molecular orbitals can be formed by combining the $1s$ atomic orbital of hydrogen with the $2p_z$ atomic orbitals on the fluorine atoms. The other atomic orbitals and the electrons associated with them are assumed to have no part at all in the formation of the molecular orbitals. We have three atoms joining and the linear combination of the atomic orbitals will result in three molecular orbitals, one bonding orbital, one nonbonding orbital, and one antibonding orbital (see Fig. 12.11). The two electron pairs involved in bonding are placed accordingly—two in the

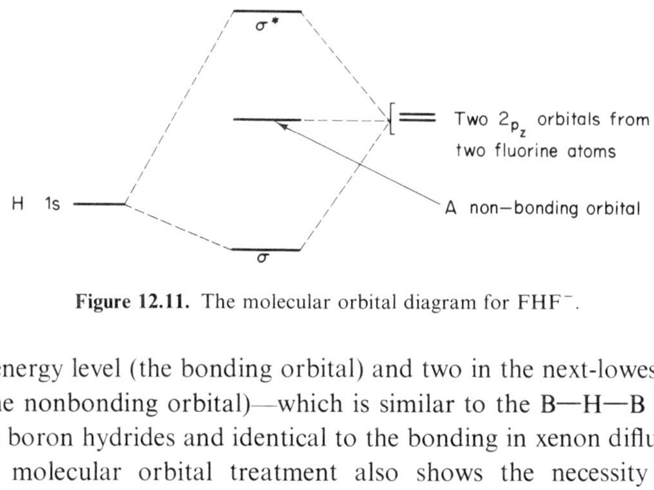

Figure 12.11. The molecular orbital diagram for FHF^-.

lowest energy level (the bonding orbital) and two in the next-lowest energy level (the nonbonding orbital)—which is similar to the B—H—B bridging bond in boron hydrides and identical to the bonding in xenon difluoride.

The molecular orbital treatment also shows the necessity for the statement that A and B should be of equal or nearly equal electronegativity. If either A or B is distinctly more electronegative than the other, the nonbonding orbital will be localized almost entirely on the more electronegative atom. The bonding orbital will then be essentially two centered, and the hydrogen bond will not form.

The effect of hydrogen bonding on the boiling and melting points of hydrides has been pointed out. The structure of chemical compounds is also affected by hydrogen bonding. Acetic acid and many dibasic organic acids exist in the liquid phase as dimers, due to hydrogen bonding. This association causes unexpected values for the molecular weight when determined experimentally.

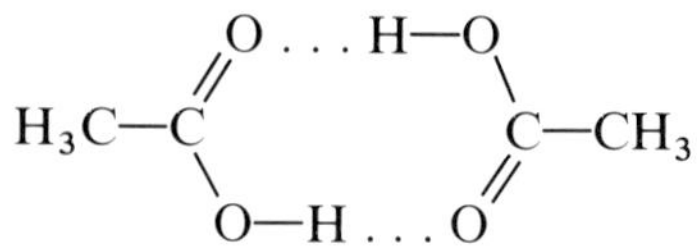

As water cools and finally solidifies to ice, a surprising fact is obvious. Ice floats and hence must be less dense than liquid water. The curiously open structure of ice results from the tetrahedral arrangement of electron pairs (both bonding and nonbonding) around the oxygen atom. As water solidifies, four hydrogen atoms arrange themselves around an oxygen atom; two of these hydrogen atoms are bound by electron pairs in a normal covalent bond, whereas the other two atoms are bound to the same oxygen atom by hydrogen bonding, being attracted to the lone pairs of electrons on the oxygen atom. On freezing, the decrease in density of the ice is due to the molecular movement to obtain proper positions for tight hydrogen bonding. Note that solid hydrogen sulfide is not less dense than liquid hydrogen sulfide.

In the case of a hydrogen bond containing fluorine, a distinction must be made between the polymer formed by HF and the single hydrogen bond formed by species such as KF. The polymer formed by HF has the structure

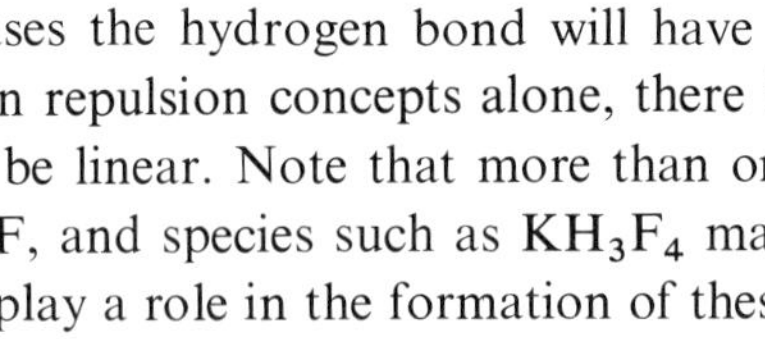

the HFH angle being about 140°. This angle is interesting, for it shows that at least in some cases the hydrogen bond will have a directional nature. Otherwise, based on repulsion concepts alone, there is no reason why the species should not be linear. Note that more than one HF molecule may be associated to KF, and species such as KH_3F_4 may be formed. Lattice energy effects may play a role in the formation of these higher members.

Solubility of nonionic organic species in water can also be attributed to hydrogen bonding. For example, dimethyl ether, $(CH_3)_2O$, is miscible with water, whereas dimethyl sulfide, $(CH_3)_2S$, is only slightly soluble. This may be attributed to the fact that there is much more hydrogen bonding in dimethyl ether, due to the greater electronegativity of oxygen, than there is in dimethyl sulfide. Some authors (Douglas and McDaniel) have stated that the solubility of organic species in water is strongly dependent upon the ratio of carbon atoms to oxygen atoms in the organic species. A 3:1 ratio appears to be the highest limit for good solubility. Thus, $(CH_3)_2O$ and

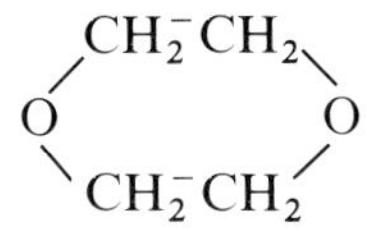

dioxane are miscible with water, whereas $(C_2H_5)_2O$ is only soluble to the extent of 7.5 g per 100 g of water.

Analysis of many compounds is based on hydrogen bonding. The determination of carbon and hydrogen in organic compounds by combustion is based on the fact that ascarite absorbs water. Ascarite is potassium hydroxide deposited on asbestos, and the potassium hydroxide absorbs water by the formation of hydrogen bonds. The sample containing carbon and hydrogen is burned, forming carbon dioxide and water vapor. The water vapor is absorbed on ascarite and the gain in weight is measured. Thus, the weight and percentage of hydrogen may be determined.

When fluorine is prepared by the electrolysis of hydrogen fluoride, it nearly always is contaminated with hydrogen fluoride. The hydrogen fluoride can be removed by passing the mixture over potassium fluoride:

$$F^- + HF \longrightarrow (FHF^-) \qquad [12.86]$$

The reaction evolves about 58 kcal/mole. The hydrogen fluoride can then be regenerated by heating the KHF_2.

Suggested Readings

1. Ashby, E. C., *The Chemistry of Complex Aluminohydrides,* vol. 8 in *Advances in Inorganic Chemistry and Radiochemistry*, H. J. Emeleus and A. G. Sharpe, eds., Academic Press Inc., New York, 1966.
2. Cotton, F. A., and G. Wilkinson, *Advanced Inorganic Chemistry,* 2nd ed., Interscience Publishers, New York, 1966.
3. Douglas, B. E., and D. H. McDaniel, *Concepts and Models of Inorganic Chemistry*, Blaisdell Publishing Co., New York, 1965.
4. Green, M. L. H., and D. J. Jones, *Hydride Complexes of Transition Metals*, vol. 7 in *Advances in Inorganic Chemistry and Radiochemistry*, H. J. Emeleus and A. G. Sharpe, eds., Academic Press Inc., New York, 1965.
5. Hurd, D. T., *Chemistry of the Hydrides*, John Wiley & Sons, Inc., New York, 1952.

6. Jolly, W. L., *The Inorganic Chemistry of Nitrogen,* W. A. Benjamin, Inc., New York, 1964.

7. Libowitz, G. G., *Solid State Chemistry of Binary Metal Hydrides,* W. A. Benjamin, Inc., New York, 1965.

8. Lipscomb, W. N., *The Boron Hydrides*, W. A. Benjamin, Inc., New York, 1963.

9. Massey, A. G., *Boron, Sci. Am.*, **210,** 88 (1964).

10. Muetterties, E. L., ed., *The Chemistry of Boron and its Compounds,* John Wiley & Sons, Inc., New York, 1967.

11. Muetterties, E. L., and W. H. Knoth, *The Polyhedral Boranes, Chem. Eng. News,* 87, May 9 (1966).

12. Phillips, S. G., and R. J. P. Williams, *Inorganic Chemistry*, Oxford University Press, New York, 1965.

13. Sebera, D. K., *Electronic Structure and Chemical Bonding,* Blaisdell Publishing Co., New York, 1964.

14. Stone, F. G. A., *Hydrides of the Group IV Elements*, Prentice-Hall, Inc., Englewood Cliffs, N. J., 1962.

Study Questions

12.1. A group of workers investigated the reaction of ethylenediamine, $H_2N—CH_2—CH_2—NH_2$, with diborane. Ethylenediamine was found to absorb diborane in a 1:1 mole ratio to form a white solid compound. The empirical formula from analysis was $BNCH_7$; experimental molecular weight was 88 ± 7. Another group of workers prepared exactly the *same* compound by the reaction of ethylenediamine dihydrochloride with $NaBH_4$. As you might have expected, both groups of workers have assigned different structures to this compound. The ^{11}B NMR spectrum of this compound has been reported [*Inorg. Chem.*, **2,** 226 (1963)] and consists in a single quadruplet having peaks of relative intensity 1:3:3:1. Write two possible structures that could be assigned to this compound, and from the evidence presented, show which structure best represents the compound in question.

12.2. The proton NMR spectrum of a compound of composition $C_4H_5O_2N$ consists of two single peaks at 6.51τ and 6.18τ, in the integrated ratio of 2:3. Deduce a reasonable structure for the compound.

12.3. On inspection of the infrared spectrum of borazole, one finds the B—N stretching frequency at 1465 cm^{-}1. However one finds that the B—N stretching frequency of B-trichloroborazole is 1443 cm^{-1}. How can this frequency shift be explained?

12.4. You are given the skeletal structure for diborane, B_2H_6,

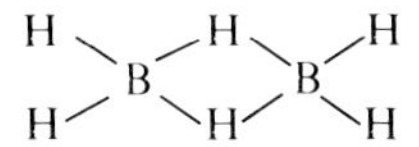

and the following experimental evidence for tetraborane (10), B_4H_{10}:

(a) Six hydrogen atoms are equivalent to each other; four hydrogens are equivalent to each other, but different from the other six.

(b) Each boron atom has its trivalence satisfied by two normal covalent bonds and by two hydrogen bridges.

Draw a skeletal structure for tetraborane and indicate which bonds are normal covalent, and which are hydrogen bridges. Predict the proton and ^{11}B NMR of this compound (see W. N. Lipscomb, *The Boron Hydrides*, W. A. Benjamin, Inc., New York, 1963).

12.5. Write balanced equations for the following:

(a) The preparation of $(DBND)_3$ from D_2O and any other species (which do not contain deuterium).

(b) The preparation of $Ca(AlCl_4)_2$. State the experimental conditions.

(c) Preparation of hydrazine.

(d) Phosphonium iodide + water.

(e) Commercial preparation of H_2O_2.

12.6. Which would you expect to be more soluble in water—ammonia or arsine? Explain.

12.7. In the text it was stated that *hydrogen peroxide is an endothermic compound*. Explain the meaning of this statement and gather data to show that it is true.

12.8. Stibine, SbH_3, is prepared by the reaction of an acid on a zinc-antimony alloy. BiH_3 is prepared by the reaction of a ground mixture of bismuth and magnesium with an acid. What is the purpose of the zinc and magnesium in these reactions? Present a mechanism for both preparations.

12.9. Prepare the following from heavy water: (a) DCl; (b) D_2SO_4; (c) LiD; (d) GeD_4; (e) SiD_4; (f) ND_3.

12.10. Predict the reaction of atomic hydrogen with the following species: (a) sulfur; (b) arsenic; (c) silver nitrate solution; (d) tungsten.

12.11. Give balanced equations for the reaction of:

(a) phosphine and hydronium ion;

(b) potassium hydride with heavy water;

(c) lithium hydride and hydrogen sulfide;

(d) sodium hydride with methyl bromide;

(e) potassium hydride with formaldehyde ($\mathrm{H{-}\overset{\displaystyle H}{\overset{|}{C}}{=}O}$) in an inert solvent.

CHAPTER 13

Oxygen and the Oxides

Because we are creatures living in a sea of oxygen diluted mainly by nitrogen gas, many of the properties of oxygen are well known to us, and the element needs little introduction. Oxides of most elements are known, and it seems that only the lighter noble gases form no oxides.

Of the elements in Group VI of the periodic table, only oxygen is a gas under normal conditions. Its isotopic composition is ^{16}O (99.7%), ^{18}O (0.2%) and ^{17}O (0.1%). Oxygen occurs in the two allotropic forms O_2 and O_3; O_3 is ozone.

Oxygen as O_2 is paramagnetic in the gaseous, liquid, and solid phases. The bond dissociation energy is 117 kcal/mole. The valence bond description is inadequate, since the :O̤=O̤: form would account for the bond strength but not for the paramagnetism, whereas the :Ȯ̤—Ȯ̤: form would account for the paramagnetism but not for the bond strength. However, the molecular orbital (MO) approach to the bonding in the molecule accounts for both these properties. The MO description is

$$O_2 \quad KK(\sigma_{2s})^2(\sigma^*_{2s})^2(\pi_p)^4(\sigma_p)^2(\pi^*_p)$$

It is immediately obvious that the bond order in the oxygen molecule is two, and the two unpaired electrons in the $2\pi^*_p$ orbitals account for the paramagnetism of oxygen.

Molecular orbital theory may also be used to describe the superoxide ion, O_2^-, and the peroxide ion, O_2^{2-}, by simply adding electrons to the preceding scheme,

$$O_2^- \quad KK(2\sigma)^2(2\sigma^*)^2(2\sigma_p)^2(2\pi_p)^4(2\pi_p^*)^3$$
$$O_2^{2-} \quad KK(2\sigma)^2(2\sigma^*)^2(2\sigma_p)^2(2\pi_p)^4(2\pi_p^*)^4$$

The corresponding bond orders for O_2^- and O_2^{2-} are 1.5 and 1. The molecular orbital approach to ozone is a bit more complicated, but extended π bonding can account for its stability. The relative bond orders of O_2, O_2^-, and O_2^{2-} should give the reader some a priori indication of the relative stability of these species.

The binary compounds of oxygen are called oxides. These oxides may be classed conveniently in terms of their reactions with water and/or acids and bases: (1) basic (ionic) oxides, whose aqueous solutions are basic; (2) oxides that are acidic (covalent); (3) amphoteric oxides, oxides that react with both acids and bases: and (4) oxides that are neutral in the sense that they have no apparent reaction with water, such as nitrogen(II) and nitrogen(I) oxides and carbon monoxide. Let this classification be restricted to the representative elements (see Fig. 13.1).

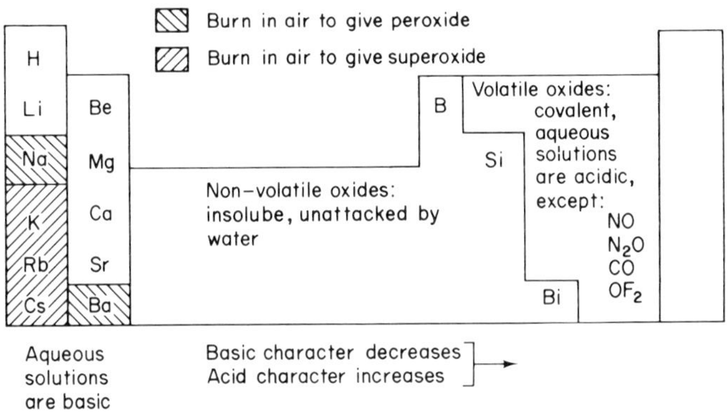

Figure 13.1. Classification of the oxides as to acid or base type.

For the most part, the transition metal oxides are nonvolatile and insoluble. Because the transition metals may assume various oxidation states, it is quite possible for a transition metal to form various oxides. It should not be at all surprising that the various oxides of a given element should exhibit acidic, basic, and amphoteric properties (see Fig. 13.7).

The formation of oxides is usually achieved by direct combination of the elements. However, sodium and barium burn in air to form peroxides, and potassium, rubidium, and cesium preferentially form superoxides.

Oxide ion, peroxide ion, and superoxide ions are strong bases, and they will remove H^+ from water or will displace OH^- from water:

$$:\ddot{\underset{..}{O}}:^{2-} + H^+\text{——}OH^- \rightarrow 2OH^- \qquad [13.1]$$

$$O_2^{2-} + H^+\text{——}OH^- \rightarrow O_2H^- + OH^- \qquad [13.2]$$

$$2O_2^- + H^+\text{——}OH^- \rightarrow O_2 + O_2H^- + OH^- \qquad [13.3]$$

From the preceding reactions, especially [13.1], which is typical of the very metallic oxides, we see that the oxide ion is a *basic anhydride*. By implication, this type of oxide should react with dilute acid, say HCl:

$$MgO\,(s) + 2HCl\,(aq) \rightarrow Mg^{2+}\,(aq) + 2Cl^-\,(aq) + H_2O \qquad [13.4]$$

This is a simple acid-base type of reaction.

The covalent oxides are generally *acid anhydrides*. These oxides of the nonmetals give acidic aqueous solutions [see equation (13.5)]:

$$O{=}C{=}O\,(g) + H_2O\,(l) \leftrightarrows H_2CO_3\,(aq) \leftrightarrows H^+\,(aq) + HCO_3^-\,(aq) \qquad [13.5]$$

(water drawn as H–:O:–H)

Metallic oxides in which the metal is in a high oxidation state usually react with water also to form acids, and hence are acid anhydrides, e.g., Mn_2O_7:

$$Mn_2O_7\,(s) + H_2O\,(l) \rightarrow 2HMnO_4\,(aq) \rightarrow 2H^+\,(aq) + 2MnO_4^-\,(aq) \qquad [13.6]$$

Once again, this is a simple acid-base reaction.

The amphoteric oxides of metals have the facility of reacting with both acids and bases. This property of the oxide seems dependent upon the fact that the central metal ion has the ability to form a complex ion of the type $M(OH)_x^{n-}$. The following will typify the reactions of an amphoteric oxide:

$$ZnO\,(s) + 2HCl\,(aq) \rightarrow Zn^{2+}\,(aq) \cdot 2Cl^-\,(aq) + H_2O \qquad [13.7]$$

$$ZnO\,(s) + 4OH^-\,(aq) \rightarrow Zn(OH)_4^{2-}\,(aq) \qquad [13.8]$$

These complex hydroxides are also formulated as oxyanions. See Table 13.1.

It is now imperative to say a few words about the hydroxide ion. For the alkali metals and the alkaline earth metals, the hydroxide ion actually exists in the crystal lattice. When such an hydroxide dissolves in water the ions become solvated:

$$M^+OH^-\,(s) + xH_2O \rightarrow M^+\,(aq) + OH^-\,(aq) \qquad [13.9]$$

TABLE 13.1 ***Formulation and Names of Complex Hydrides of Some Amphoteric Oxides***

Name	Hydroxy Formula	Oxyanion
Zincate	$Zn(OH)_4^{2-}$	ZnO_2^{2-}
Aluminate	$Al(OH)_4^-$ or $Al(OH)_6^{3-}$	AlO_2^- or AlO_3^{3-}
Stannite	$Sn(OH)_4^{2-}$	SnO_2^{2-}
Stannate	$Sn(OH)_6^{2-}$	SnO_3^{2-}
Plumbite	$Pb(OH)_4^{2-}$	PbO_2^{2-}
Plumbate	$Pb(OH)_6^{2-}$	PbO_3^{2-}

However, when the M—O bond is highly covalent, solvolytic dissociation is more than likely to occur in this fashion:

$$MOH + H_2O \rightarrow MO^- (aq) + H^+ (aq) \qquad [13.10]$$

$$B(OH)_3 + H_2O \rightarrow (HO)_2BO^- (aq) + H^+ (aq) \qquad [13.11]$$

This type of hydroxide must obviously be considered as acidic. The student should now be able to write the equations illustrating the behavior of amphoteric hydroxides as well as an amphoteric oxide.

Finally, note that oxygen should be the more negative element in all its binary compounds, save the oxygen compounds of fluorine. The reason is, of course, that except for fluorine, oxygen is the most electronegative element.

An alternate classification of oxides and hydroxides is according to their gross structures. That is, they may be classified as ionic, polymeric, or covalent (see Fig. 13.2).

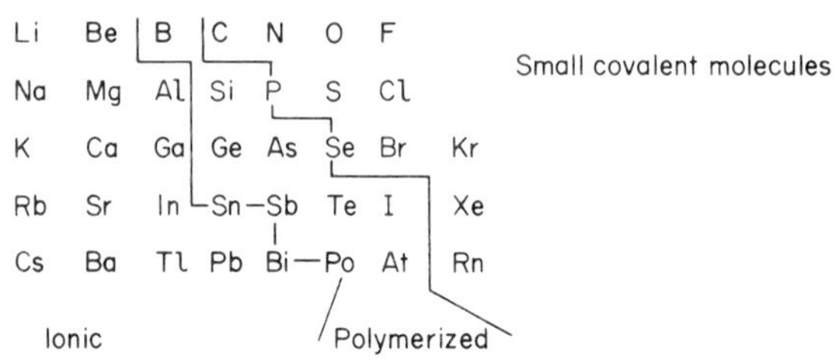

Figure 13.2. Classification of oxides with reference to type of molecule formed.

To organize the chemistry to be presented in the following pages, the student should select his own favorite method of classification. The presentation contained herein will be similar to the chapter on hydride chemistry—discussion of the oxides of the representative elements will be followed by a discussion of the oxides of the transition metals.

REPRESENTATIVE ELEMENTS

Group I-A

They include

Oxides	*Peroxides*	*Superoxides*	*Ozonides*
Li_2O	Li_2O_2		
Na_2O	Na_2O_2	NaO_2	
K_2O	K_2O_2	KO_2	KO_3
Rb_2O	Rb_2O_2	RbO_2	
Cs_2O	Cs_2O_2	CsO_2	

When lithium is burned in air, lithium oxide is formed. However, the major product of sodium metal burning in air is sodium peroxide, Na_2O_2. Potassium, rubidium, and cesium form mostly superoxides (MO_2) when burned in air.

Sodium oxide may be prepared by heating sodium metal with sodium carbonate:

$$Na_2CO_3 + Na \longrightarrow Na_2O + CO_2 \qquad [13.12]$$

The oxides of potassium, rubidium, and cesium are prepared by heating the metal in a limited amount of air and distilling off the excess metal under vacuum. The Group I-A oxides react readily with water, those of the heavier alkali metals doing so with explosive violence.

The *peroxides* of alkali metals are not so difficult to prepare. Lithium peroxide is prepared by dehydrating the hydrated peroxide with phosphorus pentoxide. Sodium peroxide is obtained by simply burning the metal in air, and the peroxides of potassium, rubidium, and cesium are prepared by heating the metal in a limited amount of air and distilling off the excess metal under vacuum. Controlling the amount of air is important in these reactions.

Sodium superoxide, a bright orange solid, may be prepared by bubbling oxygen gas through a solution of sodium amide in liquid ammonia, or by heating sodium peroxide in a pressurized oxygen system:

$$Na_2O_2\,(s) + O_2\,(g) \xrightarrow[300\text{ atm}]{500^\circ C} 2NaO_2 \qquad [13.13]$$

The remaining members of the family form superoxides by merely burning in air. Lithium superoxide has not been reported.

The only well-characterized ozonide, potassium ozonide (KO_3), is prepared by heating KOH in a stream of ozonized oxygen:

$$3KOH + 2O_3 \longrightarrow KO_3 + KOH\cdot H_2O + \tfrac{1}{2}O_2 \qquad [13.14]$$

Another report relates that KO_3 may be prepared by passing ozonized air through a suspension of KO_2 in Freon between -60 and $-100°C$.

Group II

They are

Oxides	*Peroxides*
BeO	BeO_2
MgO	MgO_2
CaO	CaO_2
SrO	SrO_2
BaO	BaO_2

The electronic configuration of the metals of Group II is ns^2; hence, the expected oxide formula is MO, the expected peroxide formula is MO_2, and the expected hydroxide formula is $M(OH)_2$.

The oxides of the metals of Group II-A of the periodic table are solid, refractory, and very stable materials. They may be prepared by various methods. Beryllium oxide and magnesium oxide may be prepared by heating their hydroxides strongly, but magnesium oxide may be made more efficiently by heating the nitrate or sulfate. The easiest preparation of magnesium oxide is simply to burn the metal in air. The oxides of calcium, barium, and strontium are best prepared by heating a carbonate with carbon. These oxides assume a colored hue due to lattice defects in the crystal structure.

The solution chemistry of these oxides is limited because of their slight solubility. The only amphoteric oxide in the group is beryllium oxide. When beryllium oxide is heated strongly, it becomes insoluble in acids. Magnesium oxide and the remaining oxides of the group are all basic in their reactions. These oxides, when placed in water, form hydroxides.

The hydroxides of Group II-A are all insoluble in water, with the exception of those of barium and strontium, which are soluble to a limited extent. Beryllium hydroxide reacts with excess base to form the beryllate ion, $Be(OH)_4^{2-}$ or BeO_2^{2-}, which is soluble in water. The other members of this family do not form complex hydroxides. With the exception of barium, heating of the hydroxide produces the oxide. When barium hydroxide is heated, the yield of barium oxide is very low.

The foregoing facts may be summarized in terms of trends. First, it might be more correct to consider the Group II-A hydroxides as hydrated oxides. Second, as the group is descended, the oxide and the hydroxide become more basic; or, we may say that as the group is descended, the hydroxide becomes less acidic. The student should be capable of rationalizing these facts.

The peroxides of the Group II-A metals have been reported but are not at all well characterized. Beryllium peroxide has been reported; it is prepared by bubbling ozone through a slurry of beryllium hydroxide in Freon-12. The action of hydrogen peroxide on magnesium hydroxide produces magnesium peroxide. An alternate preparation of magnesium peroxide is the reaction of hydrogen peroxide on diethyl magnesium in an ether solution:

$$Mg(C_2H_5)_2 + H_2O_2 \xrightarrow{\text{ether}} MgO_2 + 2C_2H_6 \qquad [13.15]$$

Similarly, calcium peroxide may be prepared by adding hydrogen peroxide to lime water if the temperature of the system is maintained below 10°C. If strontium oxide is heated to 400°C under 100 atm of oxygen gas, strontium peroxide results. The best-known peroxide of the group is barium peroxide. This peroxide may be prepared by heating barium oxide in air or by bubbling oxygen gas through a solution of barium in liquid ammonia.

The chemistry of the Group II-A peroxides is rather limited. Obviously, they would be oxidizing agents.

Group III

They include

Oxides
B_2O_3
Al_2O, Al_2O_3
Ga_2O, Ga_2O_3
In_2O, In_2O_3
Tl_2O, Tl_2O_3

The Group III-A oxides and hydroxides exhibit several interesting properties. In this series of elements, multiple valence is exhibited. The general electronic configuration of the valence electrons is ns^2np^1. It is not at all unreasonable to assume that the elements in this family could give up the lone p electron and assume the (+1) oxidation state. As a matter of fact, aluminum(I), gallium(I), indium(I), and thallium(I) compounds are known. The compounds of thallium that are best characterized and most stable are the thallous (Tl^{+1}) compounds. The +1 oxidation state increases in stability as the group is descended. That is, the ns^2 electrons become less reactive as the group is descended. When such a situation exists, as in thallium, where the ns^2 electrons are relatively unreactive, these two electrons are referred to as an *inert pair*.

Boric oxide is a polymeric oxide of boron that has characteristics belonging to both ionic and covalent species. Boric oxide may be prepared by the ignition of boric acid, $B(OH)_3$ or H_3BO_3. In turn, boric acid can be obtained by the hydrolysis of the boron hydrides and boron halides. The most efficient preparation of boric oxide is the treatment of borax with a strong acid. The basic unit of structure of B_2O_3 is the planar BO_3 arrangement:

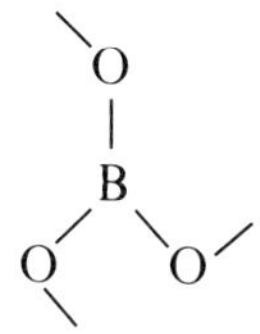

$(BO_3)^{3-}$

The structures of other anionic species, such as $B_2O_5^{4-}$ and $(BO_2^-)_n$, may be rationalized from this basic unit.

On heating boric acid, properly known as orthoboric acid, condensation may occur, producing metaboric acid:

$$H_3BO_3\,(s) \longrightarrow HBO_2\,(s) + H_2O\,(l) \qquad [13.16]$$

Boric acid is also the hydroxide of boron. It is a moderately soluble white solid that behaves as a monoprotic acid in water. The student should regard this reaction mechanistically as a nucleophilic attack by water on $B(OH)_3$:

$$H_2\ddot{O}: + B(OH_3) \rightarrow B(OH)_4^- + H^+ \qquad [13.17]$$

Derivatives of boric acid may be produced by replacing either hydrogen or hydroxyl groups:

$$B_2O_3 + 3CaF_2 + 3H_2SO_4 \rightarrow 2BF_3 + 3CaSO_4 + 3H_2O \qquad [13.18]$$

Equation [13.18] illustrates the *replacement* of the OH in boric acid by F^-. Being an acid, $B(OH)_3$ may undergo esterification, and a borate ester, $B(OR)_3$, may be produced by treatment of the acid with the appropriate alcohol.

Aluminum oxide exhibits several different structural forms but the stoichiometric composition is always the same, Al_2O_3. If a nitrate, a hydroxide, or a sulfate of aluminum is subjected to a very high temperature, above the melting point of the oxide, crystalline corundum is formed. Corundum is a form of aluminum oxide that is soluble in neither acid nor base. On the other hand, if $Al(OH)_3$ is heated gently, an amorphous form of aluminum oxide is formed which is amphoteric, dissolving in both acids and bases:

$$Al(OH)_3 \rightarrow Al^{3+} + 3OH^- \qquad K_b = 5 \times 10^{-33} \qquad [13.19]$$

$$Al(OH)_3 \rightarrow AlO_2^- + H^+ + H_2O \qquad K_a = 4 \times 10^{-13} \qquad [13.20]$$

The order of magnitude of K, where $Al(OH)_3$ is behaving as a base, indicates that it is a very weak base indeed. This fact would indicate that when an aluminum salt is dissolved in water, extensive hydrolysis takes place. Therefore, the carbonate, acetate, cyanide, and formate of aluminum cannot be prepared in aqueous solution because aluminum hydroxide precipitates. However, aluminum salts of strong acids are stable in solution because the anions will not hydrolyze. A convenient preparation of aluminum hydroxide is to add ammonium hydroxide to a solution of an aluminum salt.

The oxide of gallium(III), Ga_2O_3, is a white, polymeric solid prepared by heating an hydroxide, nitrate, or sulfate of gallium(III). The ignition of these salts produces an insoluble oxide that reacts with neither acid nor base. Thie behavior is similar to that of aluminum oxide.

The hydroxide of gallium(III) is a colorless, gelatinous, and amphoteric material:

$$Ga(OH)_3 \rightarrow Ga^{3+} + 3OH^-, \qquad K_b = 5 \times 10^{-37} \qquad [13.21]$$

$$Ga(OH)_3 \rightarrow GaO_2 + H^+ + H_2O \qquad K_a = 1 \times 10^{-15} \qquad [13.22]$$

Compare these values with the values of the equilibrium constants of aluminum hydroxide. What are your conclusions and how do you defend them?

Finally, we shall comment on the oxides and hydroxides of Group III-A elements. Indium(I) oxide is a fairly stable compound. It does not react with water but does react with dilute hydrochloric acid, with the evolution of hydrogen. No indium(I) hydroxide has been reported. Indium(III) hydroxide is a polymeric material having both a covalent and an ionic character. It is not well characterized. Indium(III) hydroxide is a white gelatinous material having some amphoteric character. It does dissolve in base, but indium(III) hydroxide reprecipitates from the solution on standing. Both indium(III) oxide and indium(III) hydroxides are soluble in acids. Thallium(I) oxide (thallous oxide) is a black solid that dissolves in water, forming thallous hydroxide. The thallium hydroxide solution is readily oxidized by air forming thallic oxide [thallium(III) oxide], which precipitates out of solution. Thallous hydroxide and oxide, as well as thallic oxide, all dissolve in acids.

Group IV

These are

Oxides
CO, CO_2
SiO, SiO_2
GeO, GeO_2
SnO, SnO_2
PbO, PbO_2

The electronic configuration of the valence shell of these atoms in the ground state is ns^2np^2. The $+2$ oxidation state may be rationalized by the loss of the np electrons.

Carbon monoxide, CO, is formed by burning carbon or a hydrocarbon in a limited supply of air. Purer samples of CO may be obtained by thermally decomposing nickel tetracarbonyl,

$$Ni(CO)_4 \xrightarrow{200^\circ C} Ni + 4CO \qquad [13.23]$$

and by heating carbon dioxide and carbon at 1000°C.

Some of the reactions of carbon monoxide are summarized in the following list:

$$CO\,(g) + NaOH\,(s) \xrightarrow[\text{pressure}]{160^\circ C} HCOONa\,(s) \qquad [13.24]$$

$$CO\,(g) + Cl_2\,(g) \xrightarrow{\text{catalyst}} COCl_2\,(g) \qquad [13.25]$$

$$CO\,(g) + Ni\,(s) \xrightarrow{80^\circ C} Ni(CO)_4\,(l) \qquad [13.26]$$

$$CO\,(g) + Fe\,(s) \xrightarrow{200^\circ C} Fe(CO)_5\,(l) \qquad [13.27]$$

$$8CO\,(g) + S_8\,(s) \xrightarrow{\text{vapor}} 8COS\,(g) \qquad [13.28]$$

$$2CO\,(g) + B_2H_6\,(g) \xrightarrow{100^\circ C} 2OC{:}BH_3\,(s) \qquad [13.29]$$

This list of reactions clearly points out that in most cases carbon monoxide acts as a Lewis base—i.e., the carbon atom is donating a pair of electrons to a Lewis acid. This fact alone should attract our attention to the structure of carbon monoxide. At first glance, one would suspect that the electrons on the oxygen would be donated, because oxygen (being more electronegative) would be richer in electron density and hence be the part of the molecule acting as the Lewis base.

Carbon monoxide has the ground state term $^1\Sigma$, a bond dissociation energy of 255.8 kcal/mole, and a dipole moment of 0.112 D. Pauling makes use of the following resonance structures to account for the very large resonance energy of 83 kcal/mole:

$$^{+}:\mathrm{C}:\underset{\cdot\cdot}{\ddot{\mathrm{O}}}:^{-} \longleftrightarrow :\mathrm{C}::\ddot{\mathrm{O}}: \longleftrightarrow :\mathrm{C}::\underset{\cdot\cdot}{\mathrm{O}}: \longleftrightarrow {}^{-}:\mathrm{C}:::\mathrm{O}:^{+}$$

(a) (b) (c) (d)

To account for the dipole moment of 0.112 D, Pauling states that structure (d) contributes 60% to the real structure.

The qualitative molecular orbital approach to the problem leads to the electronic configuration for carbon monoxide:

$$KK(\sigma_{2s})^2(\sigma_{2s}^*)^2(\pi_p)^4(\sigma_{2p})^2$$

From this configuration, a bond order of 3 and a ground state term of $^1\Sigma$ are predicted. Coulson modifies this treatment to account for the dipole moment of carbon monoxide, and for the fact that carbon donates the pair of electrons or has the partial negative charge. The following points are a summary of the Coulson modification (assume carbon and oxygen are joined along the z axis):

(a) Carbon hybridizes its $2s$ and $2p_z$ orbitals; the $2p_x$ and $2p_y$ orbitals remain unhybridized.
(b) The inner shells of electrons, $1s^2$ for carbon and the $1s^2$ and the $2s^2$ for oxygen, do not participate in bonding.
(c) Carbon and oxygen bond along the z axis, the $2p_z$ of oxygen and one of the sp_z hybrid orbitals of carbon overlapping to form a sigma bond. Result $\overset{+}{\mathrm{C}}\overset{-}{\mathrm{O}}$.
(d) The $2p_x$ and $2p_y$ orbitals of carbon and oxygen overlap side-to-side to form two π bonds. Result: $\overset{+}{\mathrm{C}}\overset{-}{\mathrm{O}}$.
(e) The other sp_z hybrid orbital on carbon contains a lone pair of electrons. Result: $^{2-}\mathrm{CO}^{+}$.

Point (e) shows a large excess of negative charge on carbon, which more than compensates for the positive charge accumulated in (c) and (d). Therefore, the negative end of the carbon monoxide molecule lies on the carbon atom. This interpretation is also supported by calculation. Hence, the Lewis base donor properties lie with the carbon atom. (This argument is presented diagrammatically in Chapter 4.)

In compounds such as nickel tetracarbonyl, where nickel exists in the zero oxidation state, *synergic bonding* is defined to account for the stability of such compounds (see Fig. 13.3). Here, one of the orbitals of the metal with the proper symmetry overlaps with the *sp* hybrid orbital of carbon monoxide containing the lone pair of electrons to form a sigma bond between the metal and carbon monoxide, carbon monoxide being the

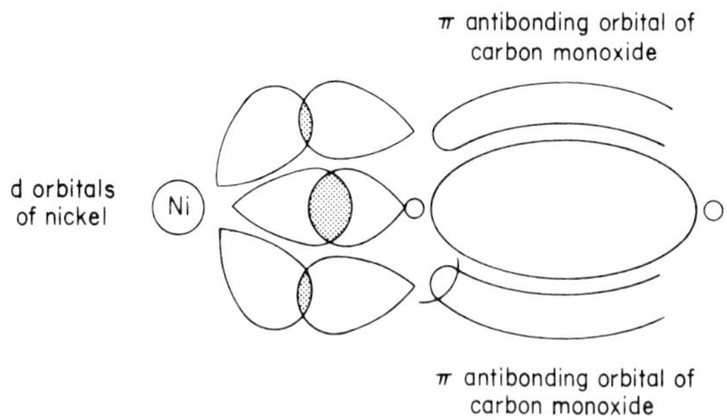

Figure 13.3. Synergic bonding in nickel tetracarbonyl.

donor through the carbon atom. At the same time, electrons in the *d* orbitals of the metal are donated to the relatively low-lying π antibonding orbitals of the carbon monoxide. Here the nickel atom is the donor of electrons. In nickel tetracarbonyl, there is a mutual donation and acceptance of electrons by nickel and carbon monoxide. This is synergic bonding.

Carbon dioxide is an easily compressible, colorless gas, the solid form subliming at $-78°C$. Carbon dioxide is easily produced by burning carbon in an excess of oxygen. If sufficient air is present, combustion of a hydrocarbon will also give carbon dioxide as a product.

The following four resonance forms are those postulated to account for the resonance energy of 33 kcal/mole:

$$:\ddot{O}::C::\underset{\cdot\cdot}{O}: \;\leftrightarrow\; :\underset{\cdot\cdot}{O}::C::\ddot{O}: \;\leftrightarrow\; :\overset{+}{O}:::C:\underset{\cdot\cdot}{\ddot{O}}: \;\leftrightarrow\; {}^{-}:\underset{\cdot\cdot}{\ddot{O}}:C:::\overset{+}{O}:$$

Each structure contributes equally to the resonance energy. The molecular orbital treatment of CO_2 has already been presented (see Chapter 4).

Neither $C(OH)_2$ nor $C(OH)_4$ has been reported, although esters of orthocarbonic acid are known. The organic carboxylic acids, already familiar to the student, have the general formula *R*COOH, where *R* is H or an organic radical. The simplest carboxylic acid is HCOOH, formic acid. These acids will not be discussed. When carbon dioxide is bubbled into water, a small amount of carbonic acid, H_2CO_3, is formed, whereas the greater amount of carbon dioxide simply exists as hydrated carbon dioxide and nothing more. Carbonic acid is a very weak acid; $K_1 = 3.5 \times 10^{-7}$

and $K_2 = 4.4 \times 10^{-11}$. The existence of this acid is based on the fact that carbonates are formed from aqueous solutions of carbon dioxide. Carbonic acid itself has not been isolated.

Most carbonates (except those of the alkali metals) are insoluble in water, but most bicarbonates are fairly soluble in water. Carbonates of many metal ions do not form (e.g., aluminum) due to the hydrolysis of carbonate ion. When sodium carbonate is added to a solution of a strongly acid salt of aluminum, aluminum hydroxide is formed:

$$2Al(OH_2)_6^{3+} + 3CO_3^{2-}\,(aq) \rightarrow 2Al(OH)_3 + 9H_2O + 3CO_2\,(g) \quad [13.30]$$

Unlike carbon monoxide, silicon monoxide is an amorphous brown powder that is pyrophoric in air. Like carbon monoxide, silicon monoxide is a strong reducing agent. If heated to 1000°C, silicon monoxide develops a face-centered cubic crystal lattice. It is generally prepared by heating silicon dioxide and silicon in vacuuo at 1450°C.

On the other hand, silicon dioxide may be made rather easily by the hydrolysis of silicon tetrachloride or by acidifying a solution of sodium silicate. Silicon dioxide is a stable substance but is attacked by base and hydrofluoric acid.

$$SiO_2\,(s) + 2NaOH\,(aq) \rightarrow Na_2SiO_3\,(aq) + H_2O \quad [13.31]$$

$$SiO_2\,(s) + 4HF\,(aq) \rightarrow SiF_4\,(g) + 2H_2O \quad [13.32]$$

Silica itself is a neutral species, and the fundamental unit of structure is the tetrahedral

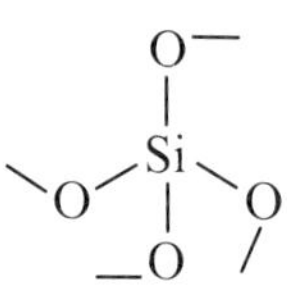

Simple salts containing this SiO_4^{4-} anion are called orthosilicates.

Silicates exist in many different forms. If silica can exist as a conglomeration of SiO_4 tetrahedra, it is not unreasonable that anionic species composed of aggregates of these units can exist. One of the simplest aggregates is the species $O_3Si{-}O{-}SiO_3^{6-}$. Such simple species are rare. The SiO_4^{4-} tetrahedra may cluster in a ring formation, as in the anionic species $Si_3O_9^{6-}$ and $Si_6O_{18}^{12-}$ (see Fig. 13.4). The student should convince himself of the charges on these species. The SiO_4^{4-} unit may also extend itself in the form of a chain (see Fig. 13.5). The repeating unit is marked

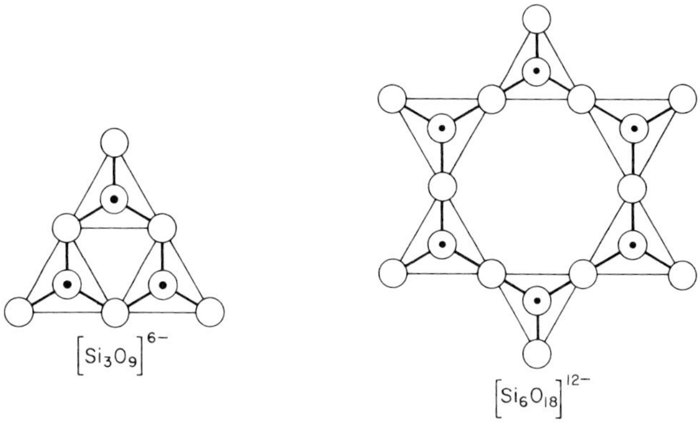

Figure 13.4. Ring-type silicate ions. Each one of the smaller triangles represents an SiO_4^{4-} tetrahedron. The larger circles represent oxide ions while the very small circles represent silicon(IV) ions enclosed in the tetrahedra.

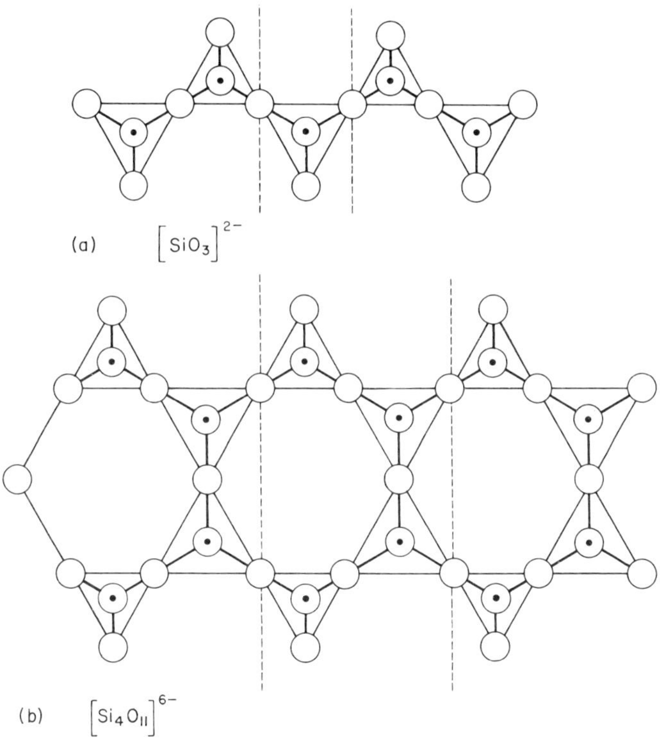

Figure 13.5. (a) Silicate chain: SiO_4^{4-} tetrahedra arranged in chains. The unit is SiO_3^{2-} and is called metasilicate. (b) Silicate bands: silicates may be arranged in sheets and also in three-dimensional lattices.

off by dotted lines. We see that silicon, which is in the +4 oxidation state, possesses two whole oxide ions (−4) and shares two oxide ions, −1 each. Therefore, the metasilicate chain unit, SiO_3^{2-}, has a charge of −2. Double bonds may also exist (see Fig. 13.5). The repeating unit would be the $Si_4O_{11}^{6-}$ anion. The reader should convince himself of the assignment of ionic charge.

Sodium silicates are prepared by fusion of SiO_2 with sodium carbonate:

$$Na_2CO_3 + SiO_2 \xrightarrow[1300\ C]{fusion} silicates + CO_2 \qquad [13.33]$$

Sodium silicate solutions hydrolize to give all of the previously described species and other polymeric species, depending on the concentration of the solution.

Germanium(II) oxide, GeO, a black powder, may be prepared in the following ways:

$$H_2O + GeCl_2 \longrightarrow GeO + 2HCl \qquad [13.34]$$

$$GeO_2 + CO \xrightarrow{900\ C} GeO + CO_2 \qquad [13.35]$$

$$Ge + CO_2 \xrightarrow{700\text{–}900\ C} GeO + CO \qquad [13.36]$$

Germanium(II) oxide is really oxidized to germanium(IV) oxide by heating in air, or by the use of oxidizing agents. Germanium(IV) oxide may also be formed by the oxidation of germanium metal, by the reaction of nitric acid on germanium(IV) sulfide, and by the basic hydrolysis of germanium tetrachloride. Germanium(IV) oxide is not readily attacked by acid or alkali, but it does react violently with hydrofluoric acid.

Germanium(IV) hydroxide, $Ge(OH)_4$, has been reported and is soluble in hot water. It is a weak acid: $K = 2 \times 10^{-9}$. It precipitates from solution as a colloidal precipitate. As with hydroxides in this group, generally, germanium hydroxide is best considered to be $GeO_2 \cdot 2H_2O$.

Sodium metagermanate, Na_2GeO_3, has been reported. Beryllium orthogermanate, Be_2GeO_4, is isomorphous with $GeSiO_4$, germanium silicate.

Tin(II) oxide, SnO, is a brown powder but is blue when crystalline. It may be prepared by heating tin(II) hydroxide at a temperature below 100°C in a closed system. If tin(II) hydroxide is heated in air, tin(IV) oxide is eventually formed.

Tin(II) hydroxide may also be formed by treating solutions of tin(II) chloride or tin(II) sulfate with ammonium hydroxide or sodium hydroxide. Tin(II) hydroxide has no definite composition; it precipitates in colloidal form. This material is amphoteric, and its acidic and basic properties are about equal. However, in concentrated alkali, tin(II) oxide disproportionates:

$$2SnO + 2KOH \rightarrow Sn + K_2SnO_3 + H_2O \qquad [13.37]$$

Tin(IV) oxide, SnO_2, is very well characterized. It is prepared by the reaction of nitric acid on tin metal:

$$3Sn + 4HNO_3 \rightarrow 3SnO + 4NO + 2H_2O \qquad [13.38]$$

If tin(IV) oxide is heated strongly, it becomes impervious to all reagents except hot concentrated acids and alkali.

When amphoteric tin(IV) oxide is allowed to react with alkali, a stannate is formed. If a stannate is treated with acid, a colloidal precipitate is formed, and this substance is sometimes written as $Sn(OH)_4$. But, again, it would be better written $SnO_2 \cdot 2H_2O$. Tin(IV) hydroxide is amphoteric and is a much stronger acid than it is a base.

Lead(II) oxide, PbO, may be formed by heating lead in air. This oxide may exist in two forms. A yellow powder is formed initially, but if the yellow powder is heated at 500°C, it cools to a red form. Both powders have the empirical formula PbO.

The slightly soluble lead(II) hydroxide gives a solution that turns red litmus blue. Lead(II) hydroxide reacts with excess base to form plumbites.

The dark brown or black lead(IV) oxide may be prepared by fusion of lead(II) oxide with sodium chlorate or nitrate. It may also be prepared by allowing red lead to react with nitric acid. The insoluble lead(IV) oxide is filtered off. This oxide is amphoteric; it reacts with excess base to form plumbates. Lead(IV) oxide is also a strong oxidizing agent. If heated with concentrated hydrochloric acid, it produces chlorine. Sulfur dioxide reacts with lead(IV) oxide to form lead sulfate.

Finally, we note that red lead, Pb_3O_4, is a mixed oxide and is equivalent to $2PbO \cdot PbO_2$. This oxide may be prepared by heating lead(II) oxide in air at 400°C.

At this point in the presentation, it might be well to point out a few trends:

1. All elements in this group form oxides that may be formulated as XO and XO_2. They differ greatly.
2. All XO_2 oxides dissolve in alkali to give oxyacid salts, the anion being XO_3^{2-}.
3. As we descend the group, we note the following:
   (a) The metallic character of the elements increases, although not regularly.

(b) The lower oxidation state of the element becomes more stable. For example, lead commonly occurs in the +2 oxidation state. This observation has earned the two 6*s* electrons in lead the appellation of the *inert pair*. This effect of the inert pair may be exemplified by the fact that tin(IV) oxide may be prepared by the action of nitric acid on tin(II) oxide, but lead(IV) oxide may not be prepared in this manner.

(c) The acid character of the oxide and hydroxide decreases.

4. In reference to Sn and Pb, we should note that the oxides XO and XO_2 are both amphoteric:

$$XO + 4OH^- \rightarrow X(OH)_4^{2-} \text{ or } XO_2^{2-} + 2H_2O \qquad [13.39]$$

$$XO_2 + 6OH^- \rightarrow X(OH)_6^{2-} \text{ or } XO_3^{2-} + 3H_2O \qquad [13.40]$$

The Pb^{4+} ion cannot exist in solution. If it is produced, it immediately precipitates as PbO_2.

5. The higher the oxidation state of a given element, the more acidic and the less basic it is.
6. There are some less obvious diagonal relationships in the periodic table. For instance, both lithium and magnesium burn in air to give the normal oxide. Also, both of these elements form unstable nitrates and carbonates; they form chlorides that are soluble in organic solvents.

Similarities also exist between beryllium and aluminum and between boron and aluminum. Both beryllium and aluminum dissolve in concentrated alkali with the evolution of hydrogen, and their carbides (BeC_2 and Al_4C_3) yield methane on hydrolysis.

Group V

These include

Oxides				
N_2O	P_4O_6	As_4O_6	Sb_4O_6	Bi_2O_3
NO	P_4O_{10}	As_4O_{10}	Sb_4O_{10}	Bi_2O_5
NO_2				
N_2O_4				
N_2O_3				
N_2O_5				

The electronic configuration of the valence electrons of the elements in this group of the periodic table is ns^2np^3. The oxidation states expected of these elements would be +3, +5, and −3. The nitrogen atom, however, assumes all of the oxidation states from -3 to $+5$, thereby distinguishing it from the other members of the group. Table 13.2 lists some nitrogen-oxygen species and names them. Nitrogen assumes only positive oxidation states with oxygen because oxygen is more electronegative than nitrogen.

TABLE 13.2 ***Binary Nitrogen-Oxygen Species***

Formula	Name	Formula	Name
N_2O	Nitrogen(I) oxide	NO_2^+	Nitrosonium ion
NO	Nitrogen(II) oxide	NO_2^-	Nitrite ion
$—NO$	Nitroso group	NO_2^{2-}	Nitroxylate ion
NO^+	Nitronium ion	N_2O_3	Dinitrogen trioxide
NO^-	Nitrosyl ion	N_2O_5	Dinitrogen pentoxide
NO_2	Nitrogen(IV) oxide	NO_3^-	Nitrate ion
$—NO_2$	Nitryl group		

Ammonia, NH_3, exemplifies nitrogen in the -3 oxidation state; hydrazine, H_2NNH_2, the -2 oxidation state; hydroxylamine, NH_2OH, the -1 oxidation state; and N_2 the 0 oxidation state.

Table 13.2 readily reveals the fact that nitrogen-oxygen chemistry is rather complicated, and a complete presentation would be a formidable task. We shall here present a mere outline to give the student a feeling for the behavior of the nitrogen atom in different chemical environments. This should help him appreciate the complexity of the study and aid him in making comparisons and observations with regard to the other elements of Group V and the other groups of the periodic table.

Nitrogen(I) oxide, N_2O, is a colorless, sweet-smelling gas that may be prepared by heating a solution of hyponitrous acid, $H_2N_2O_2$. This acid may be obtained by treating its sodium salt with a solution of sulfuric acid. The nitrogen(I) oxide thus prepared is very pure and need only be dried. When nitrogen(I) oxide is prepared by heating ammonium nitrate *gently*, a variety of products is formed, including nitrogen(II) and nitrogen(IV) oxides as well as hydrogen and oxygen. An aqueous solution of nitrogen(I) oxide has the conductivity of pure water. By itself, nitrogen(I) oxide is relatively inert, but mixtures of nitrogen(I) oxide and hydrogen gas are explosive. Sodium azide, NaN_3, may be formed by heating sodium amide with nitrogen(I) oxide.

Nitrogen(II) oxide, NO, is a colorless gas. A simple counting of electrons will reveal that nitrogen(II) oxide is a free radical because there is an odd number of electrons and one must be unpaired. It is relatively inert when compared with the activity of other free radicals. Nitrogen(II) oxide may be prepared by allowing copper to react with a dilute solution of nitric acid.

This oxide is comparatively stable thermally, beginning to decompose into nitrogen and oxygen at 1000°C. When compressed or liquified, it forms the dimer N_2O_2, which slowly forms nitrogen(III) oxide and nitrogen(I) oxide. Nitrogen(II) oxide is more reactive than nitrogen(I) oxide. With base, it forms nitrite ion, nitrogen(I) oxide, and water. Nitrogen(II) oxide reacts with oxygen and chlorine to form nitrogen(IV) oxide and nitrosyl chloride (NOCl), respectively.

Nitrogen(IV) oxide, NO_2, is a paramagnetic species because it has an unpaired electron; when prepared, it immediately forms an equilibrium mixture of NO_2 and N_2O_4. The dimeric species is diamagnetic. For the

$$N_2O_4 \rightarrow 2NO_2, \qquad \Delta H = -12 \text{ kcal/mole} \qquad [13.41]$$

reaction the enthalpy is -12 kcal/mole: consequently, there is almost always some of each species present in any sample of NO_2. At 27°C, $\frac{1}{3}$ of the mixture is NO_2; at 100°C, 94% of the mixture is NO_2. Above 100°C, the gas is essentially 100% NO_2: at this temperature, the gas begins to decompose into nitrogen(I) oxide and oxygen.

In the liquid state, one would expect NO_2 to be almost totally N_2O_4. The liquid does not ionize, but seems to behave chemically as nitronium and nitrate ions. This liquid reacts with sodium metal to produce nitrogen(I) oxide and sodium nitrate, whereas potassium chloride reacts with dinitrogen tetroxide to form potassium nitrate and nitrosyl chloride.

Nitrogen(IV) oxide may be prepared by heating a heavy metal nitrate, such as lead nitrate, or by the reaction of concentrated nitric acid on copper. As previously mentioned, the action of dilute nitric acid on copper produces nitrogen(I) oxide, but nitrogen(I) oxide is difficult to observe because its exposure to air produces nitrogen(IV) oxide. The nitrogen(IV) oxide gas dimerizes and the brown gas is what is observed.

The NO_2, N_2O_4 mixture behaves as an oxidizing agent, similar in strength to aqueous bromine. Regarding the mixture as NO_2, we find that it reacts with copper to form copper(I) oxide and nitrogen(II) oxide. At room temperature, nitrogen(IV) oxide reacts with carbon monoxide to produce carbon dioxide and nitrogen(I) oxide.

Nitrogen(III) oxide, N_2O_3, is a blue solid that decomposes above -20°C into a mixture of nitrogen(II) oxide and nitrogen(IV) oxide. Equation [13.42] represents the common preparation of this oxide:

$$4H_2O + 4HNO_3 + As_4O_6 \rightarrow 4H_3AsO_4 + 2N_2O_3 \qquad [13.42]$$

A water solution of dinitrogen trioxide [nitrogen(III) oxide] is nitrous acid, HNO_2. This solution is the only stable form of HNO_2 ($K = 4.5 \times 10^{-4}$).

This acid is both an oxidizing agent and a reducing agent. It is oxidized to nitric acid by potassium permanganate and hydrogen peroxide: it is reduced to nitrogen(II) oxide by ferrous ion, mercury, and hydrogen iodide. An alternate formulation of nitrous acid is (HO)NO. There is no doubt at all about the acid character of this oxide (N_2O_3) and hydroxide (HONO) of nitrogen.

Dinitrogen pentoxide, N_2O_5, is a colorless solid melting at 41°C. It is prepared by the action of ozone on the lower oxides of nitrogen. It may also be obtained by the distillation of a mixture of nitric acid and phosphorus pentoxide. This reaction immediately tags nitrogen(V) oxide as the anhydride of nitric acid, because phosphorus pentoxide is a well-known dessicant.

Nitric acid may also be prepared simply by heating concentrated sulfuric acid with potassium nitrate. The resulting colorless, fuming mixture, boiling at 86°C, is considered as concentrated HNO_3 that is only partially hydrated.

The reactions of nitric acid are varied. With metals above hydrogen in the electromotive series, hydrogen, ammonia, and hydroxyl amine may be produced. With metals below hydrogen in the electromotive series, nitric acid tends to react to form the nitrate of the metal and nitrogen(II) oxide or nitrogen(IV) oxide, depending on the concentration of the acid.

An alternate formulation of nitric acid is $(HO)NO_2$. The oxide and hydroxide of this state of the nitrogen atom are definitely acidic.

There are two oxides of phosphorus: P_4O_6 or P_2O_3 and P_4O_{10} or P_2O_5. The P_4 formulation is structurally correct and is based on the fact that phosphorus itself consists in four phosphorus atoms arranged in a tetrahedron. Bridging each of the P—P bonds with an oxygen atom will give P_4O_6. Addition of an oxygen doubly bonded to each of the phosphorus atoms in P_4O_6 results in P_4O_{10}. Molecular weight determinations support these formulations. These phosphorus oxides are pictured in Fig. 13.6.

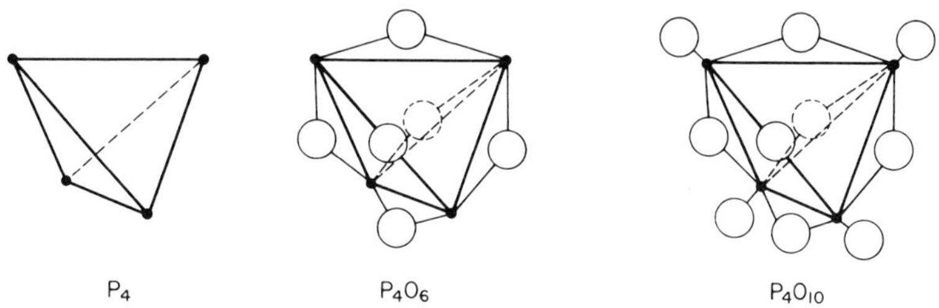

Figure 13.6. Tetrahedral arrangement of P_4 and two oxides of phosphorus.

Burning phosphorus in a limited amount of oxygen produces the white solid P_4O_6, which melts at 23.8°C. It has a garliclike odor and is very poisonous. When dissolved in water, it produces phosphorus acid H_3PO_3 or $HPO(OH)_2$. This acid may be isolated as a white solid having a melting point of 73.6°C. The acid constants for this acid are $K_1 = 1.16 \times 10^{-2}$ and $K_2 = 7 \times 10^{-7}$. When H_3PO_3 is prepared by the hydrolysis of PCl_3, it should be moderated because of the exothermicity of the reaction. This acid is generally a reducing agent.

Phosphorus(V) oxide, P_4O_{10} or P_2O_5, is a white solid that has a great affinity for water. It is one of the finest dessicants available. Oxidation of P_4O_6 produces P_4O_{10}. When P_4O_{10} reacts with water, orthophosphoric acid, H_3PO_4 or $(OH)_3PO$, is formed. This acid is a white crystalline material, melting point 42°C, that is very deliquescent. Phosphoric acid may also be produced by the direct action of nitric acid on red phosphorus, viz.,

$$P_4 + 10HNO_3 + H_2O \rightarrow 4H_3PO_4 + 5NO + 5NO_2 \qquad [13.43]$$

The acid constants for this acid are $K_1 = 9 \times 10^{-3}$, $K_2 = 6 \times 10^{-8}$, and $K_3 = 1 \times 10^{-12}$. It is quite possible to form several types of phosphates by removal of one or more hydrogen atoms. These are exemplified by the following sodium compounds:

NaH_2PO_4	sodiumdihydrogen phosphate, a primary phosphate
Na_2HPO_4	disodium hydrogen phosphate, a secondary phosphate
Na_3PO_4	trisodium phosphate, a tertiary phosphate

Apparently all primary phosphates are soluble, but only the alkali metal salts of the tertiary and secondary phosphates are soluble.

There are several *condensed* acids of phosphoric acid. They are called condensed acids because their formulas may be rationalized by removal of a water molecule from one or more molecules of the parent acid. The structure of orthophosphoric acid, and its condensed acids may be formulated as follows:

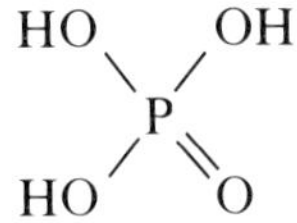

orthophosphoric acid

[13.44]

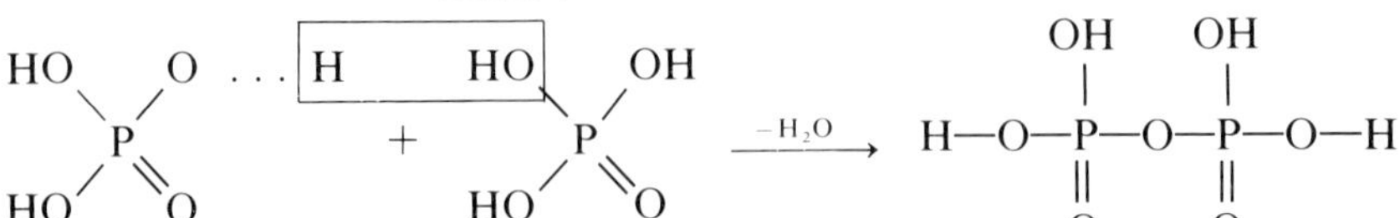

pyrophosphoric acid

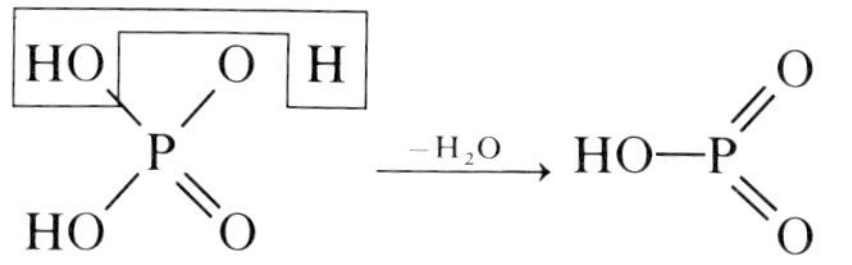

metaphosphoric acid

[13.45]

The term *pyro* usually signifies that a molecule of water is lost from two molecules of the parent acid, whereas *meta* signifies that the condensed acid contains the least amount of water that it can contain. The acid constants are listed below as follows:

Orthophosphoric Acid	*Pyrophosphoric Acid*	*Metaphosphoric Acid*
$K_1 = 9 \times 10^{-3}$	$K_1 = 1.4 \times 10^{-1}$	
$K_2 = 6 \times 10^{-8}$	$K_2 = 1.1 \times 10^{-2}$	No data are available because this acid
$K_3 = 1 \times 10^{-12}$	$K_3 = 2.1 \times 10^{-7}$	rapidly hydrolyzes to pyrophosphoric
	$K_4 = 4.1 \times 10^{-10}$	acid.

Inspection of these acid constants will lead to the conclusion that the condensed acids are stronger than the parent acid. This is a general trend. Theoretically, the process of condensation should be applicable to all oxyacids, resulting in pyro and meta condensates.

Arsenic also forms two oxides: As_4O_6 or As_2O_3, and As_4O_{10} or As_2O_5. Arsenic(III) oxide, As_4O_6, which may be prepared by burning arsenic in a limited amount of air, occurs in two crystalline forms. Both are white, poisonous solids. The trioxide is sparingly soluble in water and its solution is slightly acidic, probably due to the formation of H_3AsO_3 or $As(OH)_3$, arsenous acid. This oxide is amphoteric; it reacts with alkali and carbonates to form arsenites (AsO_3^{3-}), whereas on the other hand it reacts with concentrated sulfuric acid to form $(AsO)_2SO_4$, arsonium sulfate. Arsenic(III) oxide also reacts with sulfur trioxide at 100°C to form arsenic(III) sulfate. Arsenic(III) oxide may be oxidized to arsenic(V) oxide by ozone, hydrogen peroxide, halogens, or nitric acid.

Arsenic(V) oxide, As_4O_{10}, a white deliquescent solid, is best prepared by heating the hydrated arsenic acid [see equation (13.46)].

$$4(H_3AsO_4)_3 \cdot H_2O \rightarrow 3As_4O_{10} + 22H_2O \qquad [13.46]$$

This solid decomposes on heating to red heat to form arsenic(III) oxide. Arsenic(V) oxide is the anhydride of orthoarsenic acid. This oxide of arsenic behaves as an oxidizing agent,

$$As_4O_{10} + 10PCl_5 \rightarrow 10POCl_3 + AsCl_3 + Cl_2 \qquad [13.47]$$

Arsenous acid, H_3AsO_3 or $(HO)_3As$, a weak acid ($K_1 = 6 \times 10^{-10}$), is known only in solution. This acid dissolves in hot sodium hydroxide solution giving Na_3AsO_3, sodium arsenite, on evaporation. Further evidence for the existence of this acid may be found in the fact that it reacts with alcohols to produce esters, $As(OR)_3$, where R may be a methyl, ethyl, or propyl group.

Orthoarsenic acid, H_3AsO_4 or $(HO)_3AsO$, is a white solid melting at 36°C, and it is quite soluble in water. The acid constants are $K_1 = 5.6 \times 10^{-3}$, $K_2 = 1.7 \times 10^{-7}$, and $K_3 = 3 \times 10^{-12}$. Salts of this acid are called arsenates.

Condensed acids of orthoarsenic acid may also be formed. It should also be pointed out that methyl derivatives of orthoarsenic acid exist. Structural formulas and names are given below as follows:

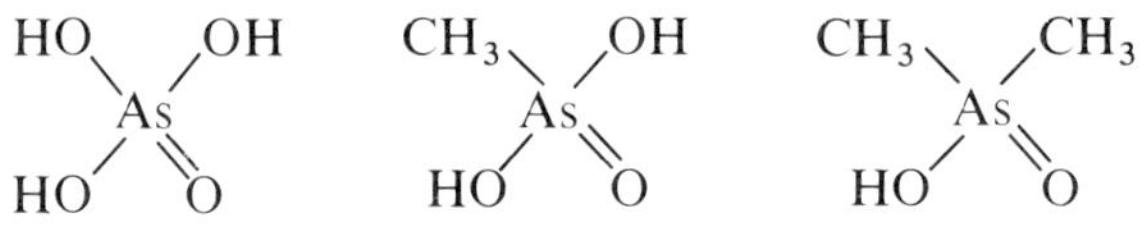

orthoarsenicacid methylarsonicacid dimethylarsinicacid

Antimony also, as would be expected, forms two oxides Sb_4O_6 or Sb_2O_3 and Sb_4O_{10} or Sb_2O_5. Antimony(III) oxide, Sb_4O_6, is a white solid formed by burning antimony in air. This oxide may also be formed by heating antimony(III) sulfide in air or by the hydrolysis of antimony trichloride in sodium carbonate solution. This oxide is amphoteric. It dissolves in concentrated acids to give antimonyl salts such as $(SbO)_2SO_4$, antimonyl sulfate, and $SbONO_3$, antimonyl nitrate. These salts completely hydrolyze in hot water. This same oxide reacts with alkali to give antimonite salts.

Antimony(V) oxide, Sb_4O_{10} or Sb_2O_5, may be prepared by igniting the residue of the reaction of nitric acid on antimony or antimony(III) oxide. The temperature must be kept below 440°C, because at this temperature antimony(IV) oxide, Sb_2O_4, is formed. Antimony(V) oxide is readily reduced by hydrogen or carbon to antimony(III) oxide. Antimony(V) oxide is nearly insoluble in water, but it does redden damp litmus paper. This oxide is completely unaffected by dilute or concentrated sulfuric acid, but, if fused with potassium hydroxide, an *antimonate* (SbO_4^{3-} or SbO_3^-) is formed. An acid of antimony(V) that is formulated as $HSb(OH)_6$ has not been isolated itself, but the potassium salt, $KSb(OH)_6$, has been isolated.

If bismuth metal, bismuth(III) hydroxide, bismuth basic carbonate, $(BiO)_2CO_3$, or bismuth nitrate is heated to redness in air, a pale yellow powder having both ionic and covalent character and polymeric in form is produced. This powder is bismuth trioxide or bismuth(III) oxide, Bi_2O_3.

The hydrated form of bismuth(III) oxide is prepared by adding ammonium hydroxide to a solution of bismuth(III) nitrate. The solid formed is formulated as bismuth(III) hydroxide and is basic in nature. If bismuth hydroxide is heated, BiO(OH) is formed, whereas continued heating generates bismuth(III) oxide. Note that bismuth(III) oxide is soluble in acids, but insoluble in alkaline solutions.

Bismuth(V) oxide, Bi_2O_5, is a red powder that looses oxygen at 100°C. The bismuth(V) oxidation state is not common, although $NaBiO_3$, sodium bismuthate, is well known and the acid, $HBiO_3$, has been reported.

In summary, we might again stress, in looking at the oxides of Group V, that the higher oxidation state of the element becomes less stable as we descend the group, and the metallic character of the elements in a given oxidation state increases with an increase in atomic number.

Group VI

These are

Oxides		
SO	TeO	
S_2O_3		
SO_2	TeO	SeO_2
SO_3	TeO_3	SeO_3
SO_4		

The ground state electronic configuration of the valence shell of the elements of Group VI is ns^2np^4. The Lewis structure may be written $:\ddot{\underset{..}{E}}.$ It could be imagined that the elements of this group could donate a pair of electrons to oxygen in a Lewis acid-base reaction, and thus rationalize the oxides EO, EO$_2$, and EO$_3$. Obviously, this explanation is oversimplified, but it is consistent with the chemistry of the oxides of Group VI. In similar terms, the student should be able to rationalize the formation of S_2O_3 and SO_4 in terms of Lewis acid-base reactions.

Sulfur monoxide, SO, is formed by the reaction of the vapors of sulfur and sulfur dioxide under the influence of an electrical discharge. Sulfur(II) oxide is a very unstable species, reacting with water to form $H_2S_2O_4$, dithionic acid.

Sulfur dioxide, SO_2, is a gas that condenses at -10°C and has a freezing point of -75°C. Because of its liquid temperature range, it has been used as a solvent (see Chapter 8). Sulfur dioxide may be prepared by simply burning sulfur in air, roasting sulfides, or by the action of acids on sulfites. If a water solution of sulfur dioxide is heated, it breaks down to form sulfur and sulfur trioxide, the sulfur trioxide immediately reacting with the water to form sulfuric acid.

When sulfur dioxide is dissolved in water, it forms an acid, H_2SO_3 or $(HO)_2SO$, only to a limited extent. The acid constants are $K_1 = 1.3 \times 10—^2$ and $K_2 = 1.0 \times 10^{-7}$. The acid (which may be $SO_2 \cdot 7H_2O$) exists only in solution and will react with magnesium to produce hydrogen. Light causes the acid to decompose. Sulfur dioxide may act as an oxidizing agent. It reacts with moist hydrogen sulfide to produce water and sulfur. In a neutral or moderately acid solution, sulfur dioxide will reduce manganese(VIII) to manganese(II), iodine to iodide, and mercury(II) to mercury.

Sulfur trioxide, SO_3, is a colorless liquid freezing at 15°C and boiling at 45°C. It is prepared by reacting sulfur dioxide with oxygen. Platinum, vanadium pentoxide, or iron(III) oxide speed up the kinetics of this thermodynamically favored reaction. Sulfur trioxide may also be produced by the distillation of pyrosulfuric acid. Chemically, sulfur trioxide oxidizes bromide to bromine, iron to iron(II), and zinc to zinc(II). It reacts with amines to form adducts of the type $R_3N:SO_3$. Sulfur trioxide is an acidic oxide, reacting with water to form sulfuric acid. The measured dipole moment of sulfur trioxide is zero. indicative of a planar molecule.

Sulfur tetroxide has been reported, and it is the anhydride of peroxymonosulfuric acid, H_2SO_5.

Sulfuric acid, H_2SO_4 or $(HO)_2SO_2$, is familiar to all students. Industrially, it is prepared by the contact process and the lead chamber process. The concentrated form of this acid is usually a 98% solution with a constant boiling point of 338°C.

Sulfuric acid has several modes of dissociation:

$$2H_2SO_4 \leftrightharpoons H_3SO_4^+ + HSO_{4-} \quad [13.48]$$

$$H_2SO_4 \leftrightharpoons SO_3 + H_2O \quad [13.49]$$

Subsequently, these species may recombine to form several new species and different equilibria:

$$SO_3 + H_2SO_4 \leftrightharpoons H_2S_2O_7 \quad [13.50]$$

$$H_3SO_4^+ + H_2O \leftrightharpoons (H_3SO_4 \cdot H_2O)^+ \quad [13.51]$$

$$HSO_4^- + H_2O \leftrightharpoons (HSO_4 \cdot H_2O)^- \quad [13.52]$$

$$H_2SO_4 + H_2O \leftrightharpoons H_2SO_4 \cdot H_2O \quad [13.53]$$

Phase studies of sulfuric acid have shown the existence of many species:

$$\begin{array}{ll} H_2S_4O_{13} & H_2SO_4 \cdot H_2O \\ H_2S_3O_{10} & H_2SO_4 \cdot 2H_2O \\ H_2S_2O_7 & \quad\vdots \qquad \vdots \\ & \quad\vdots \qquad \vdots \\ & \quad\vdots \qquad \vdots \\ H_2SO_4 & H_2SO_4 \cdot 8H_2O \end{array}$$

Thus, the sulfuric acid system is not a simple one.

Concentrated sulfuric acid is an oxidizing agent, oxidizing copper to copper(II), silver to silver(I), lead to lead(II), and mercury to mercury(II), whereas the acid is reduced to sulfur dioxide. The dilute acid is not an oxidizing agent but acts as a normal acid, reacting metals above hydrogen in the electromotive series to release hydrogen. Concentrated sulfuric acid is also a good dessicant. Care must be taken in diluting sulfuric acid because of the great exothermicity of the reaction of the concentrated acid with water.

Chemical substitutions may be made on the sulfuric acid molecule. Sulfonic acids are those in which one of the hydroxyl groups on the acid is replaced by an organic radical, as in $R(OH)SO_2$. The compound $C_6H_5(HO)SO_2$ is called phenylsulfonic acid. Salts of these acids would be called sulfonates; $NaC_6H_5SO_3$ is sodium phenyl sulfonate.

Peroxysulfuric acids are known. These acids have the peroxide linkage; $H—O—O—S(OH)O_2$ is monoperoxysulfuric acid. There also exists a class of compounds called polythionic acids, whose general formula is $H_2S_nO_6$, with n as great as six. Dithionic acid has the structure

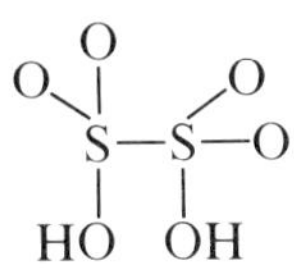

$H_2S_2O_6$

These acids are stable only in solution, but their salts have been isolated.

Selenium forms two stable oxides. Selenium(IV) oxide, SeO_2, is a colorless crystalline material that sublimes at 315°C. It may be prepared by burning selenium in air or by the reaction of nitric acid on selenium. In the vapor state, the molecular composition seems to be $(SeO_2)_3$, whereas the solid form seems to be polymeric, $(SeO_2)_x$.

Selenous acid, H_2SeO_3 or $(HO)_2SeO$, may be isolated when selenium(IV) oxide is dissolved in water and the solution is evaporated. The acid takes the appearance of colorless crystals that are very hygroscopic. The crystals tend to revert to selenium(IV) oxide on exposure to air. The acid constants are $K_1 = 2.7 \times 10^{-3}$ and $K_2 = 4.8 \times 10^{-9}$. A solution of the acid is easily oxidized to selenic acid, H_2SeO_4, and easily reduced to selenium.

Selenium(VI) oxide, SeO_3, is a white deliquescent solid with a melting point of 118°C. Selenium(VI) oxide is prepared by the electrical discharge of selenium(IV) oxide vapor with oxygen at low pressures. It may also be prepared by the dehydration of a selenic acid solution.

Selenic acid is best prepared by the following reaction:

$$Ag_2SeO_3 + H_2O + Br_2 \longrightarrow H_2SeO_4 + 2AgBr \qquad [13.54]$$

This acid is a stronger oxidizing agent than sulfuric acid. Phase studies have revealed the following moieties:

$$H_2Se_2O_7, \quad H_2SeO_4 \cdot 2H_2O$$

$$H_2SeO_4, \quad H_2SeO_4 \cdot 4H_2O$$

Selenic acid is a colorless crystalline solid that melts at 57°C. The crystals are hygroscopic.

Tellurium(II) oxide, TeO, a black powder, has been reported. When dissolved in sulfuric acid it gives a red solution, and tellurium precipitates out of solution on standing.

Tellurium(IV) oxide, TeO_2, is prepared by burning tellurium in air or oxidizing tellurium with a strong oxidizing agent. Tellurium(IV) oxide is composed of colorless crystals that sublime. When heated, tellurium(IV) oxide decomposes into tellurium(II) oxide and oxygen. Tellurium(IV) oxide also forms adducts with the composition $TeO_2 \cdot 3HCl$ and $TeO_2 \cdot 2HCl$.

Tellurium(VI) oxide, TeO_3, is an orange-yellow powder resulting from the thermal decomposition of telluric acid, H_6TeO_6. The oxide decomposes to tellurium(IV) oxide and oxygen on being heated to red heat. This oxide may be used as an oxidizing agent and it disproportionates in concentrated hydrochloric acid.

$$TeO_3 + HCl\text{ (conc)} \longrightarrow Cl_2 + TeO + TeCl_4 \qquad [13.55]$$

Tellurium(VI) oxide does not react with water to form an acid.

Tellurous acid, H_2TeO_3, is a weak acid ($K_1 = 6 \times 10^{-6}$). The acid itself has not been isolated, but tellurites, salts of the acid such as sodium tellurite, Na_2TeO_3, have been prepared by dissolving tellurium(IV) oxide in alkali solution. In solution, tellurites are easily oxidized to tellurates. Tellurites are readily reduced to tellurium by sulfur dioxide, tin, zinc, copper, or mercury.

The formula for orthotelluric acid is H_6TeO_6. This acid is prepared by oxidizing tellurium with nitric acid or hydrogen peroxide and evaporating the solution to obtain the crystalline acid. The acid constants are $K_1 = 1.53 \times 10^{-8}$ and $K_2 = 4.7 \times 10^{-11}$. When crystals of orthotelluric acid are heated, tellurium(VI) oxide and water are formed, and continued heating will produce tellurium(IV) oxide and oxygen.

Group VII

They include

Oxides			
OF_2	Cl_2O	Br_2O	$(I_4O_9)_x$
O_2F_2	ClO_2	$(BrO_2)_x$	$(I_2O_4)_x$
	ClO_3	$(BrO_3)_x$	I_2O_5

Oxygen difluoride, OF_2, which may be explosive and which is more poisonous than fluorine alone, is prepared by bubbling fluorine through a 2% solution of sodium hydroxide. It slowly decomposes into fluorine and oxygen. This oxide is a neutral oxide in the sense that when it is dissolved in water it does not give an oxygenated fluorine acid. Its reaction with water is

$$OF_2 + H_2O \rightarrow O_2 + 2HF \qquad [13.56]$$

Dioxygen difluoride, O_2F_2, is prepared by an electrical discharge in a mixture of oxygen and fluorine at liquid air temperatures. This orange solid decomposes at $-100°C$.

An aqueous solution of hydrogen fluoride is hydrofluoric acid. No oxyacids of fluorine have been isolated. However, hypofluorites, such as trifluoromethyl hypofluorite, CF_3OF, have been reported. It might be argued that hypofluorous acid is just as valid as carbonic acid, which is known only in its derivatives. Some other oxyanions of fluorine have been reported in salts such as KFO and $AgFO_3$, but the reports have been questioned, and the reactions attributed to these species have actually been caused by the impurities in the starting materials from which they were synthesized. One rationalization for oxyacids of fluorine not being isolated is related to the octet restriction of the second-row elements. The orbitals on the next energy level are too high in energy to be utilized to expand the fluorine octet. However, this has been questioned!

Chlorine(I) oxide, Cl_2O, a yellowish gas, is prepared by allowing chlorine to react with mercury(II) oxide:

$$2Cl_2 + 2HgO \rightarrow HgO \cdot HgCl_2 + Cl_2O \qquad [13.57]$$

If chlorine(I) oxide is heated, it explodes. This oxide of chlorine is a strong oxidizing agent and is the anhydride of hypochlorous acid.

The yellowish gas, ClO_2, chlorine(IV) oxide, may be prepared in two ways:

$$2AgClO_3 + Cl_2 \rightarrow 2AgCl + 2ClO_2 + O_2 \qquad [13.58]$$

$$3KClO_3 + 3H_2SO_4 \rightarrow 3KHSO_4 + H_2O + 2ClO_2 \qquad [13.59]$$

This oxide is very explosive. It is stable in carbon tetrachloride solution and has been found to be paramagnetic. Chlorine(IV) oxide is soluble in water, but light causes its decomposition into hydrochloric and chloric acids.

Another oxide of chlorine, ClO_3, chlorine(VI) oxide, decomposes at 0°C. It is prepared by the action of ozone on chlorine(IV) oxide. This species is diamagnetic even in solution, and hence appears to dimerize. In most cases, it is formulated as Cl_2O_6 and named dichlorine hexoxide. This oxide reacts with water to yield chloric acid and perchloric acid.

Chlorine(VII) oxide, Cl_2O_7, a colorless, oily liquid that detonates when struck or brought into contact with a flame, is prepared by the action of phosphorus pentoxide on anhydrous perchloric acid at −10°C for 24 hours. The structure consistent with vapor phase Raman and infrared data is $O_3Cl{-}O{-}ClO_3$.

The oxyacids (hydroxides) of chlorine are extensive. Hypochlorous acid, HClO(HOCl), is known only in solution. It is generally prepared by the reaction of chlorine with water,

$$Cl_2 + H_2O \rightarrow HCl + HOCl \qquad [13.60]$$

This acid is weak, $K = 3 \times 10^{-8}$. Hypochlorous will oxidize phosphorus, arsenic, antimony, sulfur, and selenium, whereas the acid is reduced to chloride ion and oxygen.

Chlorous acid, HOClO ($HClO_2$), is also known only in solution. It is prepared from the reaction of a barium salt and sulfuric acid,

$$Ba(ClO_2)_2 + H_2SO_4 \rightarrow 2HClO_2 + BaSO_4 \qquad [13.61]$$

Its acid constant is 10^{-2}. Although the acid has not been isolated, many chlorites have been.

Chloric acid, $HOClO_2$ ($HClO_3$), is also known only in solution. It is prepared by the action of su!furic acid on a chlorate:

$$Ba(ClO_3)_2 + H_2SO_4 \rightarrow BaSO_4 + 2HClO_3 \quad [13.62]$$

Heat causes this acid to decompose:

$$3HClO_3 \rightarrow HClO_4 + 2O_2 + Cl_2 + H_2 \quad [13.63]$$

Metal chlorates may decompose, under heat, in two ways:

$$4KClO_3 \rightarrow KCl + 3KClO_4 \quad [13.64]$$

and with managanese dioxide as catalyst, the thermal decomposition produces potassium chloride and oxygen.

Perchloric acid, $HOClO_3$ ($HClO_4$), is a colorless, mobile liquid. It may be prepared by the action of sulfuric acid on potassium perchlorate and then separated by distillation. This acid is a strong oxidizing agent. Metallic perchlorates are known. Esters of this acid are highly explosive.

Bromine(I) oxide, Br_2O, a brown liquid, is prepared by the action of bromine on a slurry of mercuric oxide in carbon tetrachloride at 0°C. This liquid is an excellent oxidizing agent.

Bromine(IV) oxide, $(BrO_2)_x$, is a yellow solid produced in an electrical discharge of oxygen and bromine vapor. It is stable at −40°C but decomposes rapidly at 0°C into its elements. This yellow solid reacts with sodium hydroxide to produce NaBrO.

Bromine(VI) oxide $(BrO_3)_x$, a white solid that decomposes violently at 10°C, is stable at −80°C. It is generally prepared by the action of ozone on the lower oxides of bromine.

Hypobromous acid, HOBr, is known only in solution ($K = 2 \times 10^{-9}$). It may be prepared by the action of bromine on water. It thermally decomposes to bromine and bromic acid,

$$5HOBr \rightarrow HBrO_3 + 2H_2O + 2Br_2 \quad [13.65]$$

Bromic acid, $HOBrO_2$ ($HBrO_3$), also known only in solution, is prepared by the action of bromine on silver bromate:

$$3H_2O + 5AgBrO_3 + 3Br_2 \rightarrow 6HBrO_3 + 5AgBr \quad [13.66]$$

Potassium bromate may be prepared by the action of bromine on potassium hydroxide.

Until 1968, perbromates proved to be illusive. The initial preparation was based on making a selenate with ^{83}Se.

$$^{83}SeO_4^{2-} \rightarrow \, ^{83}BrO_4^- + \beta^- \qquad [13.67]$$

Later, electrolysis of a solution containing perchloric acid, lithium bromate (^{82}Br), and lithium perchlorate yielded rubidium perbromate. A new and more convenient synthesis consists in the oxidation of bromate with fluorine in alkaline solution. Perbromic acid is a strong monobasic acid that is a sluggish oxidant.

Two oxides of iodine have no definite structure: $(I_2O_4)_x$ and $(I_4O_9)_x$. The first, $(I_2O_4)_x$, is prepared by allowing sulfuric acid to react with iodic acid. This solid reacts with both acids and bases:

$$(I_2O_4)_x + KOH \rightarrow KI + KIO_3 + 3H_2O \qquad [13.68]$$

$$(I_2O_4)_x + HCl \rightarrow 3Cl_2 + ICl + 4H_2O \qquad [13.69]$$

The nature of $(I_4O_9)_x$ is uncertain.

Iodine(V) oxide, I_2O_5, is produced by the reaction of nitric acid with iodine or the dehydration of iodic acid. This oxide is the anhydride of iodic acid. It acts as a strong oxidizing agent that sometimes detonates in the presence of organic substances.

Iodous acid, HOI, is known only in dilute aqueous solution and is unstable. It is usually prepared by mixing aqueous iodine with mercuric oxide. It is a very weak acid and a very weak base:

$$HOI \rightleftharpoons H^+ + OI^-, \quad K = 4.5 \times 10^{-13} \qquad [13.70]$$

$$HOI \rightleftharpoons OH^- + I^+, \quad K = 3 \times 10^{-10} \qquad [13.71]$$

Iodic acid, $HOIO_2$ (HIO_3), is a colorless solid that is very soluble in water and is prepared by dissolving iodine(V) oxide in water. This acid bleaches litmus. Structurally, it is believed that iodate ions are held together by hydrogen bonds. Iodates are usually prepared by reacting I_2 with base:

$$3I_2 + 6OH^- \rightarrow IO_3^- + 5I^- + 3H_2O \qquad [13.72]$$

The iodates are generally oxidizing agents.

There are several forms of periodic acid. Metaperiodic acid has the formula $HOIO_3(HIO_4)$. This acid is a solid, colorless crystal. Several forms of the tetrahedral ion are hydrated: $IO_4(H_2O)^-$ or $H_2IO_5^-$; $IO_4(H_2O)_2^-$ or $H_4IO_6^-$. The $H_4IO_6^-$ ion exists in strong acid solution as H_5IO_6, which is known as paraiodic acid. Complicated pH-dependent equilibria are associated with these acids.

TRANSITION METAL OXIDES

Although a very important class of compounds, limited space will be devoted to the transition metal oxides. The outstanding feature of the transition metals is that they exhibit a great degree of variable valence. In Fig. 13.7, the transition metals are listed according to groups, and their oxides are distinguished by their oxidation states. Some classification of these species is given in Fig. 13.7.

Group Number	Electronic Configuration	Elements	Oxidation States +1	+2	+3	+4	+5	+6	+7	+8
III-B	d^1	Sc Y La			M_2O_3					
IV-B	d^2	Ti Zr Hf		$\underline{(MO)}$	(M_2O_3)	MO_2				
V-B	d^3	V Nb Ta		$\underline{(MO)}$	M_2O_3	MO_2	M_2O_5			
VI-B	d^5	Cr Mo W	Basic	$\underline{(MO)}$	M_2O_3	MO_2	M_2O_5	MO_3		
VII-B	d^5	Mn Tc Re	Oxides	MO	M_2O_3	MO_2	M_2O_5	MO_3	M_2O_7	
	d^6	Fe Ru Os		$\underline{MO}$	M_2O_3	MO_2				MO_4
VIII-B	d^7	Co Rh Ir		MO	M_2O_3	(MO_2)	Acidic Oxides			
	d^8	Ni Pd Pt		MO	(M_2O_3)	(MO_2)				
I-B	d^{10}	Cu Ag Au	M_2O	MO	(M_2O_3)					
II-B	d^{10}	Zn Cd Hg		MO						

Amphoteric Line

← "Ionic" or Polymerized Oxides → | ← Small, Covalent Molecules →

Figure 13.7 Oxides of the transition metals. Adapted from C. S. G. Phillips and R. J. P. Williams, *Inorganic Chemistry*, Oxford University Press, Oxford, 1965. Oxides in parentheses are considered very unstable; those underlined are nonstoichiometric, often having metallic properties.

Perusal of Fig. 13.7 will lead one to certain conclusions relative to the nature of the transition metal oxides. In traversing a transition metal period from left to right (read down Fig. 13.7), the +2 oxidation state becomes more stable. In terms of structure, this would mean that the initial elements would tend to give up electrons to achieve a more stable configuration.

They would be reducing agents. Vanadium(II) oxide and chromium(II) oxide are strong reducing agents. Can the reader rationalize the fact that manganese(II) oxide is stable and not a reducing agent?

In traversing a transition metal period from left to right, the +3 oxidation state becomes less stable. Manganese(III) oxide is an oxidizing agent and cobalt(III) oxide oxidizes water to oxygen; copper and zinc do not possess a +3 oxidation state. The highest oxidation state achieved by any element usually corresponds to the group number. Table 13.3 contains standard heats of formation for comparison purposes.

TABLE 13.3 ***Standard Heats of Formation of Transition Metal Oxides*** **(-kcal/mole)**

Formula	ΔH_f°	Formula	ΔH_f°
ScO		Sc_2O_3	(411)[a]
TiO	123.9	Ti_2O_3	362.9
VO	−100	V_2O_3	296
CrO		Cr_2O_3	272.7
MnO	92.1	Mn_2O_3	229.2
FeO	63.8	Fe_2O_3	196.8
CoO	57	Co_2O_3	
NiO	57.3	Ni_2O_3	
CuO	37.6	Cu_2O_3	

[a] Parentheses indicate that variation in the value is possible.

As a transition metal group is descended, the higher oxidation state becomes more stable with respect to conversion to the lower oxidation state; or, as the transition metal group is ascended, the lower oxidation state becomes more stable. Osmium(VIII) oxide has been isolated, but iron(VIII) oxide, FeO_4, has not been. Ruthenium(VIII) oxide has been formed but decomposes on standing. Osmium(IV) oxide has been synthesized, but simple heating in air tends to revert it to osmium(VIII) oxide.

As to type of oxide formed, there are some relationships. First, the oxides of higher oxidation state are acidic. When chromium(VI) oxide is dissolved in water, chromic acid(H_2CrO_4) results. Manganese(VII) oxide is the recognized anhydride of permanganic acid, $HMnO_4$, which exists only in solution—but permanganates are known. Likewise, iron(VI) oxide has not been isolated, but ferrates, FeO_4^{2-}, do exist.

The lower oxidation states of the metals are basic oxides. They dissolve in acids to give salts in which the metal is the cation. From Fig. 13.7 it can be seen that the basic character of the oxides decreases on traversing a period of the transition metals from left to right.

In Fig. 13.7, one can observe the diagonal amphoteric line. The intermediate oxides in the vicinity of this line have amphoteric character. Manganese dioxide may be dissolved in fused alkali to form a manganite ion, MnO_3^{2-}, and at 0°C manganese dioxide reacts with concentrated hydrochloric acid to give manganese tetrachloride. Iron(III) oxide is attacked by concentrated alkali to give ferrite ion, FeO_2^-.

Finally, we should point out that the transition metals also form oxides of different types. These may be classified as *mixed oxides*, nonstoichiometric oxides, and spinels.

Mixed oxides may be represented by Fe_3O_4 and Mn_3O_4. These oxides contain the same metal ion in two different oxidation states. The oxide Fe_3O_4 is equivalent to $FeO \cdot Fe_2O_3$, and Mn_3O_4 is equivalent to $2MnO \cdot MnO_2$.

Nonstoichiometric oxides are those in which the oxygen-to-metal ratio may not be expressed in whole numbers. For instance, FeO is often short of containing its proper percentage of iron. This situation may be explained in terms of structure. When the oxide ions pack together to form a lattice, there are two types of holes formed in the lattice structure, octahedral holes and tetrahedral holes (depending on how the oxide ions are packed together). The cations, which are generally smaller, occupy these holes. To account for the nonstoichiometry, we simply admit that not all the octahedral or tetrahedral sites are occupied, as they should be. A compound may also be deficient in the amount of anion it should have; silver(I) oxide and copper(I) oxide are always deficient in oxide ions. To account for the electrical neutrality of the system, we postulate the movement of electrons.

A special type of mixed oxide is the *spinel*. A spinel structure is characterized by the fact that the oxygen atoms constitute a face-centered cubic lattice, the unit cell containing 32 oxide ions. One special type of spinel, which may be represented by the general formula AB_2O_4, is called a 2:3 spinel. This simply means that A is in the $+2$ oxidation state and B is in the $+3$ oxidation state. The compound $ZnFe_2O_4$ is a normal spinel—i.e., the trivalent ions occupy the octahedral holes and the divalent ions occupy the tetrahedral holes. The compound $Fe(Fe_2)O_4$ is an inverse spinel. In an inverse spinel, the trivalent ions occupy the tetrahedral holes. The 4:2 spinels are always inverse, but 2:3 spinels may be either normal or inverse. A priori predictions as to the type of spinel that may be formed by transition metal ions is one of the applications of crystal field theory (see Chapters 6 and 9).

Suggested Readings

1. Appleman, E. H., *Inorg. Chem.*; **8,** 223 (1969).
2. Appleman, E. H., *J. Am. Chem. Soc.*, **90,** 1900 (1968).
3. Cotton, F. A., and Wilkinson, G., *Advanced Inorganic Chemistry*, 2nd ed., John Wiley & Sons, Inc., New York, 1966.
4. Coulson, C. A., *Valence*, Clarendon Press, Oxford, 1952.
5. Dasent, W. E., *Non-Existent Compounds*, Marcel Dekker, Inc., New York, 1965.
6. Durant, P. J. and B. Durant, *Introduction to Advanced Inorganic Chemistry*, Longmans, Green & Company, Ltd., London, 1964.
7. Larsen, E. M., *Transition Elements*, W. A. Benjamin, Inc., New York, 1965.
8. Sanderson, R. T., *Inorganic Chemistry*, Reinhold Publishing Corp., New York, 1967.

Study Questions

13.1. Hypophosphorus acid, H_3PO_2, is a weak acid ($K_1 = 1 \times 10^{-2}$). How could you experimentally determine K_1 for this acid? Explain, making use of the Ricci approximation, whether hypophosphorus acid should be represented as a monobasic or a dibasic acid.

13.2. Which compound would you expect to act as the weaker acid—$Sn(OH)_2$ or $Sn(OH)_4$? Explain.

13.3. Which one of the following compounds would you expect to exhibit the most acidic properties—As_40_{10}, As_40_6, or Bi_20_3? Explain.

13.4. Which of the following compounds would have the highest pK_1 value—H_3BO_3 or H_2CO_3? Explain.

13.5. Which compound would you expect to be the strongest oxidizing agent—MnO_2, K_2MnO_4, or $KMnO_4$? Explain.

13.6. Which of the following would have the highest pK_A value associated with it—HO^-, HS^-, or HTe^-? Explain.

13.7. Arrange the following acids in order of intrinsic increasing acid strength: H_3PO_4, H_2SeO_4, H_3PO_3, ClOH, H_3AsO_4, IOH. Explain.

13.8. Arrange the following conjugate bases in order of increasing basicity: ClO_2^-, $HSeO_4^-$, ClO_4^-, $H_2BO_3^-$. Explain.

13.9. Nitrogen(IV) oxide, N_2O_4 (dinitrogen tetroxide), appears unique among simple molecules in the ease with which it may be dissociated homolytically (by heat) or heterolytically (by reagents) in several ways. Discuss the following:

(a) $[(CH_3)_4N]^*NO_3$, containing ^{15}N in the nitrate ion, equilibrates with liquid N_2O_4 in 36 hours. The reaction is promoted by bases.

(b) A variety of complexes of the type $N_2O_4 \cdot B$ and $N_2O_4 \cdot 2B$ are known where B is a base. Indicate their probable structures.

(c) BF_3 forms the stable complex $F_3B \cdot N_2O_4$. The complex is noncrystalline, insoluble in all inert solvents, is readily dissociated at elevated temperatures, and cannot be melted under pressure below 300°C. Chemically, it is a powerful nitrating agent, particularly for aromatic hydrocarbons. Give a probable structure of the complex and write down the most likely acid promoted dissociation of N_2O_4 in solution.

13.10. Give balanced equations for the following reactions. Show steps involved where possible:

(a) Be + NaOH solution

(b) PuO_3 + fused KOH

(c) Preparation of H_3PO_5

13.11. Which would be more extensively hydrolyzed in solution—K_2SiO_3 or K_2SnO_3? Explain.

13.12. Peroxynitric acid is made by the action of N_2O_5 on hydrogen peroxide. Give the structure of this acid based on the reaction.

13.13. When ^{14}NO and $^{15}NO_2$ are mixed in an infrared cell, the infrared monitoring of the system indicates that there is a complete exchange of nitrogen atoms. Deduce a structure for N_2O_3 consistent with these data and explain completely how the structure that you deduce is consistent with the given data.

13.15. Orthoarsenic acid is heated at 100°C. The resulting mixture contains some of the original acid and some condensed acids. Name and derive the structural formulas of each condensed acid. If these acids were separated and isolated, how would you confirm the presence of each?

13.16. Write balanced equations to show the separate reactions of hot, concentrated sulfuric acid on silver, lead, and mercury.

13.17. Arrange the following in order of increasing bond energy:
O_2^{2-}, O_2^+, O_2^-, O_2

CHAPTER 14

The Halogens

OCCURRENCE

The halogens form only a small percentage of the earth's crust; F = 0.1%, Cl = 0.1%, Br = 0.001%, and I = 0.001%. Due to their vigorous reactivity, the halogens do not exist in elemental form in nature but occur in minerals or in solution in the sea. Fluorine occurs largely in the minerals cryolite (Na_3AlF_6) and fluorspar (CaF_2). Another source is the mineral apatite [$Ca_5(PO_4)_3F$]. Apatite is the hardening agent of tooth enamel; therefore, fluoride dentifrices and fluoridation of water are used to help prevent tooth decay. Due to the high concentration of calcium ions in sea water and the low solubility product of calcium fluoride (4×10^{-11}), the amount of fluorine in solution in sea water is very low. Chlorine and bromine, however, are found in their greatest abundance as halide ions in the sea. Bromine is also found in waste salt brines from oil wells in southern California, at a concentration of from 10 to 135 ppm (in contrast to roughly 65 ppm in sea water). Iodine occurs as sodium iodate ($NaIO_3$) in nitrates found in northern Chile, and it also occurs to some extent in sea water as iodides. Certain forms of marine life have the ability to take up this iodine and concentrate it (0.1 to 2% of their body weight). Astatine is a radioactive element with a relatively short half-life (8.3 hours for the longest-lived isotope) which has never been found in nature. It can be formed by the beta decay of polonium, but since polonium is rare and will generally undergo alpha rather than beta decay, astatine is generally made by bombarding ^{209}Bi with alpha particles.

$$^{4}_{2}He + ^{209}_{83}Bi \rightarrow ^{211}_{85}At + 2^{1}_{0}N \qquad [14.1]$$

PREPARATION

Discussion of the preparation of the elemental halogens is perhaps best approached by consideration of the appropriate Latimer diagram (Table 14.1). The reduction potential of fluorine is much too positive to allow its

TABLE 14.1 ***Partial Latimer Diagrams for the Halogens***

E°_B

$$F_2^\circ \overset{2.87}{\overline{\qquad\quad}} F^-$$

$$ClO_3^- \overset{0.50}{\overline{\qquad\quad}} ClO^- \overset{0.40}{\overline{\qquad\quad}} Cl_2^\circ \overset{1.36}{\overline{\qquad\quad}} Cl^-$$

$$BrO_3^- \overset{0.52}{\overline{\qquad\qquad\qquad\qquad}} Br_2^\circ \overset{10.07}{\overline{\qquad\quad}} Br^-$$

$$IO_3^- \overset{0.23}{\overline{\qquad\qquad\qquad\qquad}} I_2^\circ \overset{0.53}{\overline{\qquad\quad}} I^-$$

E°_A

$$BrO_3^- \overset{1.51}{\overline{\qquad\qquad\qquad\qquad}} Br_2^\circ \overset{1.07}{\overline{\qquad\quad}} Br^-$$

preparation by the use of chemical oxidizing agents. Elemental fluorine is prepared by the electrolysis of a fused mixture (250°C) of hydrogen fluoride and potassium fluoride ($K^+HF_2^-$) with amorphous carbon electrodes. The electrodes must be replaced frequently because of attack by the very active fluorine. Any hydrogen fluoride evolved with the elemental fluorine is removed by adduct formation similar to that used in the electrolysis cell. Moisture must be eliminated from the cell as water will react *very* vigorously with fluorine to produce hydrogen fluoride and oxygen.

The reduction potential for the formation of chlorine from chloride ion is much less positive than that for the same process in fluorine; therefore, chlorine can be prepared by chemical means, such as the reaction of solid manganese dioxide with concentrated hydrochloric acid. Commercially, chlorine is prepared by

$$MnO_2 + 4HCl \rightarrow Cl_2 + MnCl_2 + 2H_2O \qquad [14.2]$$

the electrolysis of brine or fused chlorides. If the chlorine is prepared by the electrolysis of brine, hydroxide ions are formed at the cathode. Special precautions must be taken so that the chlorine and hydroxide ion do not mix, since chlorine disproportionates in base to form chloride and hypochlorite (in a hot solution, the hypochlorite ion disproportionates further to form chloride and chlorate; see Table 14.1).

Bromine is generally prepared from bromide in sea water, with chlorine as the oxidizing agent. The sea water is first acidified to prevent disproportionation of the bromine to be formed, and then chlorine is passed through the solution. The bromine formed is swept out with a current of air and led into a basic solution, where it disproportionates into bromide and bromate ions. When the concentration of bromine in this solution is sufficiently high, the solution is acidified to recover the elemental bromine (see Latimer diagram again for $Br^- + BrO_3^- \rightarrow Br_2$).

Iodine is also prepared by the replacement of iodide in soluble iodides by the more active chlorine, or by the reduction of iodates with sulfites:

$$2IO_3^- + 2HSO_3^- + 3SO_3^{2-} \rightarrow 5SO_4^{2-} + I_2 + H_2O \qquad [14.3]$$

GENERAL PROPERTIES

Changes in physical and chemical properties occur in a fairly regular manner as the halogen group is descended. The properties of the fluorides differ from the remainder of the halides. These distinctions will be discussed as they occur in the compounds of the halogens.

Fluorine is an extremely reactive element, and it forms compounds with almost all the elements of the periodic table. In addition to this ability to react with most elements, fluorine also has a tendency to cause the elements with which it combines to assume their highest oxidation states. This is a consequence of fluorine's small size and high electronegativity. The high electronegativity is needed to place the high positive charge or high oxidation state on the elements. In a covalent fluoride, the small size allows more fluorine atoms to be attached to an element without steric repulsions, and, in ionic fluorides, the small size of the fluoride ion favors a high lattice energy. Table 14.2 contains several general properties of the halogens that may be compared.

TABLE 14.2 ***General Properties of the Halogens***

	Physical State	Radii		Electro-negativity*	Electron Affinity (kcal/mole)	Bond Energy (kcal/mole)
		Ionic	Covalent			
Fluorine	Pale-yellow gas	1.36	0.64	4.0	81.0	38
Chlorine	Green-yellow gas	1.81	0.99	3.0	84.8	58
Bromine	Red liquid	1.96	1.14	2.8	79.1	46
Iodine	Black solid	2.20	1.33	2.5	72.1	36

* Pauling's method.

The extremely high reactivity of fluorine is related to its low bond energy and high electron affinity. The low bond energy of the fluorine molecule indicates a relatively easy transformation to fluorine atoms, whereas the high electron affinity of the atomic state is indicative of a great tendency to form the fluoride ion. The higher bond energies of the other halogens contribute significantly to their lesser reactivity.

Electron affinity and bond energy decrease as Group VII-A is descended, with the exception that the values for fluorine are lower than those for chlorine (similar effects are noted when electron affinities and *single* bond energies of nitrogen and oxygen are compared with those of phosphorus and sulfur, respectively). This lower electron affinity may be accounted for by considering the effect of the very small size of the fluorine atom as compared with the other halogen atoms. The very compact fluorine atom is unable to accept a complete electron without considerable electron-electron repulsions. These repulsions decrease to a considerable extent the energy released when an electron is completely accepted to form the fluoride ion. Similarly, when two fluorine atoms are joined, the compactness is seen in the effects of repulsion of the two interpenetrating electron clouds. This repulsion weakens the fluorine-fluorine bond and increases the bond length from the 1.28 A expected (doubling the covalent radius) to 1.42 A (observed). This type of effect is not noted in the other halogens because they are less compact.

The physical states and colors of the halogens vary as expected in view of van der Waals and charge transfer effects on the larger, more polarizable atoms of the family. Colors of solutions of the elements, especially iodine, vary with the solvent. Solutions of iodine in hydrocarbons, alkali halides, and carbon disulfide are violet, whereas iodine in alcohols, ethers, esters, and acids gives brown solutions. The brown solutions occur with solvents possessing an electron pair that can be donated to the iodine. The solubility of iodine in such solvents is generally greater than in the violet solutions, where it is essentially in the free, dissolved condition.

THE REPRESENTATIVE ELEMENT HALIDES

Group I

The alkali halides can be formed by burning the alkali metals in the appropriate halogens. The alkali halides are typical ionic compounds, and, in general, their properties reflect this ionic nature—very stable, relatively unreactive solids. The low charge and fairly large size of the alkali metal ions prevent their halides from occurring as hydrates. Their boiling and melting points are high and in general decrease with the larger cations or anions (Table 14.3). The lithium halides and sodium iodide have lower melting points than the next lower alkali metal halide, and they do not follow the general trend. The deviation occurs for the small cation-large anion combinations. In these cases, anion-anion contact may occur in the lattice. The consequent repulsions weaken the lattice and lead to low boiling and melting points. Note the rather low enthalpy of formation values for these cases.

TABLE 14.3 ***Melting Points and Enthalpies of Formation of the Alkali Halides***

	Melting Points, °C (ΔH_f in kcal/mole)				
	Li	Na	K	Rb	Ca
F	845 (−146)	995 (−136)	856 (−134)	775 (−131)	682 (−127)
Cl	610 (−98)	808 (−98)	772 (−104)	717 (−103)	645 (−103)
Br	547 (−84)	757 (−87)	748 (−94)	682 (−94)	636 (−94)
I	446 (−65)	661 (−69)	677 (−78)	642 (−79)	621 (−80)

All alkali halides, except cesium chloride, bromide, and iodide, crystallize with the sodium chloride or rock salt structure (coordination number 6). These cesium halides prefer the more dense cesium chloride structure (coordination number 8). As expected from Le Chatelier's principle, other alkali halides (notably those of rubidium and potassium, except for the fluorides) can be converted to the cesium chloride structure at high pressure. In a high temperature (above 445°C) modification, cesium chloride itself exhibits the rock salt structure. Radius ratio considerations are not generally satisfactory for determining the structures of these halides, as is evidenced by the observation that only about half the alkali halides that prefer the rock salt structure have radius ratios falling between the 0.414 to 0.732 calculated as optimum for this structure (Table 14.4). The radius ratios for those halides that naturally prefer the cesium chloride structure are, however, all above the 0.73 required for cubic coordination. With respect to structures, in many cases the structures of the halide salt and of the corresponding alkaline earth chalcenocide (Group VI-A) salt are similar.

Group II

The alkaline earth halides can generally be prepared by reacting the pure metal with the halogen or hydrogen halide. Some of the beryllium halides,

TABLE 14.4 ***Radius Ratios for the Alkali Halides*** $\left(\frac{R\ cation}{R\ anion}\right)$

	Li	Na	K	Rb	Cs
F	0.50	0.70	0.98	0.93*	0.82*
Cl	0.38	0.52	0.74	0.81	0.93
Br	0.35	0.48	0.68	0.75	0.85
I	0.31	0.43	0.60	0.67	0.76

* Calculated on the basis of $\frac{R \text{ anion}}{R \text{ cation}}$.

however, are prepared by quite different methods; for example, beryllium fluoride is prepared by heating ammonium tetrafluoroberylate, $(NH_4)_2BeF_4$, in an inert gas stream, and beryllium iodide is prepared by heating beryllium carbide, Be_2C, with hydrogen iodide. The ability of the smaller (as compared with alkali metals), highly charged alkaline earth cations to bind water strongly in their first coordination sphere is exhibited by their tendency to form as hydrates or by their deliquescent ability. The ability to form hydrates is especially noted with the lighter metals.

The regularity in boiling and melting points of the alkaline earth halides is not as great as that observed for the alkali metal halides (Table 14.5). Beryllium and magnesium halides generally differ in their properties from halides of calcium, barium, and strontium due to the increased covalent nature caused by the great ability of beryllium and magnesium to polarize attached atoms.

For both the alkali metal and alkaline earth halides, if, with a given cation, the family of halogens is descended, the enthalpy of formation of these compounds decreases. This can be explained on the basis of increasing anion-anion repulsion and decreasing coulombic attraction, due to the larger interionic distances. In general, for both series of halides, if with a given halide the alkali metal or alkaline earth family is descended, the enthalpy of formation becomes larger. In this situation, the decreasing anion-anion repulsion is the major factor. Note that the change becomes

TABLE 14.5 ***Melting Points and Enthalpies of Formation of the Alkaline Earth Halides***

	Be	Mg	Ca	Sr	Ba
F	(−227)*	1263 (−268)	1418 (−290)	1400 (−290)	1320 (−287)
Cl	405 (−122)	714 (−153)	782 (−190)	875 (−198)	962 (−206)
Br	(−88)	711 (−124)	760 (−161)	643 (−171)	847 (−180)
I	480 (−51)	(−86)	740 (−128)	460 (−136)	(−144)

* Values in parentheses refer to enthalpy of formation in kcal/mole.

smaller as the larger cations are reached and is much greater for the smaller alkaline earth metals as compared with the alkali metals.

The structures of the alkaline earth halides are not as simple as those of the alkali metal halides, even though calcium fluoride is the prototype for one of the simple structural classes (fluorite). Strontium and barium fluoride also assume the fluorite structure; however, the other halides of this group (with the exception of the beryllium halides) have crystalline structures that cannot be assigned to one of the simple classes.

Beryllium fluoride, chloride, and bromide form chain structures with halogen bridges. Beryllium chloride can be considered to be composed of tetrahedral $BeCl_4$ groups sharing edges.

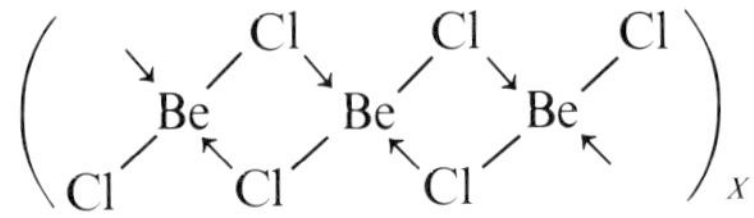

Beryllium fluoride consists of tetrahedra joined at the corners in a definite geometrical pattern (β-crystobalite structure). In the β-crystobalite structure, no two fluorine atoms share the same two beryllium atoms.

Group III

In view of the great differences in properties between the boron halides and the halides of the rest of the group, we shall consider them first. Such differences are not unexpected, since boron is the only nonmetal of this group.

Boron. Numerous boron halides have been prepared. All of the boron trihalides and diboron tetrahalides are known; in addition, a large number of higher boron-chloride species have been discovered. Although mixed boron halides have been observed spectroscopically, only boron dibromide-fluoride and boron difluoridebromide have actually been isolated. Temperatures as low as $-50°C$ are insufficient to slow down the exchange between boron tribromide and boron trichloride and to allow the separation of the mixed halides observed spectroscopically.

The boron trihalides can be prepared by treating elemental boron with the appropriate halogen, Table 14.6 lists some of their properties. Boron trifluoride is better prepared, however, by heating a mixture of fluorspar (calcium fluoride), concentrated sulfuric acid, and boric oxide:

$$B_2O_3 + 3H_2SO_4 + 3CaF_2 \rightarrow 2BF_3 + 3CaSO_4 + 3H_2O \qquad [14.4]$$

The boron trihalides are symmetrical planar molecules with the boron presumably exhibiting sp^2 hybridization. Their Lewis acid nature is evident from the vacant p orbital remaining on boron.

Their order as Lewis acids is generally

$$BI_3 > BBr_3 > BCl_3 > BF_3$$

which is accounted for on the basis of $(p \rightarrow p)$ π bonding in the boron trihalides (refer to section on Lewis acids and bases for discussion). The effect of this π bonding is seen in the short B—X bond lengths observed, as compared with the bond length calculated from simple addition of covalent radii, in these molecules. Fluorine *nmr* chemical shifts in such molecules as boron trifluoride, methylborondifluoride, and the adduct ethylborondifluoride trimethylamine, also show these π bonding affects. As fluorine is

TABLE 14.6 ***Properties of Some Boron Halides***

	M.P. (°C)	B.P. (°C)	ΔH_f (kcal/mole)
BF_3	−127	−100	−270
BCl_3	−107	13	−95.7
BBr_3	−46	90	−52.8
BI_3	50	212	−27.6
B_2F_4	−56	−34	−342
B_2Cl_4	−93	66	−118
B_2Br_4	0.5 to 1.5		−50
B_2I_4			20
B_4Cl_4	Solid with 2 mm pressure at room temperature		

replaced by an alkyl group, the remaining fluorines undergo more extensive π bonding, become more deshielded, and undergo resonance at a lower field strength. The formation of the adduct eliminates all π bonding, causing the ^{19}F resonance to appear at a higher field strength than in the boron trifluoride.

The boron trihalides are stable but reactive molecules, although boron triiodide has a comparatively low thermal stability, probably caused by steric overcrowding. One reason for their high reactivity is their coordination unsaturation. The boron halides form adducts with many donor molecules. In general, if the donor molecule has hydrogen attached to the actual donor atom, hydrogen halide is eliminated from the adduct. If there is no hydrogen attached to the donor atom, the adduct itself is generally stable. Thus, dimethylether forms stable adducts with all the trihalides. Methyl alcohol forms a stable adduct with boron trifluoride, but with boron trichloride, tribromide, and triiodide, cleavage occurs to form hydrogen halide and trimethoxyboron.

$$(CH_3)_2O + BCl_3 \rightarrow (CH_3)_2O{:}BCl_3 \qquad [14.5]$$

$$3CH_3OH + BCl_3 \rightarrow B(OCH_3)_3 + 3HCl \qquad [14.6]$$

The greater strength of the B—F bond in contrast to the strength of the other boron-halogen bonds may account for the stability of the boron trifluoride-methyl alcohol adduct. Boron trifluoride is not completely hydrolyzed by water, whereas the other halides are completely hydrolyzed. The partial hydrolysis of boron trifluoride places fluoride ions in solution. These fluoride ions react with the borontrifluoride to form the fluoroborate anion, BF_4^-, reducing the possibility of hydrolysis of these anions.

$$4BF_3 + 3H_2O \rightarrow 3BF_4^- + B(OH)_3 + 3H^+ \qquad [14.7]$$

However, the major reason for the incomplete hydrolysis of boron trifluoride must again lie essentially in the strength of the boron-fluorine bonds. The strength of these boron-fluorine bonds allow the formation of hydrates below room temperature, $BF_3 \cdot nH_2O$ ($n = 1$ or 2).

Diboron tetrachloride is prepared by passing boron trichloride through a microwave discharge,

$$2BCl_3 \rightarrow B_2Cl_4 + Cl_2 \qquad [14.8]$$

or by passing boron trichloride over heated boron monoxide

$$B_2O_2 + BCl_3 \xrightarrow[\text{450 mm}]{\text{200–236 C}} B_2Cl_4 \qquad [14.9]$$

Diboron tetrafluoride is formed by fluorinating either diboron tetrachloride with antimony trifluoride or boron monoxide with sulfur tetrafluoride. Diboron tetrabromide and diboron tetraiodide are prepared by passing the trihalide through the appropriate discharge.

The structure of dibromotetrachloride is dependent on the physical state. In the crystal, diboron tetrachloride is a planar molecule (D_{2h}) with

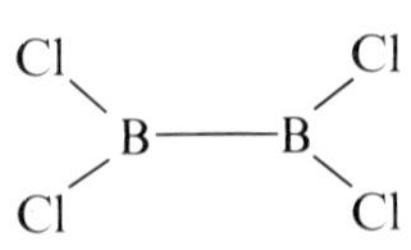

rather short B—Cl distances, therefore indicating possible boron-chlorine $(p \rightarrow p)$ π bonding. In the liquid phase (above $-98°C$), the two $B—Cl_2$ units are probably at right angles to one another (D_{2d}). Diboron tetrafluoride has the same structure as the tetrachloride, except that in the gaseous state (above $-56°C$), there is a question as to whether the BF_2 groups are staggered or in free rotation. In both molecules, the boron-boron bond is a bit longer than that expected for a single bond. Boron-boron bond lengths are listed in Table 14.7.

TABLE 14.7 ***Boron-Boron Bond Lengths (in A)***

B_2	1.59
B_2F_4	1.67
B_2Cl_4	1.75

The chemistry of the diboron tetrahalides is that of the boron-boron and the boron-halogen bonds. Much of the chemistry of the boron-halogen bonds in diboron tetrahalides can be inferred from the chemistry of the boron trihalides. The tetrahalides are much less stable than the trihalides, with diboron tetrafluoride being the most thermally stable (8% decomposition in one day at room temperature). Thermal decomposition leads to a complex variety of polyboron halides. Both the tetrafluoride and tetrachloride, being coordinatedly unsaturated, have been shown to add 2 moles of donor molecules for 1 mole of tetrahalide. The adduct is more thermally stable than the parent tetrahalide.

$$B_2X_4 + 2(CH_3)_3N \rightarrow [(CH_3)_3N]_2B_2X_4 \qquad [14.10]$$

Displacement studies have been performed on the tetrahalides to determine the relative Lewis acid strengths of the trihalides and tetrahalides. Boron trifluoride displaced diboron tetrafluoride from its adduct with trimethylamine and was without action on the diboron tetrachloride adduct of trimethylamine. Steric effects and other complicating factors do not allow a simple interpretation of these results.

The tetrahalides react with oxygen or air, in some cases igniting. Diboron tetrafluoride reacts explosively with oxygen at room temperature. The tetrahalides all appear to be hydrolyzed by water; when water is in excess, the products involve a rupture of the boron-boron bond. Many other substances can also cause rupture of this bond. Halogens generally rupture the boron-boron bond to form boron trihalides. Use of thermochemical data from reactions of this type allows calculation of the boron-boron bond energies, which all appear to be within ± 5 kcal of 78 kcal/mole.

Tetraboron tetrachloride was the first electron-deficient boron halide to be discovered. It is a yellow solid that ignites in dry air and is one of the products formed when boron trichloride is reduced in a mercury arc. Its structure consists in a tetrahedron of boron atoms, each with a terminal chlorine attached. The boron-boron bonding can be treated as multi-centered, involving all four boron atoms. Tetraboron tetrachloride does not form an adduct with diethyl ether but hydrolyzes in water.

Aluminium, Gallium, Indium, and Tallium. The trihalides of these elements are prepared by the action of the halogen or the hydrogen halide on the metal or the metallic oxide. When the metallic oxide is used, carbon is often added to aid in the reduction:

$$M_2O_3 + 3Cl_2 + 3C \rightarrow 2MCl_3 + 3CO \qquad [14.11]$$

Indium trifluoride is not prepared in the general manner, but by heating ammonium hexafluoroindiate in hydrogen fluoride.

Most of the trihalides of this group have a dimeric structure (M_2X_6: M = Al, Ga, In; X = Cl, Br, I, with the exception of GaI_3) in the liquid or vapor phase and a layer structure in the solid phase. In donor solvents and at high temperatures, the dimers generally convert to monomers. In view of the structures of these molecules, it is not surprising that they react

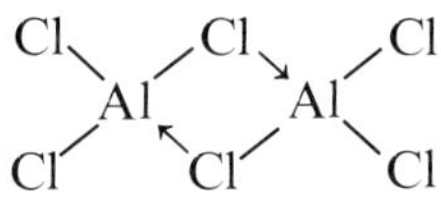

with donor molecules to form complexes such as R_3NAlCl_3 and that they also hydrolyze readily. It is interesting to recall that the boron halides do not exhibit this bridged type of structure. This fact could be due in part to the small size of the boron (steric effects) and to the ability of the boron to engage in π bonding, thereby removing its electron deficiency.

The trifluorides have crystalline structures in which the metal atom assumes a coordination number of 6. The extended array of metal and fluorine atoms in such a structure leads to high melting and boiling points for the materials involved (Table 14.8). Thallium trifluoride is the only trifluoride of this group that hydrolyzes.

The thallium trihalides are the least stable trihalides of the group and convert fairly readily to the monohalides. Thallium triiodide is not a true case of tervalent thallium but is actually, $Tl^+(I_3^-)$. The ease of conversion of the thallium trihalides to the monohalides is an example of the so-called *inert pair effect*. This name is given to this pair of electrons because it has been observed that as a family of representative elements is descended, the *s* electrons have less tendency to be used in bonding. Therefore, the element tends to assume the $N - 2$ oxidation state (N = group number).

TABLE 14.8 ***Melting Points, °C, and Enthalpies of Formation of MX_3****

	Al	Ga	In	Tl
F	1290 (−356.3)	950 sub.	1170	500 dec.
Cl	192/1700 mm (−167)	77.9 (−125)	586 (−128)	40 dec. (−84)
Br	97.5 (−132)	121.5 (−92)	436 (−97)	(−59)
I	179.5 (−75)	212 (−51)	210 (−55)	(−30 for TlI)

* Enthalpy values in parentheses, kcal/mole.

The label "inert pair" is misleading, and an explanation of the effect is probably best found in a comparison of the promotion energies of the electrons needed for bonding vs. the strengths of the resulting bonds as the table is descended. Monohalides of the other members of the group can be prepared by high temperature reactions of the metal with the metallic trihalides. Thallium monochloride is found to have some of the properties of silver chloride; it discolors in light and is insoluble in water (compare sizes and electronic configurations).

Molecules with the apparent formulas of dihalides (MX_2) have been prepared, but it appears that they are salts of the type $M(\mathrm{I})M(\mathrm{III})X_4$. Many complexes of this type and of the type $M(\mathrm{I})_3M(\mathrm{III})X_6$ have been prepared. The $M(\mathrm{I})M(\mathrm{III})X_4$ complexes are prepared in nonaqueous solvents, since in complexes of this type the metal is coordinately unsaturated and, therefore, susceptible to hydrolysis.

Group IV

All the tetrahalides of this group, with the exception of lead tetrabromide and tetraiodide, are known. In addition, many mixed halides and catenated halides (of carbon and silicon) have been observed. Some properties of these halides are listed in Table 14.9.

The lack of success in preparation of (or lack of stability of) these two lead tetrahalides probably stems from a combination of the reluctance of lead to assume or remain in the IV oxidation state, as well as the ease with which bromide and iodide can be oxidized. Carbon tetraiodide is also rather unstable, losing iodine above 100°C.

TABLE 14.9 ***Melting Points (°C) and Enthalpies of Formation of Group IV Tetrahalides***

(ΔH_f kcal/mole in parentheses; M.P., °C)

Some Tetrahalides:

	C	Si	Ge	Sn	Pb
F	−184 (−221)	−90 (−370)	−15	705 sub.	600 (−222)
Cl	−23 (−32)	−68 (−153)	−49.5 (−130)	−33 (−130)	−15
Br	94 (−0.5)	5.2 (−95)	26 (−79)	33 (−97)	
I	171 (40)	120 (−32)	144 (−31)	146	

Some Dihalides:

	F	Cl	Br	I
Sn	210	247 (−84)	232 (−61)	320 (−34)
Pb	822	298 (−86)	373 (−66)	412 (−42)

With the exception of tin and lead tetrafluorides, which are partially ionic and have octahedral coordination around the metal, the tetrahalides are covalent tetrahedral molecules. Preparation of the tetrahalides is generally accomplished by reaction of the element or oxide with the proper halogen or hydrogen halide. In some cases, such as carbon and tin tetrafluoride, they are prepared by fluorination of the chlorides. Carbon tetrachloride can be prepared by the action of chlorine on carbon disulfide.

Some interesting comparisons can be made of the properties of these Group IV tetrahalides. The melting points and boiling points of the tetrafluorides increase as the molecular weight increases. Rather high values are observed for the more ionic tin and lead tetrafluorides. For the other halides, however, the carbon compounds are found to have the highest melting and boiling points. We account for this by considering the effect of the partial multiple bonding than can exist in silicon, germanium, or tin tetrahalides due to the presence of low energy *d* orbitals on the central element. This partial multiple bonding leads to a decrease in *inter*molecular forces and thus to lower melting and boiling points.

The carbon tetrahalides are much more resistent to hydrolysis than are the other tetrahalides of this group because of the availability of *d* orbitals and the larger size of the atoms below carbon in Group IV. Both of these factors make the Group IV element more vulnerable to attack by nucleophiles, such as water or alcohols. Therefore, most of the Group IV tetrahalides, other than those of carbon, hydrolyze rather readily. Silicon tetrafluoride and tin tetrachloride do not hydrolyze completely due to formation of some MX_6^{2-} anion with the halide ion formed in the hydrolysis reaction. Recall that a similar effect is noted for boron trifluoride. The hydrolysis of germanium and tin tetrachloride is reversible, as might be expected from the amphoteric nature of the oxide products.

As was mentioned previously, lead tetrabromide and tetraiodide are not known. However, lead will exist in a formal IV oxidation state in combination with bromine in compounds of the type M_2PbBr_6. In these species, the lead(IV) is stabilized by the coordination of the two additional bromide ions with their donation of negative charge. Lead tetrachloride is also rather unstable, and the formal (IV) charge on lead is stabilized by the formation of the hexachloroplumbate(IV) anion, $PbCl_6^{2-}$.

As expected in view of the *inert pair effect*, the stability of the Group IV dihalides increases as the group is descended. Lead prefers to form the dihalide and must be forced to form the tetrahalide. Tin forms both the di- and tetrahalides, with tin(II) chloride being noted as a fairly good reducing agent. Germanium prefers the IV state but can be obtained in the II state as the chloride, bromide, or iodide. Neither carbon nor silicon forms stable unreactive dihalides. Silicon, however, forms very reactive SiX_2 or $(SiX_2)_x$ species by passing the silicon tetrahalide over elemental silicon at high temperatures and low pressures. The SiF_2 species has been extensively studied and is found to form a plastic polymer $(SiF_2)_x$ on condensation at low temperatures. When heated, this polymer decomposes to give perfluorosilanes, Si_nF_{2n+2}. Species with n values up to at least 14 have been observed in the mass spectrograph; perfluorotrisilane, Si_3F_8, and perfluorotetrasilane, S_4F_{10}, have been isolated. In addition, the $(SiF_2)_x$ polymer readily reacts with other species condensed onto it to form silicon-containing derivatives.

Carbon and silicon are the only members of the group to form catenated halides (except for the compound Ge_2Cl_6). The fluorocarbons are more stable than the chlorocarbons, whereas the chlorosilanes appear to be more stable than the fluorosilanes. Chlorosilanes up to $Si_{25}Cl_{52}$ are known and fluorosilanes up to $Si_{10}F_{22}$ have been isolated, whereas $Si_{16}F_{31}^+$ has been observed in the mass spectrum of the thermal decomposition products of $(SiF_2)_x$. The relatively small number of fluorosilanes known may then be due to difficulties in synthesis rather than to a lack of stability.

Group V

All the trihalides of this group have been prepared either as pure compounds or as ammonia adducts, and many of the mixed trihalides are also known. The properties of some of these are listed in Table 14.10. The members of this group, with the exception of nitrogen, form the pentafluorides. Phos-

TABLE 14.10 *Melting Points (°C) and Enthalpies of Formation of Group V Trihalides*
(ΔH_f in kcal/mole in parentheses)

	N	P	As	Sb	Bi
F	−208 (−30)	−152 (−219)	−6 (−218)	290 (−219)	727 (−212)
Cl	*d*71 (55)	−92 (−76)	−16 (−73)	73 (−91)	232 (−91)
Br		−42 (−44)	31 (−47)	97 (−62)	218 (−59)
I		61 (−11)	143 (−14)	171 (−24)	408 (−24)

phorus also forms the pentachloride and pentabromide, whereas antimony also forms the pentachloride. A few mixed pentahalides are also known.

The trihalides are pyramidal molecules containing a lone pair of electrons and are generally prepared either by heating an excess of the element in the appropriate halogen or by the action of the concentrated hydrohalic acid on the oxide. Nitrogen trifluoride, however, is prepared by the electrolysis of anhydrous ammonium bifluoride, NH_4HF_2. Nitrogen trifluoride is a rather unreactive molecule, which is insoluble in water and does not hydrolyze. It will react with hydrogen when ignited to give hydrogen fluoride and nitrogen. A by-product of the electrolysis of ammonium bifluoride is dinitrogen tetrafluoride, N_2F_4, which may also be prepared by the action of copper metal on nitrogen trifluoride. Dinitrogen tetrafluoride is a very reactive molecule generally reacting as if it were the free radical $\cdot NF_2$. Due to the weakness of the nitrogen-nitrogen single bond (−38 kcal/mole) and the high electronegativity of the attached fluorine atoms, this type of behavior is not unexpected. When mercury and nitrogen trifluoride are subjected to a silent electric discharge, both dinitrogen tetrafluoride and dinitrogen difluoride, F—N=N—F, are formed. Dinitrogen difluoride is also a reactive molecule and exists as cis-trans isomers that are separable by gas chromatography. In general, these N-F species are more thermally stable than are the other nitrogen halides. This stability has been explained as being due to the strong nitrogen-fluorine bond.

In contrast to the stability of nitrogen trifluoride, nitrogen trichloride (prepared by electrolysis of saturated ammonium chloride solutions) is a very unstable molecule and explodes violently when warmed. This is consistent with its enthalpy of formation value of + 55 kcal/mole. (The other two nitrogen trihalides, NBr_3 and NI_3, are also thermally unstable.) Nitrogen trichloride also readily hydrolyzes.

$$Cl_2N\text{—}Cl + OH_2 \rightarrow \underset{\displaystyle H\text{—}O\text{—}H}{\underset{\downarrow\ \uparrow}{Cl_2N\text{—}Cl}} \rightarrow NH_3 + HOCl \quad [14.12]$$

Nitrogen trifluoride has no orbitals available for attack by water, whereas the chlorine atoms on nitrogen trichloride contain empty *d* orbitals that may be looked on as the point of nucleophilic attack from the oxygen of the water molecule, thus leading to the observed hydrolysis products of ammonia and hypochlorous acid.

At this point, we observe that hydrolysis of the other trihalides of this group does not proceed to give the same type of product. Phosphorus trichloride can serve as a general example. In this molecule, both of the atoms, halogen and Group V, have empty *d* orbitals that can be attacked by the oxygen in water. Since this attack is nucleophilic, it is most likely to take place on the less electronegative atom, the Group V atom. Thus, the hydrolysis of phosphorus trichloride produces phosphorus acid and hydrogen chloride. The hydrolysis of the rest of the members of this family proceeds in a similar fashion. Antimony trichloride and bismuth trichloride precipitate as the oxychlorides, SbOCl and BiOCl, before hydrolysis is complete.

$$Cl_2P{-}Cl + OH_2 \rightarrow \underset{\substack{\uparrow\ \downarrow \\ H{-}O{-}H}}{Cl_2P{-}Cl} \rightarrow (HO)_3P + HCl \qquad [14.13]$$

Nitrogen tribromide has not been isolated as a pure compound at room temperature. Nitrogen tribromide decomposes above $-80°C$, and even at that temperature, it may be in the form of an adduct with ammonia. Nitrogen triiodide, formed by adding iodine to ammonia, exists in the form of the adduct $NH_3 \cdot NI_3$ and is extremely sensitive to shock.

Phosphorus trifluoride is also not prepared by one of the general methods but is best prepared by the fluorination of phosphorus trichloride with a fluorinating agent, such as arsenic trifluoride.

$$PCl_3 + AsF_3 \rightarrow PF_3 + AsCl_3 \qquad [14.14]$$

It hydrolyzes more slowly than does phosphorus trichloride. Phosphorus trifluoride forms many complexes, and in this respect behaves much as does carbon monoxide in stabilizing low oxidation states. Phosphorus trichloride and tribromide may act as either Lewis acids or Lewis bases, depending on the other molecule involved. Phosphorus in the trivalent state has a great tendency to be oxidized, readily forming PX_5, PSX_3, and POX_3.

The chemistry of the oxychlorides is somewhat similar to the chemistry of the trichlorides. Nitrogen and arsenic also form the oxyfluorides,

F_3MO(M = N, As). Trifluoramine oxide is an interesting molecule, since infrared indicates that the N—O bond contains appreciable double bond character and that the nitrogen-oxygen bond is stronger than the nitrogen-fluorine bond. Consequently, in this molecule, nitrogen can be considered to have expanded its octet.

In many respects, the trihalides of arsenic, antimony, and bismuth behave like the phosphorus trihalides. They all hydrolyze, and, in keeping with the greater metallic character as Group V is descended, the hydrolysis becomes reversible. Arsenic and antimony trifluorides are used as mild fluorinating agents.

Several tetrahalides in addition to dinitrogen tetrafluoride are known. Diphosphorus tetrafluoride is prepared by the reaction of PF_2I with mercury. The phosphorus-phosphorus bond in this molecule is stronger than the nitrogen-nitrogen bond in dinitrogen tetrafluoride. The other tetrahalides are P_2Cl_4, P_2I_4, and As_2I_4. Diphosphorus tetraiodide is formed by the action of white phosphorus and iodine in benzene or carbon tetrachloride. Diphosphorus tetrachloride is formed by the reaction of hydrogen and phosphorus trichloride in an electric discharge. Diarsenic tetraiodide is formed by heating arsenic and iodine in a hydrogen atmosphere. These molecules readily react with air and water and decompose on standing to give the trihalide and a solid.

TABLE 14.11 ***Melting Points (°C) and Enthalpies of Formation of Group V Pentahalides***

(ΔH_f in kcal/mole in parentheses; M.P., °C)

	P	As	Sb	Bi
F	−94 (−228)	−80	8.4	725
Cl	160 (−106)		2.8 (−105)	
Br	100 (−64)			

Many of the pentahalides can be formed by the reaction of the element with excess halogen. Phosphorus pentafluoride, however, is probably best prepared by fluorinating phosphorus pentachloride with arsenic trifluoride. The stability of the pentahalides decreases for the heavier halogens, with the loss of X_2 occurring more readily.

The preparation of the pentahalides is governed by several factors. The large metals can accommodate more halides but are better oxidizing agents in their highest oxidation state. Table 14.11 lists the known pentahalides. Arsenic, antimony, and bismuth pentafluorides are used as fluorinating agents.

The pentahalides are trigonal bipyramidal molecules in the gaseous state. The hybridization hypothesized here is sp^2 for the trigonal planar portion of the molecule and *dp* for the two axial halogens. This hybridization is consistent with the different bond lengths observed for equitorial or axial halogens. Phosphorus pentafluoride shows only one type of fluorine at room temperature according to NMR. This is probably due to rapid intramolecular scrambling rather than to actual equivalence of all the fluorine atoms.

The high viscosity of antimony pentafluoride shows the ability of these molecules to act as electron pair acceptors in forming polymers. This viscosity can be accounted for by fluorine atoms donating to adjoining antimony pentafluoride molecules to give antimony the coordination number of 6. This ability to act as electron pair donors is also evident in the different structures some of these exhibit in the solid state. Phosphorus pentachloride is found to exist as $[PCl_4^+][PCl_6^-]$, whereas phosphorus pentabromide exists in the solid state as $[PBr_4^+][Br^-]$. In addition, many mixed species of the following types may also be formed:

$$[PCl_4^+][PCl_6^-] + 2AsF_3 \rightarrow [PCl_4^+][PF_6^-] + 2AsCl_3 \qquad [14.15]$$

Sublimation of the mixed species produces a mixed pentahalide

$$[PCl_4^+][PF_6^-] \rightarrow PCl_4F + PF_5 \qquad [14.16]$$

Other mixed halides are formed by the action of a halogen on the trihalide

$$PF_3 + Cl_2 \rightarrow PCl_2F_3 \qquad [14.17]$$

In some cases, the pentahalides hydrolyze faster than the trihalides, which can be accounted for in the following manner. Hydrolysis occurs via the empty *d* orbitals on the Group V element. In the pentahalides, one of these *d* orbitals is already used in bonding, and, therefore, the other *d* orbitals are also of a proper energy to be used in bonding. In the trihalides, the empty *d* orbitals are of a higher energy than the bonding orbitals and consequently are reluctant to be used. The pentahalides then react more readily with water. Partial hydrolysis produces the oxyhalide.

$$PCl_5 + H_2O \rightarrow 2HCl + O{=}PCl_3 \qquad [14.18]$$

An interesting series of Group V halides are the phosphonitrilic halides

$(PNX_2)_n$. They can be prepared in general by heating the appropriate phosphorus pentahalide with the ammonium halide,

$$PX_5 + NH_4X \rightarrow (PNX_2)_n + 4HX \qquad [14.19]$$

where X = chlorine, bromine, and iodine. The fluorides, however, are not prepared in this manner.

The phosphonitrilic halides are cyclic or linear polymers, and generally have good thermal stability. This thermal stability makes them desirable from an industrial viewpoint. The resistance to hydrolysis is dependent to some extent on the size of the polymer.

The cyclic trimer has the structure shown.

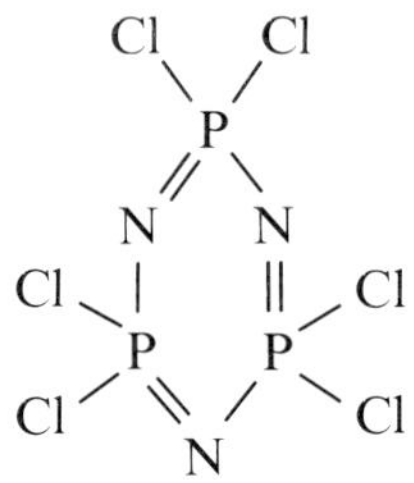

Aromatic character in the ring is suggested by the high stability of the trimer and by the maintenance of the ring structure in a variety of derivatives. Calculations have been made that support the concept of aromatic character in this ring.

Group VI

Oxygen forms more varied halides than do the other members of the group. Apart from tellurium, it is the only member of its family that forms stable compounds with all the halogens.

Oxygen-fluorine compounds are best classified as fluorides rather than oxides, since in these species the oxygen has a formal positive charge. Some properties of these oxides are listed in Table 14.12. The first member

TABLE 14.12 ***Melting Points and Enthalpies of Formation of Oxygen Fluorides***

	OF_2	O_2F_2	O_3F_2	O_4F_2
M.P. (°C)	−224	−164	−190	183 (dec)
ΔH_f (kcal/mole)	−7.6	4.7	6.2	

of the series, oxygen difluoride, OF_2, is prepared by the electrolysis of either potassium bifluoride in water or of 80% aqueous hydrofluoric acid. It is a fairly stable molecule, decomposing at from 250 to 280°C, and is less active than elemental fluorine as a fluorinating agent. Oxygen difluoride reacts slowly with water at room temperature to form oxygen and hydrogen fluoride, although the slow rate reaction is due less to an inherent thermodynamic stability than to a fairly high activation energy for reaction or decomposition. Dioxygen difluoride, O_2F_2, on the other hand, is unstable and decomposes rapidly into its elements at its boiling point. It is prepared by the glow discharge of its elements and has a structure similar to that of H_2O_2. The bonding can be explained simply by beginning with the elementary molecular orbital diagram for the oxygen molecule. This has one electron in each of the two π^* orbitals. These half-filled π^* orbitals can overlap with the half-filled p orbitals on fluorine. This retains the double bond between the two oxygen atoms and considers the oxygen-fluorine bond to be a three-center, two-electron bond. The oxygen-fluorine bond is electron deficient and is expected to be relatively long. Its ease of decomposition is not unexpected. Ozone fluoride or trioxygen difluoride, O_3F_2, is prepared by the reaction of nitrogen trifluoride and oxygen in a glow discharge. It quantitatively decomposes above −158°C to form dioxygen difluoride. Along with the dioxygen difluoride, trioxygen difluoride is one of the strongest fluorinating agents known. Tetraoxygen difluoride has also been reported and appears to be even more unstable than the other oxygen fluorides. Both the trioxygen and tetraoxygen difluorides may be mixtures rather than pure compounds.

The oxygen fluorides are all difluorides with an increasing number of oxygen atoms. The oxygen-chlorine compounds or chlorine oxides, some properties of which are given in Table 14.13, that have been isolated do not follow this pattern. Chlorine monoxide, Cl_2O, is obtained by passing chlorine over mercuric oxide. It is extremely unstable and explodes on shock or impact. It is soluble in water and produces hypochlorus acid, of which it is the anhydride. Solution in basic water will obviously produce the hypochlorite ion. Chlorine dioxide, ClO_2, is prepared by the action of sulfuric acid on potassium chlorate in the presence of oxalic acid as a

TABLE 14.13 ***Melting Points and Enthalpies of Formation of Oxygen Chlorides***

	Cl_2O	ClO_2	Cl_2O_6	Cl_2O_7
M.P. (°C)	−116	−50	3.5	−91.5
ΔH_f (kcal/mole)	−18.3	23.5		

reductant. It is very unstable and may even explode in the gas phase. The oxalic acid used in the preparation also forms carbon dioxide, which dilutes the chlorine dioxide and thereby increases the safety of the preparation. Chlorine dioxide is an odd-electron molecule, and in contrast to nitrogen dioxide it does not dimerize. The inability to dimerize has been explained by the fact that the valence bond treatment gives resonance structures (shown below) which delocalize the unpaired electron and therefore make it

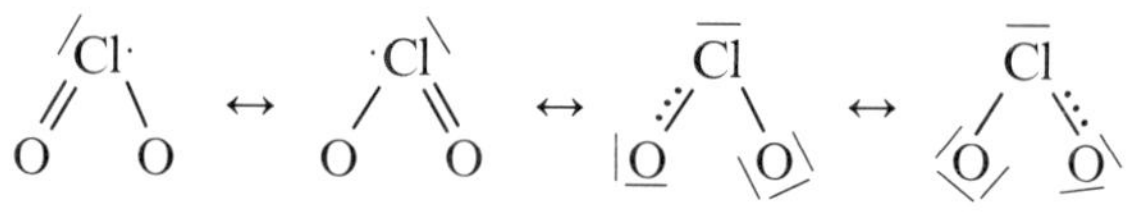

unavailable to be used in the formation of the dimer. Because hydrolysis of chlorine dioxide in alkaline water gives chlorite and chlorate as products, it can be considered as the mixed anhydride of chlorous and chloric acids. Dichlorine hexoxide is prepared by the action of ozone on chlorine under red light. The dissociation energy is only 1.5 kcal/mole ($Cl_2O_6 \rightleftharpoons 2ClO_3$), and it exists as a dimer in the liquid phase and as a monomer in the gas phase. It reacts violently with liquid water and is a strong oxidizing agent. In a number of reactions, it acts as if it were $ClO_2^+ClO_4^-$, chloryl perchlorate. In one such reaction, perchloric acid is produced.

$$Cl_2O_6 + HF \rightarrow HClO_4 + ClO_2F \qquad [14.20]$$

Dichlorine heptoxide is the anhydride of perchloric acid and as such can be produced by the dehydration of perchloric acid with phosphorus pentoxide (note that N_2O_5, the anhydride of nitric acid, can be prepared

$$4HClO_4 + P_4O_{10} \rightarrow 4HPO_3 + 2Cl_2O_7 \qquad [14.21]$$

in the same manner). The heptoxide is more stable than the monoxide and dioxide; however, it is still fairly reactive and shock sensitive. The chemistry of the various chlorine-oxygen compounds can generally be obtained from the appropriate Latimer diagram (Table 10.3).

Dibromine monoxide, Br_2O, is the anhydride of hypobromous acid, just as dichlorine monoxide is the anhydride of hypochlorus acid. It is a product of the thermal decomposition of bromine dioxide, BrO_2. Bromine monoxide decomposes at its melting point. Bromine dioxide is prepared by the glow discharge of bromine and oxygen and decomposes when warmed above $-40°C$. There are numerous higher oxides of bromine. Dibromine heptoxide does not appear to have been prepared; the fact that perbromic acid is known may mean that a suitable preparation for dibromine heptoxide has not yet been found.

A number of iodine oxides are also known. Diiodine pentoxide, the anhydride of iodic acid, is the best characterized of these oxides. It may be obtained from iodic acid by dehydration. It is a stable compound that may be heated to 300°C before it decomposes. A number of mixed halogen oxides, XO_nF, are also known.

Fluorides of Sulfur, Selenium, and Tellurium. Sulfur, selenium, and tellurium form octahedral hexafluorides by direct fluorination of the elements. Sulfur hexafluoride is a very inert molecule, unaffected by the attack of oxygen, halogens, or water. It is also inert in high intensity electrical fields and therefore is used as an insulating gas in electrical equipment. This inertness is a kinetic rather than a thermodynamic effect and appears to lie in a large activation energy for reaction. The sulfur is coordinately saturated and in addition is shielded from chemical attack by the fluorine atoms. Selenium hexafluoride is slowly hydrolyzed by water, and tellurium hexafluoride is completely hydrolyzed in 24 hours. This increase in reactivity occurs as the size of the central element increases and becomes less shielded by the fluorine atoms attached to it. The increased vulnerability to attack is also noted, in that tellurium hexafluoride can act as a Lewis acid and forms amine addition compounds, $TeF_6 \cdot 2(CH_3)_3N$. Efforts to prepare the same adducts with selenium hexafluoride were unsuccessful. This result agrees with the lower reactivity of selenium hexafluoride as compared with tellurium hexafluoride. Sulfur and tellurium are also hexacovalent in the decafluorides, M_2F_{10} (M = S, Te). Disulfur decafluoride is a very toxic liquid (B.P. = 29°C) and is more reactive than sulfur hexafluoride. There appears to be no inherent reason why diselenium decafluoride should not be formed. Some properties of the Group VI halides are given in Table 14.14.

TABLE 14.14 *Melting Points and Enthalpies of Formation of Some Group VI Halides*

(ΔH_f in kcal/mole in parentheses; M.P., °C)

Group VI	Atom	S	Se	Te
Halogens				
F	6	−51 (−262)	−35 (−267)	−38 (−315)
	4	−121 (−174)	−10	130
Cl	4	*d*−31 (−14)	sub. 191 (−46)	225 (−78)
	2	−78 (−12)		208
Br	4		*d*	308 (−50)
I	4			259

The tetrafluorides have a distorted tetrahedral structure with a lone pair of electrons (C_{2v}) and are all more reactive than the analogous hexafluorides. The difference in reactivity is especially noticeable between sulfur hexafluoride and tetrafluoride. The sulfur-fluorine bond is slightly stronger in sulfur tetrafluoride than in sulfur hexafluoride; however, sulfur tetrafluoride is coordinately unsaturated, and this vulnerability to attack makes it more reactive. Sulfur tetrafluoride is prepared by the action of sodium fluoride on sulfur dichloride in acetonitrile. Selenium tetrafluoride is formed by fluorinating selenium tetrachloride with silver fluoride. Tellurium dioxide reacts with selenium tetrafluoride to form tellurium tetrafluoride.

Sulfur forms two additional gaseous fluorides—sulfur difluoride, SF_2, and sulfur monofluoride, S_2F_2. Sulfur monofluoride exists as two isomers:

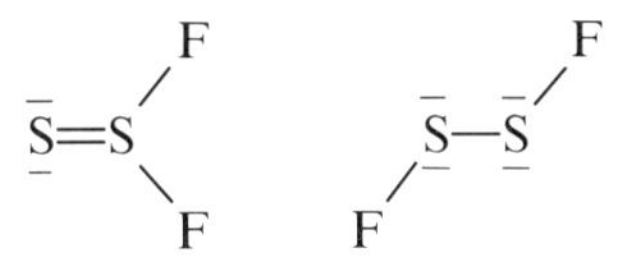

They are formed by the action of silver fluoride on excess sulfur at 200°C. Sulfur monofluoride decomposes to form free sulfur and the difluoride. It is difficult to obtain these species in the pure state.

The neutral hexachlorides of sulfur, selenium, or tellurium are not known. In the case of sulfur, this may be due to the larger size and lower electronegativity of chlorine, as compared with fluorine. Sulfur tetrachloride (the tetrachloride is the highest chloride of these elements) results from the chlorination of sulfur dichloride at low temperatures. Selenium and tellurium tetrachloride are prepared by chlorination of the element with elemental chlorine. The stability of the tetrachlorides increases as the size of the central element increases. At toom temperature, sulfur tetrachloride is completely dissociated into sulfur dichloride and elemental chlorine. Selenium tetrachloride is stable at its sublimation temperature of 191°C, and the tellurium analog is stable at its boiling temperature, 390°C. The instability of sulfur tetrachloride may be due to crowding of chlorine atoms around the sulfur. Selenium and tellurium form complex halides of the type M_2SeCl_6 and M_2TeCl_6, thereby indicating no steric problems for the formation of the tetrachlorides. The hexachlorodetellurate(IV) and hexachloroselenate(IV) anions are regular octahedrons rather than distorted octahedrons, as electron repulsions would predict. This has been rationalized in the following manner. The selenium- or tellurium-chlorine bond lengths are longer than predicted from the sum of the covalent radii, which would be anticipated if the lone pair electrons on selenium or tellurium were forced into an inner, spherically shaped *s* orbital. It is necessary to force the electron pair on these atoms into an inner orbital, since its presence in the distorted octahedral structure otherwise predicted would force the chloride ions closer together than their van der Waals radii. The electron pair in this inner orbital would increase the observed bond length and would have no stereochemical effect on the molecule. Similar effects are noted for the hexabromoantimonate(III) and hexabromotellurate(IV) anions. The anion complex, $SbBr_6^-$, which does not have this lone pair of electrons, is observed to have normal bond lengths.

Sulfur, selenium, and tellurium all form dichlorides, and, in addition, sulfur and selenium form the monochlorides. The sulfur mono- and dichlorides are formed by the reaction of sulfur and chlorine. Hydrogen polysulfide with sulfur mono- or dichloride yields a number of polysulfanes, S_nCl_2. Sulfur dichloride decomposes to the monochloride and chlorine. Therefore, it is difficult to obtain the pure materials. Both of these species hydrolyze in water.

Selenium and tellurium form di- and tetrabromides. Selenium tetrabromide exists only in the solid state and is completely dissociated in the vapor state, whereas tellurium tetrabromide is a stable molecule boiling above 400°C. The chemistry of these species is similar to the chemistry of the chlorides. The only proven iodide of this group is tellurium tetraiodide, which is prepared by heating tellurium tetrabromide and ethyl iodide in a sealed tube at 100°C.

A large number of oxyhalides of this group have been prepared. The most important are probably the thionyl halides, SOX_2, and the sulfuryl halides, SO_2X_2. The reaction of phosphorus pentachloride with sulfur dioxide produces thionyl chloride, which may then be fluorinated to yield thionyl fluoride. Sulfuryl chloride is prepared by the catalytic reaction of sulfur dioxide and chlorine. This may be fluorinated with sodium fluoride at 400°C to produce sulfuryl fluoride. Their structures are those expected from electron repulsion considerations; the thionyl compounds are pyramidal and the sulfuryl compounds are distorted tetrahedra. The thermal and hydrolytic stability of both types of compounds decreases with increasing atomic weight of the halogens; SO_2F_2, stable; SO_2Cl_2, partly decomposed at 100°C; SO_2Br_2, decomposes at 59°C. The chlorides are excellent chlorinating agents.

Group VII

The Group VII halides or interhalogens can be divided into three main classes: neutral molecules; polyhalide complex cations, such as FCl_2^+;

polyhalide complex anions, such as ICl_2^- (prepared by addition of X_2 to MX'). Only the neutral interhalogen molecules will be considered here.

The neutral interhalogen molecules have the general formula XX'_n, where $n = 1, 3, 5$, or 7, as expected from a consideration of the valence electrons available on the central element originally, and once unpairing begins. The attached halogen, X', is the more electronegative halogen, and to obtain the maximum number of attached atoms the electronegativity and size differences should be as great as possible. These interhalogens can be prepared by direct union of the elements with the specific conditions varied to produce the desired species.

TABLE 14.15 *Properties of Interhalogen Molecules*

Interhalogen	ClF	BrF	IF	BrCl	ICl	IBr
M.P. (°C)	−156	−33		−54	27	40
ΔH_f (kcal/mole)	−13.4	−14.7	−22.7	3.5	−7.7	−2.5
Interhalogen	ClF_3	BrF_3	I_2Cl_6	BrF_5	IF_5	IF_7
M.P. (°C)	−83	9		−60	9	4.5
ΔH_f (kcal/mole)	−38.8	−75	(−21.1)	−127	−205	−244

Table 14.15 lists the XX' interhalogen molecules. Iodine monofluoride has been observed only spectroscopically. The synthesis of pure bromine monofluoride is difficult also. Both these monofluorides have a great tendency to disproportionate into their more stable fluorides: iodine monofluoride into iodine, iodine pentafluoride, and even a little iodine heptafluoride; bromine monofluoride into bromine, bromine trifluoride, and bromine pentafluoride. The preference of bromine and iodine fluorides to exist as polyatomic molecules is probably due to the effect of the highly electronegative fluorine in making the *d* orbitals on the central halogen suitable for bonding. Such an effect would be greater with larger electronegativity differences between halogens. This trend is so great that iodine pentafluoride is the lowest (n value) stable iodine fluoride and chlorine trifluoride is the highest stable chlorine fluoride.

The XX' interhalogens can be prepared by direct combination of the elements. The observed order of bond strength of these interhalogens is IF > BrF > ClF > ICl > IBr > BrCl. This is the same order as for the electronegativity difference between the halogens involved and suggests that the gain in polar character at the beginning of the series strengthens the XX' bond. Chlorine monofluoride is probably the most common monofluoride. It is a very reactive molecule, even though its bond strength is greater than that of chlorine or iodine.

The XX'_3 species that have been studied are chlorine and bromine trifluoride and iodine trichloride (or, more properly, diiodine hexachloride). The stability of these compounds with respect to decomposition to their elements is $BrF_3 > ClF_3 > (ICl_3)_2$. Again, this is the order of decreasing electronegativity differences. Chlorine trifluoride has a *bent T* structure, with two of the bonds—(*a*)—weaker than the other bonds—(*b*).

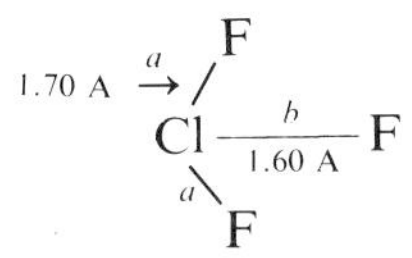

The bonding can be considered in the following manner. The *b* bond can be considered to be the normal covalent bond. A filled *p* orbital perpendicular to this normal bond can then be used to form a three-centered bond to the two fluorine atoms. Since the three-centered bond consists of, in order, a bonding orbital, a nonbonding orbital, and an antibonding orbital, and there are only four electrons to be used here, the bonds to the two additional fluorine atoms should be shorter and weaker. The same considerations apply to the bromine trifluoride molecule. Both are highly reactive molecules. Chlorine trifluoride reacts with almost all substances at low temperatures, and at high temperatures even the protective fluoride coatings on metals will not protect them from further attack. Bromine trifluoride finds use as a nonaqueous solvent in which many interesting species may be prepared (Chapter 8). Iodine trichloride is found as the planar dimer, I_2Cl_6.

All the pentafluorides are known. Again, the stability increases with increasing electronegativity difference. Chlorine pentafluoride is highly unstable and has not been completely characterized. It might be expected that such a molecule might not be too stable due to the high positive charge it would tend to place on chlorine and also due to relative size considerations. Their structures appear to be as given in Fig. 14.1.

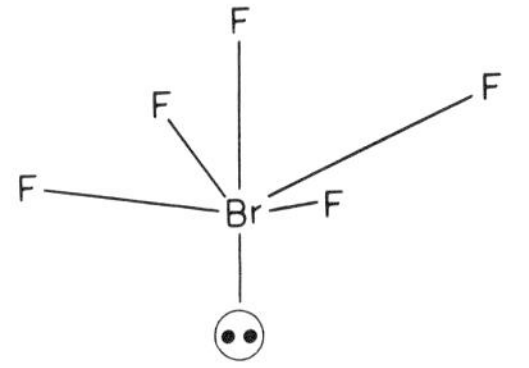

Figure 14.1. The square pyramidal structure of BrF_5 with one shorter bromine-fluorine bond (1.68 A) and four longer bromine-fluorine bonds (1.78 A).

Note that again the three-centered bonds (formed from two perpendicular filled *p* orbitals) are longer and presumably weaker than the ordinary covalent bond. Bromine pentafluoride is highly reactive. Iodine pentafluoride, however, is generally less reactive than the other interhalogens.

Iodine heptafluoride, the only representative of the XX'_7 class, is prepared by the fluorination of iodine pentafluoride. It appears to have a pentagonal bipyramidal structure and is a very reactive molecule, reacting with all the metals except those of the platinum group.

Group O

Once Bartlett showed that the so-called inert gases react, through his preparation of xenon hexafluoroplatinate(V), $XePtF_6$, it was not long before a vast number of xenon compounds were prepared. The properties of some of these compounds are listed in Table 14.16.

TABLE 14.16 ***Rare Gas Halides***

	XeF_2	XeF_4	XeF_6	KrF_2	$XeOF_4$	XeO_2F_2
M.P. (°C)	127	114	49.5	sub. −68	(−46?)	30.8
ΔH_f (kcal/mole)	−28.5	−68	−86	14.4		35

Xenon was found to react with fluorine to produce xenon di-, tetra-, and hexafluorides, depending on the reaction conditions. If excess xenon is heated with fluorine and the product condensed out to prevent conversion to higher fluorides, xenon difluoride can be produced. Xenon tetrafluoride is nearly quantitatively produced by the reaction of xenon and fluorine in a 1 to 5 ratio at 400°C and 6 atm, and xenon hexafluoride is nearly quantitatively formed by the reaction of xenon with fluorine in a 1 to 2 ratio at 700°C and 200 atm. The reaction conditions do not always have to be so vigorous, as a good yield of xenon difluoride was observed when a mixture of xenon and fluorine was allowed to stand in the sun.

These fluorides are stable when dry and pure. All of them hydrolyze in water.

$$XeF_2 + 2OH^- \rightarrow Xe + \tfrac{1}{2}O_2 + 2F^- + H_2O \qquad [14.22]$$

$$3XeF_4 + 6H_2O \rightarrow XeO_3 + 2Xe + \tfrac{3}{2}O_2 + 12HF \qquad [14.23]$$

$$XeF_6 + 3H_2O \rightarrow XeO_3 + 6HF \qquad [14.24]$$

Controlled hydrolysis yields oxyfluorides such as xenon oxide tetrafluoride, $XeOF_4$, and xenon dioxide difluoride, XeO_2F_2. Xenon oxide tetrafluoride is thermally stable, whereas xenon dioxide difluorine is thermally unstable and must be treated cautiously.

The structures of xenon difluoride and xenon tetrafluoride are linear ($D_{\infty h}$) and square planar (D_{4h}), respectively, as predicted by consideration of electron pairs in the valence shell. Their bonding can be simply considered as three-centered, four-electron bonds, as for the interhalogens discussed previously. Such a concept predicts the formation of the three halides as is observed and is consistent with the observed geometry.

When one considers xenon hexafluoride via the three-center, four-electron system, one predicts a regular octahedral structure (O_h), and when one considers this molecule on the basis of electron pair repulsions, one predicts a distorted octahedral structure. Experimental evidence points to a nonoctahedral structure or to an octahedral ground state with a very low barrier to distortion. An unambiguous determination of its structure is therefore of interest. Xenon hexafluoride can also act as a donor molecule, forming $XeF_6 \cdot BF_3$, or as a fluoride ion acceptor, forming Cs_2XeF_8. It is the most vigorous fluorinating agent of the three binary fluorides.

Krypton difluoride is prepared by the electric discharge of krypton and fluorine taken in a 1 to 2 ratio. It is a linear molecule that is thermally unstable at room temperature. There is some evidence for the existence of a radon fluoride, but the intense radioactivity of radon makes this study difficult and dangerous.

At least two studies have been made on the formation of xenon chlorides. Evidence for xenon dichloride has been found by passing xenon and chlorine through a microwave discharge and then condensing the product on an optical window at 20°K and studying the infrared spectrum. The decay of $^{129}ICl_4^-$ has been shown by Mossbauer spectroscopy to form xenon tetrachloride (see Chapter 11). Neither of these two chlorides would appear to have any long-term stability.

TRANSITION METAL HALIDES

The transition metal fluorides have been studied in more detail than have the other transition metal halides; therefore, many of the following generalizations apply most often to the fluorides. The study of transition metal halides is made difficult by the generally low volatility and high susceptibility to hydrolysis of these species. Due to this tendency to hydrolyze, oxyhalides readily form; consequently, some of the early work on the binary halides is suspect, since they may have been contaminated with oxyhalides.

Some general aspects of the chemistry of these transition metal halides can be noted (see Table 14.17). In most cases, the 3*d* transition elements tend to exist in a low oxidation state. Thus, chromium through zinc are common as stable difluorides, and almost all the possible dihalides are known. On the other hand, the 4*d* and 5*d* transition elements tend to exist in higher oxidation states. Molybdenum through iridium form hexafluorides that are stable at room temperature. The highest oxidation states occur with fluorine, and the highest oxidation states with other halogens

TABLE 14.17 ***Reported Transition Metal Halides*** **(MX_n)***

Group		F	Cl	Br	I
	Sc	3	3	3	3
III–B	Y	3	3	3	3
	La	3	3	3	3
	Ti	2, 3, 4	2, 3, 4	2, 3, 4	2, 3, 4
IV–B	Zr	2, 3, 4	2, 3, 4	2, 3, 4	2, 3, 4
	Hf	4	2, 3, 4	2, 3, 4	3, 4
	V	2, 3, 4, 5	2, 3, 4	2, 3, 4	2, 5
V–B	Nb	3, 4, 5	3, 4, 5	2, 3, 4, 5	2, 3, 4, 5
	Ta	3, 5	(?2), 3, 4, 5	3, 4, 5	3, 4, 5
	Cr	2, 3, 4, 5	2, 3, 4	2, 3	2, 3
VI–B	Mo	3, 4, 5, 6	2, 3, 4, 5, 6	2, 3, 4, 5	2, 3, 4
	W	4, 6	2, 4, 5, 6	2, 3, 4, 5, 6	2, 3, 4, 5
	Mn	2, 3, 4	2, 3	2	2
VII–B	Tc	5, 6	4, 6		
	Re	4, 5, 6	3, 4, 5, 6	3, 4, 5	2, 3, 4
	Fe	2, 3	2, 3	2, 3	2
	Ru	3, 4, 5, 6	2, 3, 4	3	3
	Os	4, 5, 6	2, 3, 4	3, 4	2, 3
	Co	2, 3	2, (3?)	2	2, 3
VIII–B	Rh	2, 3, 4, 5, 6	2, 3	3	3
	Ir	3, 4, 5, 6	2, 3, 4	2, 3	2, 3
	Ni	2	2	2	2
	Pd	2, 3, 4	2	2	2
	Pt	2, 4, 5, 6	2, 3, 4	2, 3, 4	2, 3, 4
	Cu	2	2	2	2
I–B	Ag	2			
	Au	3	2, 3	3	3

*Monohalides not included. The number refers to the oxidation state of the metal.

are lower and decrease with increasing atomic weight of the halogen. Note that this tendency for higher oxidation states lower in the periodic table is the opposite of the trend observed in the representative elements in which the lower oxidation states become more stable as the table is descended.

Although a number of rather general methods can be used for the preparation of the transition metal halides, in many cases synthetic methods specific for the halide desired must be employed. Some of the more general methods follow.

(a) *Direct halogenation of the metal.* This method generally produces the compound with the metal in the highest oxidation state possible in combination with that particular halogen. The reaction is performed in a flow system; adjustment of temperatures and flow rates are used to help determine the product formed. When the reaction is run at a high temperature, quenching of the product is sometimes used to prevent its decomposition. As might be expected, direct halogenation has somewhat limited use for the preparation of iodides.

Hydrogen halides also react with some transition metals to form the transition metal halides. In contrast to those halides prepared by the use of pure elemental halogen, the hydrogen halides generally produce the lower valence halides.

(b) *Reaction of halogenating agents with metals or metallic oxides.*

$$Nb_2O_5 + 5SOCl_2 \rightarrow 2NbCl_5 + 5SO_2 \qquad [14.25]$$

Fluorides are often prepared by the use of reactive interhalogens, such as bromine or chlorine trifluorides. Some problems arise here, however, since many of the interhalogens have a tendency to form adducts with the products desired. Aluminum trihalides are extensively used in this reaction with oxides. Aluminum triiodide is especially powerful. In these cases, the driving force has been considered to be the high lattice energy of aluminum oxide. Similarly, thionyl chloride is extensively used, because it forms volatile by-products.

The metallic oxides can also be halogenated with elemental halogen. In some cases, this method forms the oxide halides. Carbon is therefore often added to remove the oxygen and ensure complete halogenation.

(c) *Reduction of higher halides.* This method has been used with either the metal itself or with an active metal, such as aluminum, as the reducing agent.

$$3ZrCl_4 + Zr \rightarrow 4ZrCl_3 \qquad [14.26]$$

(d) *Special methods for fluorides.* Disproportionation of fluorides, especially MF_5, has been used to produce some fluorides otherwise prepared with difficulty. Halogen exchange with hydrogen fluoride has also been used. Very pure hydrogen fluoride is essential for the success of this method.

The chemistry of the transition metal halides is dependent to a large extent on the number and kind of halogens involved and on the structure of the halide. The following is a brief look at some of the binary halides based mainly on the number of halogens in the halides.

Univalent Halides

The most common representatives of this type of halide occur in Group I-B. Some rather unexpected monohalides, such as iridium chloride and bromide and the iodides of osmium and rhenium, have also been reported. Zirconium and hafnium monochlorides have also been reported recently and are surprisingly stable.

The silver halides are perhaps the best known examples of this class. The photosensitive nature of silver chloride, bromide, and iodide is responsible for their extensive use in the photographic process. Silver chloride, due to its light sensitivity, is also being used in glass materials that darken on exposure to light and lighten again once the light has been removed. Silver fluoride is the only water soluble silver halide. Recall also that the solubility of the silver halides in liquid ammonia is just the reverse of that in water (see Chapter 8).

Copper and gold monohalides can be made; however, they are generally not extremely stable. Table 14.18 lists some of their properties. In some

TABLE 14.18 ***Melting Points (°C) and Enthalpies of Formation (kcal/mole, in parentheses) of Univalent Halides***

	Cu	Ag	Au
F	908	435 (−48.5)	
Cl	430 (−32.2)	455 (−30.4)	185*d* (−8.3)
Br	488 (−25.3)	430 (−24.0)	212*d* (−3.3)
I	588 (−16.4)	557 (−14.8)	177*d* (0.3)

cases, such as for copper chloride and bromide, the extreme insolubility of the salt in water allows its preparation. The great tendency some of these have to form complexes with excess halide ion also increases their stability. The potential for the reaction

$$2Cu^+ \rightarrow Cu^{2+} + Cu^\circ, \qquad E^\circ = 0.368 \text{ V}$$

indicates the instability of the copper(I) ion in aqueous solution. Removal of this ion from aqueous solution by precipitation or complexation increases the stability of the monovalent form.

Divalent Halides

A very large number of transition metal dihalides have been observed (see Table 14.17). Approximately two-thirds of the possible dihalides have been observed. They can generally be prepared by the action of the halogens or hydrogen halide on the metal, or in some cases on the metallic oxide. The hydrogen halide is probably used more often than is the free halogen, since the free halogen has a tendency to put the metal into a higher oxidation state. Sometimes the dihalide is obtained by reduction of a higher halide, either with an active metal or with hydrogen. Depending on the relative sizes of the cation and anion, they usually have either the fluorite or rutile structure. In general, they are rather inert high-melting solids. The species that are not hydrolyzed by water often occur as hydrates.

The second and third row dihalides, especially the fluorides, have not been prepared or are not as well characterized as those of the first row. This is not unexpected, in view of the tendency for the 4*d* and 5*d* transition elements, especially in combination with fluorine, to prefer higher oxidation states.

Molybdenum and tungsten dihalides, particularly the chlorides, form hexameric metal *cluster compounds*. The name *cluster compound* has been given to discrete units containing at least three transition metal atoms in which strong metal-metal bonding occurs. At present, a large number of cluster compounds have been observed and studied. The general formula of the cluster compound depends upon the transition metal involved.

The formula for the cluster ion of molybdenum dichloride derivatives is $[Mo_6Cl_8]^{4+}$ (see Fig. 5.17). The structure is either an octahedron of metal atoms in a cube of chloride atoms or a face-centered cubic structure with molybdenum atoms in the faces and chlorine atoms at the corners of the cube. Each chlorine atom then bridges between three molybdenum atoms. After considering the electrons necessary for the molybdenum chlorine bonds, we see that there are 24 electrons remaining for the 12 metal-metal bonds. This gives a bond order of one for each metal-metal bond. Molybdenum dichloride itself has the same general $Mo_6Cl_8^{4+}$ unit. Two of the remaining four chlorines are attached to two molybdenum atoms trans to each other. The remaining chlorines bridge between the four remaining coplanar molybdenum atoms and neighboring clusters, giving a layer structure infinite in two dimensions. This unit persists undestroyed through many chemical reactions.

Tervalent Halides

Excluding Group II-B from consideration as transition metals, over 78 of the possible 108 transition metal trihalides have been reported (see Table 14.17). These, like the dihalides, are generally inert, high-melting solids, some of which hydrolyze rather readily. They can be prepared in the same manner as the dihalides, except that the free halogen is probably used more often than is the hydrogen halide. Some authorities in the field suggest that perhaps some of the compounds listed as trihalides might actually be complex oxyhalides; they should perhaps be reinvestigated.

The structure of iron trichloride is interesting. In the solid state, it exists in a layer structure in which the iron is octahedrally surrounded by six chlorine atoms, whereas in the vapor state it exists as a dimer, Fe_2Cl_6, with the same structure as $Al_2Cl_6(D_{2h})$.

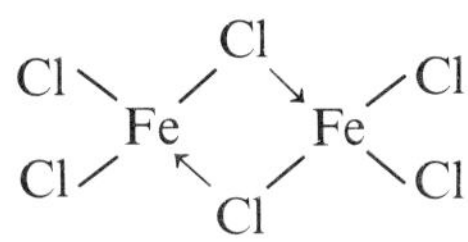

A number of the trivalent halides form complexes of the type A_3MX_6 (A = alkali metal), in which the metal atom can be in an octahedral configuration. Rhenium(III) forms a most interesting complex that is not of this A_3MX_6 type but has the formula $[Re_2Cl_8]^{2-}$. This is not a cluster compound since it does not have the required three transition metal atoms; however, significant metal-metal bonding is involved. The structure is shown to be

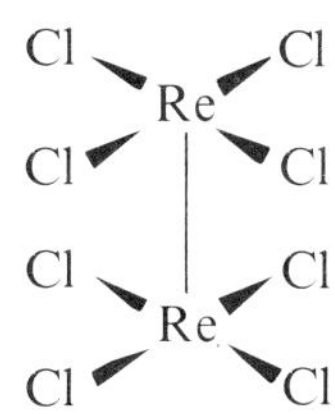

The rhenium-rhenium bond length is very short and the chlorine atoms are eclipsed rather than staggered, as might be expected. Simple molecular orbital considerations show that a quadruple bond (δ bond) explains these factors. The $d_{x^2-y^2}$, p_x, p_y, and s orbitals in the valence shell of the rhenium atom are set aside for bonds to the chlorine atom. Of the remaining orbitals on rhenium (containing one electron each), either the d_z^2 orbital or a p_z–d_z^2 hybrid on each rhenium atom can overlap to form a σ bond, the d_{xz} and d_{yz} orbitals on one rhenium can overlap with the same orbitals on the other rhenium atom in π symmetry, and the d_{xy} orbitals can overlap with δ symmetry. Since there are eight electrons available, the ground state configuration is $\sigma^2\pi^4\delta^2$. The δ bond explains the eclipsed configuration of the ion. It was estimated that 7 kcal/mole for the δ bond would be enough to account for the eclipsed configuration. Calculations based on the spectrum of the species indicates that the δ bond may be around 90 kcal/mole. A number of other species have since been found to be explained on the same basis.

Quadrivalent Halides

Table 14.17 indicates that this type of molecule is rather well known—however, not as extensively as are the di- and trihalides. Because of the higher formal charge on the metal, the amount of ionic character is less in these halides than in the lower halides. Thus, although most of these halides still have high melting points, titanium tetrachloride is a volatile liquid that will undergo many of the reactions of the covalent halides of the representative elements to the right in the periodic table. Most of these tetrahalides will react with water. Disproportionation of some of the tetrahalides can be used as a method of preparation for some of the pentahalides. In the solid state, most of the structures are coordinated so as to give octahedral coordination to the metal.

Pentavalent Halides

These halides are more common with the 4*d* and 5*d* transition elements and with the more electronegative halogens. The preparations are fairly specific for the pentahalide derived. Their melting points are generally low enough, in most cases being 200°C or less, so that vacuum sublimation is often used for purification. In general, they are fairly susceptible to hydrolysis.

The metals tend to assume an octahedral coordination. Niobium and tantalum pentachlorides in the solid state form dimers with bridging chlorine atoms similar to the bridging observed in the ferric chloride. The octahedra around the metal atoms in the pentafluorides are distorted, depending on the number of *d* electrons on the metal. Metals with many *d* electrons do not accept the electron density from the fluorine too well and are more distorted.

Complexes of the type AMX_6 are well known. The molybdenum and tungsten hexafluoride complexes are extremely sensitive to moisture. Since a color change is involved in their reaction with water, they have been considered as possible irreversible moisture indicators.

Hexavalent Halides

These are essentially all fluorides except for molybdenum, tungsten, technecium, and rhenium hexachlorides and tungsten hexabromide. The hexafluorides are generally prepared by the action of elementary fluorine on either the metal or a lower fluoride. They attack glass in the presence of moisture.

The second-row hexahalides are generally more reactive than the third row, and reactivity generally increases as we go to the right in the periodic table, shown by the following order: $OsF_6 > IrF_6 > PtF_6 > RuF_6 > RhF_6$. The last three hexafluorides are thermally unstable at room temperature, whereas the earlier ones are stable at room temperature. Pi bonding effects are used to explain the relative stabilities. Fluorine to metal $(p \rightarrow d)$ π bonds stabilize the molecule. As the number of *d* electrons increases to the right, π bonding, and therefore stability, also decrease. Since the 5*d* metals are approximately the same size as the 4*d* metals but have a much larger nuclear charge, they bond more effectively with the fluorine atoms.

These hexahalides are volatile materials, generally with very short liquid ranges. Tungsten hexafluoride is a gas at room temperature (M.P. 2°C; B.P. 18°C) and is said to be the most dense of all known gases.

Oxyfluorides of the hexavalent metals are also well known. These oxyfluorides often behave as if they contained only fluorine atoms (MOF_4 behaves as if it were MF_5), since the similarity in size between oxygen and fluorine allows the oxygen to behave as if it were a fluorine atom.

Only a few heptahalides are known.

Suggested Readings

1. Canterford, J. H., and R. Cotton, *Halides of the Second and Third Row Transition Metals*, John Wiley & Sons, Inc., New York, 1968.
2. Cotton, F. A., *Inorg. Chem.,* **4,** 334 (1965). A discussion of the bonding in $[Re_2Cl_8]^{2-}$.
3. Emeleus, H. J., *The Chemistry of Fluorine and Its Compounds*, Academic Press, Inc., New York, 1969.
4. Fox, W. B., et. al., *Inorg. Chem.,* **7,** 2064 (1968). A discussion of trifluoramine oxide.
5. George, J. W., "Group V and VI Halides and Oxyhalides," *Progr. Inorg. Chem.,* **2,** 33 (1960), F. A. Cotton, ed., Interscience Publishers, New York.

6. Holloway, J. H., *Noble Gas Chemistry*, Methuen & Co., Ltd., London, 1968.

7. Kolditz, L., "The Halides of Phosphorus, Arsenic, Antimony, and Bismuth," in vol. 7, *Advances in Inorganic Chemistry and Radiochemistry,* H. J. Emeleus and A. G. Sharpe, eds., Academic Press, Inc., New York, 1965, p. 1.

8. Massey, A. G., "The Halides of Boron," in vol. 10, *Advances in Inorganic Chemistry and Radiochemistry*, H. J. Emeleus and A. G. Sharpe, eds., Academic Press, Inc., New York, 1967, p. 1.

9. Paddock, N. L., and H. T. Searle, "The Phosphonitrilic Halides and Their Derivatives," in vol. 1, *Advances in Inorganic Chemistry and Radiochemistry*, H. J. Emeleus and A. G. Sharpe, eds., Academic Press Inc., New York, 1959, p. 347.

10. Peacock, R. D., "Fluorine Compounds of the Transition Metals," *Progr. Inorg. Chem.,* **2,** 194 (1960), F. A. Cotton, ed., Interscience Publishers, New York.

11. Schmeiseer, M., and K. Brandle, "Oxides and Oxyfluorides of the Halogens," in vol. 5, *Advances in Inorganic Chemistry and Radiochemistry*, H. J. Emeleus and A. G. Sharpe, eds., Academic Press Inc., New York, 1963, p. 41.

12. Timms, P. L., et. al., *J. Am. Chem. Soc.*, **87,** 2824 (1965). Preparations of $(SiF_2)_x$.

13. Wells, A. F., *Structural Inorganic Chemistry*, 3rd ed., Oxford University Press, Oxford 1962.

14. Weiberga, E. H., E. E. Havinga, and K. H. Boswijk, "Interhalogen Compounds and Polyhalides," in vol. 3, *Advances in Inorganic Chemistry and Radiochemistry*, H. J. Emeleus and A. G. Sharpe, eds., Academic Press Inc., New York, 1961, p. 133.

Study Questions

14.1.

$$Cl^- \xrightarrow{-1.3595} Cl_2 \xrightarrow{-1.63} HClO \xrightarrow{-1.42} ClO_3 \xrightarrow{-\ -1.19} Cl0^{4-}$$

$$Br^- \xrightarrow{-1.07} Br_2 \xrightarrow{-1.59} HBrO \xrightarrow{-1.49} BrO_3^-$$

$$I^- \xrightarrow{-0.5355} I_2 \xrightarrow{-1.45} H1O \xrightarrow{-1.14} IO_3^- \xrightarrow{-1.7} H_5IO_6$$

$$H_2O \xrightarrow{-1.229} O_2$$

Discuss the relative stability of these species in standard *acid* or *basic* solution, with regard to disproportionation and liberation of oxygen.

14.2. Give and explain the detailed stereochemistry of the following: $T1Cl_3$, ICl_3, SbF_6^-, $SeBr_6^{2-}$, $C10_2F$.

14.3. It has been found that $(C_2H_5)_2A1F$ is unusually viscous. It had very weak acceptor properties and formed no adduct with ethers but amine adducts were identified. Explain [Lanbengayer, A. W. and G. F. Lengnick, *Inorg. Chem.*, **5,** 503 (1966)].

14.4. P_2I_4 has been found to form the adducts $P_2I_4 \cdot BBr_3$ and $P_2I_4 \cdot 2BBr_3$ No other adducts were observed, and, in addition, BF_3 did not form an adduct with P_2I_4. Discuss these results. [(Coroley and Cohen, *Inorg. Chem.,* **4,** 1200 (1965.]

14.5. Calculate ΔH_f for NF_3 from the data given in chapter 2. Compare this with the experimental value given in this chapter.

14.6. Explain why SF_6 is inert whereas SF_4 hydrolyzes, even though the SF_4 bond energy is greater than that in SF_6.

14.7. Calculate ΔH_f for $CuCl_2$; $CuCl_2$ is known, but apparently ΔH_f has not been determined. Do your calculations agree with the existence of $CuCl_2$? Why or why not? Explain any assumptions. (*Hint*: Assume same lattice energy as $MgCl_2$. Why?)

14.8. Calculate ΔH_f for Ne^+F^-. Would such a compound be expected to be formed? What factors make it feasible or nonfeasible? State any assumptions made.

14.9. *Explain*: In covalent compounds, a bromide generally has a higher B.P. than a corresponding chloride, but in ionic compounds, the reverse is generally true.

14.10. Brencic, J. V. and F. A. Cotton, [*Inorg. Chem.*, **8**, 7 (1969)], give the following qualitative MO diagram for the metal-metal bond portion of an M_2X_8 species of symmetry D_{4h} (σ_n' and σ refer to nonbonding orbitals):

————	$b_{1g}(\sigma^*)$
════	$e_g(\pi^*)$
————	$b_{1u}(\delta^*)$
————	$a_{2u}(\sigma_n')$
————	$a_{1g}(\sigma_n)$
————	$b_{2ag}(\delta)$
════	$e_u(\pi)$
————	$a_{1g}(\sigma)$

For the species $Mo_2Cl_8^{4-}$, neglect Mo—Cl bonds and the electrons in them, and fill in the diagram with the remaining electrons on the metal atoms that can be used for metal-metal bonding.

Discuss the resultant in terms of number and type of bonds formed, including the atomic orbitals necessary to form them, and the effect of these on bond distances and structure. Give the structure for this ion based on your reasoning and explain how you arrived at this structure.

14.11. Fluoride is not found to any significant extent in sea water. Why? Fluoridation of drinking water (perhaps 0.01 mole/liter of fluoride ion) is used to help prevent tooth decay. Discuss the problems that would arise in fluoridating hard water (about 1.32×10^{-5} mole/liter of Ca^{2+}).

14.12. Write the equations and give the potentials for the reactions employed in obtaining elemental bromine from sea water.

14.13. Explain the relative bond energies (kcal/mole) of the following halogens: $F_2 = 38$; $Cl_2 = 58$; $Br_2 = 46$; $I_2 = 36$.

14.14. Explain the general decrease in melting point of the alkali metal halides when either the cation or anion is increased in size while the size of the other is held constant.

14.15. Compare the structures of boron and aluminum trichlorides and explain any differences or similarities.

14.16. *Explain*: Carbon tetrachloride is stable in water, whereas silicon tetrachloride immediately hydrolyzes. Both of these molecules are thermodynamically unstable with respect to hydrolysis.

14.17. Write equations for the following reactions:

(a) Boron trifluoride with water.
(b) Thallium(III) ion with iodide ion to form thallium triiodide (write in ionic form).
(c) The preparation of silicon difluoride.
(d) The hydrolysis of phosphorus tribromide.
(e) The preparation of phosphonitrilic chloride trimer.
(f) The preparation of dichlorine heptoxide.
(g) The preparation of xenon trioxide.

14.18. The structure of chlorine trifluoride is

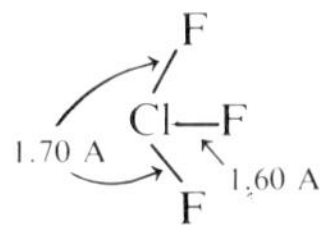

Explain why the bond lengths are different.

14.19. Ammonium halides sublime on heating by a mechanism that involves their dissociation to ammonia and HX (where X = halogen) followed by recombination to form NH_4X on the cooler surface of the system. Predict which halide would be expected to sublime at the lower temperature—NH_4F or NH_4Br. Explain.

14.20. Give balanced equations for the following reactions:

$FeCl_3 \cdot H_2O$ + heat
$BeCl_2$ + excess Na_2CO_3 solution

14.21 It has been reported [Demitras, G. C., and A. G. MacDiarmid, *Inorg. Chem.*, **6,** 1903 (1967)], that 5.540 mmole of PF_5 react with 3.250 mmole of $CH_3N[Si(CH_3)_3]_2$, yielding 6.350 mmole of $(CH_3)_3SiF$ (which is 97.7% of the expected yield) and a new compound that is analyzed as follows: C, 10.50%; H, 2.77%; N, 11.70%; P, 26.37%; F, 48.25%. The mass spectrum of the new compound gives a molecular ion peak at 234.0 m/e. The infrared spectra indicates the presence of methyl groups, C—N and P—F stretching. The proton NMR of this new compound gives a simple triplet 2.42 ppm downfield from TMS. Determine a structure for this new compound consistent with the data presented.

CHAPTER 15

Organometallic Chemistry

INTRODUCTION

Although the term *organometallic* is often loosely applied to any compound containing carbon and a metal atom, it will be limited in the present discussion to those compounds containing a metal-to-carbon bond. Such compounds would thus include, in addition to various hydrocarbon derivatives, carbonyls, certain chelate complexes, carbides, and related compounds. The question of what elements may be considered *metallic* is somewhat difficult. Boron and phosphorus are normally excluded in the scope of organometallic compounds, but the metalloids silicon, arsenic selenium, and tellurium are sometimes included when comparison to neighboring elements is made.

Since the discovery of the alkyls of zinc and mercury over a century ago, a number of important discoveries have been made. The discovery of Grignard reagents (organomagnesium halides) in 1900 provided versatile intermediates for a variety of organic syntheses. More recently, the synthesis of ferrocene, $(C_5H_5)_2Fe$, in 1951, and the later determination of its structure, opened up a new field of research (π complexes), which has contributed greatly to our understanding of chemical bonding.

PROPERTIES

The primarily covalent character of most organometallic compounds is reflected in their physical properties. They are typically low-melting solids, liquids, or gases at ordinary temperatures and are usually soluble in non-polar or weakly polar organic solvents, such as alkanes, toluene, ether, or even alcohols. Their chemical properties, but especially their thermal stability, vary widely. For example, tetramethylsilane (Me_4Si) is unchanged after many days at 500°C, whereas tetramethyltitanium decomposes rapidly at room temperature. Reaction rates of these compounds also vary widely; e.g., dimethylmercury is not attacked at room temperature by oxygen in the air, whereas dimethylzinc is spontaneously inflammable.

CLASSIFICATION

Organometallic compounds may be classified by two convenient methods. One method, based on the nature of the bonding, can be correlated well with position of the metal in the periodic table (Table 15.1). The very electropositive alkali metals form ionic organoderivatives as expected, since they must bond to carbon, which is fairly electronegative. The formation

TABLE 15.1 ***Types of Organometallic Compounds and the Periodic Table****

Covalent Cluster Structures (→ Be, Mg; → B, Al)

Li	Be											B	C	N	O	F
Na	Mg											Al	Si	P	S	Cl
K	Ca	Sc	Ti	V	Cr	Mn	Fe	Co	Ni	Cu	Zn	Ga	Ge	As	Se	Br
Rb	Sr	Y	Zr	Nb	Mo	Tc	Ru	Rh	Pd	Ag	Cd	In	Sn	Sb	Te	I
Cs	Ba	La	Hf	Ta	W	Re	Os	Ir	Pt	Au	Hg	Tl	Pb	Bi	Po	*Zi*
Fr	Ra	Ac														

Ionic (alkali metals)

Nature of metal-carbon bonding unclear (Be–Ra)

d-block transition elements form both *M*—C bonds and complexes with π-bonding ligands.

Volatile covalent *M*—C bonding

*Taken with permission from Coates, G. E., M. L. H. Green, P. Powell, and K. Wade, *Principles of Organometallic Chemistry*, Methuen & Co., Ltd., London, (reprinted 1970).

of ionic compounds is especially favored when the hydrocarbon anion may be stabilized, as, for example, when the negative charge may be delocalized over several carbon atoms in an aromatic or unsaturated system. The high stability of cyclopentadienyl derivatives, such as $K^+C_5H_5^-$, in which six π electrons produce a delocalized aromatic system, is a prime example. In acetylene derivatives, such as $Na^+C^-{\equiv}CH$, stabilization is mainly the result of the greater electronegativity of *sp* relative to sp^3 hybridized carbon atoms.

Volatile Covalent

The less electronegative metals and metalloids form volatile organic derivatives in which the metal-carbon bond consists of a single covalent electron pair linkage. The stabilities of these compounds are very similar to the wide range of stabilities found for other nonmetallic organic compounds.

Electron Deficient

Like the boron hydrides, certain of the organic derivatives of the metals lithium, beryllium, magnesium, and aluminum contain electron-deficient multicentered bonds; e.g., Me_4Li_4, $(Me_2Be)_n$, and $Me_3(Al)_2$.[1] These metals form strongly polarizing cations because of their high charge densities; this is considered to be an important factor in their forming compounds with electron-deficient structures.

Transition Metals

A third classification of organometallic compounds includes those of the transition metals. These compounds may involve σ, π, or sometimes δ bonding, or a combination of these types. This ability of transition metals to form multiple bonds is usually associated with the presence of electrons in the *d* orbitals of the metal, which are of suitable energy for overlap, especially with unsaturated organic ligands.

Group II-A Metals

The more metallic elements of Group II-A form organic derivatives in

[1] To conserve space and hopefully make the organometallic formulas more simple, the following abbreviations will be used: Me = methyl; Et = ethyl; Ph = phenyl; Bu = butyl; *t*-Bu = tertiary butyl; *n*-Bu = normal butyl; *i*-Bu = isobutyl; AcO^- = acetate; *i*-Pr = isopropyl; *o*-phen = orthophenanthroline.

which the nature of the metal-carbon bonding is not clearly established, although at least in some instances it is predominantly ionic in character—e.g., the cyclopentadienyl derivatives of calcium, strontium, and barium.

Another method of classifying organometallic compounds involves describing the number of electrons formally available from the neutral hydrocarbon or hydrocarbon radical. Such a classification is independent of the bonding character, and the method of formation is used primarily with the transition metals. Thus, with increasing number of available electrons, one has

1-electron donors:	alkyls
2-electron donors:	olefins
3-electron donors:	allyls
4-electron donors:	conjugated dienes
5-electron donors:	dienyls, such as the cyclopentadienyl radical
6-electron donors:	trienes and arenes
7-electron donors:	trienyls, such as the cycloheptatrienyl radical

In many complexes involving multielectron donor ligands, notably with cyclopentadiene in the formation of ferrocene, $Fe(C_5H_5)_2$, aromatic character is found in the system although none existed in the reactants prior to the reaction. Delocalization of electrons occurs to varying degrees in complexes involving donors of three electrons or more, such as conjugated dienes, and it is an important factor in their unusual stability.

BONDING AND STABILITY CONSIDERATIONS

As noted previously in Chapter 2, the character of a chemical bond is reflected to a large degree by its bond strength. Although thermochemical data on organometallic compounds are not generally available and precise comparisons cannot be made, sufficient measurements have been made to lead to the following order of stability for organometallics containing various organic groups attached to a given metal:

$$\text{Et—}M < \text{Me—}M < \text{Ph—}M < \text{CF}_3\text{—}M$$

This order has been suggested to be valid for most metals. Higher primary alkyls are generally of the same order of stability as ethyl, but secondary and tertiary alkyls are less stable. The differences are due at least in part to certain steric factors that prevent association and hence stabilization of the molecules by intermolecular secondary bonding forces.

The greater stability of aryls compared with alkyls is probably restricted to those metals with *d* orbitals, either filled and capable of back donating electrons into the antibonding (π^*) orbitals of the aromatic nucleus, or empty and hence able to overlap and accept electrons from the aromatic π orbitals. Such π bonding increases the bond multiplicity and strengthens what would otherwise be a normal covalent bond of rather moderate strength.

Back donation is particularly important for carbon monoxide and nitrile ligands and is essential for stability in their complexes. These ligands are quite similar in their bonding characteristics. For CO (Chapter 5), we may consider that the free molecule has two π bonds as well as a σ bond linking the two atoms, and, in addition, a lone pair of electrons in a *p* orbital on carbon. Donation of the latter to a neutral metal atom results in the bonding condition depicted in Fig. 15.1 (in which only the π orbital in the plane of the paper is shown). If *M* is a transition metal (the only

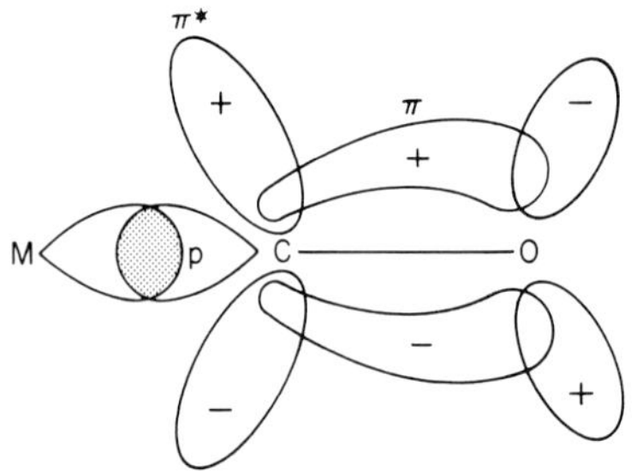

Figure 15.1. σ-bonding system in a metal carbonyl.

type of metal that forms stable compounds of this type), a filled *d* orbital of the metal can now overlap with the vacant π^* orbital of the carbonyl group. This is illustrated in Fig. 15.2.

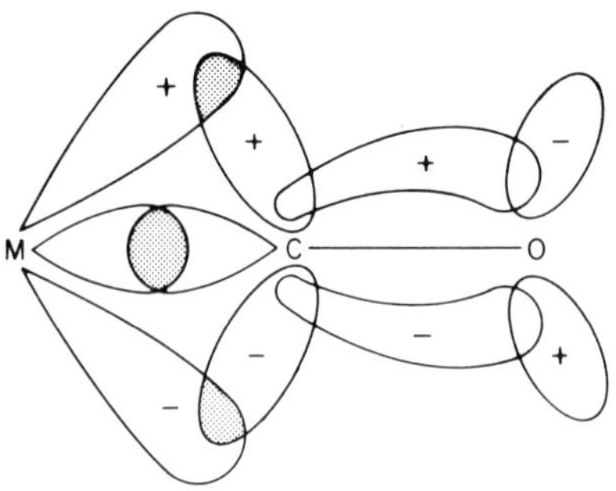

Figure 15.2. Back-bonding in a transition metal carbonyl.

The extremely weak basic character of carbon monoxide is evidenced by its inactivity toward hydrogen ion to form HCO^+. This weak basic character is also exhibited in its reactions with metals and metal ions. The accumulation of positive charge on the ligand with pure σ bonding is too unfavorable to permit the formation of a strong bond with the metal without back donation from the metal to the ligand. Although back donation is adequate in some positive metal ions [e.g., $Mn(CO)_6^+$ is stable], more filled orbitals on the metal become available as one reduces the oxidation number. Thus, carbonyls are favored in the zero [e.g., $Ni(CO)_4$] or negative [e.g., $Fe(CO)_4^{2-}$] oxidation states of the metal.

The greater basicity of isonitriles (isocyanides), CNR, as compared with CO, results in higher oxidation states being favored [$Fe(CNR)_6^{2+}$] with correspondingly weaker back donation. This difference can be attributed to the lower electronegativity of nitrogen compared with oxygen.

The cyanide ion, CN^-, is isoelectronic with carbon monoxide, and in at least the majority of its compounds it bonds to metals through carbon. As a negative ion it will generally form a strong simple covalent bond with metallic ions. Back donation cannot be a significant factor in such cases, except in certain anionic complexes in which the metal is in a low oxidation state. The existence of $[Ni(CN)_4]^{4-}$ containing Ni(O) is an example, and it is directly comparable to $Ni(CNR)_4$ and $Ni(CO)_4$.

The bonding in metal olefins provides a somewhat different picture. Whereas the carbon of CO uses its lone pair to form the σ bond, in the case of the olefin, the π electrons themselves are involved. These electrons are thus shared between at least three atoms, and the bond would be quite weak without back donation from the transition metal d to the π^* orbital of the olefin, as shown in Fig. 15.3.

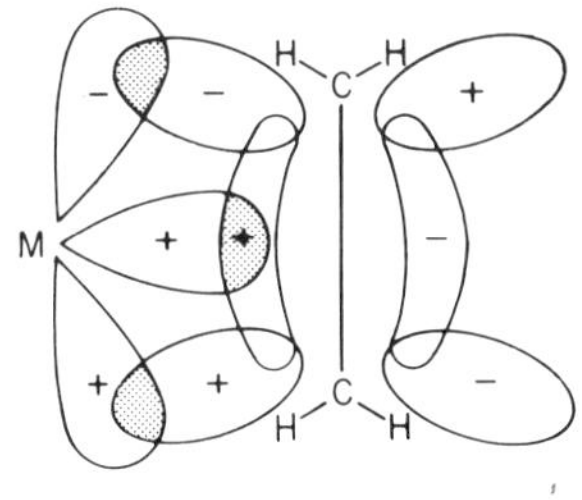

Figure 15.3. Bonding in a metal-olefin.

In contrast to carbonyls, where the metal lies in the nodal plane of the π electrons, bonding in olefin complexes is at right angles to the nodal plane. To distinguish between the two types, the former is referred to as a π bond, the latter as a μ bond. However, since π orbitals are utilized in an important way in both types, both are correctly referred to as π complexes without specifying the precise bond type.

KINETIC STABILITY

Many aspects of thermodynamics and kinetics must be considered when one discusses the stability of an organometallic compound, whether one speaks of resistance to thermal decomposition or to chemical attack.

As described previously (Chapter 3), the thermodynamic stability of a compound is really measured by its standard free energy of formation, ΔG°_f. However, as is true for many other compounds, standard entropy data for organometallic compounds are not commonly available at this time; thus we need to use standard enthalpy of formation values to indicate approximately the thermodynamic stability of organometallic compounds (Table 15.2 and Fig. 15.4).

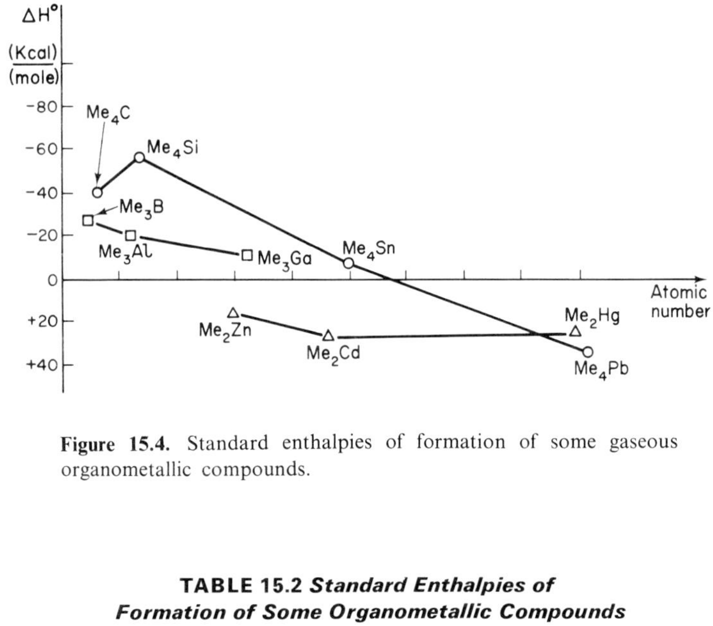

Figure 15.4. Standard enthalpies of formation of some gaseous organometallic compounds.

TABLE 15.2 *Standard Enthalpies of Formation of Some Organometallic Compounds*

Compound	ΔH_f° (kcal/mole)	Compound	ΔH_f° (kcal/mole)
EtLi (g)	13.9 ± 1.4	Me_2Zn (g)	13.1 ± 2
Me_3B (g)	-29.3 ± 5.5	Me_2Cd (g)	26.2 ± 0.3
Me_3Al (g)	-21.0 ± 2	Me_2Hg (g)	22.3 ± 1
Me_3Ga (g)	-9.2 ± 4	Me_4Si (g)	-57
Me_3P (g)	-23.0 ± 1.5	Me_4Sn (g)	-4.6 ± 0.6
Me_3As (g)	3.7 ± 1.2	Me_4Pb (g)	32.6 ± 0.3
Me_3Sb (g)	7.4 ± 3.2	Me_6Al_2 (g)	-62 ± 4

As may be seen, particularly from Fig. 15.4, the organoderivatives of most of the lighter elements are thermodynamically stable at room temperature with respect to decomposition to the elements, whereas those of the heavier metals are unstable to such decomposition; they are endothermic compounds.

An example of the latter is dimethyl mercury, for which ΔH_f° for the reaction in equation (15.1) is 30 kcal/mole

$$2C + 3H_2 + 2Hg \longrightarrow Me_2Hg \qquad [15.1]$$

whereas ΔH for the process

$$Me_2Hg\,(l) \rightarrow Me_2Hg\,(g) \qquad [15.2]$$

is 7.7 kcal/mole. The stability of this compound and other thermodynamically unstable organometallics can be attributed to the necessity of a reaction path of sufficiently low energy being available for the decomposition. That is, the decomposition reaction may be kinetically controlled.

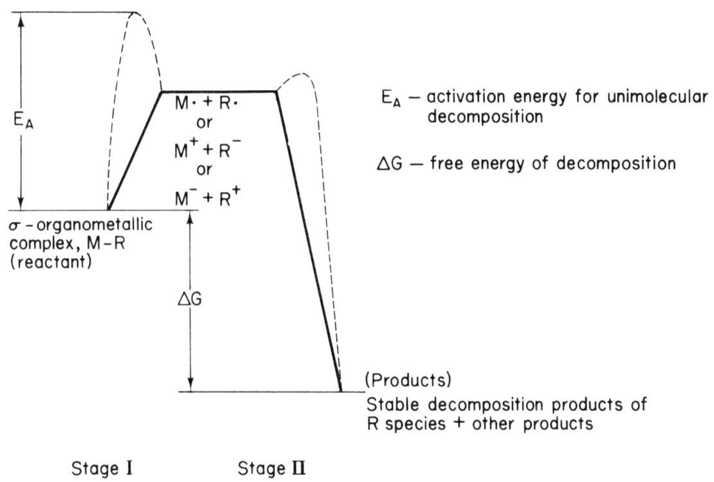

Figure 15.5. Schematic representation of the thermal decomposition of a σ-organo complex, MR.

As seen from Fig. 15.5, for decomposition to proceed, the activation energy must be sufficiently low for the rate-controlling step in the decomposition. It is likely that this rate-controlling step for many organometallics involves dissociation of the *M*—C bond to form radicals (*M*· and *R*·) or ions, such as M^+ and R^-. Thus, the stronger is the bond, the greater is the activation energy needed for dissociation, and the less readily it will dissociate. For example, the thermal decomposition of Group IV-A organometallics generally is favored kinetically as well as thermodynamically, as one goes from carbon to lead; the greater ease of decomposition is consistent with the greater relative strength of M—C bonds when M is carbon as compared with lead.

Stability to Oxidation

All organometallic compounds are thermodynamically unstable to oxidation, the driving force being provided by the large negative free energies of formation of the product metal oxide, carbon dioxide, and water. Many are also kinetically unstable to oxidation at room temperature.

Although Me_2Hg and Group IV-A derivatives are inert, most derivatives of Group A elements are rapidly oxidized. Kinetic instability to oxidation may be associated with the presence either of low-lying orbitals, e.g., $4p$ in Me_3Ga, or of a lone pair of electrons, e.g., Me_3Bi. In contrast, the Group IV alkyls possess neither of these features and behave as saturated compounds.

Stability to Hydrolysis

Whereas the organoderivatives of Groups IV-A and V-A are kinetically stable to attack by water, those of Groups I-A and II-A and of zinc, cadmium, aluminum, gallium, and indium are readily hydrolyzed. This is consistent with the concept that these hydrolysis reactions usually involve the nucleophilic attack by water and is thus facilitated by the presence of empty low-lying orbitals in the metal atom (e.g., Al), which attract the electron pair of the nucleophile. In the tetraalkyls of Group IV-A and trialkyls of Group V-A, the metals contain filled orbitals and nucleophilic attack is not favored. The rate of hydrolysis is dependent on the polarity of the metal-carbon bond; where this is high (e.g., Me_3Al), rapid hydrolysis occurs, whereas Me_3B is unaffected by water at room temperature, despite the empty $2p$ orbital on the boron atom. The majority of the neutral organoderivatives of the transition metals are inert to hydrolysis.

Causes of Kinetic Stability

One of the most striking differences between reaction rates of the elements and their compounds in the second row of the periodic table and the heavier elements is the much greater kinetic instability in the latter in many compounds. If one contrasts carbon with silicon, for example, the greater kinetic stability of carbon compounds is well documented and can be attributed largely to the full use of the four sp^3 valence orbitals, resulting in a maximum coordination number of 4. The high energy of the empty antibonding orbitals into which electrons could be donated to initiate thermal decomposition is largely responsible for this inability of carbon to expand its coordination number beyond 4.

In silicon, on the other hand, the empty $3d$ orbitals lie relatively close to the $3s$ and $3p$ valence electrons and may sometimes contribute significantly to bonding, e.g., $(SiH_3)_3N$ and SiF_6^{2-}, or, more often, be functional in forming a transition state in reactions of silicon compounds. However, expansion of the coordination number above 4 in silicon and heavier elements ordinarily occurs only when highly electronegative groups, such as halogen atoms, are attached to the metal atom (e.g., SiF_6^{2-}). The occurrence of coordination numbers above 4 and resulting kinetic instability to hydrolysis of silicon tetrachloride as compared with carbon tetrachloride, may be associated with the presence of these low-lying $3d$ orbitals on silicon. When such highly electronegative groups are absent (e.g., tetramethylsilane), there is generally no tendency of the metal atom to expand above 4, with the result that tetramethylsilane is *kinetically* inert to hydrolysis, oxidation, and thermal decomposition at room temperature. *Thus, kinetic stability of organometallic compounds may be associated with a lack of availability of empty and low-energy levels around the metal atom.*

In the case of the transition metals, however, empty *d valence* orbitals are often available, and the elements are inherently electron deficient producing much higher kinetic instability. The spontaneous decomposition of tetramethyltitanium at room temperature, in contrast to the stability of tetramethylsilane up to 500°C, is an example.

METHODS OF FORMATION

The many methods that have been used to synthesize organometallic compounds can be organized into three general reaction categories: substitution, addition, and miscellaneous. The systemization of a particular reaction into either of the first two categories is usually ascertained quite readily. The third category encompasses all other reactions. A study of preparative methods is important in revealing the relative reactive behavior of organometallic compounds as well as in providing other useful information of a synthetic nature. We shall consider representative reactions in each of these categories with a view toward providing insight into why certain preparative reactions are feasible and/or why they are more or less valuable than others.

Substitution Reactions

Substitution methods for preparing organometallic compounds may be further subdivided according to the following general types:

$$\begin{array}{llll} \text{(A)} & RX + M & \rightarrow & RM + X \\ \text{(B)} & RY + 2M & \rightarrow & RM + MY \\ \text{(C)} & RY + MZ & \rightarrow & RM + YZ \\ \text{(D)} & A + MY & \rightarrow & AM + Y \end{array}$$

Type A Reactions. They involve those compounds containing an acidic hydrogen or a less electropositive metal than the substituting metal. In acetylenes, the terminal hydrogen has been made acidic by virtue of the greatly increased electronegativity of the *sp* hybridized carbon compared with the more common sp^3 or sp^2 states, and reaction involves liberation of hydrogen:

$$R\text{C}{\equiv}\text{CH} + \text{Na} \rightarrow R\text{C}{\equiv}\text{C}^-\text{Na}^+ + \tfrac{1}{2}\text{H}_2 \qquad [15.3]$$

Resonance stabilization of the anion can also produce acidic character in a hydrocarbon. Cyclopentadiene, reacting with potassium, involves this type of stabilization:

$$\text{C}_5\text{H}_5 + \text{K} \rightarrow \text{C}_5\text{H}_5^-\text{K}^+ \qquad [15.4]$$

Also, triphenylmethane will react with alkali metals in liquid ammonia:

$$\text{Ph}_3\text{CH} + \text{Na} \xrightarrow{\text{NH}_3\text{(l)}} \text{Ph}_3\text{C}^-\text{Na}^+ + \tfrac{1}{2}\text{H}_2 \qquad [15.5]$$

As a general principle, reactions of type A, in which X is a metal less electropositive than M, proceed spontaneously toward the formation of RM. The preparative value of this reaction is very limited, but it is probably the best source of highly pure alkali and alkaline earth metal alkyls and commonly uses the easily purifiable and stable mercury derivatives as starting materials. Again, if we use the approximation that $\Delta H \approxeq \Delta G$, the

ΔH value for the reaction may be used as an indicator of reaction spontaneity. Thus, zinc will replace mercury completely according to

$$Zn + Me_2Hg \rightarrow Me_2Zn + Hg, \qquad \Delta H = -8 \text{ kcal/mole} \qquad [15.6]$$

However, in the reaction

$$Cd + Me_2Hg \rightarrow Me_2Cd + Hg, \qquad \Delta H = +2.7 \text{ kcal/mole} \qquad [15.7]$$

a mixture of Me_2Hg and Me_2Cd is obtained.

Type B Reactions. These are the most widely used and most important of all methods for the synthesis of organoderivatives of the most electropositive metals, they are seemingly the cornerstone of organometallic chemistry. Incuded here are the preparation of zinc alkyls discovered in 1849 by Frankland,

$$EtI + Zn \rightarrow EtZnI \qquad [15.8a]$$

$$2EtZnI \rightarrow Et_2Zn + ZnI_2 \qquad [15.8b]$$

the famous Grignard reaction,

$$2RBr + 2Mg \xrightarrow{\text{ether}} 2RMgBr \rightarrow MgBr_2 + R_2Mg \qquad [15.9]$$

and important methods utilizing lithium or sodium metal with an alkyl or aryl halide,

$$2Li + EtI \rightarrow EtLi + LiI \qquad [15.10]$$

$$2Na + C_6H_5Cl \xrightarrow{\text{pentane}} C_6H_5Na + NaCl \qquad [15.11]$$

For less electropositive metals, alloys with an alkali metal have been found to be extremely useful in producing products with positive ΔH values. In this regard, the highly exothermic nature of the portion of the reaction involving the alkali metal acts as a driving force for the reaction. Thus, the alkyls and aryls of Hg, Tl, Sn, Pb, and Bi can be made—e.g.,

$$2Hg + 2MeBr \rightarrow Me_2Hg + HgBr_2 \qquad \Delta H \doteqdot 0 \text{ kcal/mole} \qquad [15.12a]$$

$$HgBr_2 + 2Na \rightarrow Hg + 2NaBr \qquad \Delta H = -132 \text{ kcal/mole} \qquad [15.12b]$$

$$Hg + 2MeBr + 2Na \rightarrow Me_2Hg + 2NaBr \qquad \Delta H = -132 \text{ kcal/mole} \qquad [15.12c]$$

Although a compound such as benzyl lithium is not available by reaction of benzyl chloride, $C_6H_5CH_2Cl$, it is available readily from reaction of the corresponding alkoxy derivative:

$$C_6H_5CH_2OMe + 2Li \xrightarrow{THF} C_6H_5CH_2Li + LiOMe \qquad [15.13]$$

Also, cleavage of diaryl ethers, ordinarily quite difficult, proceeds smoothly in liquid ammonia:

$$C_6H_5OC_6H_5 + 2Na \xrightarrow{NH_3(l)} C_6H_5Na + C_6H_5ONa \qquad [15.14]$$

Type C Reactions. These reactions include a range of diverse processes that may be subdivided into at least six somewhat different groups. In each case, there is a double displacement of nonmetallic, inorganic groups bonded to an alkyl group and a metal. The driving force in the reaction may be the base strength of the group being exchanged, the electrophilic nature of M^+, stability of the products, or some combination of these factors.

(1) *Metalation* will be considered to be the reaction of an acid with a salt (usually weaker) of an acid. The term is usually restricted to metal-hydrogen exchange reactions. The acidic species is most commonly a hydrocarbon, in which negative charge is delocalized to produce an acidic site. Cyclopentadiene and its derivatives indene 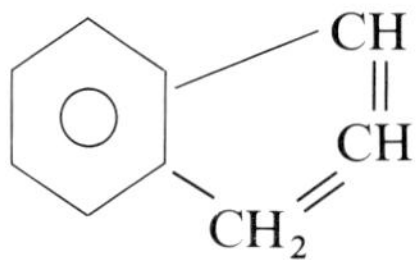 and fluorene 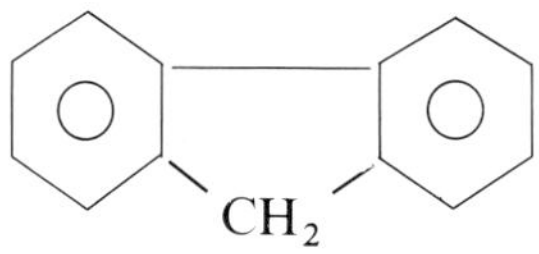 are examples, as are most aromatic systems and acetylenes. Such species react readily with bases such as alkoxides:

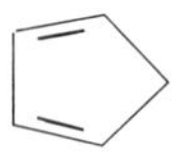 $+ RONa \rightarrow$  $Na^+ + ROH$ [15.15]

This reaction goes to completion with the more highly branched alkoxides, such as *t*-butyl.

The evolution of hydrogen when NaH is used ensures completion and product purity:

$$RC{\equiv}CH + NaH \rightarrow RC{\equiv}CNa + H_2 \qquad [15.16]$$

Purity can also be ensured in liquid ammonia by using $NaNH_2$:

$$RC{\equiv}CH + NaNH_2 \xrightarrow{NH_3(l)} RC{\equiv}CNa + NH_3 \qquad [15.17]$$

Although metalation of benzene is of no preparative value, that of the more acidic substituted benzenes frequently is, particularly when an orthodirecting group is present on the ring (CF_3, *R*O, etc):

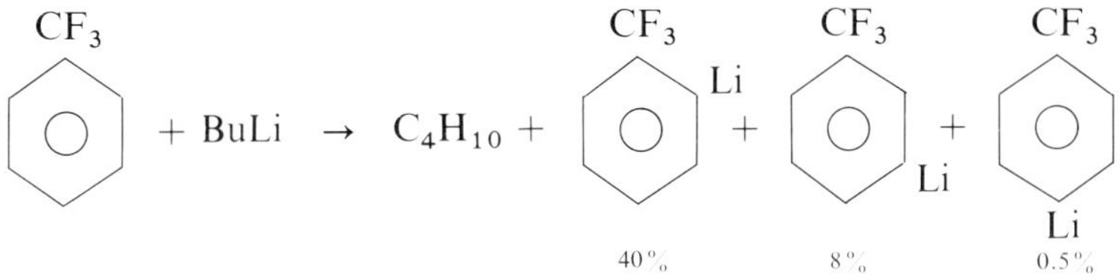

[15.18]

Organolithium compounds have been found to be most effective in the preceding reaction. In contrast, Grignard reagents are relatively inefficient.

Although in a formal sense the mercuration of aromatic compounds is a metalation reaction, in a mechanistic sense it is different, since it involves electrophilic attack by species such as $AcOHg^+$, whereas the previously described metalations are considered to depend upon proton abstraction by the base as the rate-controlling step. However, the same relative order of reactivity of the aromatic substrate is obtained in the two cases.

Benzene and toluene are mercurated smoothly by mercury(II) acetate in ethanol:

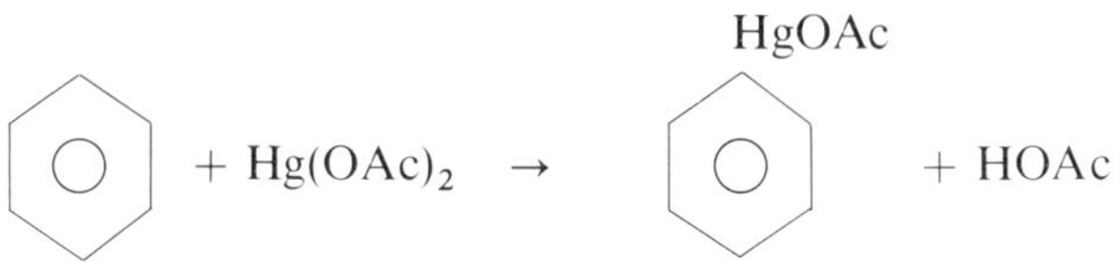

[15.19]

Aromatic compounds, such as thiophene and furan, which are more susceptible to electrophilic attack, are attacked more readily and yield the di- and tetraderivatives, respectively:

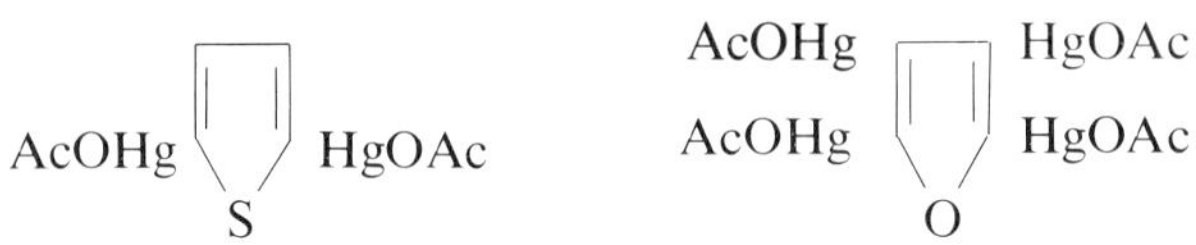

Kinetic studies on aromatic mercuration reactions reveal that the reaction is first order both in the aromatic compound and in mercury(II) acetate:

$$\text{Rate} = k[\text{ArH}][\text{Hg(OAc)}_2] \qquad [15.20]$$

The reaction is catalyzed by strong acids, such as $HClO_4$, up to a 1:1 ratio of $HClO_4$ to $Hg(OAc)_2$, suggesting the formation of a complex intermediate, $[HgOAc]^+ [ClO_4]^-$.

The relative ease of mercuration of nitrobenzene and other deactivated aromatic systems has been explained by the free radical mechanism:

$$Hg(OAc)_2 \rightleftarrows AcOHg\cdot + AcO\cdot \qquad [15.21a]$$

[15.21b]

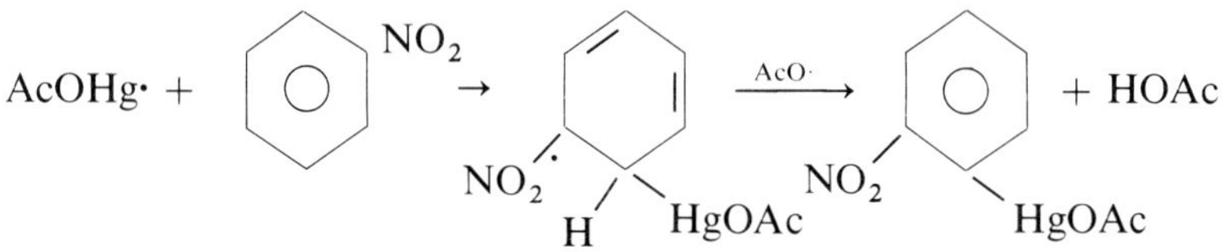

The catalytic action of strong acid favors the ionic mechanism with more reactive aromatic compounds. Thallium and gold(III) salts have been shown to metalate aromatic compounds similar to $Hg(OAc)_2$:

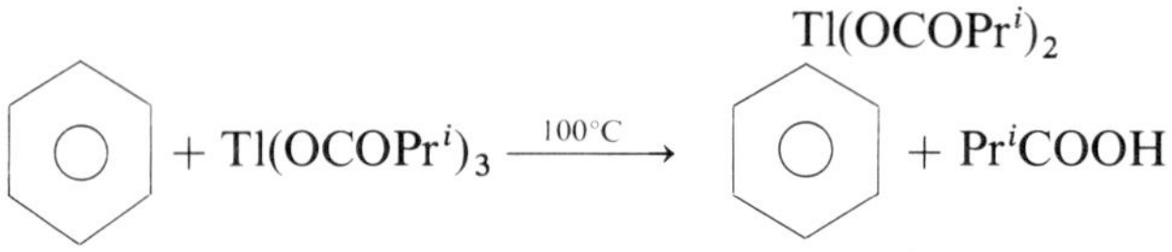

[15.22]

(2) *Metal-halogen exchange reactions* are best illustrated by reaction of a halogenated phenol with *n*-butyllithium:

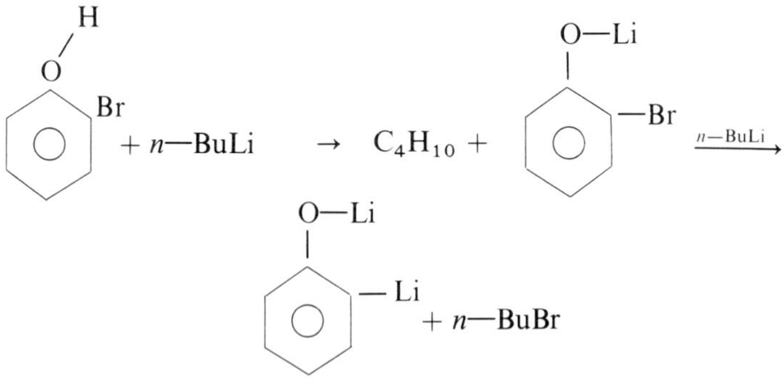

[15.23]

This method is favored over metalation for metal substitution in aromatic compounds because it ensures formation of the orthoderivative that is obtainable in better yields than other derivatives. If lithium metal is used, the reaction stops at the formation of the lithium salt; hence, the use of normal butyllithium is essential. The metal atom becomes preferentially attached to the more electronegative organic radical, or the radical which would form the more stable carbanion.

(3) *Reactions of organometallic compounds with metal salts* are of the type

$$RY + MZ \rightarrow RM + YZ \qquad [15.24]$$

where X is a metal or metal-containing group. They are generally the most useful preparative methods for all those organometallic compounds that cannot be obtained by direct reaction of the metal with hydro- or halocarbon. Such replacement will occur whenever Y is a more electropositive species than M. An important contributing factor to this restriction is the ΔH_f° of the YZ product, which is usually a halide. The effect is thus to combine the more electropositive element Y with the halogen and the organic radical with the more electronegative metal M.

Thus, although sodium will displace mercury from $R_2\text{Hg}$,

$$R_2\text{Hg} + 2\text{Na} \rightarrow 2R\text{Na} + \text{Hg} \qquad [15.25a]$$

and dialkylmercury is formed when alkylsodium reacts with HgCl_2,

$$2R\text{Na} + \text{HgCl}_2 \rightarrow R_2\text{Hg} + 2\text{NaCl} \qquad [15.25b]$$

the most convenient and easily available starting materials for such reactions are the Grignard reagents and organolithium compounds. Thus, the alkyls of cadmium, mercury, beryllium, and aluminum are best prepared in this way, according to

$$2R\text{MgCl} + \text{CdCl}_2 \rightarrow R_2\text{Cd} + 2\text{MgCl}_2 \qquad [15.26]$$

$$2R\text{MgCl} + \text{HgCl}_2 \rightarrow R_2\text{Hg} + \text{MgCl}_2 \qquad [15.27]$$

$$2R\text{MgCl} + \text{BeCl}_2 \rightarrow R_2\text{Be} + 2\text{MgCl}_2 \qquad [15.28]$$

$$3R\text{MgCl} + \text{AlCl}_3 \rightarrow R_3\text{Al} + 3\text{MgCl}_2 \qquad [15.29]$$

The Group IV-A metals tend to yield complex mixtures, such as RMX_3, RMX_2, R_3MX, and R_4M, and a large excess of Grignard reagent is needed to obtain the fully substituted derivative R_4M. Mixed alkyl-metal halides can sometimes be isolated, as from the reaction

$$2\text{PhLi} + \text{SnCl}_4 \rightarrow \text{Ph}_2\text{SnCl}_2 + 2\text{LiCl} \qquad [15.30]$$

and then used to permit further transformation, according to the reaction

$$Ph_2SnCl_2 + 2RLi \rightarrow Ph_2SnR_2 + 2LiCl \qquad [15.31]$$

The transition metals, including the lanthanide elements, either do not yield simple alkyls or form unstable products. The presence of other ligands can greatly stabilize the alkyls or aryls of transition metals. Thus, solvent-free triphenylchromium cannot be isolated but the (tris) tetrahydrofuranate, $Ph_3Cr(OC_4H_3)_3$, is relatively stable. Phosphines and similar π-bonding ligands can stabilize the aryls of Group VIII, yielding complexes such as $(Et_3P)_2NiPh_2$. Ligands that are more strongly π bonding than phosphines give particularly stable alkyls and aryls; e.g.,

$$XMn(CO)_5 + RMgX \rightarrow RMn(CO)_5 + MgX_2 \qquad [15.32]$$

Reaction of bimetallic compounds with halides. In the reaction type under consideration [equation (15.24)], Y may be a second metal or metal-containing group. Many derivatives of the Group IV-A metals and metal carbonyls are best prepared by this method. For lead and other Group IV-A metals, the method is usually used to introduce a fourth organic group into a trialkyl or triaryl metal derivative. For example, Ph_3PbLi, as prepared from

$$PbCl_2 + 3PhLi \rightarrow Ph_3PbLi + 2LiCl \qquad [15.33]$$

can then be treated with benzyl chloride to produce Ph_3PbCH_2Ph:

$$Ph_3PbLi + PhCH_2Cl \rightarrow Ph_3PbCH_2Ph \qquad [15.34]$$

In the metal carbonyl series, the reaction of the alkali metal derivative with an alkyl or aryl halide has been used:

$$MeI + NaMn(CO)_5 \rightarrow MeMn(CO)_5 + NaI \qquad [15.35]$$

Hydrocarbon-metal-carbonyl anions react similarly—e.g.,

$$2C_5H_5Fe(CO)_2Na + ClCO(CH_2)_3COCl \rightarrow C_5H_5(CO)_2FeOC(CH_2)_3COFe(CO)_2C_5H_5 + 2NaCl \qquad [15.36]$$

Type D Reactions. They are of the type where A and Y are ligands, and the reaction may be described as simple ligand exchange.

(1) *Exchange of one neutral ligand for another.* Neutral ligand exchange includes examples of the displacement of phosphorus and other heteroatoms by carbon, as well as many examples of the displacement of one carbon ligand by another. The extent of displacement varies with the

(a) *Metal.* Thus, isonitriles displace carbon monoxide to differing degrees depending on the metal:

$$Ni(CO)_4 + 4PhNC \rightarrow Ni(CNPh)_4 + 4CO \qquad [15.37]$$

$$Cr(CO)_6 + 3PhNC \rightarrow Cr(CO)_3(CNPh)_3 + 3CO \qquad [15.38]$$

(b) *Type of ligand.* In contrast with the reaction [15.37], ethyl isocyanide only displaces three CO groups from $Ni(CO)_4$:

$$Ni(CO)_4 + 3EtNC \rightarrow Ni(CO)(EtNC)_3 + 3CO \qquad [15.39]$$

(c) *Nature of solvent.* With boiling THF as the solvent in the reaction

$$BrMn(CO)_5 + 5PhNC \xrightarrow{\text{THF}} BrMn(CNPh)_5 + 5CO \qquad [15.40]$$

all five carbonyls are displaced; but in ethanol solution, reactions stop at the formation of $BrMn(CO)_3(CNPh)_2$.

The reverse of this reaction has been clearly demonstrated in several cases; e.g., a reverse reaction similar to

$$Cr(CO)_6 + C_6H_5CH_3 \xrightarrow[\text{excess toluene}]{\text{reflux in}} (CH_3C_6H_5)Cr(CO)_3 + 3CO \qquad [15.41]$$

has been developed into a practical synthesis of chromium hexacarbonyl:

$$(C_6H_6)_2Cr + 6CO \rightarrow Cr(CO)_6 + 2C_6H_6 \qquad [15.42]$$

As will be discussed later, both unimolecular dissociation, S_N1, and bimolecular displacement, S_N2, mechanisms may be involved in ligand exchange reactions. Many such reactions, which are catalyzed by photolysis, proceed by a dissociative S_N1 mechanism, whereas thermal reactions may proceed by either S_N1 or S_N2 mechanisms. Thus, photochemical techniques have been used to effect substitution when thermal substitution by an S_N2 is difficult. For example,

$$Fe(CO)_5 + \begin{matrix} CH\text{—}CO_2Et \\ \| \\ CHCO_2Et \end{matrix} \rightarrow \left[Fe(CO)_4 \leftarrow \begin{matrix} CHCO_2Et \\ \| \\ CHCO_2Et \end{matrix} \right] + CO \qquad [15.43]$$

in which the rate-controlling step is believed to be

$$Fe(CO)_5 \xrightarrow{h\nu} Fe(CO)_4 + CO \qquad [15.44]$$

(2) *Exchange of neutral ligands for negative ligands.* Many complexes containing neutral ligands such as carbon monoxide, alkylisonitrile, and olefins may be formed by displacing negative groups from a complex anion, as exemplified by

$$K_2[PtCl_4] + 2CO \rightarrow Pt(CO)_2Cl_2 + 2KCl \qquad [15.45]$$

When the starting material is a neutral halide, a Lewis acid, e.g., an aluminum halide, capable of complexing with the halide ion and thus assisting its removal, may be needed. This is the case in the reactions of halocarbonyls with carbon monoxide or olefins:

$$Mn(CO)_5Br + CH_2 = CH_2 + AlBr_3 \rightarrow [Mn(CO)_5C_2H_4]^+AlBr_4^- \qquad [15.46]$$

Alternatively, complex ions such as perchlorate or tetraphenylborate will promote such reactions; e.g.,

$$C_5H_5Fe(CO)_2Cl + CO + NaBPh_4 \rightarrow [C_5H_5Fe(CO)_3]BPh_4 + NaCl \qquad [15.47]$$

(3) *Exchange of negative ligands for neutral ligands.* Some examples of this type are

$$[Fe(CN)_5OH_2]^{3-} + CN^- \rightarrow [Fe(CN)_6]^{4-} + H_2O \qquad [15.48]$$

and

$$[Fe(o\text{-phen})_3]^{2+} + 2KCN \rightarrow Fe(o\text{-phen})_2(CN)_2 + 2K^+ + o\text{-phen} \qquad [15.49]$$

both of which can be rationalized by crystal field theory concepts.

(4) *Exchange of one anionic ligand for another.* The driving force for these reactions is the insolubility of one of the products—e.g., KSCN in liquid NH_3:

$$K_2[Pt(SCN)_4] + 4KC{\equiv}CR \xrightarrow{NH_3(l)} K_2\,Pt(C{\equiv}CR)_4 + 4KSCN \qquad [15.50]$$

(5) *Type D Variation.* A variation of the type D method involves a reactant that adds in two positions as it displaces the ligand *Y*. Only perfluorinated or allylic halides have been shown to react in this way, as, for example, with iron pentacarbonyl:

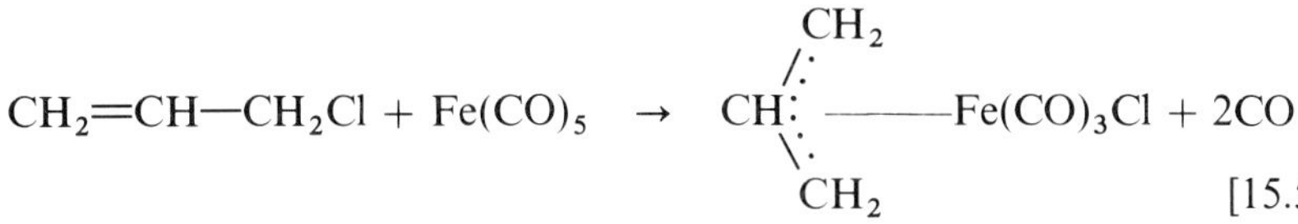

[15.51]

Addition Methods

Addition methods for the preparation of organometallic compounds are of two fundamental types: (1) the reaction of the elemental metal with various species, such as radicals and olefins; and (2) insertion reactions which may be the addition of metal hydrides and alkyls to unsaturated hydrocarbons. Each type is distinctive and has other useful aspects, sometimes more important than that of synthesis.

Reactions of Metals

(1) *Reaction with carbon monoxide.* A very limited method but one that is quite satisfactory for nickel and iron is the reaction of the free metal with carbon monoxide to give the equivalent complex:

$$Ni + 4CO \rightarrow Ni(CO)_4 \qquad [15.52]$$

This is the Mond process, used in the purification of nickel. Certain phosphorus and arsenic compounds are known to react directly with nickel in a similar way.

(2) *Reaction with unsaturated hydrocarbons.* Although some covalent π complexing reactions of metals are known, such as

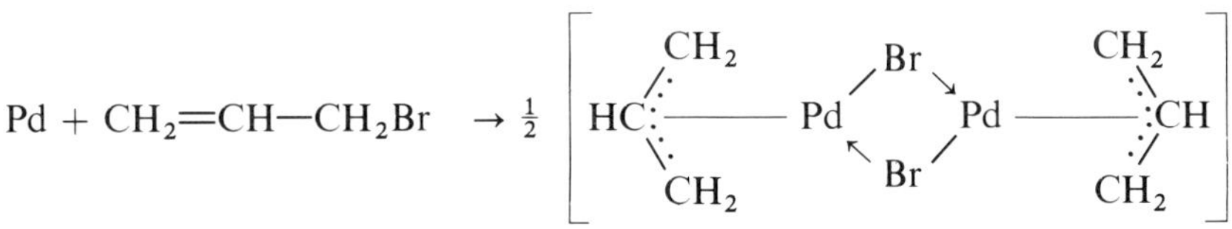

[15.53]

the formation of ionic complexes by the direct reaction of alkali metals is an important preparative method for certain aromatic organometallic compounds. The direct addition of alkali metals to unsaturated hydrocarbons appears to be limited to highly substituted aromatic olefins and certain aromatic hydrocarbons containing two or more aromatic rings. Addition products are of two types—those that can be regarded as hydrocarbon salts and those that cannot. The addition of sodium to 1,1-diphenylethylene by a one-electron transfer followed by dimerization

$$Ph_2C{=}CH_2 + Na \rightarrow (Ph_2C{-}\dot{C}H_2)Na^+ \rightarrow (Ph_2C{-}CH_2{-}CH_2{-}\bar{C}Ph_2)Na_2^{2+} \qquad [15.54]$$

is commonly known as the *dimerizing addition* of sodium to an olefin. The second type of addition product involves the addition of one or more electrons to the lowest vacant molecular orbital of an aromatic hydrocarbon. Such compounds as biphenyl, naphthalene, or 1,4-diphenylbutadiene react with alkali metals without loss of hydrogen. An example is anthracene:

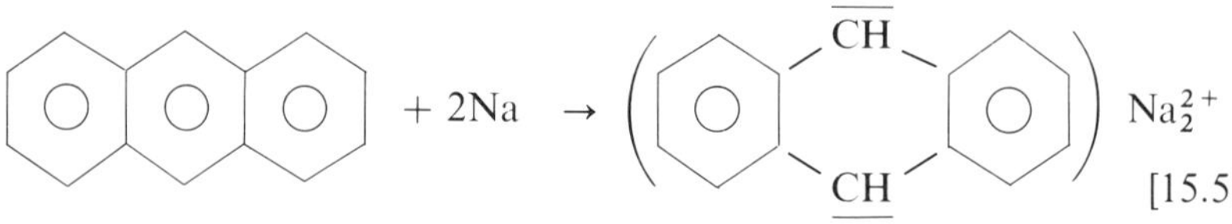

[15.55]

Such addition compounds are all highly colored, and their formation is greatly facilitated in basic solvents, such as tetrahydrofuran or 1,2-dimethoxyethane.

(3) *Reaction with radicals.* The general reaction

$$R\cdot + M \rightarrow R{-}M \qquad [15.56]$$

is more practically useful as a detection method for free radicals than as a preparative method, although it could conceivably find application in the preparation of otherwise inaccessible miscellaneous compounds. The same reaction occurs with stable free radicals if the metal is highly electropositive:

$$Ph_3C\cdot + Na \rightarrow Ph_3C^-Na^+ \qquad [15.57]$$

Insertion reactions. Insertion reactions are of two basic types. The first involves the insertion of olefins and acetylenes into metal-hydrogen bonds. The reaction may be represented in general by

$$M\text{—}\mathrm{H} + \mathrm{\gt C{=}C\lt} \;\rightarrow\; M\text{—}\overset{|}{\underset{|}{\mathrm{C}}}\text{—}\overset{|}{\underset{|}{\mathrm{C}}}\text{—}\mathrm{H} \qquad [15.58]$$

The reaction is particularly important for aluminum hydride, as it can be made readily to prepare important compounds such as triethylaluminum from aluminum metal, hydrogen, and ethylene:

$$3\mathrm{CH_2 = CH_2 + Al + \tfrac{3}{2}H_2} \;\rightarrow\; 3\mathrm{Et_2AlH} \qquad [15.59]$$

which may be followed by

$$\mathrm{Et_2AlH + CH_2{=}CH_2} \;\rightarrow\; \mathrm{Et_3Al} \qquad [15.60]$$

Similarly to boron hydrides, the corresponding addition of GaH bonds to olefins is known:

$$\mathrm{(HGaCl_2)_2} + R\mathrm{CH{=}CH_2} \xrightarrow{\text{ether}} (R\mathrm{CH_2CH_2GaCl_2)_2} \qquad [15.61]$$

Also, germanes, stannanes, and plumbanes react readily with acetylene without the need of a catalyst, *tris*(*n*-butyl) lead hydride adding rapidly even at 0°C:

$$\mathrm{Ph_3SnH + HC{\equiv}CPh} \xrightarrow{90^\circ\mathrm{C}} \mathrm{Ph_3SnCH{=}CHPh} \qquad [15.62]$$

$$\mathrm{Bu_3PbH + PhC{\equiv}CH} \xrightarrow{0^\circ\mathrm{C}} \textit{trans}\text{-}\mathrm{PhCH{=}CHPbBu_3} \qquad [15.63]$$

The increased ease of addition from Ge to Pb correlates well with the decrease in strength of the M—H bonds. Metal carbonyl hydrides, especially $\mathrm{HCo(CO)_4}$, add to olefins in the hydroformylation reaction. The addition to ethylene is believed to proceed by a dissociative mechanism:

$$\begin{array}{ccccc} & & \mathrm{HCo(CO)_4} & \leftrightharpoons & \mathrm{CO + HCo(CO)_3} \\ \mathrm{HCo(CO)_3 + CH_2{=}CH_2} & \leftrightharpoons & \mathrm{H{-}Co(CO)_3} & \leftrightharpoons & \mathrm{CH_3CH_2Co(CO)_3} \\ & & \uparrow & & \mathrm{CO}\downarrow\uparrow\mathrm{-CO} \\ & & \mathrm{CH_2{=}CH_2} & & \mathrm{CH_3CH_2Co(CO)_4} \end{array} \qquad [15.64]$$

Other insertion reactions can be represented in general by

$$A\text{—}B + X{=}Y \;\rightarrow\; A\text{—}X\text{—}Y\text{—}B$$

where A is a metal, B is an element other than hydrogen, carbon, nitrogen, phosphorus, oxygen, or halogen, and XY is an olefin or acetylene. This method is the basis of the Ziegler polyethylene process, in which triethylaluminum adds to ethylene in an indefinite number of steps similar to

$$Et_2Al + CH_2{=}CH_2 \rightarrow Et_2AlCH_2CH_2Et \qquad [15.65]$$

The ease of addition of ethylene increases in the order $1° < 2° < 3°$ alkyls, and only the latter two add ethylene at ordinary pressure:

$$Pr^iLi + CH_2{=}CH_2 \xrightarrow{0°C} Pr^iCH_2CH_2Li \qquad [15.66]$$

The reaction stops here, since the product is a primary alkyllithium, which will only add further (and polymerize) at high pressure. The reaction occurs most readily when the polar $A{-}B$ bond can add to unsaturated species bearing electron-withdrawing substituents. These substituents either increase the polarity of the C=C or C≡C bond or stabilize a negative charge in the transition state.

The reaction of n-butyllithium with butadiene does not stop with the first step:

$$n\text{-BuLi} + CH_2{=}CH{-}CH{=}CH_2 \rightarrow n\text{-BuCH}{-}CH{=}CH{-}CH_2Li + \underset{\displaystyle Li}{BuCH_2{-}\overset{}{CH}{-}CH{=}CH_2} \qquad [15.67]$$

Further addition leads to polymerization. Polyaryl olefins and acetylenes react by simple addition:

$$Ph{-}C{\equiv}CPh + BuLi \rightarrow cis\text{-}\underset{\displaystyle Bu}{PhC}{=}\underset{\displaystyle Li}{CPh} \qquad [15.68]$$

Oxidative Addition. A unique and interesting type of addition reaction is one in which simultaneous oxidation of the metal accompanies the increase in coordination number. Such oxidative-addition reactions are generally found among low-spin transition metal complexes, and, more particularly, among those metals with nearly filled d orbitals in which the preferred coordination number is lower than 6. Thus, as electrons are lost in oxidation, additional ligands may be incorporated into the primary coordination sphere of the metal (see subsequent discussion of the 18-electron rule).

Some examples of these reactions involve five-coordinate (d^7) complexes,

$$2[Co^{II}(CN)_5]^{3-} + X\text{—}Y \rightarrow [Co^{III}(CN)_5X]^{3-} + [Co^{III}(CN)_5Y]^{3-} \quad [15.69]$$

where $X\text{—}Y = H_2$, Br_2, HOH, CH_3I, ICN, etc.; four-coordinate (d^8) complexes,

$$[Ir^{I}(CO)(PPh_3)_2Cl] + X\text{—}Y \rightarrow [Ir^{III}(X)(Y)(CO)(PPh_3)_2Cl] \quad [15.70]$$

where $X\text{—}Y = H_2$, Cl_2, HCl, CH_3I, RSO_2Cl, RHgCl, R_2SiH, etc.; and two-coordinate (d^{10}) complexes,

$$Pt^0(PPh_3)_2 \xrightarrow{X\text{—}Y} Pt^{II}(PPh_3)_2(X)(Y) \xrightarrow{X\text{—}Y} Pt^{IV}(PPh_3)_2X_2Y_2 \quad [15.71]$$

where $X\text{—}Y = CH_3I$, CH_3Br, Ph_3SnCl, etc.

Three Miscellaneous Methods

Of the various miscellaneous methods for synthesizing organometallic compounds, the more important involve reduction, either chemically or electrolytically, and merit mention here.

Chemical Reduction. The majority of methods for forming metal carbonyls depends on a selective method of reducing the metal salt to the zero or low oxidation state in the presence of carbon monoxide. In many cases, carbon monoxide itself is a sufficiently good reducing agent, reducing nickel(II) sulfide and cobalt(II) sulfide in alkaline solution:

$$NiS + 5CO + 4OH^- \rightarrow Ni(CO)_4 + CO_3^{2-} + S^{2-} + 2H_2O \quad [15.72]$$

Evidence of the disproportionation of Ni(II) to Ni(0) and Ni(IV) followed by the reaction of the Ni(IV) with carbon monoxide to give Ni(II) and carbon dioxide has been obtained. When sulfides are unavailable, carbonyls, oxides, and halides can be employed under somewhat more drastic conditions. Thus the oxides and/or halides of cobalt, technicium, rhenium, ruthenium, osmium, iridium, platinum, and gold have been used conveniently, e.g.,

$$Tc_2O_7 + CO \xrightarrow[400 \text{ atm}]{275^\circ} Tc_2(CO)_{10} \quad [15.73]$$

$$2RuCl_3 + 5CO \rightarrow 2Ru(CO)_2Cl_2 + COCl_2 \quad [15.74]$$

For the Group VI-B metals, stronger reducing agents are frequently necessary:

$$WCl_6 + 6CO + 2Al \rightarrow W(CO)_6 + Al_2Cl_6 \quad [15.75]$$

Probably the most efficient procedure for Group V-B metals employs sodium metal in diglyme (diethylene glycol dimethyl ether), e.g.,

$$VCl_3 + 4Na + 6CO + \xrightarrow[80\%]{100^\circ C} [Na^+][V(CO)_6^-] + 3NaCl \quad [15.76]$$

Isonitriles are prepared by similar reduction techniques, e.g.,

$$ReI_3 + 6RNC \rightarrow [Re(CNR)_6]^+I_3^- \quad [15.77]$$

which occurs with disproportionation of the iodide ion. The preparation of di-arene chromium cations utilizing Al and $AlCl_3$ belongs in this category:

$$6C_6H_6 + 3CrCl_3 + 2Al + AlCl_3 \rightarrow 3(C_6H_6)_2Cr^+AlCl_4^- \quad [15.78]$$

Many related preparations have been carried out by the same method.

Electrolytic Reduction. A method that merits much fuller exploitation is electrolytic reduction using metal cathodes. Examples are

$$CH_2CHCN + NaOH \xrightarrow[\text{Sn cathode}]{\text{electrolysis}} Sn(CH_2CH_2CN)_4 \quad [15.79]$$

$$(CH_3)_2C{=}O \xrightarrow[\text{Pb or Hg cathode}]{\text{electrolysis}} (CH_3)_2CH_2Hg \text{ or } (Pb) \quad [15.80]$$

$$(Et_3Al)_2 \cdot NaF \xrightarrow[\text{Pb cathode}]{\text{electrolysis}} PbEt_4 + Al + NaF \quad [15.81]$$

Reaction (15.81) produces very pure aluminum.

PROPERTIES AND REACTIONS OF ORGANOMETALLIC COMPOUNDS

The reactions of organometallic compounds can best be described in terms of position of the metal in the periodic table. In terms of various properties, including polarity and type of bonding, reactivity toward various reagents including water, and thermal stability, such compounds may be grouped into three categories: (1) the highly polar and reactive organic derivatives of Groups I, II, and III; (2) the simple organic derivatives of Groups IV and V; and (3) the organic derivatives of the transition metals. In each of these categories, various types of reactions will be noted and discussed.

Organometallic Compounds of Groups I, II, and III

The properties and reactivity of the organometallic compounds of Groups I-A, II-A, and III-A can in general be correlated with the degree of ionic character in the metal-carbon bond—i.e., reactivity parallels the electropositive nature of the metal. The alkyls of the highly electropositive alkali metals are thus more reactive than those of the elements in Groups II-A or III-A. Also, in general, reactivity increases with increasing atomic weight in a given group of the periodic table—e.g., lithium to cesium. Across a period from left to right there is a corresponding decrease in reactivity:

$$\mathrm{Li} > \mathrm{Be} > \mathrm{B}$$
$$\mathrm{Na} > \mathrm{Mg} > \mathrm{Al}$$

In the alkali metal family, a dissimilarity exists between the properties and reactivity of organolithium compounds and the reactivity of the remaining alkali metal compounds. This dissimilarity can be attributed to the much smaller size and higher polarizing ability of lithium. Lithium compounds are thus appreciably covalent, whereas the other alkali metals form highly ionic derivatives containing a carbanion. The most useful aspect of this unique nature of the lithium derivatives is their solubility in inert solvents, such as the saturated hydrocarbons. This allows not only for their relatively easy preparation but provides a convenient medium in which a variety of reactions can be carried out. The reactivity of the lithium compounds is, at the same time, sufficiently high so that they are of ever-increasing value in chemical synthesis. Although the other alkali metal compounds are more reactive, their solubility is severely limited by a combination of the same high reactivity and highly ionic character. Thus, they are often insoluble in or reactive with the solvent. As with Grignard reagents, organolithium compounds are not isolated but used immediately (in situ) for synthetic purposes and are prepared by a similar method:

$$\text{Li} + RX \xrightarrow[\text{dry } N_2]{\text{ether}} R\text{Li} + \text{Li}X \qquad [15.82]$$

The most widely used lithium reagents for synthetic means are the *n*-butyl and phenyl derivatives, *n*-BuLi and PhLi. The use of these reagents may be illustrated by the three main types of exchange reactions involving organolithium compounds:

(1) Metal-metal exchange:

$$\text{PhLi} + \text{Ph}_3\text{SnCH}_2\text{CH}{=}\text{CH}_2 \rightarrow \text{CH}_2{=}\text{CHCH}_2\text{Li} + \text{Ph}_4\text{Sn} \qquad [15.83]$$

(2) Metal-halogen exchange:

$$n\text{-BuLi} + p\text{-BrC}_6\text{H}_4\text{NMe}_2 \rightarrow n\text{-BuBr} + p\text{-Li-C}_6\text{H}_4\text{NMe}_2 \qquad [15.84]$$

(3) Metal-hydrogen exchange:

$$\text{PhLi} + \text{PhC}{\equiv}\text{CH} \rightarrow \text{C}_6\text{H}_6 + \text{PhC}{\equiv}\text{CLi} \qquad [15.85]$$

Reactions of the more ionic alkali metal derivatives can be divided into two types:

(1) Those in which the negative charge on the organic radical cannot be dispersed over several carbon atoms, and
(2) Those in which conjugation results in dispersal of the negative charge on the organic radical.

The former compounds are characterized by intense reactivity due to the localization of negative charge on one carbon atom. As an example, dibenzfuran is rapidly demetallated by ethylsodium, and reaction with butyllithium gives the monolithium derivative:

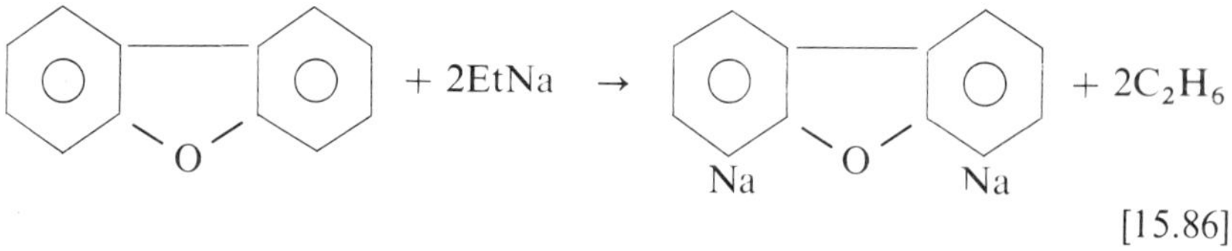

[15.86]

The alkyl sodium derivatives are generally unstable at room temperature, decomposing according to

$$C_2H_5Na \rightarrow C_2H_4 + NaH \qquad [15.87]$$

Organosodium compounds in which the negative charge is delocalized are all brightly colored and are less reactive. They are soluble in ether, without decomposition. Triphenylmethylsodium, one of the most extensively studied members of this class of compound, is generally used as a deep-red ether solution, which is hydrolyzed by water

$$Ph_3C^-Na^+ + H_2O \rightarrow Ph_3CH + NaOH \qquad [15.88]$$

and is oxidized by O_2, giving various peroxy compounds.

The organometallic chemistry of the Group II-A metals is largely the chemistry of Grignard reagents, $RMgX$. Since the use of such reagents is commonly discussed in some detail in a basic organic chemistry text, no further mention of them will be made here, except to note the constitution of these reagents, which have been in question for many years. Reliable molecular weight determinations show a tendency toward a monomeric state with increasing dilution in diethyl ether. As the concentration increases to the 1-to-2 molar range, the degree of association increases and commonly is two and sometimes more. The following structure has been indicated for *t*-butylmagnesium chloride over a wide concentration range:

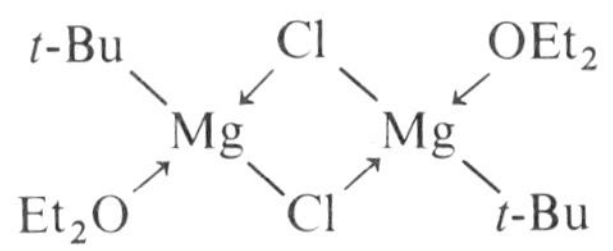

Very little is known about the organic compounds of calcium, strontium, and barium. The best-known and characterized derivatives are the ionic cyclopentadienyls, $(C_5H_5)_2M$.

Alkylberyllium hydrides, $RBeH$, are interesting because their ether or amine complexes are dimeric in benzene and show an electron-deficient BeH_2Be bridge structure.

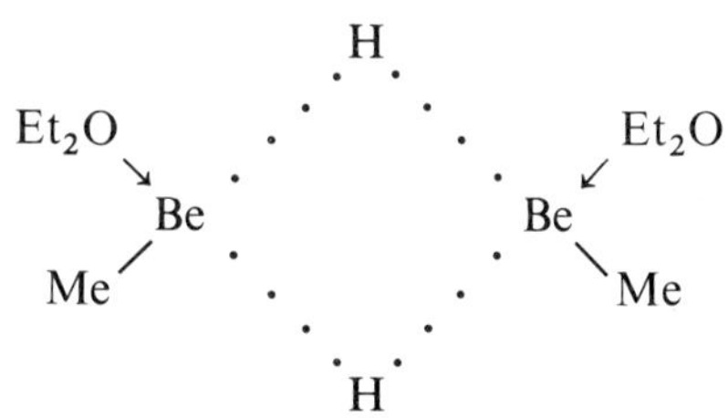

This structure differs from the similar boron compound in that it exhibits cis-trans isomerism, and the bridge bonds are not split in reaction with bases. Such alkylberyllium hydrides do react very rapidly with carbonyl bonds to yield alkoxy-beryllium alkyls:

$$4MeBeH + 4PhCHO \rightarrow (MeBeOCH_2Ph)_4 \qquad [15.89]$$

Group II-B Metals. Although ethylzinc iodide has been known since 1849, its structure was determined only in 1966. It is a polymer in which each zinc is tetracoordinated (bound to one carbon and three iodines) and each iodine is bound to three zinc atoms. The use of alkyl zinc halides as synthetic reagents has been almost entirely superseded by Grignard reagents. Many dialkyl zinc derivatives that are volatile and air sensitive are commercially available and have been found useful in many syntheses of zinc complexes involving acidic (15.90) and neutral (15.91) ligands and formation of anionic complexes (15.92):

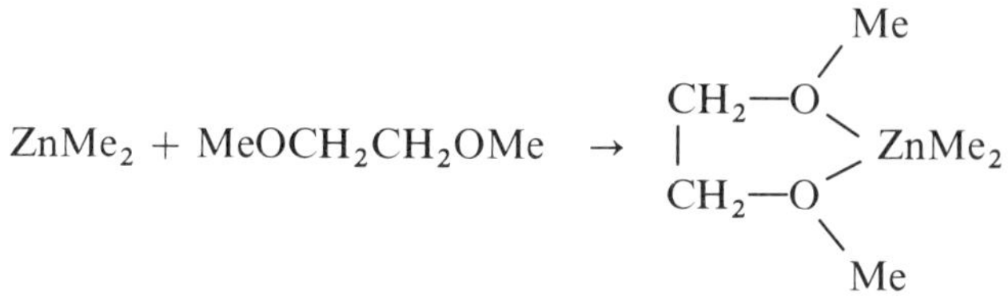
[15.90]

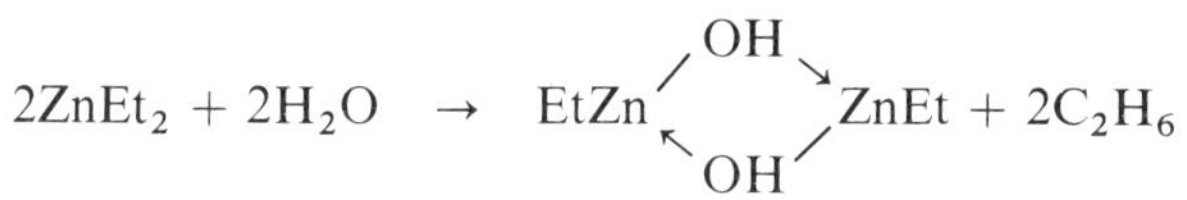
[15.91]

$$3Et_2Zn + 2Na \rightarrow 2NaEt_3Zn + Zn \qquad [15.92]$$

Little is known about alkyl cadmium halides, but dialkyl cadmium compounds have been found useful for ketone synthesis:

$$R_2Cd + R'COCl \rightarrow R'COR + CdCl_2 \qquad [15.93]$$

Alkyl and aryl mercury halides have long been known and are monomeric linear molecules. Conversion into hydroxides, cyanides, and other derivatives is easily accomplished by ion exchange, or by the use of silver salts:

$$MeHgI + AgCN \rightarrow MeHgCN + AgI \qquad [15.94]$$

The properties of $RHgX$ are largely dependent on the polar character of the HgX bond, the compounds ranging from saltlike (RHg^+X^-), when X is highly electronegative (F^-, NO_3^-,SO_4^{2-}), to nonpolar compounds, soluble in organic solvents when X is less electronegative (I^-). With the exception of Me_2Hg, dialkyl mercury compounds are thermodynamically unstable with respect to decomposition to mercury and hydrocarbons. The diaryl derivatives are usually stable indefinitely at room temperature. Both dialkyl and diaryl derivatives are less reactive than those of zinc and cadmium. They are inert to water and oxygen and do not generally form coordination complexes. Considerable study of the mechanism and kinetics of exchange reactions of the type

$$HgX_2 + HgR_2 \leftrightarrows 2HgXR \qquad [15.95]$$

has been undertaken and discussed elsewhere (Chapter 10).

A number of complexes of these alkyls have been studied. The reaction between dimethylzinc and diphenylamine yields a product, $(MeZnNPh_2)_2$, in which the unusual presence of three-coordinate zinc has been shown by X-ray diffraction. The highly symmetrical structure of the methylzinc methoxide tetramer, also determined by X-ray diffraction, can be regarded as simple cubic with Zn and O atoms alternating at the corners of the cube.

Group III Metals. A very important and significant aspect of organo-aluminum chemistry is the addition of aluminum-hydrogen bonds to olefins, carbonyls, and similar groups in the absence of ether and other bases. The reaction may be illustrated by

$$R_2Al{-}H + CH_2{=}CHMe \rightarrow R_2Al{-}CH_2{-}CH_2Me \qquad [15.96]$$

A wide range of aluminum alkyls can be obtained simply by heating i-Bu_3Al with $RCH{=}CH_2$. The reactions are believed to involve the fission of an aluminum-carbon bond giving an aluminum-hydrogen bond, followed by addition of the latter to the olefin. At high pressures, about 100 atm, and elevated temperature, greater than 100°C, a competing reaction has been observed—the insertion of olefin molecules into the aluminum-carbon bonds in the trialkyl:

$$Al\begin{cases} C_2H_5 \\ C_2H_5 \\ C_2H_5 \end{cases} \xrightarrow{C_2H_4} Al\begin{cases} (C_2H_4)_m \cdot C_2H_5 \\ (C_2H_4)_n \cdot C_2H_5 \\ (C_2H_4)_o \cdot C_2H_5 \end{cases} \qquad [15.97]$$

Such electron-deficient bonding is also presumed to be present in hydrocarbon solutions of $LiAlR_4$ and $NaAlR_4$ and is probably due to polarization by the alkali metal.

Organometallic Compounds of Groups IV-A and V-A

The elements of Groups IV-A and V-A differ from those in Groups I-A through III-A in some important respects; their bonds to carbon are appreciably less polar, and their simple organic derivatives do not normally function as acceptor molecules and are not bonded through electron-deficient linkages. They are less reactive to nucleophiles due to the absence of vacant low-energy orbitals, and most are stable in air and do not react with water. Reactivity and thermal instability both increase down the groups—however, with decreasing strength of the M—C bond.

Group IV-A Elements. The organic derivatives of the Group IV-A elements other than carbon are strikingly similar in a number of respects to the analogous carbon compounds; however, there are some important differences. The elements are almost exclusively tetravalent and form a large number of organic derivatives, particularly of the type R_nMX_{4-n}, where R = alkyl or aryl; M = Si, Ge, Sn, or Pb; X = a monovalent atom or group such as H, OR, halogen, or alkali metal; and $n = 1, 2, 3$, or 4. The fully substituted tetraalkyls are saturated compounds and are thus generally unreactive chemically, in contrast to the R_3M derivatives of Groups III-A and V-A, which have good acceptor and donor properties, respectively. A large number of catenated derivatives exist containing M—M bonds. The stability of such bonds decreases down the group with the result that only a few derivatives containing Pb—Pb bonds are known, whereas Si—Si bonds are quite common.

The effect of substituents on the thermal stability of catenated derivatives is worth noting; such stability increases in the order halide < hydride < organo. Significantly, the substituents that stabilize catenated compounds are not those which form the strongest bonds with these elements, but those which release electrons most readily. If fission of the M—M bond is critical to such stability as is likely, it is reasonable that the lower is the partial positive charge on M, the stronger is the bond and the more stable is the compound.

Although tetravalency is almost universally a feature of silicon and germanium, coordination numbers greater than 4 are common in organotin and organolead chemistry by association through groups containing fluorine and oxygen. In such cases, vacant *d* orbitals of the metal are utilized, resulting in trigonal bipyramidal or octahedral geometry. Some compounds of this type, such as the pentacoordinate trimethyltin fluoride, are polymeric with the fluorides bridging adjacent Me_3Sn groups, whereas others, such as trimethyltin monochloride, are monomeric but form five-coordinated adducts with Lewis bases such as pyridine. Lead and tin also form octahedral-type polymeric structures of the type Me_2SnF_2 and Ph_2PbCl_2, in which the halogens act as bridging groups. The presence of $(p \rightarrow d)$ π bonding (Chapter 12) has been implied in many derivatives of these elements, particularly Si and Ge, involving electronegative atoms.

The organic derivatives of the Group IV-A elements are generally unreactive with respect to cleavage of the metal-carbon bond. This is attributed to the low polarity of the metal-carbon bond and is particularly true for silicon and germanium, for which few reactions are known involving transfer of organic groups. The relative weakness of the tin-carbon bond offsets its low polarity, with the result that organotin compounds find more uses for such transfer reactions. Cleavage also occurs more readily if the organic group has electronegative substituents, since these facilitate separation of the group as R^- during nucleophilic attack on the metal. Thus, chlorinated methyl groups are quite readily removed from silicon by bases such as OH^-, with reactivity increasing in the order CH_2Cl < $CHCl_2$ < CCl_3.

The low reactivity of the metal-carbon bond influences the reactivity of compounds containing other bonds, such as metal-hydrogen, metal-oxygen and metal-nitrogen, which may be present in the same molecule. The presence of unreactive organic groups serving as blocking groups simplify the chemistry of hydride, alkoxy, or amino derivatives. Thus, trialkyl or triaryl derivatives R_3MH, R_3MOR, or R_3MNR_2 are more convenient materials to use for reasons of thermal stability, solubility, and reactivity than MH_4, $M(OR)_4$, or $M(NR)_4$. The stabilities of the hydrides of a particular element, for example, increase in the order $MH_4 < RMH_3 < R_2MH_2 < R_3MH$.

The hydrides of silicon, germanium, tin, and lead all have slightly polar metal-hydrogen bonds and can add to polar unsaturated functional groups. Extensive studies on silicon and tin have shown such addition reactions to operate by a variety of mechanisms, ranging from heterolytic fission of $M^{\delta+}H^{\delta-}$ to homolytic fission into radicals, to heterolytic fission, if polarized, of $M^{\delta-}$ $H^{\delta+}$.

The most important organosilicon compounds are the silicone polymers, a typical example of which is *methylsilicone oil*: $Me_3Si(OSiMe_2)_nOSiMe_3$;

n ranges from 20 to 500. Nearly all silicone polymers are made from products obtained by hydrolizing organochlorosilanes, R_nSiCl_{4-n}, in which R may be methyl (usually), phenyl, or hydrogen. Such oils, resins, and rubbers have been found extremely useful for many purposes requiring stability and electrical insulation at high temperatures that are not available from hydrocarbon products.

Group V-A Elements. The organic derivatives of arsenic, antimony, and bismuth are of three types, the most common being R_3M (R = alkyl, aryl, or H). These are pyramidal structures and are generally stable except for the hydrides. Saltlike *-onium* compounds ($R_4M^+X^-$), in which the coordination number is four, and compounds of the type R_5M are also known and well characterized.

In the trialkyls, Lewis base character is quite important. Arsenic is particularly capable of forming ($d \rightarrow d$) π bonds with transition elements such as platinum. Antimony and particularly bismuth show the same complexing tendency to a much weaker degree. Thermal stability in general decreases from arsenic to bismuth. Catenation is also more of a feature with arsenic than with antimony or bismuth, and many compounds containing arsenic-arsenic bonds, such as $(MeAs)_5$, are known. The *-onium* salts are known to be tetrahedral, analogous to NH_4Cl, whereas the pentacoordinated derivatives exhibit a trigonal bipyramidal structure, with the exception of Ph_5Sb, which has been shown by X-ray analysis to be square pyramidal.

Organometallic Compounds of the Transition Elements

As described previously, the transition metal organic derivatives may be classified on the basis of the multiple π or δ bonds frequently exhibited, or in terms of the type of ligand that donates electrons to the transition metal. Whatever the classification method used, however, the adherence of a great many such organic complexes of the transition metals to the 18-electron rule is quite significant. The rule *that stable compounds are formed when the valence energy level of the metal contains* 18 *electrons* applies most generally for elements in the middle of the transition series, such as vanadium, chromium, manganese, iron, and cobalt. To illustrate the rule, consider the complexes in Table 15.3.

The basis for the 18-electron rule lies in the kinetic stability that may be associated with filled sets of orbitals. From Fig. 9.13, we see that 18 electrons in all are required to fill all the bonding and nonbonding molecular orbitals in an octahedral complex, the electrons originating from both

TABLE 15.3 ***Application of the 18-Electron Rule***

Compound	Number of Electrons Contributed by Metal	Ligands	Total
$Fe(CO)_5$	8	$2 \times 5 = 10$	18
$(\pi—C_5H_5)_2Fe$	8	$2 \times 5 = 10$	18
$Co(Ph_3P)_4^-$	$9 + 1 = 10$	$2 \times 4 = 8$	18
$\pi—C_5H_5Mo(CO)_2\pi—C_3H_5$	6	$\pi—C_5H_5 = 5$, $C_3H_5 = 3$ $(CO)_2 = 2 \times 2 = 4$	18
$\pi—C_5H_5Re(CO)_2C_5H_6$	7	$\pi—C_5H_5 = 5$, $C_5H_6 = 2$* $(CO)_2 = 2 \times 2 = 4$	18

* C_5H_6 is known to attach to Re by only one of its double bonds.

the metal and the ligands. Kinetic stability will be achieved if no low-lying orbitals are available into which electrons may readily be promoted to initiate thermal decomposition, or be donated, as in nucleophilic attack. Conversely, kinetic stability is unlikely to be attained if low-lying bonding or nonbonding molecular orbitals are empty, i.e., if less than 18 electrons are present. Thus, if the energy difference, 10 Dq, is large, then the energy difference between the highest filled orbital and the lowest vacant orbital is also large. This energy difference may be increased when π back donation from metal to ligands occurs (Figs. 9.14 and 9.15). This is at least partly why the majority of complexes that obey the 18-electron rule contain ligands such as carbon monoxide, trialkyphosphorus compounds, and olefins, which are good π bond ligands, accepting electrons from the metal by back donation. Such π *acids* also assist in stability by removing charge from the metal atom, which may seldom be larger than ± 1.

Reactions and Structure. Using the classification scheme described previously based on type of ligand, we may conveniently discuss the reactions of the transition metal organic derivatives. Because of the many unusual and distinctive structures characteristic of these compounds, correlation to structure will be made frequently.

(1) *One-electron ligands.* A great many compounds involving one-electron ligands have already been described. These include all the compounds containing metal-carbon σ bonds to various alkyl and aryl groups. In this group, other σ donors, such as carbon monoxide, cyanide ion, and acetylides will be considered. Transition metal derivatives involving one-electron ligands differ from those of the main group elements in that the d orbitals in transition metals may overlap with proper ligand orbitals to produce π bonds in addition to the σ bond. In metal-alkyl bonds, little contribution by π bonding is likely, since the p_x and p_y orbitals of the carbon atom, which are of the correct symmetry for π overlap with metal

p_x, p_y, d_{xz}, or d_{yz} orbitals, are already fully involved in bonding with other carbon or hydrogen atoms, which are attached to this carbon atom.

As expected, with simple σ-type covalent bonds, the greater is the electronegativity difference between the alkyl group and the metal, the greater will be the ionic resonance energy and the stronger the bond. Although data are limited, the fact that $F_3CCo(CO)_4$ can be distilled without decomposition at 91°C, whereas $H_3CCo(CO)_4$ is stable only at −30°C or below, supports this observation.

In metal aryls, π electron donation may take place from the filled π orbitals of the ligand to empty metal orbitals, or from the metal to empty π-antibonding orbitals of the ligand. Accordingly, metal aryls are usually more stable than their alkyl analogs. Thus, complexes such as $(R_3P)_2NiAr_2$ [Ar = mesityl; sym-$C_6H_3(Me)_3$] are quite stable, whereas $(R_3P)_2NiR_2$ is so unstable it has not been isolated. The influence of ring methyl groups (e.g., mesityl) is believed to contribute to stability by sterically hindering attack on the metal-carbon bond.

Alkyl complexes, which contain the $M{-}C{\equiv}CR$ grouping, are similar to metal cyanides and carbonyls and are often more stable than the corresponding alkyls or aryls. The nickel complex, $(Et_3P)_2Ni(C{\equiv}CPh)_2$, for example, melts at from 149 to 151°C without decomposing. The ligands, $-C{\equiv}CR$, CN^-, and CO are all isoelectronic and possess both σ-donor and π-acceptor properties.

Transition metal alkyl and aryl bonds also undergo insertion reactions:

$$M{-}R + X \rightarrow M{-}X{-}R \quad (X = CO, SO_2, F_2C{=}CF_2) \qquad [15.103]$$

At least in the carbonylation reaction, it has been shown by isotopic tracer technique that a coordinated carbonyl group and not a gaseous carbon monoxide molecule is involved in the insertion process.

(2) *Two-electron ligands*. The typical two-electron ligand is an olefin such as ethylene. The ethylene-metal bond may be represented (Fig. 15.3) as consisting of a π-acceptor and σ-donor component. Rotation about the C=C bond is thus severely restricted, and geometrical isomers can be isolated.

Two-electron ligands can be formed from and can be converted into one-electron ligands:

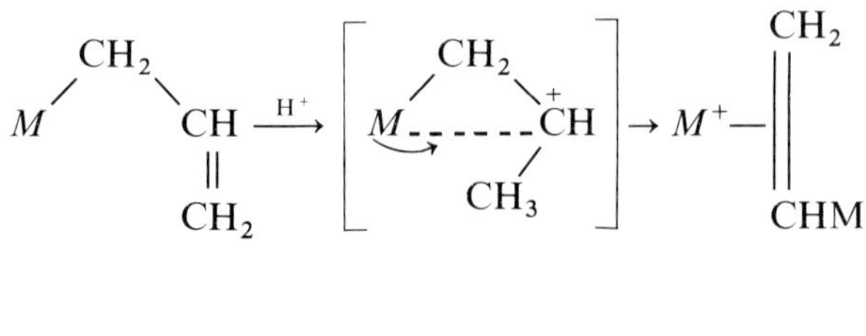

[15.104]

They can also be converted to a three-electron system:

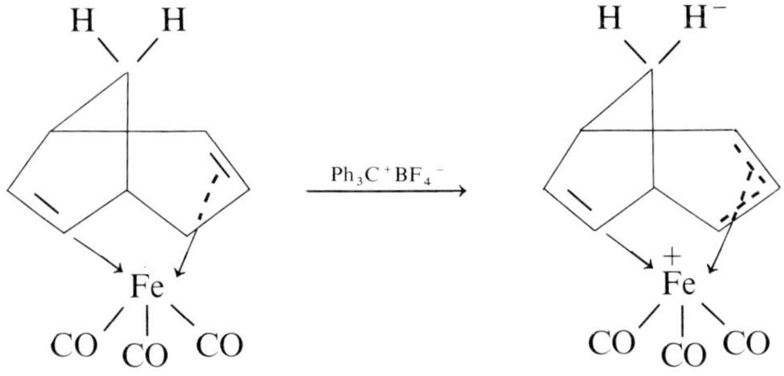

[15.105]

(3) *Three-electron ligands*. A typical π-enyl complex is *bis*-π-2-methylallylnickel. Its structure, determined by X-ray diffraction, has the nickel atom sandwiched between the parallel planes of two allyl groups:

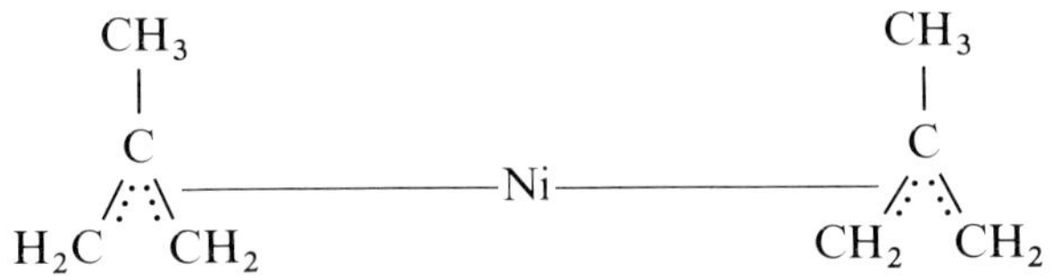

Such π-allyl complexes have been shown by proton magnetic resonance data to be of two types. Most are of the type A_2B_2X or

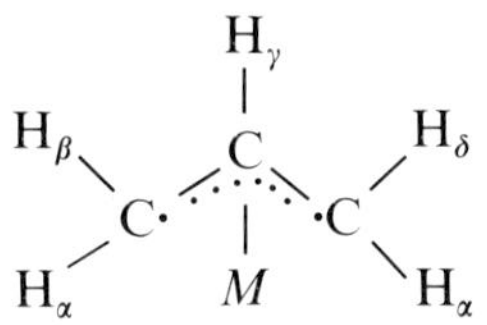

with two types of terminal hydrogens. Some are of the type A_4X at high temperatures (equivalent terminal hydrogens) and of the A_2B_2X type at lower temperatures. This suggests rapid scrambling of the terminal hydrogens at higher temperatures so that their environment averages out and a single peak is observed in the NMR spectrum. Other allyl complexes, e.g., π-allyl palladium chloride, show the A_4X spectrum only in the presence of ligands, such as R_3P, R_3As, or Me_2SO. Rapid exchange processes are thus concluded to cause the equilibration of the terminal hydrogens.

(4) *Four-electron ligands*. Conjugated olefins, such as butadiene, are typical ligands that may be considered to be four-electron donors. Butadiene tricarbonyl iron is an 18-electron complex with the structure

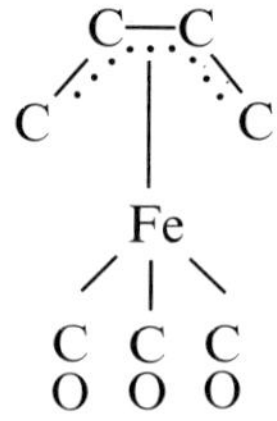

in which the *planar cis*-butadiene molecule is almost parallel to the plane formed by the three carbonyl groups. Similar to one-electron ligand systems shown, complexes of conjugated olefins may be converted by hydride abstraction [15.106] to five-electron π-dienyl ligands or to three-electron π-enyl ligands [15.107] by treatment with acids, e.g., HCl:

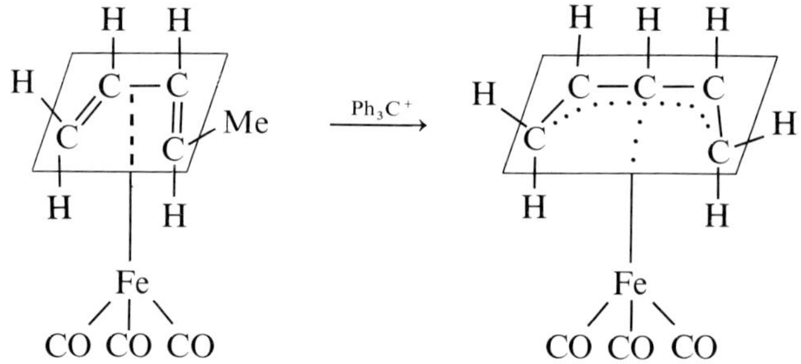

[15.106]

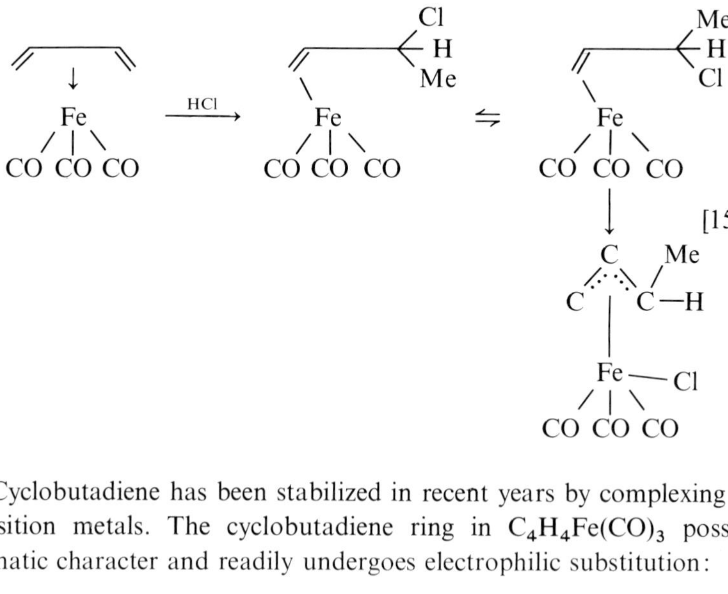

[15.107]

Cyclobutadiene has been stabilized in recent years by complexing with transition metals. The cyclobutadiene ring in $C_4H_4Fe(CO)_3$ possesses aromatic character and readily undergoes electrophilic substitution:

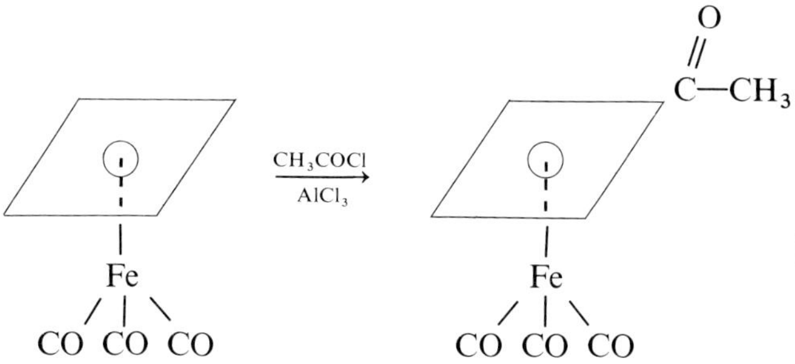

[15.108]

(5) *Five-electron Ligands.* The most familiar of the five-electron ligands is the cyclopentadienyl radical, $C_5H_5\cdot$. The most important types of compounds formed by C_5H_5 are the π-cyclopentadienyl complexes, in which the C_5H_5 ring is bonded essentially covalently to the metal with the metal lying equidistant from all five carbon atoms. The well-known *bis*-π-cyclopentadienyl sandwich compounds of the transition metals are exemplified by $(C_5H_5)_2Fe$, or ferrocene:

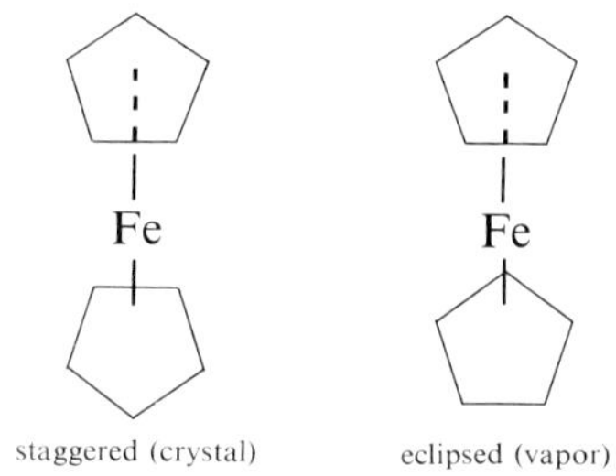

Ferrocene is isomorphous with other transition metal derivatives, $(C_5H_5)_2M$ (where M = V, Cr, Fe, Co, or Ni). Some derivatives, such as $(C_5H_5)_2Ru$, crystallize in the eclipsed form. The C_5H_5 rings are planar, all the carbon-carbon bond distances are equal, and they possess considerable aromatic character. The sandwich complexes are quite stable thermally and chemically, being resistant to hydrolysis and catalytic hydrogenation. Stability to oxygen varies, ferrocene being inert whereas chromocene is pyrophoric in air.

There are a large variety of π-cyclopentadienyl metal carbonyl complexes known, adhering with very few exceptions to the 18-electron rule. This is exemplified by the series π-$C_5H_5Co(CO)_2$, π-$C_5H_5Mn(CO)_3$, and π-$C_5H_5V(CO)_4$. These compounds are stable thermally, and most resist oxidation by oxygen at room temperature. They undergo a wide variety of reactions, including reduction with sodium amalgan in THF to yield π-cyclopentadienyl carbonyl anions by loss of carbon monoxide. Acidification of anions leads to metal hydrides containing M—H bonds.

$$[\pi\text{-}C_5H_5Mo(CO)_3]^- + H^+ \rightarrow \pi\text{-}C_5H_5Mo(CO)_3H \qquad [15.109]$$

(6) *Six-electron ligands.* Benzene and its substituted derivatives are the most familiar arenes of six-electron ligands. A wide range of *bis*-π-arene complexes, e.g., $(\pi\text{-}C_6H_6)_2M$ (where M = V and Cr) and $(\pi\text{-mesitylene})_2M$ [where M = V, Cr, Fe, Co, and their cations as well as mono-π-arenes are known in which one arene group is present along with other ligands, e.g., $\pi\text{-}C_6H_6Cr(CO)_3$] are known. They are stable thermally, although not as stable as their *bis*-π-cyclopentadienyl counterparts. Bond dissociation energy data suggest that the iron-carbon bond in ferrocene is almost twice as stable as the chromium-carbon bond in *bis*-π-benzene chromium, the corresponding 18-electron complex in the *bis*-π-arene series.

(7) *Six-, seven-, and eight-electron olefin ligands.* A number of olefins may act as multielectron donors. Cycloheptatriene molybdenum tricarbonyl can undergo hydride abstraction to give π-cycloheptatrienyl metal cations. These compounds are called *piano stool* compounds.

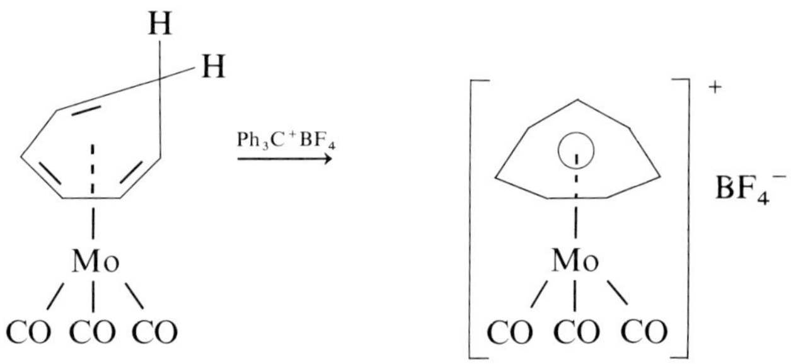

[15.110]

Mixed sandwich compounds can undergo ring replacement reactions:

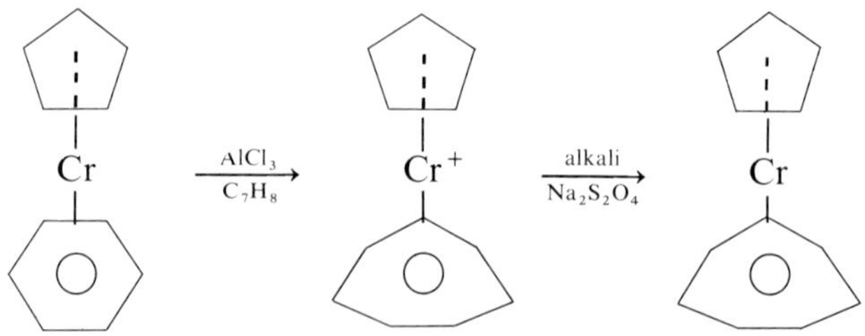

[15.111]

Cyclo-octatetraene can bond in many different ways, in which it may donate some or all of its double bonds to transition metals, e.g.,

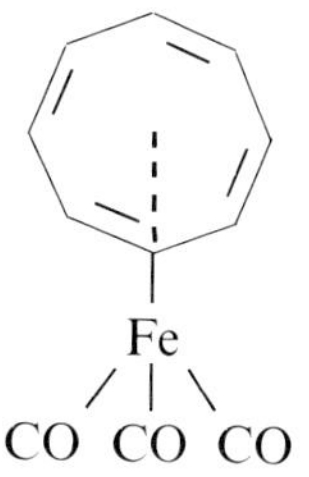

C_8H_8 ring is not planar.

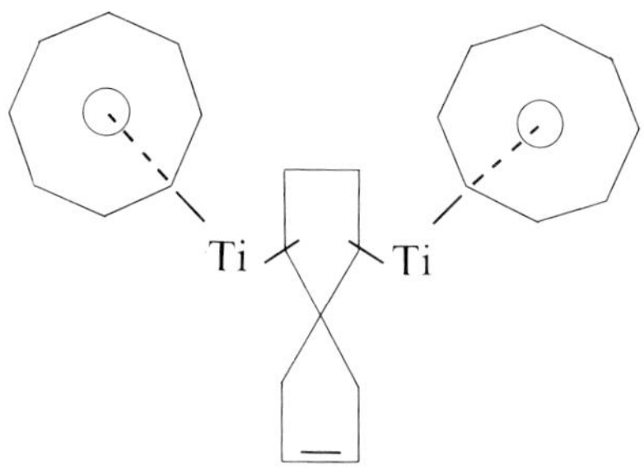

Two of the C_8H_8 rings are planar.

Suggested Readings

1. Coates, G. E., M. L. H. Green, P. Powell, and K. Wade, *Organometallic Compounds*: vol. 1, 3rd ed—*The Main Group Elements*, Methuen & Co., Ltd., London, 1967; vol. 2—*The Transition Elements*, Methuen & Co., Ltd., London, 1968.
2. Coates, G. E., M. L. H. Green, P. Powell, and K. Wade, *Principles of Organometallic Chemistry*, Methuen & Co., Ltd., London, 1968.
3. Eisch, J., and H. Gilman, *Organometallic Compounds*, vol. 2, Academic Press, Inc., New York, 1960.
4. Halpern, J., *Accounts Chem. Res.*, **3**, 386 (1970).
5. Pauson, P. L., *Organometallic Chemistry*, St. Martins Press, New York, 1967.
6. Richardson, J. W., "Carbon-Metal Bonding," in *Organometallic Chemistry*, H. Zeiss, ed., Reinhold Publishing Corp., New York, 1960.
7. Stone, F. G. A., *Hydrogen Compounds of the Group IV Elements*, Prentice-Hall, Inc., Englewood Cliffs, N. J., 1962.

Study Questions

15.1. Suggest a reason why the thermal stability of organometallic compounds varies so greatly, e.g., tetramethylsilane is unchanged after many days at 500°C. whereas tetramethyltitanium decomposes rapidly at room temperature.

15.2. Explain why a strongly polarizing cation such as Li or Be because of its high charge density has a tendency to form compounds with electron-deficient structures.

15.3. Sketch the appropriate orbitals that are suitable for both σ- and π-type bonding in a transition metal carbonyl compound.

15.4. Predict the relative kinetic stability of (a) $(CH_3)_3Ga$, $(CH_3)_4Ge$, and $(CH_3)_3As$; (b) $(CH_3)_4Ge$ and $GeCl_4$. Explain any differences.

15.5. Using appropriate starting materials, write equations for the preparation of

(a) CH_3Li (b) $CH_3C{\equiv}CK$
(c) $(CH_3)_2Cd$ (d) Ph_2GeCl_2
(e) $Mo(CO)_6$ (f) Chromocene
(g) $(CH_3CH_2CH_2GaCl_2)_2$ (h) $MnCOCH_3$
(i) $(CO)_3Fe(\pi\text{-}C_5H_7)$ (j) R_3SnOH

15.6. Predict the principal product in each of the following reactions:

(a) $Na + C_2H_5Br$ (b) $C_6H_5CH_2Cl + Li$
(c) $C_6H_5CH_3 + Hg(OAc)_2$ (d) $CH_3MgCl + HgCl_2$
(e) $AuCl_3 + CO$ (f) $BrMn(CO)_5 + PhNC$ (in ethanol)
(g) $Ph_2C{=}CPh_2 + Na$ (h) $Et_3Al + C_2H_4$ (cyclohexene)
(i) $n\text{-}BuLi + Ph_3SnCH_2CH{=}CH_2$ (j) $(CH_3)_3SnCl + (CH_3)_3N$

15.7. Explain why a reaction such as $RX + M$ proceeds spontaneously if X is a less electropositive metal than M.

15.8. What factor(s) acts predominantly as the driving force in each of the following reactions?

(a) $RC{\equiv}CH + NaH$ (b) $Ph_2NiCL + Ph_3PLi$
(c) $BuLi + PhC{\equiv}CPh$ (d) C_2H_5Na (decomposition at RT)
(e) $(CH_3)_6Al_2 + (CH_3)_3N$ (f) $(Ph_3P)_2PtCl_2 + CH_2{=}CH_2 \xrightarrow{N_2H_4}$

15.9. Explain why a ligand such as PhNC will displace CO from $Ni(CO)_4$ to produce $Ni(CNPh)_4$.

15.10. Using the soft-hard acid-base theory of Pearson, give an explanation of the reactivity of the trialkyls of Al, Ga, In toward Me_3N, Me_3P and Me_3As.

15.11. Sketch the specific orbitals on each appropriate atom that are suitable for (a) $(p \to d)$ π bonding in $(CH_3)_3GeN(CH_3)_2$, and (b) $(d \to d)$ π bonding in $(CH_3)_3AsPt(CH_3)_2$.

15.12. Explain why the thermal stabilities of the Group IV-A organoderivatives vary in the order $MH_4 < RMH_3 < R_2MH_2 < R_3MH$.

Index

C

D

P

Q

R

W

Z